Responding to Extreme Weather Events

Other Hydrometeorological Extreme Events Titles

Hydrometeorological Hazards: Interfacing Science and Policy
Edited by Philippe Quevauviller

Coastal Storms: Processes and Impacts
Edited by Paolo Ciavola and Giovanni Coco

Drought: Science and Policy
Edited by Ana Iglesias, Dionysis Assimacopoulos, and Henny A.J. Van Lanen

Facing Hydrometeorological Extreme Events: A Governance Issue
Edited by Isabelle La Jeunesse and Corinne Larrue

Hydrometeorological Extreme Events and Public Health
Edited by Franziska Matthies-Wiesler and Philippe Quevauviller

Responding to Extreme Weather Events
Edited by Daniel Sempere-Torres, Anastasios Karakostas, Claudio Rossi, and Philippe Quevauviller

Responding to Extreme Weather Events

Edited by

Daniel Sempere-Torres
Center of Applied Research in Hydrometeorology
Universitat Politècnica de Catalunya
Barcelona, Spain

Anastasios Karakostas
DRAXIS Environmental S.A., Thessaloniki, Greece

Claudio Rossi
LINKS Foundation
Turin, Italy

Philippe Quevauviller
European Commission
Vrije Universiteit Brussel
Brussels, Belgium

WILEY

This edition first published 2024

Registered Offices
John Wiley & Sons, Inc., 111 River Street, Hoboken, NJ 07030, USA
John Wiley & Sons Ltd, The Atrium, Southern Gate, Chichester, West Sussex, PO19 8SQ, UK

For details of our global editorial offices, customer services, and more information about Wiley products visit us at www.wiley.com.

Wiley also publishes its books in a variety of electronic formats and by print-on-demand. Some content that appears in standard print versions of this book may not be available in other formats.

Library of Congress Cataloging-in-Publication Data Applied for:

Hardback ISBN: 9781119741589

Cover Design: Wiley
Cover Image: © Patrick Orton/Getty Images

Set in 10/12pt Times by Straive, Pondicherry, India

Contents

List of Contributors

Stelios Andreadis
Centre for Research and Technology Hellas (CERTH), Information Technologies Institute (ITI), Thessaloniki, Greece

Edoardo Arnaudo
Dipartimento di Automatica e Informatica (DAUIN), Politecnico di Torino, Turin, Italy

Francesco Baghino
Cima Research Foundation, Savona, Italy
DIBRIS – Department of Informatics, Bioengineering, Robotics and Systems Engineering, University of Genoa, Genoa, Italy

Marc Berenguer
Center of Applied Research in Hydrometeorology, Universitat Politècnica de Catalunya (UPC), Barcelona, Spain

Giacomo Blanco
LINKS Foundation, Turin, Italy

Lorenzo Bongiovanni
LINKS Foundation, Turin, Italy

Mirko D'Andrea
Cima Research Foundation, Savona, Italy

Paolo Fiorucci
Cima Research Foundation, Savona, Italy

Antonella Frisiello
LINKS Foundation, Turin, Italy

Oliver Gebhardt
Department of Economics, Helmholtz Center for Environmental Research GmbH – UFZ, Leipzig, Germany

Ilias Gialampoukidis
Centre for Research and Technology Hellas (CERTH), Information Technologies Institute (ITI), Thessaloniki, Greece

Panagiotis Giannakeris
Centre for Research and Technology Hellas (CERTH), Information Technologies Institute (ITI), Thessaloniki, Greece

Milan Kalas
KAJO s.r.o., Bytca, Slovakia

Anastasios Karakostas
DRAXIS Environmental S.A., Thessaloniki, Greece

Denys Kolokol
KAJO s.r.o., Bytca, Slovakia

Ioannis Kompatsiaris
Centre for Research and Technology Hellas (CERTH), Information Technologies Institute (ITI), Thessaloniki, Greece

Ilias Koulalis
Centre for Research and Technology Hellas (CERTH), Information Technologies Institute (ITI), Thessaloniki, Greece

Christian Kuhlicke
Department of Urban & Environmental Sociology, Helmholtz Center for Environmental Research GmbH – UFZ, Leipzig, Germany
Institute for Environmental Sciences and Geography, University of Potsdam, Potsdam, Germany

Henny A.J. Van Lanen
Hydrology and Quantitative Water Management, Department of Environmental Sciences, Wageningen University and Research, Wageningen, the Netherlands

Annakaisa Von Lerber
Finnish Meteorological Institute, Helsinki, Finland

Gianluca Marucco
LINKS Foundation, Turin, Italy

Erika Meléndez-Landaverde
Center of Applied Research in Hydrometeorology, Universitat Politècnica de Catalunya (UPC), Barcelona, Spain

Emmanouil Michail
Centre for Research and Technology Hellas (CERTH), Information Technologies Institute (ITI), Thessaloniki, Greece

Claudia Di Napoli
School of Agriculture, Policy and Development, University of Reading, Reading, UK
Department of Geography and Environmental Science, University of Reading, Reading, UK

Joy Ommer
KAJO s.r.o., Bytca, Slovakia
Department of Geography and Environmental Science, University of Reading, Reading, UK

Shinju Park
Center of Applied Research in Hydrometeorology, Universitat Politècnica de Catalunya (UPC), Barcelona, Spain

Nicolò Perello
Cima Research Foundation, Savona, Italy
DIBRIS – Department of Informatics, Bioengineering, Robotics and Systems Engineering, University of Genoa, Genoa, Italy

Marco Pini
LINKS Foundation, Turin, Italy

Philippe Quevauviller
European Commission, Vrije Universiteit Brussel, Brussels, Belgium

Nicola Rebora
Cima Research Foundation, Savona, Italy

Claudio Rossi
LINKS Foundation, Turin, Italy

Tommaso Sabattini
KAJO s.r.o., Bytca, Slovakia

Dario Salza
LINKS Foundation, Turin, Italy

Daniel Sempere-Torres
Center of Applied Research in Hydrometeorology, Universitat Politècnica de Catalunya (UPC), Barcelona, Spain

Amin Shakya
Faculty of Geo-information and Earth Observation, University of Twente, WRS, Enschede, Netherlands

Samuel J. Sutanto
Water System and Global Change, Department of Environmental Sciences, Wageningen University and Research, Wageningen, the Netherlands

Andrea Trucchia
Cima Research Foundation, Savona, Italy

Saša Vraníc
KAJO s.r.o., Bytca, Slovakia

Stefanos Vrochidis
Centre for Research and Technology Hellas (CERTH), Information Technologies Institute (ITI), Thessaloniki, Greece

Series Preface

The increasing frequency and severity of hydrometeorological extreme events are reported in many studies and surveys, including the 5th IPCC Assessment Report. This report and other sources highlight the increasing probability that these events are partly driven by climate change, while other causes are linked to the increased exposure and vulnerability of societies in exposed areas (which are not only due to climate change but also to mismanagement of risks and 'lost memories' about them). Efforts are on-going to enhance today's forecasting, prediction and early warning capabilities in order to improve the assessment of vulnerability and risks and develop adequate prevention, mitigation and preparedness measures.

The Book Series on 'Hydrometeorological Extreme Events' has the ambition to gather available knowledge in this area, taking stock of research and policy developments at international level. While individual publications exist on specific hazards, the proposed series is the first of its kind to propose an enlarged coverage of various extreme events that are generally studied by different (not necessarily interconnected) research teams.

The Series encompasses several volumes dealing with various aspects of hydrometeorological extreme events, primarily discussing science-policy interfacing issues, and developing specific discussions about floods, coastal storms (including storm surges), droughts, resilience and adaptation, governance, and public health impacts. While the books are looking at the crisis management cycle as a whole, the focus of the discussions is generally oriented toward the knowledge base of the different events, prevention and preparedness, early warning and improved prediction systems.

The involvement of internationally renowned scientists (from different horizons and disciplines) behind the knowledge base of hydrometeorological events makes this series unique in this respect. The overall series will provide a multidisciplinary description of various scientific and policy features concerning hydrometeorological extreme events, as written by authors from different countries, making it a truly international book series.

The Series so far is made of five volumes, an introductory one on '*Hydrometeorological Hazards: Interfacing Science and Policy*' (2015, Ed. Ph. Quevauviller), a second volume dealing with '*Coastal storms: Processes and Impacts*' (2017, Ed. P. Ciavola and G. Coco), a third volume on '*Droughts: Science and Policy*' (2019, Ed. A. Iglesias, D. Assimacopoulos and H.A.J. Van Lanen), a fourth volume intitled '*Facing Hydrometeorological Extreme Events: A Governance issue*' (2020, Ed. I. La Jeunesse and C. Larrue), and a fifth one on '*Hydrometeorological Extreme Events and Public Health*' (2022, Ed. F. Matthies-Wiesler and Ph. Quevauviller). The volume '*Responding to Extreme Weather Events*' is the sixth book of this Series; it has been written by experts in the field, covering various horizons and (policy and scientific) views gather from three major international research projects funded by the European Union Horizon2020 Framework Programme. It offers the reader an overview of scientific knowledge about challenges related to responses to weather extreme events, in particular impact forecasting, use of artificial intelligence and cybertechnologies for extreme weather event's management, and communication and public warnings.

Ph. Quevauviller
Series Editor

1

The ANYWHERE Paradigm Shift in Responding to Weather and Climate Emergencies

Impact Forecasting, Dynamic Vulnerability and the Need for Citizen's Involvement

Daniel Sempere-Torres and Marc Berenguer
Center of Applied Research in Hydrometeorology, Universitat Politècnica de Catalunya (UPC), Barcelona, Spain

1.1 Disaster Risk Management in Times of Climate Change: The Need of a Proactive Approach

The world has just seen the hottest decade on record during which the title for the hottest year was beaten eight times (WMO 2023). This tendency will continue for decades, even if global and European efforts to cut greenhouse gas emissions prove effective. We also know today that **'There is no definitive way to limit global temperature rise to 1.5°C above pre-industrial levels'** *(IPCC SR 1.5)*. Even a drastic temporary decrease in emissions (the 2008 financial crisis or during COVID-19 pandemic) has proved to have little effect on the overall trajectory of global warming. Therefore, and especially after the extreme events observed worldwide during 2021 and 2022, it is widely recognized that the effects of climate change (CC) are already happening today.

Responding to Extreme Weather Events, First Edition. Edited by Daniel Sempere-Torres, Anastasios Karakostas, Claudio Rossi, and Philippe Quevauviller.

Moreover, the analysis of the impacts of natural hazards in the last 50 years (see WMO 2021) shows that the **frequency and severity of these extreme climate and weather events are increasing and exacerbating climate-related economic and social losses.** And the urgency to react to their consequences is a social priority with significant political and economic implications, as proven by the climate emergency declaration of the *EU parliament (November* 2019),[1] and several other *national and regional parliaments*[2] and *leading cities*[3].

As stated by the EU Strategy on Adaptation to Climate Change (EC 2021a, b),[4] the EU and the global community are underprepared for the increasing intensity, frequency and pervasiveness of climate change impacts, especially as emissions continue to rise. We must rapidly build our resilience to CC by moving from raising public awareness and concern to mass action on adaptation. Accordingly, the **'Adaptation to CC, including Societal Transformation'**, has become one of the five Horizon Europe Missions to push this significant societal challenge[5].

In this regard, Early Warning Systems (**EWSs**) have become a crucial instrument for disaster risk management **(DRM)**. Now promoted by the United Nations (UN) through the 'Early Warnings for All initiative'[6], EWS can be especially critical during weather/climate emergencies. However, to be effective, they must be able to trigger the intended actions for damage reduction to be undertaken by authorities, first and second responders and citizens (i.e. the earliest responders in place, also seen as the zero-order responders, Briones et al. 2019).

Nonetheless, triggering the full chain of emergency management starting with the hazard forecasts up to the emergency management actions is not a simple objective, as the *catastrophic floods of July 2021 in Germany and Belgium*[7] exemplified (over 180 deaths in just a 200 mm daily rainfall event, see Table 1.2). Currently, the available scientific and technical advancements enabling us to anticipate extreme events are not well integrated into the real-life protocols of authorities and first responders. Hence it is critical to develop and implement EWSs adapted to the local needs of authorities, first responders and the population. And be able to connect them to local/community risk management plans able to ensure that the warnings can trigger the required local actions that can effectively reduce damages and loss.

This chapter, and some of the following ones, summarizes the **paradigm shift in responding to weather and climate emergencies based in the project results and lessons learnt during the ANYWHERE innovation action.**

1.2 Adapting Risk Management to the 'New Normality': The Case of Flood Risk Management

Before describing the details of the **ANYWHERE** proposed tools and results, it is important to illustrate the challenge of what it means to consider the effects induced by the CC through a particular well-known hazard, such as floods.

Floods are the most significant natural hazards in Europe in terms of the number of events, people affected and economic losses. But it is also, together with storms, the most relevant natural hazard worldwide (CRED 2020). Hydrological hazards (floods, and heavy-rain-induced disasters) are also the natural hazard that has most increased in frequency in the last 30 years (Kron et al. 2019).

Table 1.1 Differences between riverine and coastal floods compared with pluvial and flash floods under climate change effects.

	Riverine and coastal floods	Pluvial and flash floods
Time response	Long: days	Short: several 1/4 hours
Location	We know where: • Mapping of risk can be done • Defence and structural measures are possible • **PLANNING** is CRUCIAL	Can be ANYWHERE: • The probability increases with climate change (an increase in heavy rains) • The probability increases with an increase in wildfires • Structural measures are out of the question • **REAL-TIME MANAGEMENT** of the response is crucial
What to do	**We know** what to do • River restoration • Floodplain recuperation • EVACUATION is possible	**At present we DO NOT know** what to do: • Need of a NEW PARADIGM • **CITIZENS'** involvement is crucial • **SELF-PROTECTION** Flood Risk Management Plans • Subsidiarity principle

In this context and as seen in Table 1.1, it is important to recognize the differences between what are considered 'classical or typical floods' (e.g. riverine and coastal floods) and the 'new intensified floods', episodes that are not only increasing in their frequencies but also in their intensities and amount (and level) of seen socio-economic impacts due to CC (e.g. pluvial and flash floods).

On one hand, riverine and coastal flood events have long response times (that can go from several hours to several days) and thus the time between the event starts and the main consequences is of the order of several hours or days. However, pluvial and flash floods are directly related to heavy-rains and the associated torrential phenomena have extremely short response times (usually a few quarters of an hour). Consequently, these types of events trigger emergencies that develop too quick for a reactive response based on direct observations. Thus, the only appropriate emergency response must be based on the timely anticipation of the event (at least a few hours in advance, Alfieri et al. 2016). This implies that decision-making needs to rely into trusted, but uncertain, high-resolution forecasts instead of waiting to receive direct observations (only available when the impacts are already occurring), what it is by itself a significant operational challenge.

On the other hand, for the first category we can anticipate where these kinds of events will happen (around the river flood-prone areas, or in particular areas of the coastal line). Therefore, risk cartographies can be pre-established, making possible to plan defences and structural measures, as well as evacuation plans. Thus, planning is crucial to cope with these types of floods.

Contrarily, heavy-rainfall-induced floods can happen anywhere. Moreover, given the effects of CC on the increase of the frequency and severity of heavy rains, as well as on other factors amplifying the torrential character of flash floods (such as the increase of the number and severity of forest fires, which worsens the magnitude of flash floods due to the loss of vegetation; see Lavabre et al. 1993; Versini et al. 2013; Wine and Cadol, 2016; Wagenbrenner et al. 2021), pluvial and flash floods have multiplied by 3 in the last 30 years[8] and they are at present the climate-induced hazard that has increased the most. In this context, emergency management cannot only be based on planning, and the **real-time management of the response becomes crucial.**

Furthermore, whereas in riverine and coastal floods we have enough experience with effective measures to reduce and manage the associated risks (through river restoration works, floodplain recuperation actions and evacuation plans etc . . .), in the case of pluvial and flash floods, we need to recognize that essentially the current established knowledge in flood risk planning and management turns out to be useless (as in the case of July 2021 in Germany and Belgium).

In these floods, as in any other hazards where climate change is making knowledge based on past experience irrelevant, we need to acknowledge that a change of paradigm is required. Change of paradigm that implies accepting to **move from planning-based strategies towards real-time management strategies, essentially based on EWSs adapted to the local needs**, providing actionable information able to trigger the response, not just of the local authorities, but also of the citizens. This requires a disruptive societal transformation in emergency management through the implementation of flood risk management plans, which should include as a major component the concept of **self-preparedness and self-protection actions**, previously identified and adapted to the most vulnerable points and communities, a transformation that should be supported by advanced and adapted technological tools (Gräßler et al. 2020).

To understand the urgency of such a societal transformation, Table 1.2 shows the main characteristics of recent heavy-rainfall events recorded in Europe in 2020 and 2021. Whereas during the catastrophic floods in Germany in July 2021, the quantities recorded represented the equivalent of 2 months of accumulated rainfall registered in 24 hours, we can see that this event is not extraordinary in our 'new' CC times. Thus, we urgently need to start being prepared to face events delivering these 2-month accumulated rainfall in less than 24 hours or more, such as the event on the 4 October 2021 in Rossiglione, Liguria (IT), where the European rainfall-accumulated record in 12 hours has been beaten: 740 mm in 12 hours, representing **one year mean rainfall accumulated in 12 hours**.

These and the other events in Table 1.2 can help us to understand the magnitude of the new scenarios we need to be prepared to, and the urgency with which we need to start initiating the adaptation to the consequences of climate change.

1.3 Changing the Paradigm: Impact-Based Multi-Hazard Early Warning Systems to Move from Reactive to Pro-Active Emergency Response Strategies

In this context, adapted **DRM** will require an update of the tools and methodologies to evolve our present risk assessment capacities, crisis management and preparedness strategies for the natural hazards under CC. Thus, an enhanced DRM cycle will require tools using different types of information and forecasts that can enable the anticipation of disasters, providing **Early Warnings supporting the situational awareness and rapid deployment of responders in vulnerable areas.**

Table 1.2 Some examples of recent heavy rainfall events giving us what can characterize the 'new normality' under climate change times.

	Total accumulated rainfall	Maximal intensity	In terms of average monthly rainfall
14 July 2021 Germany	200 mm in 24 h	>40 mm in 1 h	2-month rainfall in 24 h
1 September 2021 @ Alcanar (ES)	220 mm in **3 h** 260 in 24 h	**77 mm in 30 minutes**	6-month rainfall in 24 h
8 September 2021 @ Agen (FR)	130 mm in **3 h**	80 mm in 1 h	2-month rainfall in 3 h
18 December 2020 @ Cerdanyola (ES)	**300 mm** in 24 h	>100 mm in 1 h	6-month rainfall in 24 h
4 October 2021 @ Rossiglione, Liguria (IT)	**900 mm** in 24 h **740 mm in 12 h**	**>180 mm in 1 h**	**12-month rainfall in 12 h**

To that end, impact-based EWS **(IEWS)** (WMO 2021) and particularly multi-hazard impact-based EWS (**MH-IEWS**) for weather emergencies have been *promoted by the WMO and the Sendai Framework (Target G)*,[9] (Murray 2021) see Figure 1.1[10,11], as the next step to translate forecasts into information **supporting actionable decisions during emergencies** and triggering **site-specific actions based on early risk forecasts.**

Although many current initiatives are trying to develop the concept of IEWS for weather and climate-induced disasters, there are very few successful experiences in implementing and demonstrating MH-IEWS in an operational environment (Merz et al. 2020). The successful H2020 Innovation Action **ANYWHERE** (www.anywhere-h2020.eu), winner of the EC Security Innovation Resilience Award in 2022,[12] is one of them.

1.3.1 From Reactive to Proactive Emergency Response Strategies

This innovative pathway can be clarified by taking the example of the case of the floods in Germany in July 2021. For this event, a clear warning for the *river Ahr*[13] (one of the most affected areas) was available through the European Flood Awareness System (*EFAS*,[14] part of the Copernicus Emergency Management Services, *CEMS*[15]), was available **more than 24 hours in advance of the floods**, see Figure 1.2. Moreover, the *ANYWHERE A4EU system*[16] provided a high-resolution warning based on the *OPERA network radar data*[17] for the portion of the river most affected **5 hours in advance** (with enough time to take self-protection actions and reduce the number of fatalities). Consequently, the technology to activate actionable solutions through a risk management self-protection protocol was fully available and working correctly. However, the EU society has not yet the capacity to react effectively to these new climate-induced emergencies, even if we have already the technology to anticipate their occurrence and impacts,[18] as the declarations in front of the Commission of Inquiry about these floods in the Walloon Parliament have shown.[19] Thus, the main

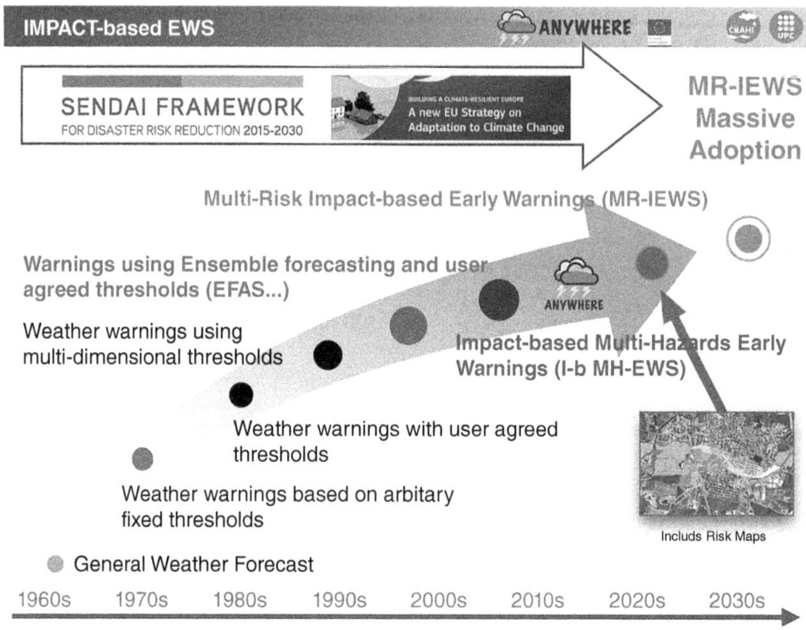

Figure 1.1 Evolution of the warning systems to support decision-making during weather and climate emergency. The initial general weather forecast has been transformed in different families of weather warnings issued by the National Meteorological Services. The advancements of the last years include the integration of probabilistic approaches using ensemble forecasting, as in the European Flood Awareness System (EFAS), or the impact-based multi-hazard early warning systems (MH-IEWS), among which *ANYWHERE* is one of the first real-time systems tested in operational environment in several Emergency Management Centres in Europe. In the next years, it is foreseen that these MH-IEWS could evolve towards new Multi-Risk Early Warning Services able to be massively adopted to support international initiatives such as the Sendai Framework for Disaster Risk Reduction, to promote the EWS4ALL initiative of the WMO as well as the international initiatives supporting climate change adaptation (CCA).

challenge is how to use this technology to empower Emergency Management Centres to transform advanced meteorological forecasts into high-resolution hazard and impact forecasting products providing information about the magnitude of the event and the expected consequences, allowing them to trigger the required actions to minimize damages and losses.

In this strategy, an important step is to understand that nowadays, the usual practices in most emergency management centres (EMCs) are still mainly reactive (first the emergency is detected, usually through 112 calls, then the reaction follows pre-established protocols, see Figure 1.3-above). There are very few exceptions of EMCs **able to act in proactive mode**, i.e. integrating forecasting capabilities or initiating the response based on the early detection of weak signals (before the emergency becomes evident). In the last years, several H2020 projects (**EMERGENT**, **ANYWHERE**, **I-REACT**, **BEAWARE**) have shown that technological developments can be of precious help for an anticipated response of first responders.

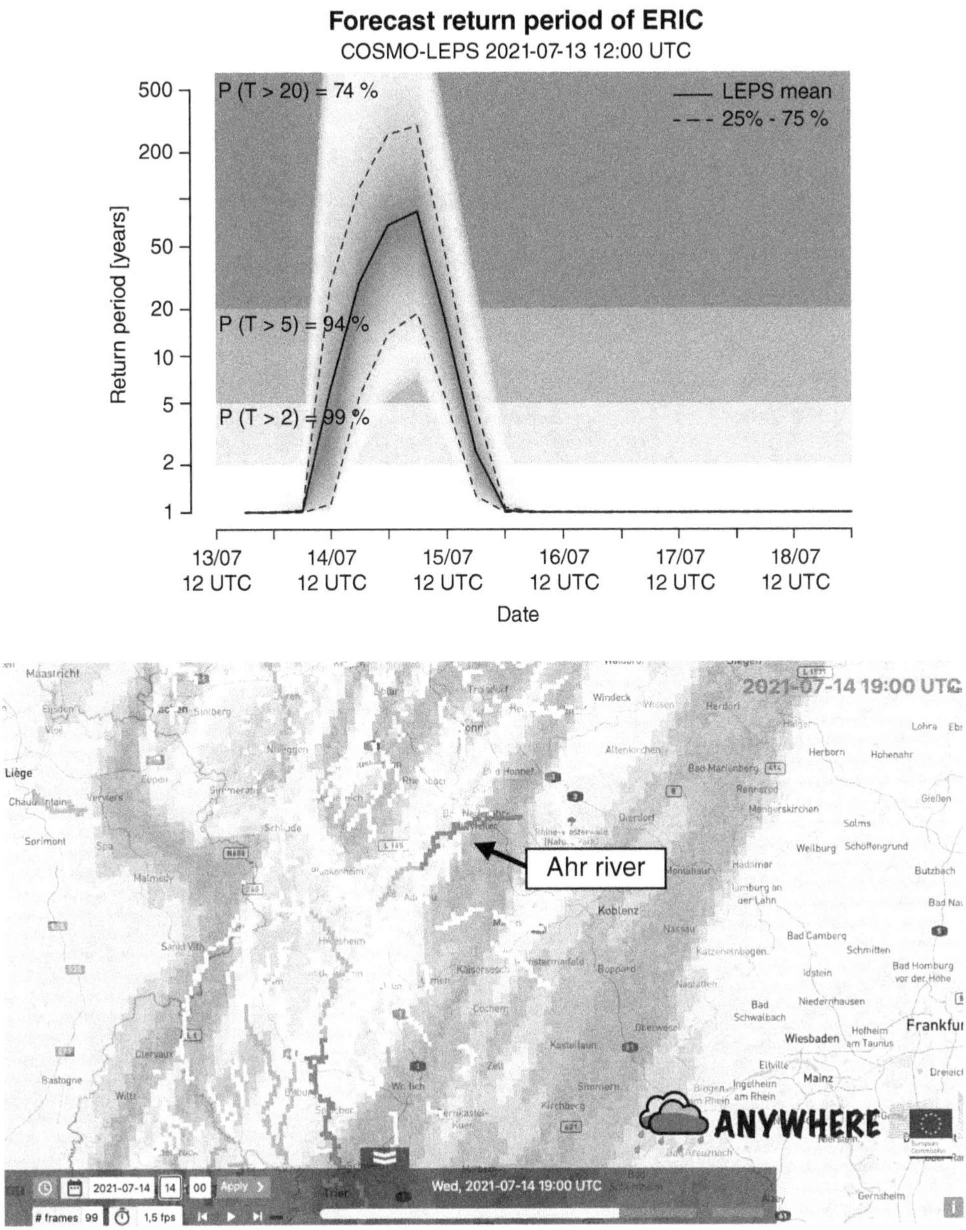

Figure 1.2 (Above) *ERIC flash flood indicator* announcing 74% probability of exceeding the highest warning level for the Ahr river (Germany) using the meteorological model forecast run on the 13 July at 12 : 00 UTC (>24 hours before the flood peak). *Source*: *EFAS*. (Below) Forecasted *ERICHA flash flood indicator* showing the maximum warning level on the Ahr river issued the 14 July at 14 hours UTC (5 hours in advance) using the rainfall nowcasts from the OPERA radars composites. *Source*: *ANYWHERE*.

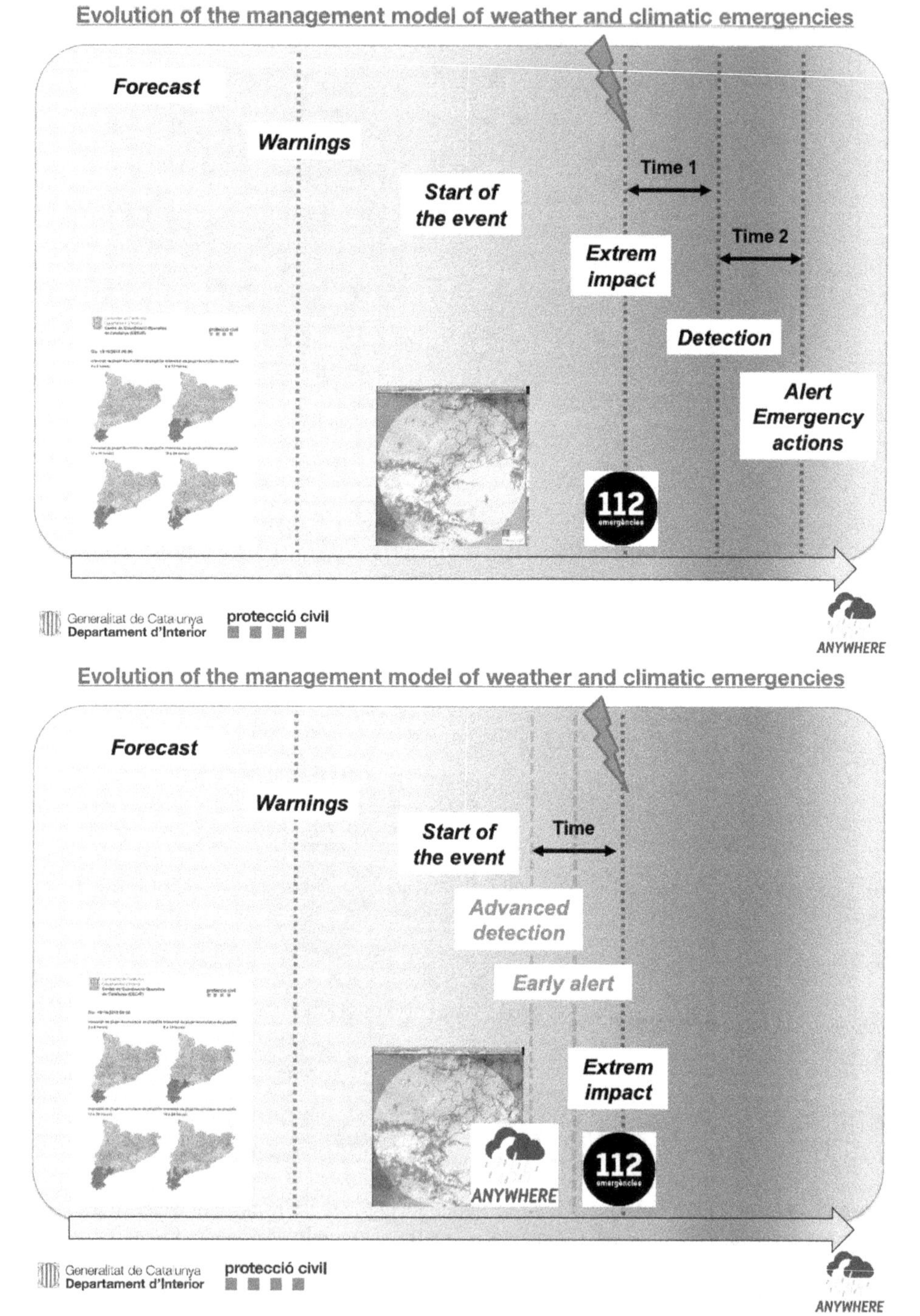

Figure 1.3 Change on the management model of weather-induced emergencies thanks to the *ANYWHERE* project developments: (above) Instead of detecting the impacts with delay time 1, and start the emergency actions with delay 2; (below) the ANYWHERE platform allow to anticipate the detection of the event and advance the response before the occurrence of the impact. *Source*: Courtesy of Sergio Delgado, Department of Civil Protection of the Generalitat de Catalunya.

In particular the *ANYWHERE* project has developed an operational **multi-hazard EWS** for extreme weather and climate events, able to translate the most advanced meteorological forecasts into *impact forecasting products* to support emergency management (Abily et al. 2020, see Section 1.3.2). The system was verified, tested and operationally demonstrated in **7 Emergency Management Centres covering the entire climatic range in the EU for 18 months,**[20] demonstrating in real time that the generalization of the proactive way of working in EMCs is now possible (see Figure 1.3-below).

These **ANYWHERE** innovations translate meteorological forecasts into anticipated impacts and **automatically connect them to critical points to trigger a set of pre-defined actions** (for instance, those of the self-protection plans), allowing civil protections and EMCs to start the response phase before the occurrence of the impacts, **reducing the damages through the concept of dynamic vulnerability** (Sempere-Torres 2019), see Section 1.4.

This capacity was tested operationally during the 50-year return period **Storm Gloria (19–23 January 2020)**, which severely affected the east coast of Spain, and in particular Catalonia in a severe way. During this event the Civil Protection of Catalonia triggered several response actions (including the management of the river Ter dams, and the confinement of tens of thousands of affected inhabitants of different cities) **based on impact forecasting early warnings for the first time in Europe**, before observations were available (saving over six hours for the operations).

1.3.2 The ANYWHERE MH-IEWS

The **impact forecasting concept** implemented by the ANYWHERE project consists of running state-of-the-art hazard-forecasting algorithms and models (driven by advanced meteorological forecasts) and combining them with the available exposure and vulnerability information to translate them into impact forecasts (see Figure 1.4).

In ANYWHERE, these algorithms and models are connected or encapsulated in a joint real-time MH-IEWS running in parallel to generate hazard forecasts for floods, flash floods, landslides and debris flows, storm surges, forest wildfires, droughts, heatwaves and weather-induced health impact, convective storms, severe winds and snowfall. The outputs of these algorithms were compiled in a catalogue of products describing the hydro-meteorological situation and forecasting the hazard level and expected impacts[21] that were served in real time by the ANYWHERE MH-IEWS to support emergency management and self-protection actions in the pilot sites of the project.

Given the differences in the characteristic scales of the different weather and climate hazards considered, the driving meteorological inputs are adapted to each hazard. These included the use of observations and radar-based precipitation nowcasts for the most local and fast-evolving hazards, such as convective storms or local flash floods and landslides (Palau 2021; Palau et al. 2020, 2023); limited-area Numerical Weather Prediction (NWP) models (driving the forecasting systems for floods, flash floods) and medium-range and seasonal forecasts (for the drought impact forecasting algorithms).

The ANYWHERE MH-IEWS is connected to the Continental-scale hazard and impact forecasts generated by the Copernicus Emergency Services (CEMS)[22]: mainly, the hydrological forecasts of the European Flood Awareness System (EFAS); the fire

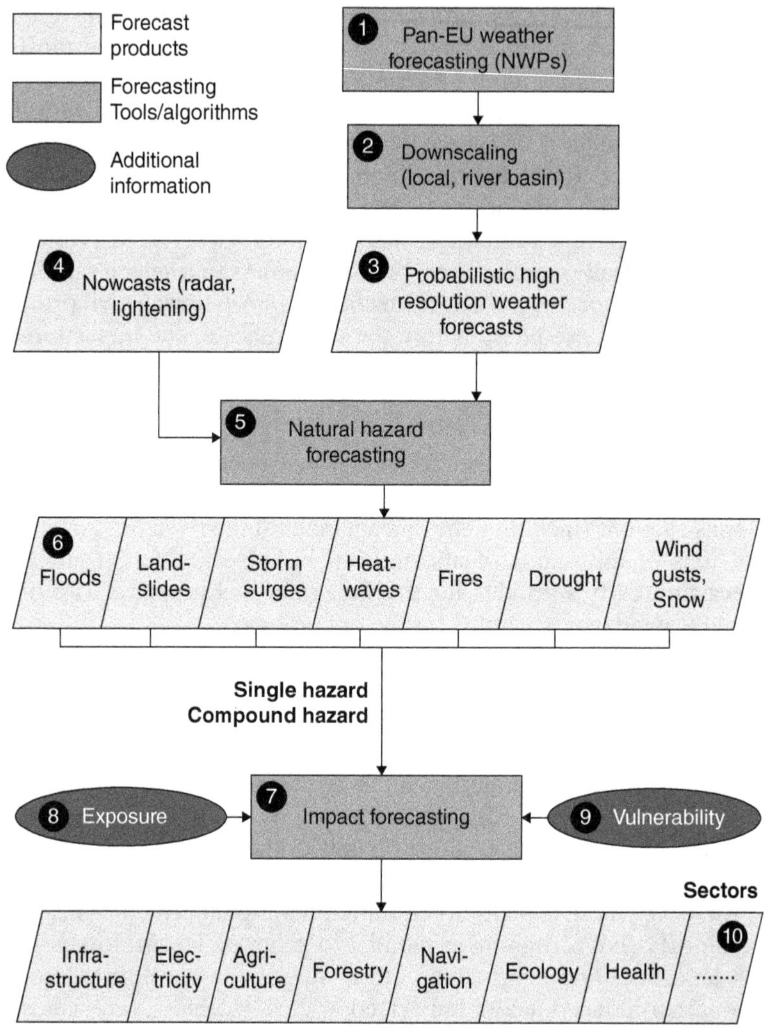

Figure 1.4 ANYWHERE multi-hazard IEWS forecasting platform: products and tools/algorithms to forecast weather-induced natural hazards and associated impacts.

of the European Forest Fire Information System (EFFIS) and the European Drought Observatory (EDO).

The EFAS flood products were complemented with flash flood hazard and impact nowcasts at Continental scale (Park et al. 2019; Ritter et al. 2021) and regional scale (Corral et al. 2019; Poletti et al. 2019; Ritter et al. 2020, 2022; Láng-Ritter et al. 2022), combining the hazard forecasts with the flood hazard and risk maps developed in the framework of the EU Floods Directive (2007); the vulnerability layers at the relevant scale to assess the expected losses and the expected impacts on population and critical points.

The storm surge forecasting models relied on a nested approach covering the European coasts at coarse resolution (EFAS-COAST, Fernández-Montblanc et al. 2019) to regional scale to forecast surge levels and waves parameters, feeding a high-resolution flood/erosion model at local scale providing flow velocity, maximum inundation depth and expected shoreline retreat (Armaroli et al. 2019; Duo et al. 2020).

The ANYWHERE MH-IEWS also integrates weather-induced health impact forecasts at the European scale by including two different types hazards: (i) those related to Air quality from the Copernicus Atmosphere Monitoring Service using a regional multi-model approach (Marécal et al. 2015), and (ii) those due to heat waves based on forecasts of the Universal Thermal Climate Index (UTCI, Di Napoli et al. 2018, 2021a, b) assessing the heat strain on the human body by combining the weather forecasts with a physiological model and an adaptive model for clothing insulation.

Complementing the products from EFFIS (based on computing fire danger indices based on medium-range weather forecasts, Di Giuseppe et al. 2016, 2017, 2018; Vitolo et al. 2018, 2019), the RISICO model (Fiorucci et al. 2008; Perello et al. 2022) is used to forecast the forest fire hazard both at European and regional scales, in the latter using higher-resolution and more accurate exposure datasets. After ignition, PROPAGATOR (Trucchia et al. 2020) is used to estimate the trajectory and extent of the forest fire given the weather conditions and identify the vulnerable areas potentially at risk.

Forecasts of drought impacts are based on transforming hazard forecasts (typically characterized with drought indices computed from medium-range and seasonal weather and hydrological forecasts; Diaz et al. 2020a, b; Sutanto et al. 2020a, b; Sutanto and Van Lanen 2021) into impact forecasts using machine-learning to train the algorithm from a database of historical reported impacts (Sutanto et al. 2019, 2020c, d).

And finally, the impacts caused by precipitation included the analysis of the impacts caused by convective cells (based on radar cell tracking; Rossi et al. 2014) and on forecasting the type of precipitation (Fehlman et al. 2018a, b, 2020; Gascón et al. 2018), particularly focusing on the impacts of snowfall on road conditions and road traffic (Cerreta et al. 2019).

Running the algorithms in parallel in the MH-IEWS allows to explore the compound and cascading effects of weather and climate events (Schauewecker et al. 2019; Sutanto et al. 2019; Láng-Ritter et al. 2022).

These real-time hazard and impact forecasting products are integrated with additional high-resolution local information about vulnerability and exposure with **artificial intelligence** and served by the **MH-IEWS** together with **regional layers of exposure and vulnerability in the emergency command centres** (see Figure 1.5).

All this information, *usually available but not interconnected*, is now processed in the single platform ANYWHERE for EU (**A4EU**)[23] to **automatically identify the affected most vulnerable points**, including their characteristics and location, and other advanced services, with the capacity to **convert the warnings into actionable decisions supporting the response** in emergency management centres (see Figure 1.6).

This is an **innovation disruption in the field of Civil Protection and Emergency Management** because the system allows the emergency response specialist to **focus on**

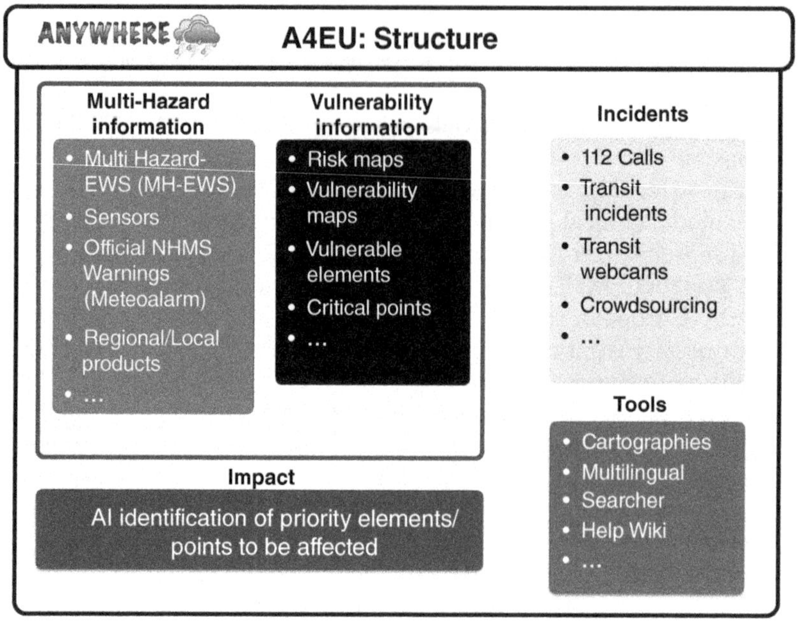

Figure 1.5 A4EU platform impact forecasting scheme: artificial intelligence is used to integrate the hazard impact forecasting products served by the MH-EWS with high-resolution local impact models and local layers of exposure and vulnerability in the emergency command centres.

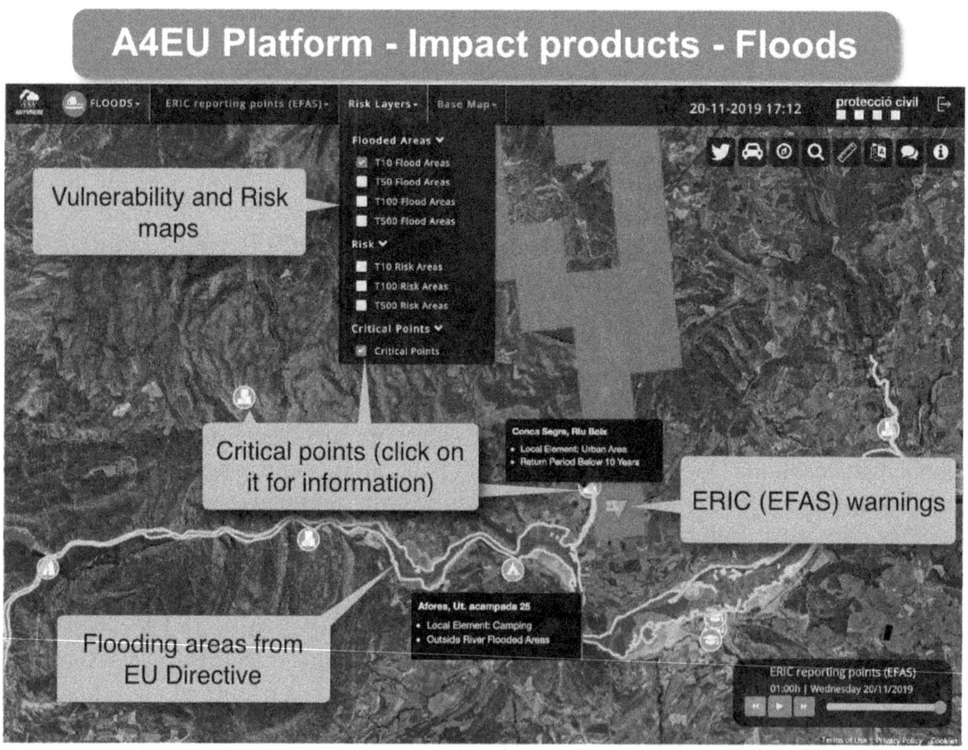

Figure 1.6 Impact forecasting implemented on the A4EU platform, showing critical points that are at risk, by automatic integration of the MHEWS forecasts with high-resolution impact models and the local vulnerability and exposure cartographies. The figure shows how the flash-flood early warning product ERIC (from EFAS) can be used to trigger an automatic warn to the predefined critical points potentially affected using the risk flood maps prepared under the Flood Directive.

local impacts, without the necessity to look in detail on the meteorological forecasts and triggers, **and on a narrow set of vulnerable locations** (i.e. Schools, Train Stations, Hospitals, Seveso facilities, among others) instead of vast regions, supporting them to magnify their response capabilities (see Figure 1.7).

1.4 The New Paradigm: Dynamic Vulnerability

The resilience of societies heavily depends on **how their citizens behave individually or collectively**. Therefore resilience, and emergency management in general, is **primarily based on the capacity to coordinate many human actions**; to share situation assessment; to make, implement and control coordinated actions; and to adapt the response to changing situations. In addition, the **human factor is essentially critical when looking at communications between first-responder authorities and citizens**, obtaining trust and confidence, avoiding false rumours and managing the psycho-social elements during the crisis and the recovery period afterwards.

In this framework, the traditional concepts of the emergency management cycle are facing a **critical paradigm shift due to the rapid change of society** by the disruption of mobile devices, IoT and technologies **transforming the world into an interconnected society**. However, the usual emergency management often disregards that both first responders and, around eighty per cent of European citizens are connected through mobile networks, carrying in our pockets what 15 years ago would have been an unimaginably sophisticated and miniaturized equipment, **wasting what should be seen as an outstanding opportunity**.

On the other hand, the increase of frequency and magnitude of extreme climate events induced by CC requires a **paradigm change in risk governance and policies** at European and global scales. The 'new normality' situation in which we are reaching new records more frequently requires acknowledging that the capacities of public response services might be exceeded and **the key role of the citizens and communities** and their **engagement throughout the phases of climate DRM** and CCA (Hügel and Davies 2020), as well as providing a methodological approach to **empower citizens as 'assets' in terms of self-protection response** (Sempere-Torres 2019).

Moreover, during an emergency situation, citizens and communities will in the first instance depend on themselves. The capacity of professional emergency services is limited and therefore prioritizes the weak and endangered persons who need them most. Thus, it is important that citizens and communities protect themselves during emergencies and this can be more efficiently achieved if **self-preparedness and protection plans** are pre-defined and incorporated into the routine of the communities. The key to the success of these self-protection plans and protocols is a multiple effort, including information (first knowing the risk in your environment, Terti et al. 2019, 2020; Weyrich et al. 2021), preparedness (taking the necessary precautionary measures to cope with emergency situations) and solidarity (reaching out to those in need before, during and after an emergency).

However, involving the citizens in DRM is a challenging societal transformation, especially in such a complex field which involves many stakeholders and organizations;

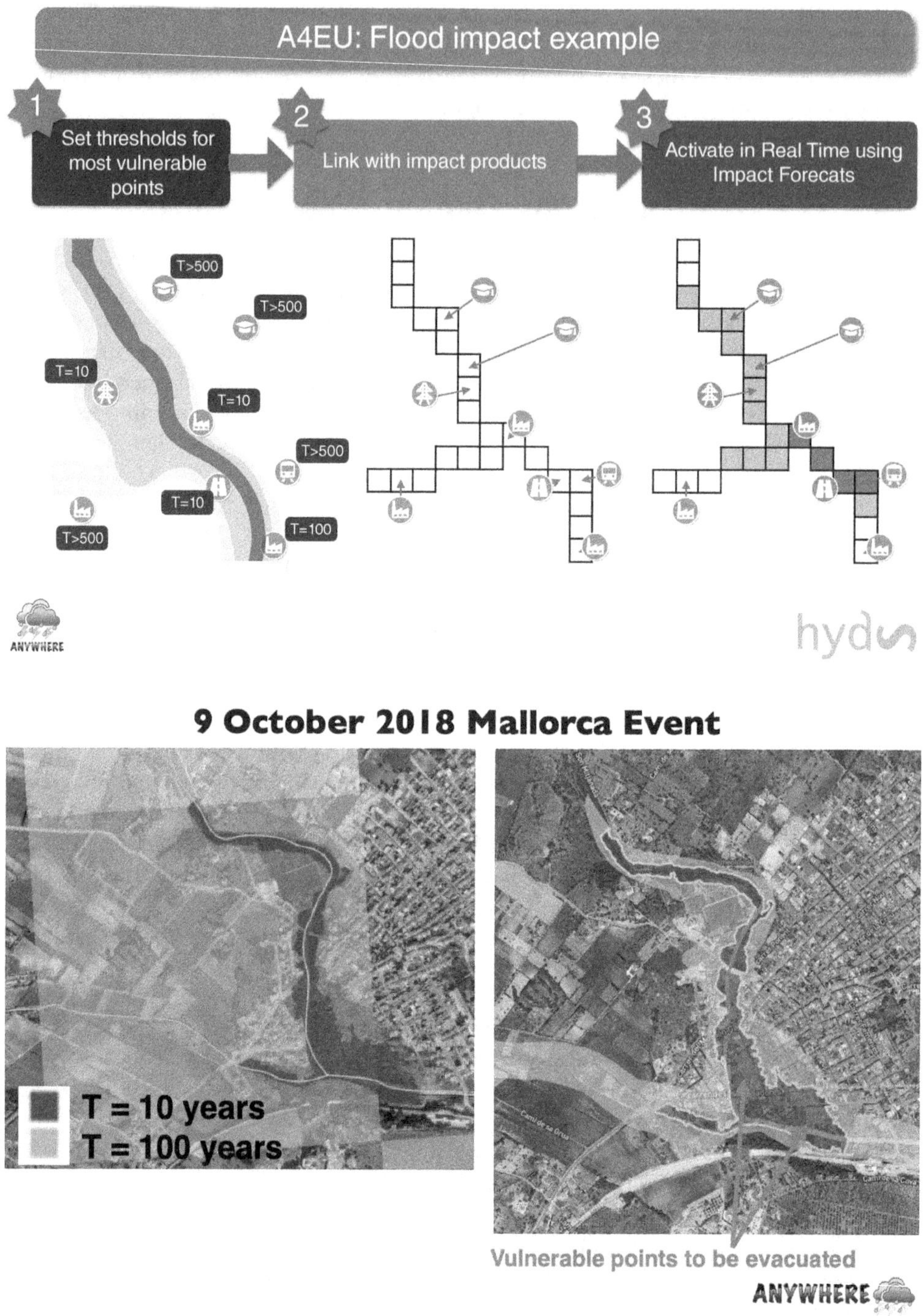

Figure 1.7 (Above) Scheme of the automatic activation of prioritary actions on the most vulnerable points for the A4EU flash flood impact forecasting. The pre-identified vulnerable points are labelled (Courtesy of HYDS). (Below) Application to the event of 9 October 2018 in Mallorca Island (ES). The orange pixel (1 × 1 km) is the trigger for a flash flood warning over 10-year return period The Flood Directive risk map allows the system to identify the associated area to be flooded and the vulnerable buildings in which the self-protection protocols should be activated.

with different objectives, procedures, reporting structures and definitions. And that involves citizens in various roles: as disaster victims, as potential sensors providing relevant information, as well as participating in generating and distributing rumours and fake news (Simon et al. 2015; Venier 2020). **Therefore, citizens can be both an important 'asset' to help first responders but may also create a lack of trust or confidence** (Díaz et al. 2016) in the emergency management since they 'use personal information and communication technology to respond to disasters in creative ways to cope with uncertainty' (Palen and Anderson 2016; Venier and Capone 2019).

At present, citizens involved in an emergency are **mainly seen by first responders as victims or as a potential nuisance or liability,** but not as a potential source of help ignoring their capacities **as an asset in crisis management**. However, in the next future, given the rising disaster risk due to population growth and CC, citizens as 'informal' volunteers (Whittaker et al. 2015) can effectively assist first/second responders by providing the much-needed additional support to react during emergencies. Changing this perception is **not just a technological challenge but a societal challenge** that requires understanding the present barriers and enablers and making a **considerable effort to show the benefits of embracing such a change.**

In this context, *ANYWHERE* has proposed the innovative concept of **dynamic vulnerability** (see Figure 1.8). First **high-resolution local information**, from predefined priority or most vulnerable points, is added to the impact-based early warnings**,** to convert them into **site-specific warnings** (SSWs, Melendez-Landaverde et al. 2020; Melendez-Landaverde and Sempere-Torres 2022). Thus, the information contained in the warnings is not only translated into the expected impacts on citizens, but can **add enriched information beyond text or voice messages**, including images simulating the impacts using augmented reality (e.g. how a specific road passing under a railway track will be

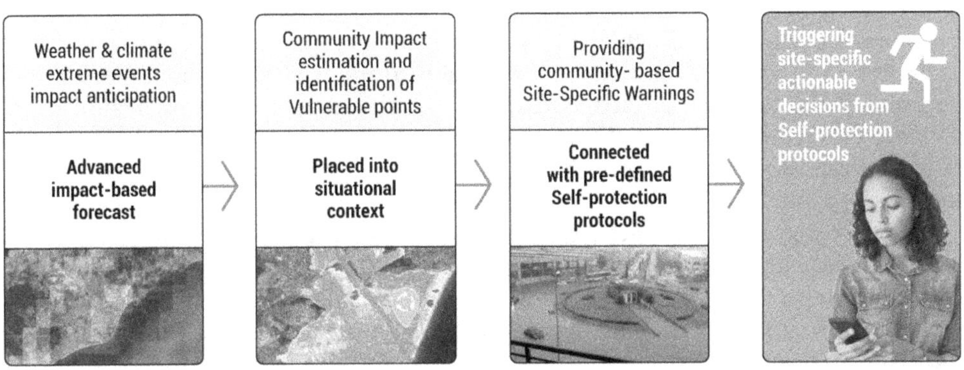

Figure 1.8 Dynamic vulnerability approach: impact forecasting IEWS are used to anticipate expected impacts in the next hours. These forecasts, reinterpreted in their social situational context, identify the most vulnerable points. The system provides SSWs for these points with pre-defined messages for the affected communities. It also enables all actors (authorities, first/second responders, engaged citizens and associations) to trigger pre-defined site-specific actions from the community self-protection protocol to reduce their vulnerability dynamically in a given location and time window.

affected); providing the best way to evacuate the place you are located; or **connecting with pre-defined actions of the self-protection plan of a given location, building or community**, even if the concerned people are just passing by.

These innovative SSWs are **the key instrument to connect the anticipated impacts with adequate actions to reduce vulnerability** as soon as the impacts are forecasted. These actions will emerge from a holistic, participatory approach (van Aalst et al. 2008) involving the civil, private sectors and the citizens **to co-design their community risk management plans and their self-protection protocols** to increase the resilience of their communities. Through these *self-protection plans*,[24] the **SSWs will be able to trigger pre-defined actions to reduce the vulnerability in a given location for a given time window** based on the local risk and response capabilities. This paradigm change in IEWS to support community adaptation is a societal transformation requiring a framework to gain **TRUST among the citizens** and **massively implement multi-risk plans of actions adapted to every city or community**. Chapter 6 of this book illustrates an example of application in the framework of the *ANYWHERE* project.

1.5 Future Work: From Multi-Hazards to Multi-Risk IEWS

The next step is moving from MH-IEWS, as the one proposed in ANYWHERE, into a **real-time multi-risk IEWS platform**, what requires supplying impact products able to identify the magnitude of the consequences (fatalities, damage and loss) triggered by weather and climate hazards. The objective is to provide **impact forecasts and/or estimates of the impacts** triggered by these hazards to generate the earliest impact information (mapping) to support the activation of community self-protection protocols, the rapid deployment of first responders in vulnerable areas, and the efficient management of the emergency.

For each hazard, this is achieved by setting the methodological approach shown in Figure 1.5: Real-time forecasts and estimates of the triggering phenomena are to be used as inputs in the hazard-assessment algorithms; the resulting hazard products are then combined with different layers characterizing exposure and vulnerability to describe the severity of the event in terms of the expected socio-economic impacts.

This new step ahead will need to capitalize on existing algorithms for hazard and impact assessment and extend them for improved impact forecasting/assessment in real-time integrating hazard forecasts with available exposure and vulnerability datasets considering factors such as scale application (from continental to regional and local) and data availability.

The **multi-risk approach** proposed (Figure 1.9) is defined as the integration of this scheme into a **unified real-time IEWS platform running algorithms for impact forecasting/estimation for the different types of weather/climate hazards**. Incorporating this impact-assessment capability for the different hazard types in the central processing core of the MR-IEWS platform supposes a significant step forward with respect to the results achieved up to now and gives direction to future works.

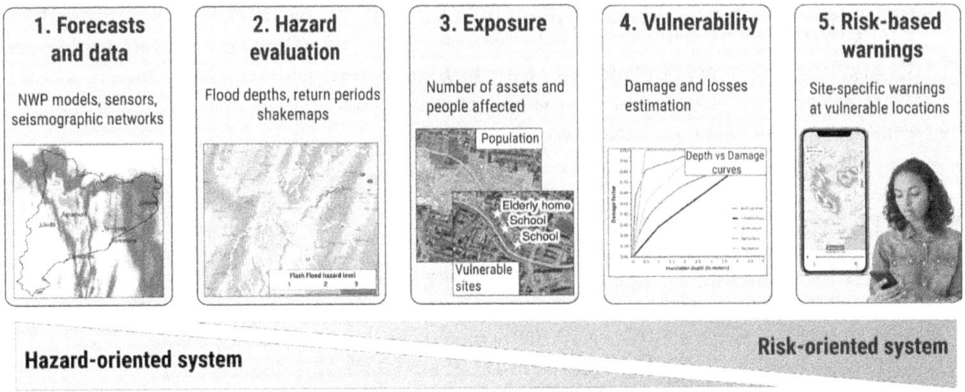

Figure 1.9 Scheme of the methodological approach to transform MH-IEWS into **multi-risk impact-based EWS** (MR-IEWS). *Source*: (5) Krakenimages.com / Adobe Stock.

Notes

1. https://www.europarl.europa.eu/news/en/press-room/20191121IPR67110/the-european-parliament-declares-climate-emergency.
2. https://www.cedamia.org/global/.
3. https://www.cedamia.org/global-ced-maps/.
4. https://ec.europa.eu/commission/presscorner/detail/en/ip_21_663.
5. https://research-and-innovation.ec.europa.eu/funding/funding-opportunities/funding-programmes-and-open-calls/horizon-europe/eu-missions-horizon-europe_en.
6. https://public.wmo.int/en/events/events-of-interest/un-global-early-warning-initiative-implementation-of-climate-adaptation.
7. https://www.thetimes.co.uk/article/germany-knew-the-floods-were-coming-but-the-warnings-didnt-work-cn99wjxzs.
8. https://www.munichre.com/en/solutions/for-industry-clients/natcatservice.html.
9. https://library.wmo.int/index.php?lvl=notice_display&id=17257#.Y_9UAi8zmZx.
10. https://www.efas.eu/en.
11. http://anywhere-h2020.eu.
12. https://home-affairs.ec.europa.eu/news/security-innovation-award-2022-2022-09-30_en.
13. https://www.dw.com/en/flooding-in-germany-before-and-after-images-from-the-ahr-and-eifel-regions/a-58299008.
14. https://www.efas.eu/en.
15. https://emergency.copernicus.eu.
16. http://anywhere-h2020.eu/services/multi-hazard-early-warning-platforms/a4eu/.
17. https://www.eumetnet.eu/activities/observations-programme/current-activities/opera/.
18. https://www.politico.eu/article/germany-floods-dozens-dead-despite-early-warnings/.
19. https://www.rtbf.be/article/comment-les-autorites-pouvaient-elles-anticiper-les-inondations-de-juillet-ce-que-disaient-les-previsions-europeennes-d-inondations-de-l-efas-10842799?id=10842799.
20. see http://anywhere-h2020.eu/services/multi-hazard-early-warning-platforms/a4eu/pilot-sites.
21. The full catalogue of products can be accessed at www.anywhere-h2020.eu/catalogue.
22. https://emergency.copernicus.eu.

23. The system is at present commercialised by ANYWHERE partners HYDS (www.hyds.es) and PREDICT SERVICES (www.predictservices.com) and is the base for a family of operational systems running in Ireland (Met Éireann), Spain (National Civil Protection), Catalonia (Civil Protection of Catalonia) and over 40 cities in Spain.
24. http://www.ingetecnia.com/en/consultori/pau.

References

van Aalst, M.K., Cannon, T., and Burton, I. (2008). Community level adaptation to climate change: the potential role of participatory community risk assessment. *Global Environmental Change* 18: 165–179. https://doi.org/10.1016/j.gloenvcha.2007.06.002.

Abily, M., Gourbesville, P., De Carvalho Filho, E. et al. (2020). Anywhere: enhancing emergency management and response to extreme weather and climate events. In: *Advances in Hydroinformatics* (ed. P. Gourbesville and G. Caignaert), 29–37. Singapore: Springer. https://link.springer.com/chapter/10.1007/978-981-15-5436-0_3.

Alfieri, L., Berenguer, M., Knechtl, V. et al. (2016). Flash flood forecasting based on rainfall thresholds. *Handbook of Hydrometeorological Ensemble Forecasting* 1–38. https://doi.org/10.1007/978-3-642-40457-3 (Full Text).

Armaroli, C., Duo, E., and Viavattene, C. (2019). from hazard to consequences: evaluation of direct and indirect impacts of flooding along the Emilia-Romagna Coastline, Italy. *Frontiers in Earth Science* 7. https://doi.org/10.3389/feart.2019.00203.

Briones, F., Vachon, R., and Glantz, M. (2019). Local responses to disasters: recent lessons from zero-order responders. *Disaster Prevention and Management* 28 (1): 119–125. https://doi.org/10.1108/DPM-05-2018-0151.

Centre for Research on the Epidemiology of Disasters (CRED) (2020). The human cost of disasters – an overview of the last 20 years 2000-2019. https://reliefweb.int/report/world/human-cost-disasters-overview-last-20-years-2000-2019 (accessed January 2023).

Cerreta, V., Tesfai, I., Poggioli, A. et al. (2019). ANYWHERE deliverable 5.2: report describing the products, services and results of the 4 case studies in Tasks 5.3-5.6. http://www.anywhere-h2020.eu/wp-content/uploads/docs/ANYWHERE_D5.2_submitted.pdf. (accessed January 2023)

Corral, C., Berenguer, M., Torres, D.S. et al. (2019). Comparison of two early warning systems for regional flash flood hazard forecasting. *Journal of Hydrology* 572: 603–619. https://doi.org/10.1016/j.jhydrol.2019.03.026.

Di Giuseppe, F., Pappenberger, F., Wetterhall, F. et al. (2016). The potential predictability of fire danger provided by numerical weather prediction. *Journal of Applied Meteorology and Climatology* 55: 2469–2491.

Di Giuseppe, F., Rémy, S., Pappenberger, F., and Wetterhall, F. (2017). Improving forecasts of biomass burning emissions with the fire weather index. *Journal of Applied Meteorology and Climatology 56* (10): 2789–2799. https://doi.org/10.1175/jamc-d-16-0405.1.

Di Giuseppe, F., Rémy, S., Pappenberger, F., and Wetterhall, F. (2018). Using the fire weather index (FWI) to improve the estimation of fire emissions from fire radiative power (FRP) observations. *Atmospheric Chemistry and Physics* 18: 5359–5370. https://doi.org/10.5194/acp-18-5359-2018.

Di Napoli, C., Pappenberger, F., and Cloke, H.L. (2018). Assessing heat-related health risk in Europe via the universal thermal climate index (UTCI). *International Journal of Biometeorology* 62 (7): 1155–1165. https://doi.org/10.1007/s00484-018-1518-2.

Di Napoli, C., Barnard, C., Prudhomme, C. et al. (2021a). ERA5-HEAT: a global gridded historical dataset of human thermal comfort indices from climate reanalysis. *Geoscience Data Journal* 8 (1): 2–10. https://rmets.onlinelibrary.wiley.com/doi/full/10.1002/gdj3.102.

Di Napoli, C., Messeri, A., Novák, M. et al. (2021b). The universal thermal climate index as an operational forecasting tool of human biometeorological conditions in Europe. In: *Applications of the Universal Thermal Climate Index UTCI in Biometeorology*, 193–208. Cham: Springer https://doi.org/10.1007/978-3-030-76716-7_10.

Diaz, P., Carroll, J.M., and Acedo, I. (2016). Coproduction as an approach to technology-mediated citizen participation in emergency management. *Future Internet 8* (3): 41. https://doi.org/10.3390/fi8030041.

Diaz, V., Perez, G.A.C., Van Lanen, H.A. et al. (2020a). Characterisation of the dynamics of past droughts. *Science of the Total Environment* 718: 134588. https://www.sciencedirect.com/science/article/pii/S0048969719345796.

Diaz, V., Corzo, G.A., Perez, H.A.J. et al. (2020b). An approach to characterise spatio-temporal drought dynamics. *Advances in Water Resources* https://doi.org/10.1016/j.advwatres.2020.103512.

Duo, E., Fernández-Montblanc, T., and Armaroli, C. (2020). Semi-probabilistic coastal flood impact analysis: from deterministic hazards to multi-damage model impacts. *Environment International* 143: 105884. https://www.sciencedirect.com/science/article/pii/S0160412020318390.

EC (2021a). A new EU strategy on adaptation to climate change. https://eur-lex.europa.eu/legal-content/EN/TXT/?uri=COM:2021:82:FIN (accessed January 2023).

EC (2021b). EU mission: adaptation to climate change. https://research-and-innovation.ec.europa.eu/funding/funding-opportunities/funding-programmes-and-open-calls/horizon-europe/eu-missions-horizon-europe/adaptation-climate-change_en (accessed January 2023).

EU Parliament (2019). Climate emergency declaration. https://www.europarl.europa.eu/news/en/press-room/20191121IPR67110/the-european-parliament-declares-climate-emergency (accessed January 2023).

Fehlmann, M., Gascón, E., Rohrer, M. et al. (2018a). Improving medium-range forecasts of rain-on-snow events in pre-alpine áreas. *Water Resources Research* 55: 7638–7661.

Fehlmann, M., Gascón, E., Rohrer, M. et al. (2018b). Estimating the snowfall limit in alpine and pre-alpine valleys: a local evaluation of operational approaches. *Atmospheric Research* 204: 136–148. https://doi.org/10.1016/j.atmosres.2018.01.016.

Fehlmann, M., Gascón, E., Rohrer, M. et al. (2020). Improving medium-range forecasts of rain-on-snow events in pre-alpine áreas. *Water Resources Research* 55: 7638–7661.

Fernández-Montblanc, T., Vousdoukas, M.I., Ciavola, P. et al. (2019). Towards robust pan-European storm surge forecasting. *Ocean Modelling* 133: 129–144. https://www.sciencedirect.com/science/article/abs/pii/S0964569119309445?via%3Dihub.

Fiorucci, P., Gaetani, F., and Minciardi, R. (2008). Development and application of a system for dynamic wildfire risk assessment in Italy. *Environmental Modelling and Software* 23: 690–702.

Gascón, E., Hewson, T., and Haiden, T. (2018). Improving predictions of precipitation type at the surface: description and verification of two new products from the ECMWF ensemble. *Weather and Forecasting* 33 (1): 89–108. https://doi.org/10.1175/waf-d-17-0114.1.

Gräßler, I., Pottebaum, J., Scholle, P., and Thiele, H. (2020). *Innovation Management and Strategic Planning of Innovative Self-Preparedness and Self-Protection Services*. Springer. https://doi.org/10.1007/978-981-15-5436-0_5.

Hügel, S. and Davies, A.R. (2020). Public participation, engagement, and climate change adaptation: a review of the research literature. *Wiley Interdisciplinary Reviews: Climate Change* 11 (4): https://doi.org/10.1002/wcc.645.

Kron, W., Löw, P., and Kundzewicz, Z. (2019). Changes in risk of extreme weather events in Europe. *Environmental Science and Policy* 100: 74–83. https://doi.org/10.1016/j.envsci.2019.06.007.

Láng-Ritter, J., Berenguer, M., Dottori, F. et al. (2022). Compound flood impact forecasting: integrating fluvial and flash flood impact assessments into a unified system. *Hydrology and Earth System Sciences Discussions* https://doi.org/10.5194/hess-2021-387. https://hess.copernicus.org/preprints/hess-2021-387.

Lavabre, J., Sempere Torres, D., and Cernesson, F. (1993). Changes in the hydrological response of a small Mediterranean basin a year after a wildfire. *Journal of Hydrology* 142: 273–299. https://doi.org/10.1016/0022-1694(93)90014-Z.

Marécal, V., Peuch, V.-H., Andersson, C., Andersson, S., Arteta, J., Beekmann, M., Benedictow, A., Bergström, R., Bessagnet, B., Cansado, A., Chéroux, F., Colette, A., Coman, A., Curier, R. L., Denier van der Gon, H. A. C., Drouin, A., Elbern, H., Emili, E., Engelen, R. J., Eskes, H. J., Foret, G., Friese, E., Gauss, M., Giannaros, C., Guth, J., Joly, M., Jaumouillé, E., Josse, B., Kadygrov, N., Kaiser, J. W., Krajsek, K., Kuenen, J., Kumar, U., Liora, N., Lopez, E., Malherbe, L., Martinez, I., Melas, D., Meleux, F., Menut, L., Moinat, P., Morales, T., Parmentier, J., Piacentini, A., Plu, M., Poupkou, A., Queguiner, S., Robertson, L., Rouïl, L., Schaap, M., Segers, A., Sofiev, M., Tarasson, L., Thomas, M., Timmermans, R., Valdebenito, Á., van Velthoven, P., van Versendaal, R., Vira, J., and Ung, A. 2015: *A regional air quality forecasting system over Europe: the MACC-II daily ensemble production*, *Geoscientific Model Development*, *8*, 2777–2813. https://doi.org/10.5194/gmd-8-2777-2015.

Meléndez-Landaverde, E.R. and Sempere-Torres, D. (2022). Design and evaluation of a community and impact-based site-specific early warning system (SS-EWS): the SS-EWS framework. *Journal of Flood Risk Management* e12860. https://doi.org/10.1111/jfr3.12860.

Melendez-Landaverde, E., Sempere-Torres, D., and Berenguer, M. (2020). Towards impact-based communication during emergencies: development of site-specific warning services in Catalonia. EGU General Assembly 2020, Online, Wien (Austria) (4–8 May 2020), EGU2020-1011. https://doi.org/10.5194/egusphere-egu2020-1011.

Merz, B., Kuhlicke, C., Kunz, M. et al. (2020). Impact forecasting to support emergency management of natural hazards. *Reviews of Geophysics* 58: e2020RG000704. https://doi.org/10.1029/2020RG000704.

Murray, V. (2021). Science and Technology Commitment to the Implementation of the Sendai Framework for Disaster Risk Reduction. *Proceedings of the 3rd Global Summit of Research Institutes for Disaster Risk Reduction* 55. https://link.springer.com/chapter/10.1007/978-981-15-8662-0_4.

Palau, R.M. (2021). Landslide and debris flow warning at regional scale. A real-time system using susceptibility mapping, radar rainfall and hydrometeorological thresholds for Catalonia (NE Spain). PhD thesis. Department of Civil and Environmental Engineering. Univeritat Politècnica de Catalunya. https://futur.upc.edu/32074725.

Palau, R.M., Hurlimann, M., Berenguer, M., and Sempere-Torres, D. (2020). Landslide and debris flow warning at regional scale. A real-time system using susceptibility mapping, radar rainfall and hydrometeorological thresholds for Catalonia (NE Spain). *Landslides* 17 (9): 2067–2083. https://doi.org/10.1007/s10346-020-01425-3.

Palau, R.M., Berenguer, M., Hürlimann, M. and Sempere-Torres, D. Implementation of hydrometeorological thresholds for regional landslide warning in Catalonia (NE Spain). *Landslides* (2023). https://doi.org/10.1007/s10346-023-02094-8.

Palen, L. and Anderson, K.M. (2016). Crisis informatics-new data for extraordinary times: focus on behaviors, not on fetishizing social media tools. *Science* 353 (6296): 224–225. https://doi.org/10.1126/science.aag257.

Park, S., Berenguer, M., and Sempere-Torres, D. (2019). Long-term analysis of gauge-adjusted radar rainfall accumulations at European scale. *Journal of Hydrology* 573: 768–777. https://doi.org/10.1016/j.jhydrol.2019.03.093.

Perello, N.; Trucchia, A.; D'Andrea, M.; Esposti, S.D.; Fiorucci, P. RISICO, an enhanced forest fire danger rating system: validation on 2021 extreme wildfire season in Southern

Italy. *Environmental Sciences Proceedings* 2022, *17*, 37. https://doi.org/10.3390/environsciproc2022017037

Poletti, M.L., Silvestro, F., Davolio, S. et al. (2019). Using nowcasting technique and data assimilation in a meteorological model to improve very short range hydrological forecasts. *Hydrology and Earth System Sciences Discussions* 102: https://doi.org/10.5194/hess-2019-75.

Ritter, J., Berenguer, M., Corral, C. et al. (2020). ReAFFIRM: real-time assessment of flash flood impacts – a regional high-resolution method. *Environment International* 136: 105375. https://www.sciencedirect.com/science/article/pii/S0160412019314485?via%3Dihub.

Ritter, J., Berenguer, M., Park, S., and Sempere-Torres, D. (2021). Real-time assessment of flash flood impacts at pan-European scale: the ReAFFINE method. *Journal of Hydrology* 603: 127022. https://www.sciencedirect.com/science/article/pii/S0022169421010726.

Ritter, J., Berenguer, M., Dottori, F. et al. (2022). Compound flood impact forecasting: integrating fluvial and flash flood impact assessments into a unified system. *Hydrology and Earth System Sciences Discussions* https://doi.org/10.5194/hess-2021-387. https://hess.copernicus.org/preprints/hess-2021-387.

Rossi, P.J., Hasu, V., Koistinen, J. et al. (2014). Analysis of a statistically initialized fuzzy logic scheme for classifying the severity of convective storms in Finland. *Meteorological Applications* **21**: 656–674. https://doi.org/10.1002/met.1389.

Schauwecker, S., Gascón, E., Park, S. et al. (2019). Anticipating cascading effects of extreme precipitation with pathway schemes – three case studies from Europe. *Environment International* 127: 291–304. https://doi.org/10.1016/j.envint.2019.02.072.

Sempere-Torres, D. (2019). EnhANcing emergencY management and response to extreme WeatHER and climate Events (ANYWHERE) Final report. D9.4. EC-H2020- H2020-DRS-01-2015-700099.

Simon, T., Goldberg, A., and Adini, B. (2015). Socializing in emergencies – a review of the use of social media in emergency situations. *International Journal of Information Management* 5 (35): 609–619. https://doi.org/10.1016/j.ijinfomgt.2015.07.001.

Sutanto, S.J. and Van Lanen, H.A. (2021). Streamflow drought: implication of drought definitions and its application for drought forecasting. *Hydrology and Earth System Sciences* 25 (7): 3991–4023. https://hess.copernicus.org/articles/25/3991/2021/hess-25-3991-2021.html.

Sutanto, S.J., van der Weert, M., Wanders, N. et al. (2019). Moving from drought hazard to impact forecasts. *Nature Communications* https://doi.org/10.1038/s41467-019-12840-z.

Sutanto, S.J., Vitolo, C., Di Napoli, C. et al. (2020a). Heatwaves, droughts, and fires: exploring compound and cascading dry hazards at the pan-European scale. *Environment International* 134: 105276. https://www.sciencedirect.com/science/article/pii/S0160412019308530.

Sutanto, S.J., Weert, M.V.D., Blauhut, V., and Van Lanen, H.A. (2020b). Skill of large-scale seasonal drought impact forecasts. *Natural Hazards and Earth System Sciences* 20 (6): 1595–1608. https://nhess.copernicus.org/articles/20/1595/2020/.

Sutanto, S.J., Wetterhall, F., and Van Lanen, H.A. (2020c). Hydrological drought forecasts outperform meteorological drought forecasts. *Environmental Research Letters* 15 (8): 084010. https://iopscience.iop.org/article/10.1088/1748-9326/ab8b13/meta.

Sutanto, S.J., Van Lanen, H.A.J., Wetterhall, F., and Llort, X. (2020d). Potential of pan-European seasonal hydro-meteorological drought forecasts obtained from a multi-hazard early warning system. *Bulletin of the American Meteorological Society* https://doi.org/10.1175/bams-d-18-0196.1.

Terti, G., Ruin, I., Kalas, M. et al. (2019). ANYCaRE: a role-playing game to investigate crisis decision-making and communication challenges in weather-related hazards. *Natural Hazards and Earth System Sciences* 19: 507–533. https://doi.org/10.5194/nhess-19-507-2019.

Terti, G., Ruin, I., Kalas, M. et al. (2020). Anycare: a serious game to evaluate the potential of impact-based and crowdsourced information on crisis decision-making. In: *Advances in*

Hydroinformatics (ed. P. Gourbesville and G. Caignaert), 103–120. Singapore: Springer https://doi.org/10.1007/978-981-15-5436-0_9.

Trucchia, A., D'Andrea, M., Baghino, F. et al. (2020). PROPAGATOR: an operational cellular-automata based wildfire simulator. *Firehouse 3* (3): 26. https://doi.org/10.3390/fire3030026.

Venier, S. (2020). A right to information relevant to disaster situations: broadening the concept beyond early warning and addressing the challenges posed by information and communication technologies. *Yearbook of International Disaster Law* https://doi.org/10.1163/26662531-01001011.

Venier, S. and Capone, F. (2019). Speaking with one or multiple voices in multi-hazard early warning systems? a survey of international and national legal and policy frameworks. *The Cambridge Handbook of Disaster Risk Reduction and International Law.* https://doi.org/10.1017/9781108564540.010.

Versini, P.-A., Velasco, M., Cabello, A. et al. (2013). Hydrological impact of forest fires and climate change in a Mediterranean basin. *Natural Hazards* 66: 609–628.

Vitolo, C., Di Giuseppe, F., and D'Andrea, M. (2018). Caliver: an R package for CALIbration and VERification of forest fire gridded model outputs. *PLoS One* 13 (1): e0189419. https://doi.org/10.1371/journal.pone.0189419.

Vitolo, C., Di Napoli, C., Di Giuseppe, F. et al. (2019). Mapping combined wildfire and heat stress hazards to improve evidence-based decision making. *Environment International* 127: 21–34. https://doi.org/10.1016/j.envint.2019.03.008.

Wagenbrenner, J., Ebel, B.A., Bladon, K.D., and Kinoshita, A.M. (2021). Post-wildfire hydrologic recovery in Mediterranean climates: a systematic review and case study to identify current knowledge and opportunities. *Journal of Hydrology* 602: https://doi.org/10.1016/j.jhydrol.2021.126772.

Weyrich, P., Ruin, I., Terti, G., and Scolobig, A. (2021). Using serious games to evaluate the potential of social media information in early warning disaster management. *International Journal of Disaster Risk Reduction* 56: 102053. https://www.sciencedirect.com/science/article/abs/pii/S2212420921000194.

Whittaker, J., McLennan, B., and Handmer, J. (2015). A review of informal volunteerism in emergencies and disasters: definition, opportunities and challenges. *International Journal of Disaster Risk Reduction* 13: 358–368. https://doi.org/10.1016/j.ijdrr.2015.07.010.

Wine, M.L. and Cadol, D. (2016). Hydrologic effects of large southwestern USA wildfires significantly increase regional water supply: fact or fiction? *Environmental Research Letters* 11: 085006.

WMO (2021). Guidelines on multi-hazard impact-based forecast and warning services – part II: putting multi-hazard IBFWS into practice. https://reliefweb.int/report/world/wmo-guidelines-multi-hazard-impact-based-forecast-and-warning-services-part-ii-putting.

WMO (2023). Past eight years confirmed to be the eight warmest on record. https://public.wmo.int/en/media/press-release/past-eight-years-confirmed-be-eight-warmest-record. (accessed January 2023)

2

Hydrometeorological Drought Forecasts
Lessons Learned from ANYWHERE and Next Steps to Improve Drought Management

Samuel J. Sutanto[1] and Henny A.J. Van Lanen[2]

[1] *Water System and Global Change, Department of Environmental Sciences, Wageningen University and Research, Wageningen, the Netherlands*

[2] *Hydrology and Quantitative Water Management, Department of Environmental Sciences, Wageningen University and Research, Wageningen, the Netherlands*

2.1 Introduction

Drought, as one of the most severe weather-related natural disasters, causes damages and losses comparable to other destructive hazards, such as floods and landslides (FAO 2017). The economic loss in all sectors due to drought events in Europe, from 1976 to 2006, was estimated at more than 100 billion EUR (EC 2007), with an average of 9 billion EUR annually (Rekovec et al. 2022). In the United States, for comparison, the total economic loss during drought and wildfires in California in 2020 was estimated at around \$10–20 billion (UNCCD 2022). In other drought vulnerable regions, such as Africa and Asia, the impacts of drought mainly affect the agricultural sectors, especially livestock and crop production. FAO (2017) estimated that agriculture accounts for more than 80% of the damage and losses due to drought worldwide. These numbers will increase threefold by 2100 if 2 °C global warming scenario is assumed and no adaptation measures are taken (EC 2021).

Due to the high losses triggered by drought in Europe, the EU Joint Research Centre (JRC) has developed the European Drought Observatory (EDO) in the framework of

Responding to Extreme Weather Events, First Edition. Edited by Daniel Sempere-Torres, Anastasios Karakostas, Claudio Rossi, and Philippe Quevauviller.

the 2007 EU Water Scarcity and Drought Communication to provide European-wide up-to-date information on the occurrence and severity of droughts (Vogt et al. 2013). EDO, in principle, produces a warning signal based on the forecasted Standardized Precipitation Index (SPI) with a three-month lead-time and soil moisture anomaly with a seven-day lead time, meaning that a meteorological drought forecast and a soil moisture drought forecast are available. These lead times, in particular the weekly one, are short, because of the creeping nature of drought hazard. Drought events may sustain from months up to more than a year (multi-year drought). Furthermore, hydrological droughts, e.g. streamflow drought and groundwater drought, are not forecasted by EDO. For these reasons, a seasonal drought early warning system (EWS) has been developed under the EU-funded ANYWHERE innovation project, i.e. ANYWHERE Drought EWS (ADEWS), which provides the forecasted drought up to seven months ahead and includes more hydrometeorological variables than precipitation and soil moisture. ADEWS is a part of the multi-hazard EWS (MH-EWS) that addresses all major weather-related natural hazards in Europe, such as floods, flash floods, landslides, storm surges, heatwaves and air quality, wildfires, convective storms, and heavy snowfall. This is in line with the implementation of the United Nations Sendai Framework for Disaster Risk Reduction 2015–2030 that highlights the need for MH-EWS as one of the important Sendai's global targets (Poljanšek et al. 2017).

As mentioned above, the ADEWS generates seasonal hydrometeorological drought forecasting products that cover most of the water cycle components (drought in precipitation, soil moisture, groundwater, runoff, and streamflow). Please note that the term hydrological drought forecasting, however, may create misunderstanding and confusion in the scientific hydrometeorological community (authors, readers, editors, and reviewers), as well as among policymakers, and in operational weather and water management services. First, forecasting of hydrological time series (e.g. streamflow) obtained from the hydrological model forced with climate prediction (not projection) cannot be classified as streamflow drought forecasts in the narrower sense. These are known as streamflow forecasts. One needs to realize that hydrological drought forecasts require an additional step in the process, which is applying a drought identification method to the forecasted time series of hydrological variables (see Sutanto et al. 2020a; Sutanto and Van Lanen 2021). In other words, forecasted hydrological drought is a derived product of hydrological forecasts. Second, there is a clear difference between low streamflow and drought in streamflow. Many scientific communities use the terms streamflow drought and low flow interchangeably. Low flow is not necessarily classified as a drought, although streamflow drought can be caused by a long period of below-normal low streamflow (e.g. see https://www.epa.gov/ceam/definition-and-characteristics-low-flows#low). According to Tallaksen and Van Lanen (2004), the concept of a drought refers to a certain threshold that distinguishes a drought event from a non-drought situation and the event has thus beginning and an end. Therefore, the general concept of drought has key aspects of the time of occurrence, drought duration, severity, and in some cases deficit volume, which is missing water in the river. Low flow characteristics, on the other hand, describe the lowest streamflow each year, i.e. the annual minimum series, which do not in all years, especially in wet years, indicate a streamflow drought (Hisdal et al. 2004; Van Loon 2015). Third, a clear distinction should be made between drought forecasts and drought projections under a warming climate.

The latter uses climatological projection data obtained from Global Climate Models with assumed future scenarios for concentrations of greenhouse gases, aerosols, and other atmospheric constituents. These data are then used to force a hydrological model, resulting in hydrological time series from the past to the future (e.g. 2050 and 2100). Drought projections are derived from these hydrological projection data sets and not from the climate forecasts.

Given the clear distinctions among different aforementioned products, we would like to stress that the ADEWS provides hydrometeorological drought forecasts, which are defined as the forecasting of below-normal hydrometeorological variables (Hisdal et al. 2004; Van Loon 2015). The experience gained during the development of ADEWS is presented in this chapter, which is anticipated to be of utmost importance for any operational hydrometeorological institution in developing DEWS. The chapter starts with a method for forecasting hydrometeorological droughts, including the forecast data and hydrological model used in the ADEWS, approaches to derive droughts from hydrometeorological forecasted time series, and the drought forecast algorithms (Section 2.2). It then continues with examples of meteorological and hydrological drought forecasts derived from the ADEWS for 2018 drought in Europe (Section 2.3). Section 2.4 describes the performance of the ADEWS. In this section, we discuss the origin of seasonal hydrometeorological drought forecast skill, which is higher than that of the forecasted hydrometeorological time series. The performance of ADEWS is also evaluated using the categorical drought classification that has been developed in the ANYWHERE project and the commonly used metric, such as Brier Skill Score (BSS; Brier 1950). The importance of catchment memory in shaping the skill of hydrological drought forecasts is presented in Section 2.5. Finally, an outlook and future improvements of hydrometeorological drought forecasting is given in Section 2.6.

2.2 Method for Forecasting Hydrometeorological Droughts

2.2.1 The Climate (ECMWF SEAS5) and Hydrological (LISFLOOD) Models

Unlike EWSs for floods, storms, and landslides that require forecasts at short lead times, such as hours and days, a drought EWS requires a long forecast time scale, from month up to (sub) season due to its creeping nature of the drought phenomenon (Tallaksen and Van Lanen 2004; Van Loon 2015). For this reason, we utilized the European Centre for Medium-Range Weather Forecasts Seasonal (ECMWF SEAS) forecasting system S5 with a lead time of seven months (Stockdale et al. 2018). The forecasting system covers the global domain and contains perturbation models for data assimilation and generation of forecast ensembles (ENS, 51 ensemble members).

LISFLOOD, the hydrological model used within European Flood Awareness System (EFAS; Pappenberger et al. 2011; Cloke et al. 2013), is a spatially distributed, raster-based rainfall-runoff model, which has been developed at JRC (van der Knijff et al. 2010; Burek et al. 2013). It is driven by gridded meteorological forcing data (precipitation, temperature, potential evapotranspiration, and evaporation rates for open water and bare soil). LISFLOOD calculates a complete daily water balance for every grid cell within the model domain. The LISFLOOD model used in our study incorporates water abstraction for irrigation, water demands for the livestock, energy production and cooling, and manufacturing industry, and 1454 surface water reservoirs across Europe (Arnal et al. 2019). Although originally the LISFLOOD model was developed for flood forecasting, the model is also capable to realistically produce drought and low flows in Europe (Forzieri et al. 2014; Wetterhall and Di Giuseppe 2018; Sutanto and Van Lanen 2020, 2022), as well as in Africa (Trambauer et al. 2013).

Two types of data sets are obtained from EFAS, i.e. proxy hydrometeorological observed data and forecast data. Gridded meteorological observation data (5 by 5 km) collected from >5000 ground observations (meteorological observed data) were used to provide a continuously updated simulation of current conditions with the LISFLOOD model. This simulation, referred to as the simulation forced with observation (SFO or proxy observed), is used (i) to provide hydrological initial conditions (HICs) for the forecasts, (ii) to monitor the current state of the European hydrological system, and (iii) as a proxy for hydrological observed data (from 1990 to present). The same model was run with the forecast meteorological data obtained from the ECMWF SEAS5 to simulate forecasted hydrological data.

2.2.2 The Drought Indices

Two well-known drought identification methods, i.e. the standardized drought approach and the threshold drought approach were implemented in the ADEWS. Standardized indices are used to indicate anomalies (drought) in different domains of the water cycle (e.g. drought in precipitation, groundwater, and runoff). The Standardized Precipitation Index (SPI-x, McKee et al. 1993), the Standardized Precipitation Evaporation Index (SPEI-x; Vicente-Serrano et al. 2010), the Standardized Runoff Index (SRI-x; Shukla and Wood 2008), and the Standardized Groundwater Index approaches (SGI-1; Bloomfield and Marchant 2013) have been implemented with accumulation period $x = 1, 3, 6$, and 12 months (temporal resolution). The standardized indices allow determination of the probability of an ongoing drought at a given time scale (temporal resolution) for any site (e.g. grid cell) based on historical hydrometeorological data (WMO 2012). However, it can also be applied in the forecasting mode (Yuan et al. 2013a; Dutra et al. 2014a), if hydrometeorological forecasts are available. The standardized drought index calculation for any grid cell is based on a long monthly hydrometeorological record that is fitted to a probability distribution, which is then transformed into a normal distribution so that the median drought index for the site and selected accumulation period is zero. The standardized indices can be used to determine drought characteristics, such as drought onset (month), termination (month), and duration (month) (Figure 2.1a). In the ADEWS, a drought event is assumed to occur when the standardized drought index is less than 0 (i.e. that the event is dryer than the median for that grid cell) and ends when the standardized drought index becomes greater or equal than 0. Each month during a drought event, the standardized drought index (X) is categorized into four classes: (i) mild drought:

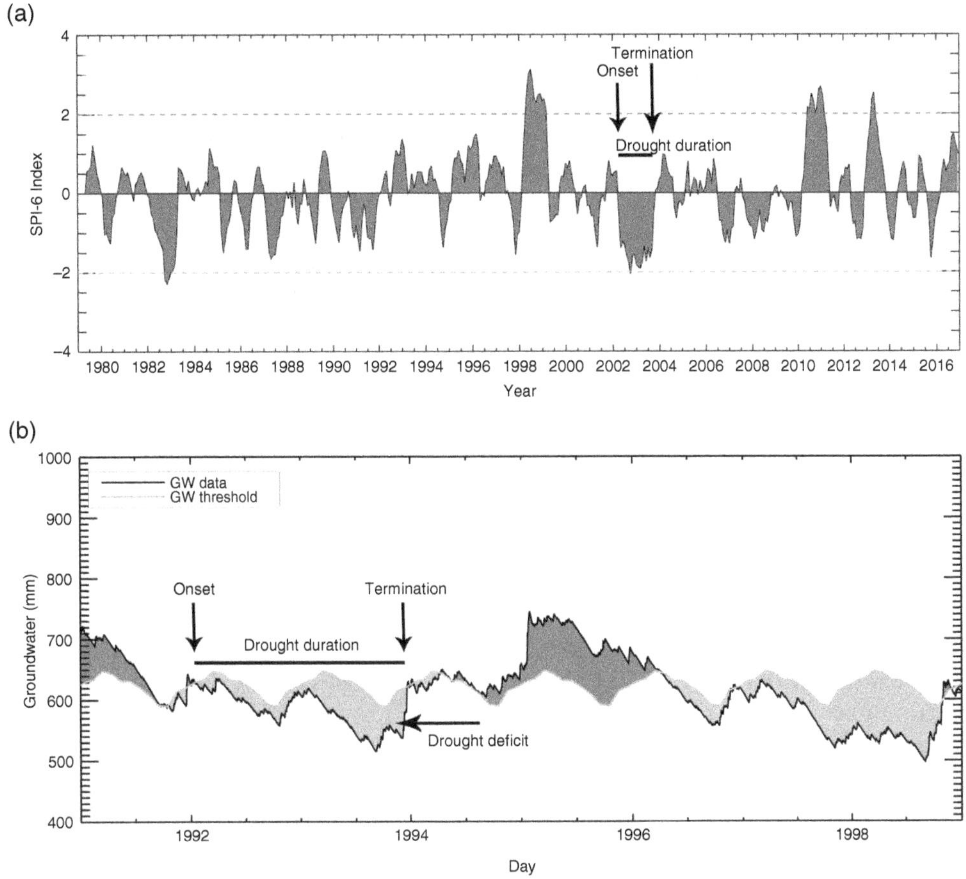

Figure 2.1 Drought characteristics derived from (a) the standardized drought method, i.e. SPI-6, and (b) the variable threshold method. Areas below zero for SPI-6 and red-shaded areas for the threshold method indicate drought event. *Source*: Modified from Sutanto and Van Lanen (2020).

$-1 \leq X < 0$, (ii) moderate drought: $-1.5 \leq X < -1$, (iii) severe drought: $-2 \leq X < -1.5$, and (iv) extreme drought: $X < -2$ (red line in Figure 2.1a). The definition of mild drought with $-1 \leq X < 0$, however, will produce many mild drought events. WMO (2012) revised the SPI categories by grouping SPI values from 0.99 to −0.99 into near-normal condition. Thus, the mild drought category, as mentioned above, was removed from the SPI classification. However, in the ANYWHERE project, we defined mild drought as $-1 \leq X < 0$, which should be revised in the future.

Similar to standardized indices, drought threshold indices are also frequently used in particular to quantify water deficits (drought) in different domains of the water cycle (precipitation, soil moisture, runoff, streamflow, and groundwater). Compared to standardized indices, the threshold approach can quantify missing water volumes that are needed to recover from the drought. It is based on defining a threshold, below which the precipitation or other hydrological variables is considered in a drought

(Yevjevich 1967; Hisdal et al. 2004). A drought event starts when the time series falls below the defined threshold value and ends when the time series rises above the defined threshold value. The following drought characteristics are determined: (i) onset and termination (day), (ii) duration (days), and (iii) deficit volume (mm) for fluxes (precipitation, runoff, streamflow) or average cumulative deviation from the threshold (mm) for state variables (soil moisture, groundwater) (Figure 2.1b). The threshold values are usually derived from the 80th percentile of the duration curve (e.g. Q80), which means that the values are equal or exceeded in 80% of the time (Hisdal et al. 2004; Van Loon and Van Lanen 2012). During a prolonged dry period, it is often observed that the data exceed the threshold level for a short period of time and thereby a large drought is divided into a number of minor droughts that are mutually dependent, which usually leads to pooling droughts (Hisdal et al. 2004; Fleig et al. 2006). We pooled minor drought by applying a centred 30-days smoothing to all forecasted hydrometeorological variables (Van Loon and Van Lanen 2012) and to reduce the number of minor droughts due to flashy behaviour. All threshold indices are calculated for median, 10th, 25th, 75th, and 90th percentiles obtained from the 51 ensemble members.

2.2.3 The Drought Forecast Algorithms

In the ADEWS, the drought identification is carried out in two steps (Figure 2.2). The first step (left part, Figure 2.2) is developing: (i) algorithms to calculate the probability distributions for each of the standardized indices for each of the 12 months (i) and accumulation levels (x), and (ii) algorithms to calculate the threshold values for each of the 12 months and each of the hydrometeorological variable. This step was carried out using the LISFLOOD SFO (proxy observed data, Section 2.2.1). The probability distributions and threshold values for each month and variable were calculated using 29 years of daily SFO data. In the second step (right part, Figure 2.2), we have developed drought algorithms to process the forecasted hydrometeorological time series into drought products using the probability distributions or threshold values that have been calculated in the first step. These algorithms are encapsulated in the ADEWS and the second-step algorithms are run every month when new forecast time series become available.

Monthly time series of proxy data from the last 11 months and daily data from the last 1 month are needed to calculate the droughts using standardized indices with different accumulation levels (i.e. $x>1$) and to calculate the droughts using threshold method, respectively. For example, for SPI-12, 11 months of proxy observed data are blended with the first month of forecasted data to predict a possible drought event in the first month ($i=1$). To predict a possible drought event in the second month ($i=2$) using the same accumulation level ($x = 12$), 10 months of proxy data and 2 months of forecasted data are blended, and so on for other months (Figure 2.3). Therefore, the calculation of drought forecasts using standardized indices for $x>1$ contains both observed data (or a proxy for observed, as in the ADEWS) and forecast data. The ratio of observed and forecast data depends on the selected temporal accumulation level and lead time. Higher accumulations level and shorter lead times require more observed data than lower accumulation levels and longer lead times (Yuan et al. 2013a; Dutra

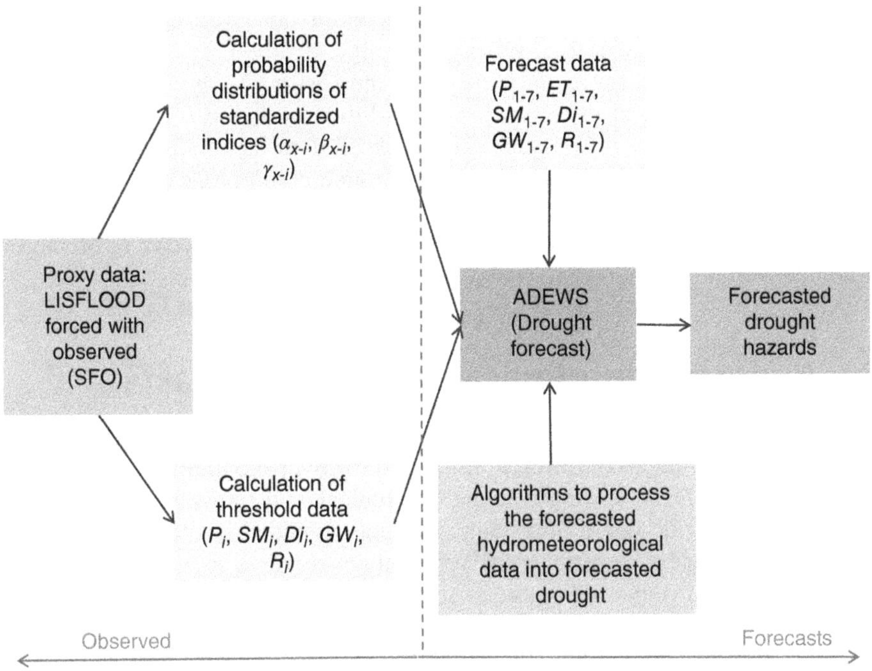

Figure 2.2 Flowchart of drought algorithms, where $\alpha_{x\text{-}i}$, $\beta_{x\text{-}i}$, and $\gamma_{x\text{-}i}$ stand for standardized indices parameters alpha, beta, and gamma. *i* stands for months (1–12 months), and *x* stands for temporal accumulation levels (1, 3, 6, and 12 months). *P* is precipitation, SM is soil moisture, *D* is streamflow, *T* is temperature, and *R* is runoff. *Source*: Adapted from Sutanto et al. (2020a).

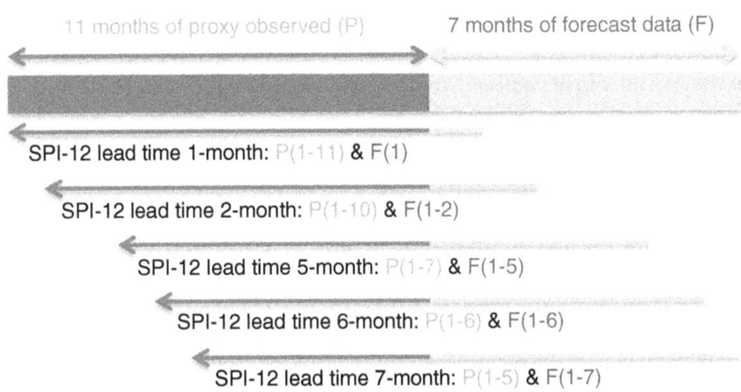

Figure 2.3 Schematic presentation of the SPI-12 calculation using a fusion of proxy observed time series data and forecast time series data. *Source*: Modified from Sutanto et al. (2020a).

et al. 2014b). For the threshold method, 15 days of observed data and 15 days of forecast data are required to predict a possible drought even on the first day. Similarly, 14 days of observed data are combined with 16 days of forecast data to predict a possible drought event on the second data and so on. This means that the prediction of day 30 does not include the observed data. The merging of data (proxy observed, SFO, and SEAS S5), which have the same climatological properties within proxy observed, reforecast (known as hindcast or past forecast), and with operational forecasts, is basically a fusion of proxy observed with seasonal forecast of precipitation/hydrological variables.

2.3 Hydrometeorological Drought Forecasts

The ADEWS provides a wide range of hydrometeorological drought forecasting products to fulfil the needs of different users and to cover the multi-faced drought phenomenon. In the ADEWS, we cover meteorological, soil moisture, and hydrological droughts. With the wide range of drought products provided in the ADEWS, different drought impacts can be investigated. Every impact requires its own specific drought index or own set of drought indices since not every drought impact can be derived from a single drought index. Thus, it is very important to deliver suitable indices for studying the targeted impact. For the agricultural sector, information on water availability for plants to grow (precipitation and soil moisture drought indices in the upcoming months) is more relevant than, for instance, drought indices addressing runoff. For water resources managers, on the other hand, the streamflow, runoff, and groundwater drought forecasting products (e.g. streamflow and runoff threshold indices, SRI-x, SGI-1) are far more important than products based on solely precipitation forecasts (e.g. SPI-x, precipitation threshold). With the wide range of different indices, accumulation periods and lead times provided by the ADEWS, the users can decide by themselves which indices are best suited for their needs. Suggestions are given by Kruse and Seidl (2013) who specify information about drought variables needed from the perspective of user groups. Here, we provide examples of meteorological drought forecasts (i.e. SPI) and hydrological drought forecasts (i.e. SGI) to illustrate the great potential of the ADEWS.

2.3.1 Meteorological Drought Forecasts

Forecasted meteorological drought events in each grid cell in Europe with lead times up to seven months can be obtained from different standardized products in different percentiles of the 51 ensemble members (median, 10th, 25th, 75th, and 90th percentile). Figure 2.4 shows some examples of the forecasted severity of European meteorological drought events derived from the SPI-3 based on the median of the 51 ensemble members for July up to September 2018. The SPI was calculated using precipitation data forecasted at the beginning of July 2018. Here, the ADEWS was proved as a useful DEWS that can forecast the drought events in Europe months ahead (see also Section 2.4). Northern, central, west European countries, and the United Kingdom were forecasted to experience severe to extreme drought events based on SPI-3 in July and

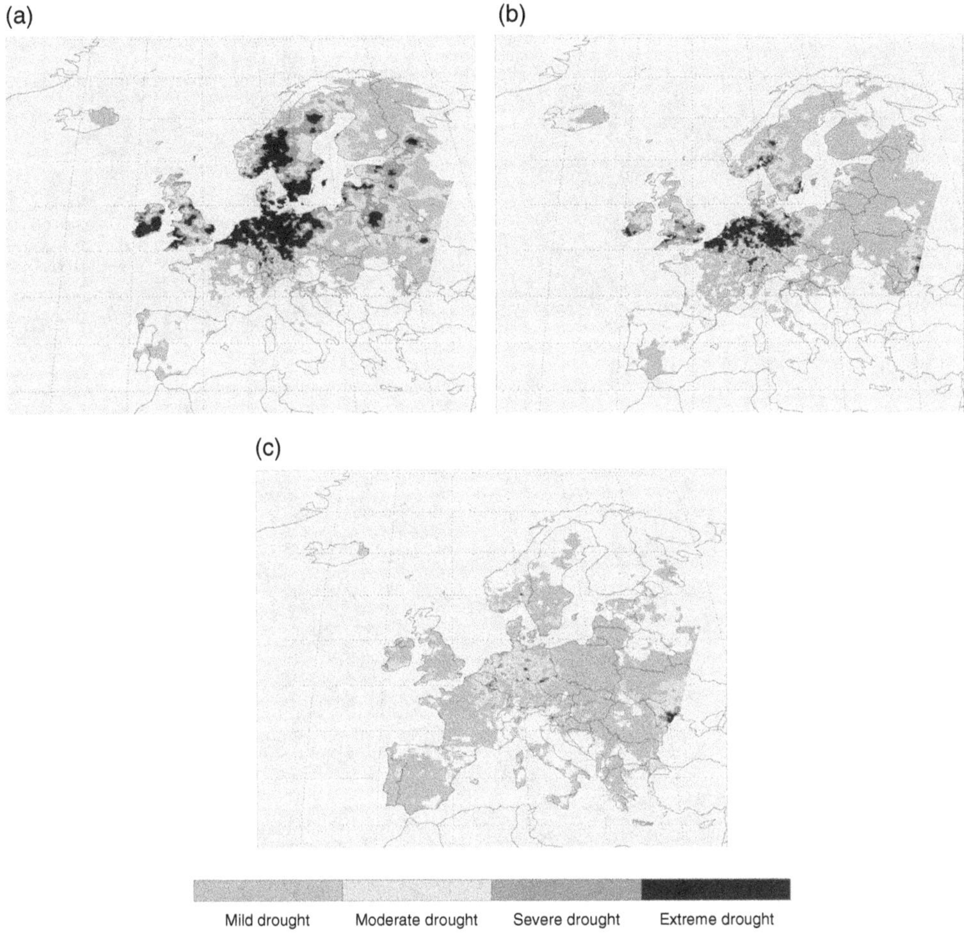

Figure 2.4 Forecasted drought severity class (median of the 51 ensemble members) using SPI-3 for the months July (a), August (b), and September 2018 (c) obtained from the forecast done on 2 July 2018.

August 2018, and droughts were expected to slowly alleviate starting from September 2018, with many regions in Europe affected by mild to moderate droughts. The 2018 drought is well known in Europe and the forecasted spatio-temporal pattern was observed in the majority of the European regions (Hari et al. 2020; Rakovec et al. 2022).

2.3.2 Hydrological Drought Forecasts

Hydrological drought forecasts issued by the ADEWS also show groundwater drought events (SGI) in most of European regions in 2018 but with less severe than the meteorological droughts (compare Figures 2.5 and 2.4). The forecasted SGI indicates mild to moderate drought for the same regions as for the SPI-3 (Figure 2.4). In general, the SPI drought class is more severe than the SGI class, indicated by more reddish colours on

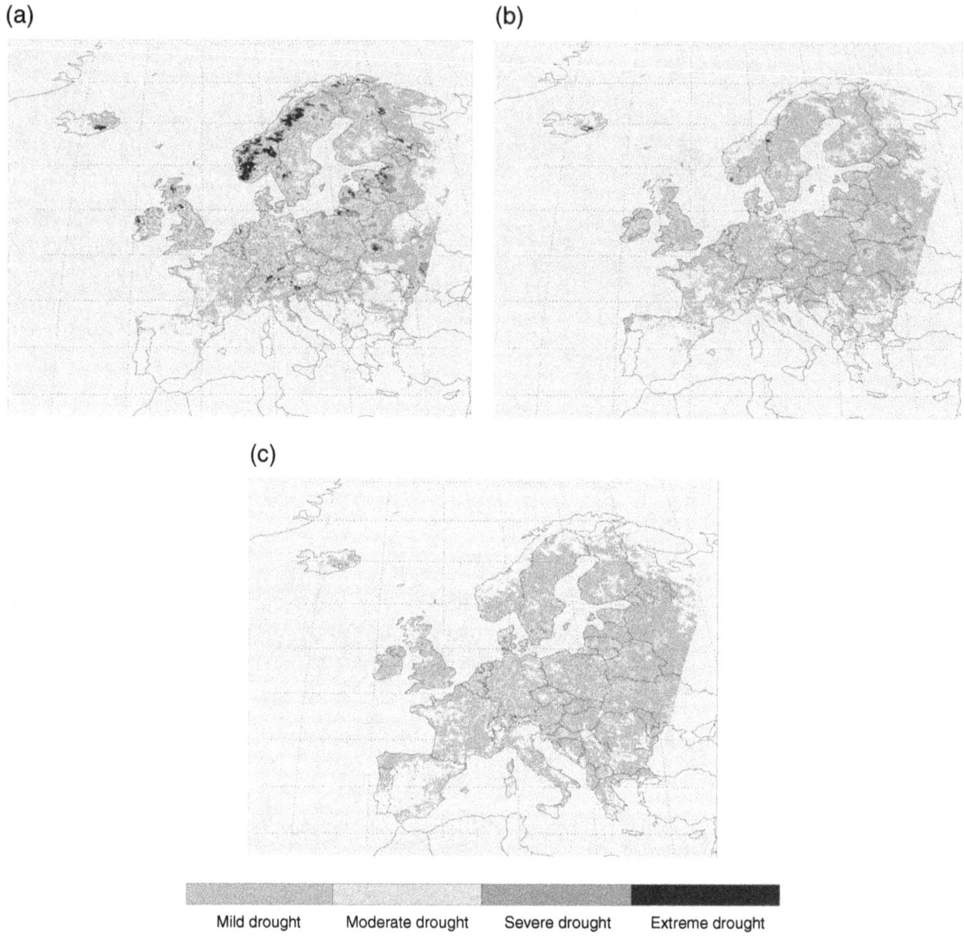

Figure 2.5 Forecasted drought severity class (median of the 51 ensemble members) using SGI for the months July (a), August (b), and September 2018 (c) obtained from the forecast done on 2 July 2018.

the maps. This might be attributed to drought propagation, meaning meteorological droughts were observed before groundwater recharge drought, and they propagate to groundwater (Peters et al. 2003; Van Loon et al. 2012). The propagation of meteorological drought to hydrological systems generally leads to less severe hydrological droughts, except for extreme and long meteorological droughts, which might generate severe hydrological drought (Van Lanen 2006).

Besides the standardized drought indices, the ADEWS also provides drought characteristics, such as the onset and termination of the longest drought (day number) within the forecast period (seven months), the total drought duration (in days rather than in months as for the standardized indices), the deficit volume (for fluxes), and average cumulative deviation from threshold (for state variables) for seven months ahead for precipitation, soil moisture, runoff, and groundwater. For streamflow drought, there are three products available, showing the number of ensemble members in drought (*n* out of 51), drought duration, and drought deficit volume in the longitudinal profile of the major European river (see Sutanto et al. 2020a for examples). In the ADEWS, drought characteristics are provided for both the standardized and threshold drought methods. The forecasted drought onset and termination give information to the users on when the drought will start and when it will end, respectively, the forecasted drought duration provides information on the cumulative length of all drought events (major and minor) in the seven months forecast period, and the forecasted drought deficit volume/average cumulative deviation from threshold is potentially useful information, especially for water resources managers. The latter applies only for the threshold droughts and not to the standardized indices (e.g., SPI, SRI, SGI).

2.4 Drought Forecast Performance

2.4.1 The Origin of Seasonal Drought Forecast Skill

During the development of ADEWS, stakeholders and end users were often questioning the performance of the ADEWS. Stakeholders and end users often ask what is the skill of a DEWS? How many months in advance can drought be forecasted? End users are typically more familiar with weather and/or hydrological forecasts, e.g. precipitation and streamflow forecasts rather than drought forecasts (Section 2.1). Many doubt that drought can be forecasted a few months ahead because they have only experience with weather forecasts that have skill only for a few days up to a week ahead (e.g. see Vitart 2004; Cui, et al. 2012). Thus, the follow-up question that may arise is why drought forecasts have a significant skill from a month up to a (sub)season, whereas weather forecasts cannot go beyond days? Why does the low skill of precipitation forecasts not translate to low-drought forecast skill? There are at least three main reasons why drought forecasts outperform weather forecasts. The first is the temporal resolution used in the drought analysis, which is typically a month, the second is the memory of the data or accumulation period using in the standardized indices and the application of moving average for daily data (Section 2.2.3), and the third is the catchment memory of the hydrological systems (Section 2.5). All play vital roles in explaining the skill of drought forecasts.

The use of monthly resolution in the drought forecasts, e.g. in the standardized indices, increases the forecast skill. Drought forecasts do not require the detailed day-to-day (or even sub-daily) evolution of weather, as demanded in short-term or medium-range weather forecasts (1–10 days forecasts). Instead, drought predictions need estimates of monthly or seasonal-mean statistics of weather, which provide information on how likely the coming season will be wetter or drier than normal (Weisheimer and Palmer 2014). Over- or underestimation of precipitation by 10 mm on a daily basis is vital for weather and associated flood forecasting. For drought forecasting, on the other hand, missing or having more precipitation of 10 mm on a certain day in a month is less significant as long as it can be counteracted by more or less precipitation on other days in the month. Thus, the predicted monthly median of precipitation derived from the ensembles is still fairly reliable. This fits well with the purpose of seasonal weather and drought forecasts that are used to identify the monthly deviation of available water from normal.

The fusion of preceding monthly observed data with forecasts to calculate drought indices with higher accumulation periods (Section 2.2.3) increases the forecast skill. Even though drought forecasts use a monthly temporal resolution, the skill of meteorological drought indices using the SPI with one-month accumulation period (SPI-1) is still quite low due to the chaotic nature of the atmosphere. The SPI-1 consists of only one-month forecasted precipitation data, and no observed data are merged in the calculation of drought indices. The skill will be improved if we incorporate the observation data into forecasts, e.g. using the SPI-x, with an accumulation period x longer than one month (Yuan et al. 2013a; Van Hateren et al. 2019; Sutanto et al. 2020b). For the threshold approach, the smoothing of the forecasted data, for instance using the 30-day moving average (30DMA), also increases the forecast skill. The application of 30DMA is commonly used to derive drought in streamflow (e.g. Yuan, et al. 2013b; Sutanto and Van Lanen 2021) and is effectively smoothing the forecasted daily streamflow data and therefore considers flow data over a 30-day period.

In addition to the accumulation period and the fusion of observed and forecasted data (see above), the high skill in hydrological drought forecasts is anticipated coming from the land surface water storages/memory (e.g. lakes, soils, groundwater) that pools, attenuates, and lengthens the effect of the driving forces (i.e. precipitation) (Van Loon et al. 2012; Van Lanen et al. 2013; Van Loon and Laaha 2015). A recent study by Pechlivanidis et al. (2020) on seasonal streamflow forecasts (note: not streamflow drought) also concludes that the predictability is higher in the slowly responding river basin, associated with large catchment memory. The impact of catchment memory in determining the hydrological drought forecast skill is discussed in Chapter 2.5.

2.4.2 Examples of Assessment of Seasonal Drought Forecast Performance

In this section, the performance of drought forecasts evaluated using two approaches is presented. The first approach is a categorical drought classification and the second is the BSS.

2.4.2.1 Categorical Drought Classification Using the categorical drought classification, the scores for forecasted hydrometeorological drought using different standardized indices (in this case drought in precipitation and groundwater) are analysed by comparing the forecasts with observations (here the SFO) (Figure 2.6). This new approach uses drought class scores, which appeared to be better aligned with the requirements of end users. In our study, we used drought classes to describe the severity of drought in the various hydrometeorological variables and to determine the forecasting score. Drought classes, such as mild, moderate, severe, and extreme (Section 2.2.2), are easier to understand by end users than a drought index (e.g. a SPI value of −1.5). In this method, we assigned a drought severity class expressed as a number as follows: 1: no drought, 2: mild drought, 3: moderate drougxht, 4: severe drought, and 5: extreme drought, for each forecast time step (month) and each grid cell. Thus, the score of the

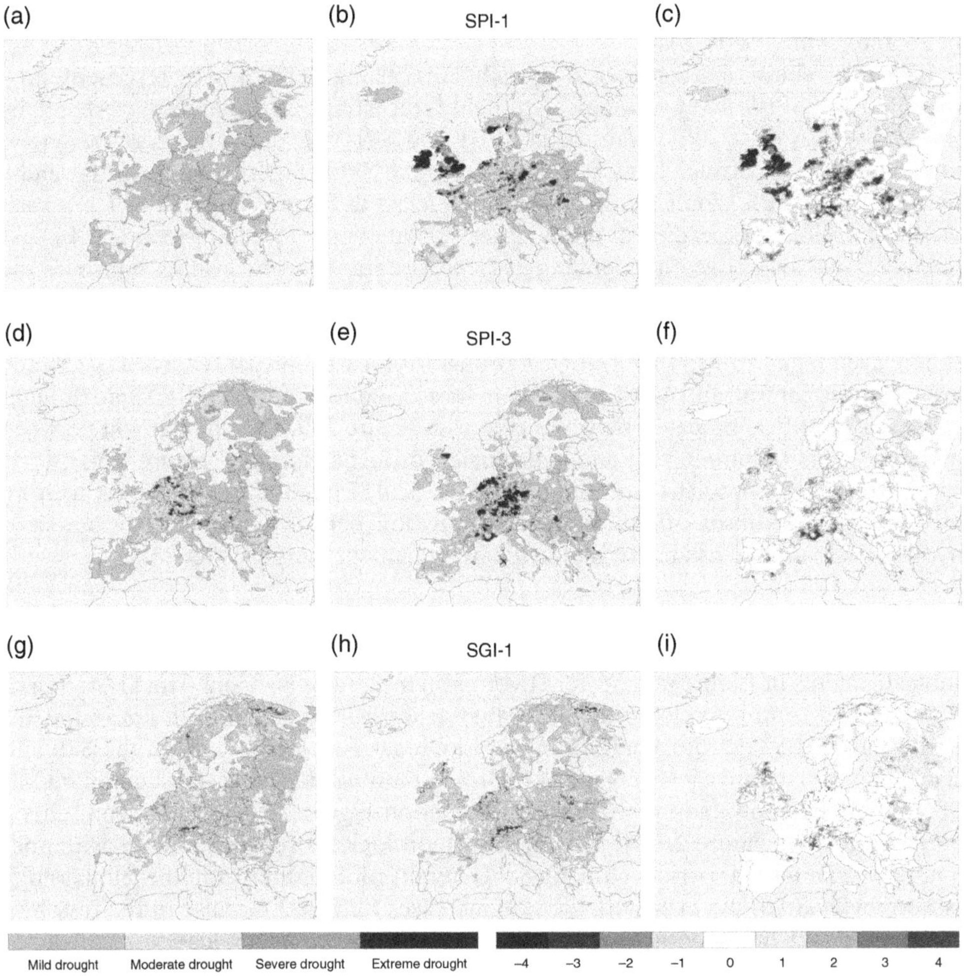

Figure 2.6 Meteorological drought for the pan-European region in August 2003 expressed as drought severity classes using the SPI-1: (a) forecasted drought done on 2 August 2003 (median of 15 ensembles), (b) drought obtained from SFO precipitation for August 2003, (c) drought forecast score expressed as difference in drought classes between the forecasts and the SFO, (d–f) as in (a–c), but for SPI-3, and (g–i) as in (a–c), but for SGI-1. Reddish colours in (c), (f), and (i) indicate that the forecast underestimates the drought class and vice versa for bluish colours. White colour in figures (c), (f), and (i) shows a perfect forecast.

forecast is identified by comparing the drought class derived from the observations with the forecasts (Figure 2.6). A perfect forecast is obtained if the drought class difference is zero (white), meaning that both forecast and SFO have the same drought class. The higher difference between drought class derived from forecast and drought class derived from SFO, either negative (forecast underestimating drought severity) or positive (forecast overestimating drought severity), shows low forecast performance. For example, a class difference of –4 or +4 (dark red or dark blue) indicates the lowest performance of drought forecasts. Detailed information about the method is presented in Sutanto et al. (2020a, b).

Figure 2.6 shows meteorological drought forecasting performance represented by the SPI-1 and SPI-3, and hydrological drought forecasting performance represented by the SGI-1 in August 2003. Forecasted SPI-1 and SPI-3 (Figure 2.6a, d, respectively) underestimate the drought classes compared to the SFO (Figure 2.6b for SPI-1 and e for SPI-3). The SPI-3 forecasts smaller drought class difference than the SPI-1, associated with fusion of preceding monthly observed data in the forecast (Section 2.4.1). As already discussed above, the blending of forecast data with observation improves the drought forecast skill. This is clearly shown in Figure 2.6f for SPI-3 where many regions in Europe have less dark reddish or bluish colours than in Figure 2.6c for SPI-1. The skill is even higher for the hydrological drought forecast represented by SGI-1. Many areas in west, central, and east Europe were forecasted to experience the same drought classes as the SFO, indicated by white colour in Figure 2.6i. The comparison between meteorological (Figure 2.6c, f) and hydrological drought forecasts (Figure 2.6i) clearly shows the higher predictive power of hydrological drought forecasts, even without inclusion of past observational data (accumulation periods). This is plausible since hydrological variable used here is affected by catchment memory (Section 2.5).

2.4.2.2 Forecast Skill Using BSS This section provides example of hydrometeorological drought forecasting skill assessment evaluated using the BSS metric for two sub-catchments of Llobregat and Ter River basin in Catalonia, Spain. The SFO streamflow is taken from a model grid cell located at the Ripoll gauging station in the Ter River and at the Guardiola station in the Llobregat River, upstream of the Sau and Baells dams, respectively. The skill of hydrometeorological forecasts as compared to SFO data for Ripoll and Guardiola is presented in Figure 2.7 (BSS = 1 represents a perfect forecast). Figure 2.7a, d show that meteorological drought forecasts with one-month accumulation period (SPI-1) do generally not outperform the climatology (shown by BSS < 0). As expected, the skill increases for SPI-3 because of the fusion of two-month and one-month observation data with forecasts for lead time 1 and 2, respectively (Figure 2.7b, e). Using SPI-3, drought can be forecasted up to two months ahead with a certain skill as the BSS is above zero. As expected, the higher the accumulation time, the higher the skill will be. The high skill for meteorological drought indices for longer accumulation periods should not be misinterpreted. Using higher accumulation periods in the SPI-x will artificially inflates the skill; it is generally comparing semi-observation data set (forecast) with observation data set. For instance, SPI-12 with an LT of 1 month includes 11 months of observed data, which is then compared with 12 months of observed data. The use of SPI-x with high accumulation periods, e.g. SPI-6 or SPI-12, is regularly used as an alternative for hydrological drought forecast, if hydrological drought forecasts are not available (WMO 2012). Short timescales of SPI, e.g. SPI-1 or SPI-3, reflect soil moisture conditions and are used to indicate drought in agriculture.

For streamflow drought, we analysed streamflow drought using the variable threshold method derived from the 80th percentile (Yevjevich 1967; Tallaksen and Van Lanen 2004). A two-sided 30-day moving average was applied to the daily forecasted streamflow to enable a fair comparison between streamflow drought forecasts and

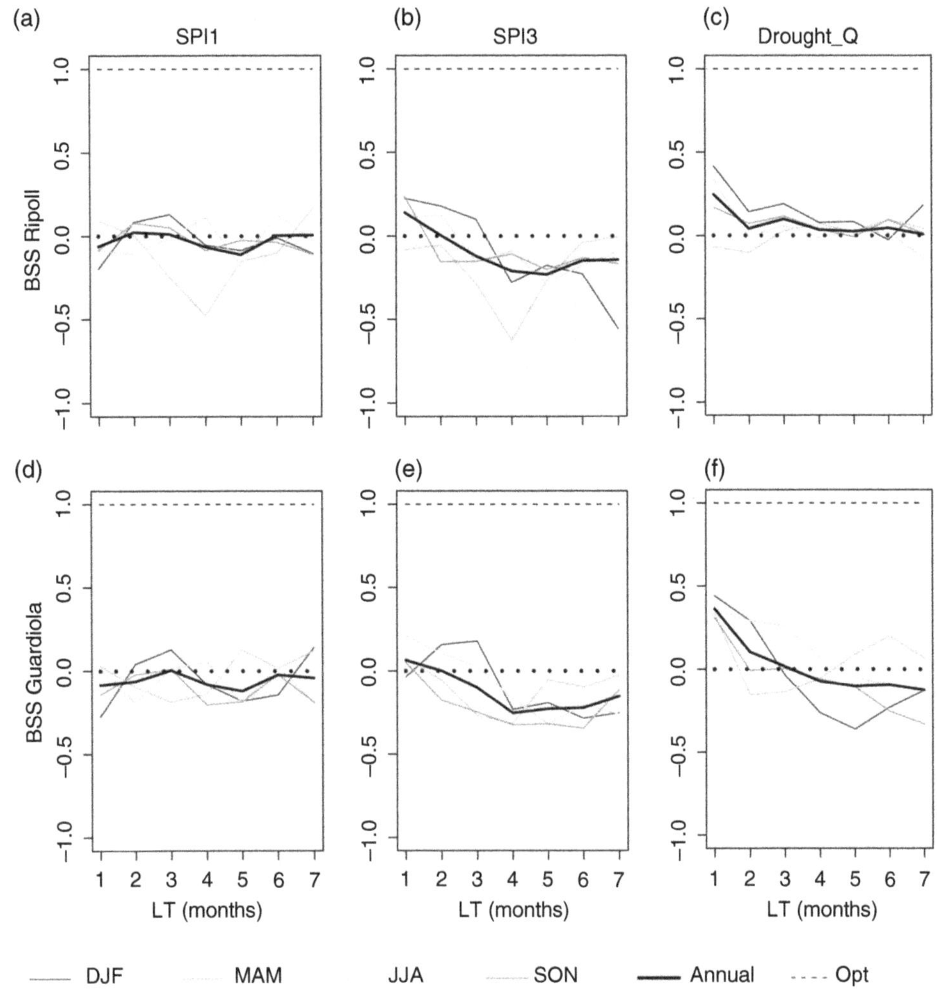

Figure 2.7 Brier Skill Scores (BSS) for the Ripoll and Guardiola (Ter and Llobregat catchments, respectively): (a) SPI1 forecasts, (b) SPI3 forecasts, and (c) streamflow drought forecast (Drought_Q) for the Ripoll. (d–f) as in (a–c) for the Guardiola. The coloured and black lines show the BSS per season and for the entire year, respectively. The dashed line shows the optimum of the forecast performance (BSS = 1) and the dots at BSS = 0 are the limit when the model skill outperforms the climatology. *Source*: Modified from Van Hateren et al. (2019).

SPI-1 forecasts, which are accumulated over one month. Applying the two-sided moving average led to missing values for the first 15 days of every forecast. The SFO streamflow data were then added to the forecast for the 15 days leading up to the first forecast day, before applying a moving average. By doing so, the forecast skill on the first month will be inflated. This is shown by the higher skill for streamflow drought with an LT of one month (Figure 2.7c, f). The skill of streamflow drought with inclusion of 15 days observed data is even higher than the SPI-3, which includes the SFO data up to two months. This proved that the memory of the catchments, that is storage capacity, which does not play a role in meteorological forecasts has a great importance in improving the hydrological drought forecasts (Section 2.5). In general, hydrological drought forecasts represented by drought in streamflow outperform the climatology up to three months ahead (shown by BSS > 0 in Figure 2.7c, f).

2.5 Importance of Catchment Memory

The previous section clearly shows that the hydrological drought forecasts have higher skill than meteorological drought ones, even beyond two months. Several studies reported the influence of catchment memory in increasing the skill of the hydrological forecasts, e.g. streamflow forecast (Pechlivanidis et al. 2020; Girons Lopez et al. 2021) (note: this is not streamflow drought). Although meteorological drought forecasts for longer accumulation periods than one month and short lead times also contain memory due to fusion with observed data (e.g. SPI-3 LT = 1 consists of two months observed), the skill of hydrological drought forecasts is still higher than the meteorological drought forecasts. Sutanto and Van Lanen (2022) have conducted a pioneering study that investigates the importance of catchment memory on the forecast performance of streamflow drought across Europe that will be summarized in this section. They used the Standardized Streamflow Index (SSI, Nalbantis and Tsakiris 2009) to identify streamflow drought, the Baseflow Index (BFI, IOH 1980) to explain the capacity of catchment to store and release water from catchment sources of water, and the groundwater Recession Coefficient (gRC) to describe storage and releases of water from subsurface water. The forecast performance was evaluated using the Brier Score (BS) approach (Brier 1950) and dichotomous forecast methods, namely forecast accuracy and success ratio (Stanski et al. 1989; Wilks 2011).

Figure 2.8 summarizes the correlation between catchment memory, represented by BFI and gRC, and forecast skills evaluated using BS, forecast accuracy, and success ratio (SR). A negative correlation between BFI and BS for all major river grid cells ($n = 10{,}106$) across Europe is seen in Figure 2.8a. Here, BS values were calculated from SSI-1 forecast data (2002–2016) with a LT = 1 compared to SSI-1 SFO data and BFIs were derived from the SFO data from 1990 to 2018. A significant negative correlation ($p < 0.05$) with Spearman's correlation value of −0.32 indicates that rivers with high BFI values yield higher streamflow drought forecast skill (low BS). High BFI associated with hydrological persistence of a river (more stable streamflow) causes high streamflow forecast performance, which is translated to streamflow drought forecasts. Harrigan et al. (2018) and Girons Lopez et al. (2021) correlated the streamflow forecast skill with BFI and they found that high skill is seen in catchments with high BFI values. Our study supports their findings that high forecast skill is obtained in the rivers with high BFI, associated with high memory.

The analysis using gRC as a second metric to study the impact of catchment memory on drought forecast performance further supports our hypothesis. A low gRC implies a slow release of groundwater caused by large catchment storage resulting in a slowly responding catchment. The BFI could be calculated for all major river grid cells (Figure 2.8a) since streamflow at a certain river grid cell represents by definition all

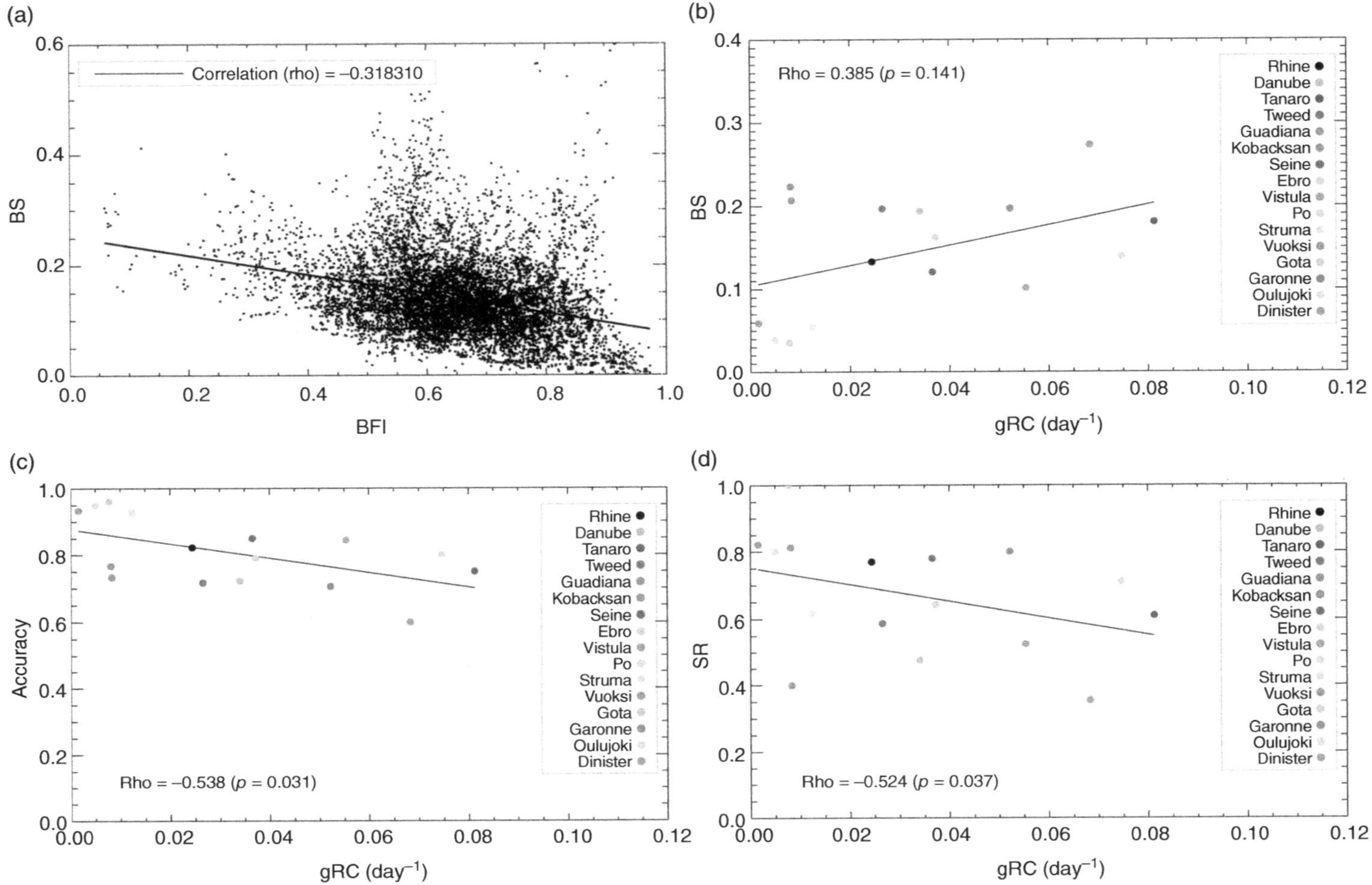

Figure 2.8 Correlation between forecast performance (BS), BFI, and gRC for 16 selected river catchments: (a) based on BS for SSI-1 LT = 1 and BFI, (b) based on BS for SSI-1 LT = 1 and gRC, (c) based on forecast accuracy for SSI-1 LT = 1 and gRC, and (d) based on forecast SR for SSI-1 LT = 1 and gRC. *Source*: Modified from Sutanto and Van Lanen (2022).

accumulated flow from upstream river grid cells. One should note that such analysis for all major grid cells across Europe was not feasible for the gRC, because the gRC value of each major river nodal point should have been obtained by spatially averaging gRCs of all land cells upstream of that major river grid cell that contribute to groundwater flow. Thus, all grid cells of a groundwater catchment (both river and land grid cells) of each over the 10,000 major river grid cells should have been investigated to obtain the gRC of the major river grid cell. This was beyond the scope of the study. Hence, we selected 16 river catchments that have a broad range of BS and gRC values to derive the correlation between BS and gRC. When using BS as a forecast skill metric (SSI-1 LT = 1), a weak positive correlation between gRC and BS is seen with Rho=0.38 (Figure 2.8b). The correlation, however, increases up to >0.7 for SSI-3 LT = 3 due to the inclusion of two months preceded observation data (not shown, please see Sutanto and Van Lanen 2022). The presence of lakes in the northern European river catchments, such as Vuoksi, Göta, and Oulujoki, explains why these rivers have low gRC and high forecast performance. The Guadiana and Struma Rivers have a slow catchment response similar to Vuoksi, Göta, and Oulujoki (gRC = 0.008 and 0.012, respectively), but they generate lower prediction performance. This might be due to the influence of anthropogenic activities to cope with water scarcity, such as the rapid development of reservoirs in the past decades. The Guadiana River, for example, has 39 man-made reservoirs located along the catchment (Canuto et al. 2019) that modify the flow regime and associated streamflow drought (Van Loon and Van Lanen 2013). Although LISFLOOD includes 1454 reservoirs across Europe (Arnal et al. 2019), these may not be sufficiently modelled and thus reduce the forecast performance.

Forecast accuracy and success ratio (SR) were used to test our hypothesis that catchment with lower gRC generates higher drought forecast performance. Unlike BS, a perfect forecast is identified if the accuracy and SR values are close to 1. Figure 2.8c, d show significant negative correlations between gRC and forecast accuracy and between gRC and SR, respectively (Rho < −0.5), which confirm our hypothesis. Results indicate that catchments with gRC values lower than 0.04, on average, yield correct forecasts up to 80% (accuracy > 0.8, Figure 2.8c) and 70% of the forecasted drought events were actually observed (SR > 0.7, Figure 2.8d). Our study proved that groundwater memory, denoted by gRC, has a strong impact in determining the drought forecast skill, independently from the evaluation metrics used in the study.

2.6 Outlook and Future Improvements

During the ANYWHERE project, we learned that the developed MH-EWS from which ADEWS is part of it is very useful to help the stakeholders coping with extreme weather-induced events. The ANYWHERE pilot sites have experienced the benefit of using the ANYWHERE platform in supporting their daily activities and decision-making to mitigate the disasters (see http://anywhere-h2020.eu/videos-page-1/). We, however, also acknowledge that the ANYWHERE MH-EWS is still under progress. For example, ANYWHERE does not forecast the impacts of hazards for all sectors and all hazards. This is also the case for ADEWS, which only forecasts the drought hazard and not the impacts. The forecast of compound and cascading (CC) hazards is another missing component in our current system. Nevertheless, we already made progress in developing approaches to forecast drought impacts, as well as CC dry hazards. These approaches are foreseen to be implemented in the next generation of the MH-EWS. In the subsections below, we will highlight these approaches as an outlook for future improvements.

2.6.1 Drought Impact Forecasts

Sutanto et al. (2019a) were the first study assessing the feasibility of forecasting drought impacts up to seven months ahead using machine-learning (ML) techniques. In their study, drought impact forecasting functions were developed using reported drought impacts obtained from the European Drought Impact Inventory from 1990 to 2017 (EDII, Stahl et al. 2016), time series of drought hazard indicators from 1990 to 2017, and machine-learning algorithms, i.e. Random Forest and logistic regression. What can we learn when developing drought impact forecasting functions using ML? Although drought impacts can be predicted months in advance, the performance of the drought impact functions depends on the severity of the drought hazard and drought duration. The higher skill is obtained from models that were trained on long-lasting droughts and having a long period of reported impacts (50 months).

An Illustration of drought impact forecasting as if it would be encapsulated in a drought EWS is given in Figure 2.9 for Germany. The median forecasted likelihood of occurrence (LIO) of an impact using forecast data for each month from January 2002 to December 2010 with LT = one to seven months for the Brandenburg (BB) region is presented in Figure 2.9a. The LIO holds for the combined impacts of drought on water quality, freshwater ecosystem, and terrestrial ecosystem (drought impact categories). Low LIO values indicate that no drought impact is forecasted and vice versa for high LIO values. There are some high LIO values forecasted in 2008, 2009, and 2011, but no impact was reported (no grey bar). Miss alarms in the DEWS could have triggered unneeded preparation measures. High LIO values that led to drought impacts, e.g. in 2003 and 2006, increased from LT $\geq$ one month, and later these gradually decreased after drought impact occurred. For drought that did not trigger impacts, LIO values are rather low and short. A reason for the miss alarms 2008, 2009, and 2011 might be the duration of these drought events that might be too short to create a noticeable impact in this region. In contrast to BB, the drought impact function for the Mecklenburg-Vorpommern (MP) region was not able to forecast the impacts for the same drought impact categories as BB in a timely manner (Figure 2.9b). Spikes in LIO are noticed in 2003, months before the impact occurred but these LIOs are too low and short. The model produces high LIOs up to 0.5 at the end of the year 2003 when the impact was already reported, which is not useful for preparing early contingency actions and drought mitigation strategies.

Although drought impact forecasting functions developed using ML techniques are able to predict drought impacts with adequate skill, the models do not always predict the impacts well in advance and have low prediction skill in some regions (see also Sutanto et al. 2020c). Here, we would like to highlight the importance of comprehensive and complete drought impact databases, which is of utmost importance to train the

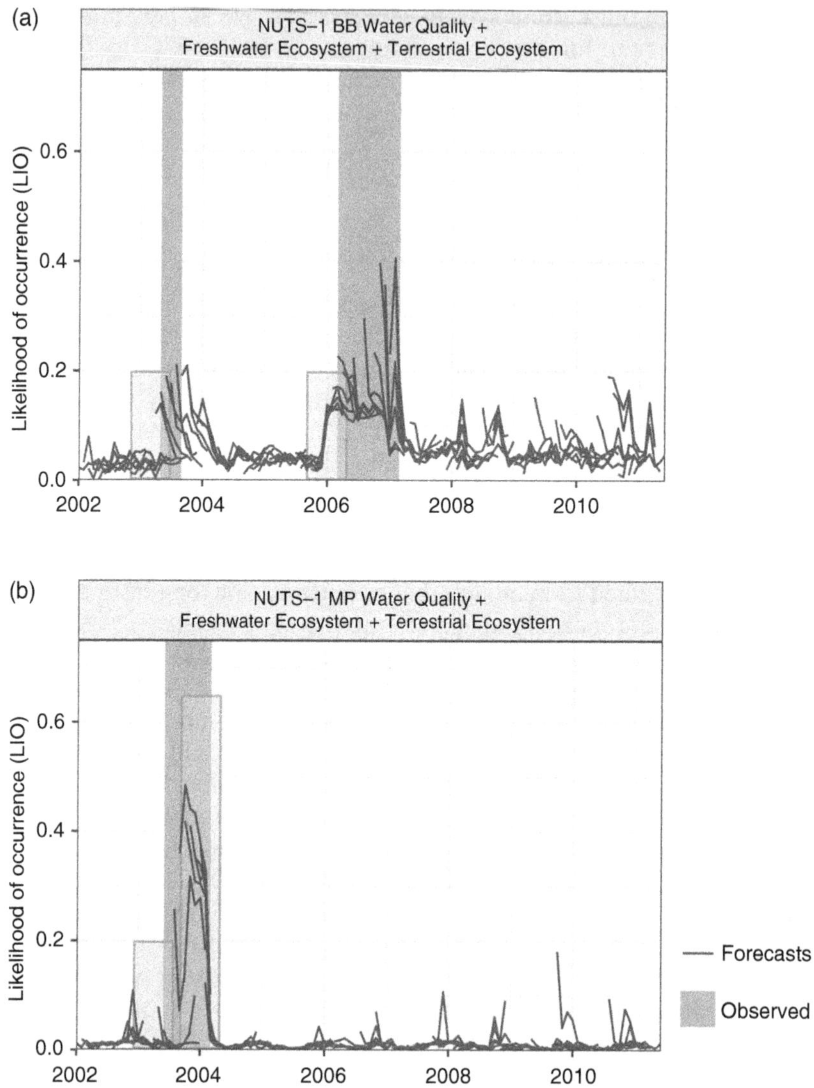

Figure 2.9 Examples of drought impact forecasting: (a) forecasted LIO and reported drought impact (EDII, grey bars) for the German NUTS-1 region Brandenburg (BB), and (b) same as (a) but for the region Mecklenburg-Vorpommern (MP). Blue lines are the forecasted median LIO done for each month from January 2002 to December 2010 with LT = 1–7 months. Red boxes show the forecasted LIOs prior and during drought events. *Source*: Modified from Sutanto et al. (2019a).

ML model. ML is a data-driven approach, in which the model performance is strongly related to the quality and quantity of the underlying data (Elshorbagy et al. 2010; Mount et al. 2016). Sutanto et al. (2020c) show that a drought impact model can be built in a region that has a total reported impact of 20 months (combination of several impact into impact categories). However, a drought impact model cannot be made when the reported impacts were analysed individually (not combined), meaning fewer reported impacts to train the model. Therefore, we highly encourage institutions that plan or use an operational DEWS to develop a drought impact database and later to move one step forward to provide drought impact forecasts. The information on predicted drought impacts can be of great value for water managers, stakeholders, and policy makers in understanding and planning drought management responses in due time.

2.6.2 Compound and Cascading (CC) Dry Hazards

The impacts of drought as a single hazard are intensified when more hazards occur simultaneously (concurrent or compound hazards) or sequentially (cascading events). For example, the 2003 drought as a single hazard was not the most severe in Europe (Spinoni et al. 2015). However, in combination with extended heatwaves and fires, it is considered the most fatal and costly (Robine et al. 2008). In the United States, the total state-wide economic cost for the 2014 California multi-hazard events, drought, heat-wave, and fire was estimated at $2.2 billion (Howitt et al. 2014). Despite the notable impacts of CC hazards, most studies have focused on forecasting single hazards due to the challenges of analysing and forecasting of a multi-hazard (Kappes et al. 2012). In the ANYWHERE, we conducted two studies related to CC dry hazard events (Vitolo et al. 2019; Sutanto et al. 2019b). The Sutanto et al. (2019b) study is pioneering in investigating spatio-temporal patterns of CC dry hazards, namely heatwaves, droughts, and fires across Europe. They proposed a method to combine three different hazards into CC events. Their study, however, was only based on historical data. Vitolo et al. (2019) were the first that forecasted the occurrence of compound heat and fire using dynamical weather forecasts (Figure 2.10).

Figure 2.10 shows the forecasted heat, fire, and compound heat and fire issued on 17 June 2017 with a lead time up to three days ahead. Many regions in Europe faced compound heat and fire (yellow), such as Spain, Portugal, France, Italy, and the Balkans.

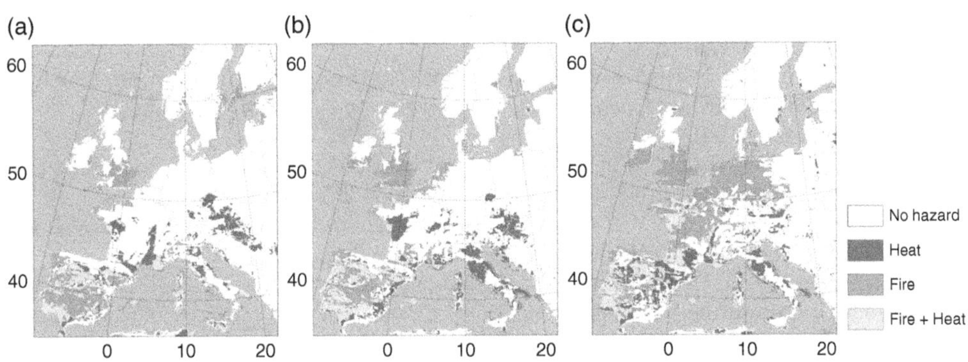

Figure 2.10 Multi-hazard hotspot forecast maps at pan-European scale issued on 17th June 2017 morning for (a) 17 June 2017 with a lead time of 1 day, (b) 18 June 2017 with a lead time of two days, and (c) 19 June 2017 with a lead time of three days. *Source*: Vitolo et al. (2019)/with permission of Elsevier/CC BY 4.0.

Furthermore, the Vitolo et al. (2019) study also provided an illustration of the percentage of Portugal's areas that have been forecasted to experience compound heat and fire hazards using 10-day forecasts issued from 1 June 2017 to 30 June 2017. Results indicate that the dramatic compound event on 17 June 2017 in Pedrógão Grande could have been forecasted 10 days in advance, with >50% of Portugal would have been affected by a high probability of compound heat and fire starting from 16 June 2017. Their study, and the one from Sutanto et al. (2019b), clearly demonstrate that compound and cascading hazards can be forecasted days in advance and their approaches can be easily implemented in an operational MH-EWS. Given that warnings can be issued in advance, the local and national authorities would have time for pre-emergency planning and allocation of resources. To the authors' knowledge, there is no study forecasting CC hazard impacts yet (note: not hazards). Forecasting CC hazard impacts is extremely challenging because of the difference in their spatial and temporal dynamics, and the need for a cross-hazard comparable method to forecast CC dry hazards including their impacts (de Ruiter et al. 2020). Future studies aiming to forecast CC hazard impacts will improve our understanding of the occurrence and interactions of CC dry hazards and associated impacts, and, in the end, will contribute to timely predicting CC dry hazards and their impacts (improved MH-EWS).

References

Arnal, L., Asp, S.-S., Baugh, C. et al. (2019). EFAS Upgrade for the Extended Model Domain. *JRC Tech. Rep. EUR29323EN. JRC, 58.*

Bloomfield, J.P. and Marchant, B.P. (2013). Analysis of groundwater drought building on the standardised precipitation index approach. *Hydrology and Earth System Sciences* 17 (12): 4769–4787.

Brier, G.W. (1950). Verification of forecasts expressed in terms of probability. *Monthly Weather Review* 78 (1): 1–3.

Burek, P., Van Der Knijff, J., and De Roo, A. (2013). LISFLOOD Distributed Water Balance and Flood Simulation Model. *JRC Tech. Rep. EUR26162EN*. 102788/24719.

Canuto, N., Ramos, T.B., Oliveira, A.R. et al. (2019). Influence of reservoir management on Guadiana streamflow regime. *Journal of Hydrology: Regional Studies* 25: 100628. https://doi.org/10.1016/j.ejrh.2019.100628.

Cloke, H., Pappenberger, F., Thielen, J., and Thiemig, V. (2013). Operational European flood forecasting. In: *Environmental Modelling: Finding Simplicity in Complexity*, 2e (ed. J. Wainwright and M. Mulligan), 415–434. Wiley.

Cui, B., Toth, Z., Zhu, Y., and Hou, D. (2012). Bias correction for global ensemble forecast. *Weather and Forecasting* 27: 396–410. https://doi.org/10.1175/WAF-D-11-00011.1.

Dutra, E., Pozzi, W., Wetterhall, F. et al. (2014a). Global meteorological drought-part 2: seasonal forecasts. *Hydrology and Earth System Sciences* 18: 2669–2678.

Dutra, E., Wetterhall, F., Di Giuseppe, F. et al. (2014b). Global meteorological drought-part 1: probabilistic monitoring. *Hydrology and Earth System Sciences* 18: 2657–2667. https://doi.org/10.5194/hess-18-2657-2014.

EC (2007). Addressing the challenge of water scarcity and droughts in the European union, communication from the commission to the European parliament and the council. European Commission, DG Environment, Brussels.

EC (2021). Overview of natural and man-made disaster risks the European union may face. European Commission, Luxembourg, https://doi.org/10.2795/1521.

Elshorbagy, A., Corzo, G., Srinivasulu, S., and Solomatine, D.P. (2010). Experimental investigation of the predictive capabilities of data driven modeling techniques in hydrology – part 1: concepts and methodology. *Hydrology and Earth System Sciences* 14: 1931–1941. https://doi.org/10.5194/hess-14-1931-2010.

FAO (2017). *The Impact of Disasters on Agriculture and Food Security*. Food and Agriculture Organization of the United Nations. ISBN: 978-92-5-108962-0.

Fleig, A.K., Tallaksen, L.M., Hisdal, H., and Demuth, S. (2006). A global evaluation of streamflow drought characteristics. *Hydrology and Earth System Sciences* 10: 535–552.

Forzieri, G., Feyen, L., Rojas, R. et al. (2014). Ensemble projections of future streamflow droughts in Europe. *Hydrology and Earth System Sciences* 18: 85–108. https://doi.org/10.5194/hess-18-85-2014.

Girons Lopez, M., Crochemore, L., and Pechlivanidis, I.G. (2021). Benchmarking an operational hydrological model for providing seasonal forecasts in Sweden. *Hydrology and Earth System Sciences* 25: 1189–1209. https://doi.org/10.5194/hess-25-1189-2021.

Hari, V., Rakovec, O., Markonis, Y. et al. (2020). increased future occurrences of the exceptional 2018–2019 central European drought under global warming. *Scientific Reports* 10: 12207. https://doi.org/10.1038/s41598-020-68872-9.

Harrigan, S., Prudhomme, C., Parry, S. et al. (2018). Benchmarking ensemble streamflow prediction skill in the UK. *Hydrology and Earth System Sciences* 22 (3): 2023–2039. https://doi.org/10.5194/hess-22-2023-2018.

Hisdal, H., Tallaksen, L.M., Clausen, B. et al. (2004). Hydrological drought characteristics. In: *Hydrological Drought. Processes and Estimation Methods for Streamflow and Groundwater. Development in Water Science*, vol. 48 (ed. L.M. Tallaksen and H.A.J. Van Lanen), 139–198. Elsevier Science B.V. http://europeandroughtcentre.com/hydrological-drought-1st-edition-file-15/.

Howitt, R.E., Medellin-Azuara, J., MacEwan, D. et al. (2014). Economic analysis of the 2014 drought for California agriculture. Center for Watershed Sciences, University of California, Davis, California, p. 20. http://watershed.ucdavis.edu.

IOH (1980). Low flow studies report, institute of hydrology, Wallingford, UK. http://nora.nerc.ac.uk/id/eprint/9093/1/Low_Flow_01.pdf (accessed 6 August 2020).

Kappes, M.S., Keiler, M., von Elverfeldt, K., and Glade, T. (2012). Challenges of analyzing multi-hazard risk: a review. *Natural Hazards* 64: 1925–1958, https://doi.org/10.1007/s11069-012-0294-2.

Kruse, S. and Seidl, I. (2013). Social capacities for drought risk management in Switzerland. *Natural Hazards and Earth System Sciences* 13: 3429–3441. https://doi.org/10.5194/nhess-13-3429-2013.

McKee, T.B., Doesken, N.J., and Kleist, J. (1993). The relationship of drought frequency and duration to time scale. *Proc. of 8th Conf. on Applied Climatology*, Anaheim, California (17–22 January 1993), pp. 179–184. Boston: American Meteorological Society.

Mount, N.J., Maier, H.R., Toth, E. et al. (2016). Data-driven modelling approaches for socio-hydrology: opportunities and challenges within the Panta Rhei science plan. *Hydrological Sciences Journal* 61 (7): 1192–1208. https://doi.org/10.1080/02626667.2016.1159683.

Nalbantis, I. and Tsakiris, G. (2009). Assessment of hydrological drought revisited. *Water Resources Management* 23: 881–897. https://doi.org/10.1007/s11269-008-9305-1.

Pappenberger, F., Thielen, J., and del Medico, M. (2011). The impact of weather forecast improvements on large scale hydrology: analysing a decade of forecasts of the European flood alert system. *Hydrological Processes* 25: 1091–1113.

Pechlivanidis, I.G., Crochemore, I., Rosberg, J., and Bosshard, T. (2020). What are the key drivers controlling the quality of seasonal streamflow forecasts? *Water Resources Research* 56: e2019WR026987. https://doi.org/10.1029/2019WR026987.

Peters, E., Torfs, P.J.J.F., Van Lanen, H.A.J., and Bier, G. (2003). Propagation of drought through groundwater – a new approach using linear reservoir theory. *Hydrological Processes* 17: 3023–3040. https://doi.org/10.1002/hyp.1274.

Poljanšek, K., Marín Ferrer, M., De Groeve, T. et al. (2017). *Science for Disaster Risk Management 2017: Knowing Better and Losing Less*,. Executive Summary, EUR 28034 EN. Luxembourg: Publications Office of the European Union. ISBN: 978-92-79-69673-2 https://doi.org/10.2760/451402.

Rakovec, O., Samaniego, L., Hari, V. et al. (2022). The 2018–2020 multi-year drought sets a new benchmark in Europe. *Earth's Future* 10: e2021EF002394. https://doi.org/10.1029/2021EF002394.

Robine, J.-M., Cheung, S.L.K., Roy, S.L. et al. (2008). Death toll exceed 70,000 in Europe during the summer of 2003. *Comptes Rendus Biologies* 331: 171–178. https://doi.org/10.1016/j.crvi.2007.12.001.

de Ruiter, M.C., Couasnon, A., van den Homberg, M.J.C. et al. (2020). Why we can no longer ignore consecutive disasters. *Earth's Future* 8: e2019EF001425. https://doi.org/10.1029/2019EF001425.

Shukla, S. and Wood, A.W. (2008). Use of a standardized runoff index for characterizing hydrologic drought. *Geophysical Research Letters* 35 (2): L02–L405.

Spinoni, J., Naumann, G., Vogt, J.V., and Barbosa, P. (2015). The biggest drought events in Europe from 1950 to 2012. *Journal of Hydrology: Regional Studies* 3: 509–524. https://doi.org/10.1016/j.ejrh.2015.01.001.

Stahl, K., Kohn, I., Blauhut, V. et al. (2016). Impacts of European drought events: insights from an international database of text-based reports. *Natural Hazards and Earth System Sciences* 16: 801–819. https://doi.org/10.5194/nhess-16-801-2016.

Stanski, H.R., Wilson, L.J., and Burrows, W.R. (1989). Survey of Common Verification Methods in Meteorology. *World Weather Watch Tech. Rep. No.8*, WMO/TD No.358, WMO, Geneva, p. 114. https://www.cawcr.gov.au/projects/verification/Stanski_et_al/Stanski_et_al.html (accessed on 1 December 2021).

Stockdale, T., Johnson, S., Ferranti, L. et al. (2018). ECMWF's new long range forecasting system SEAS5. *ECMWF Newsletter* 154: https://doi.org/10.21957/tsb6n1.

Sutanto, S.J. and Van Lanen, H.A.J. (2020). Hydrological drought characteristics based on groundwater and runoff across Europe. *Proceedings of the International Association of Hydrological Sciences* 383: 281–290. https://doi.org/10.5194/piahs-383-281-2020.

Sutanto, S.J. and Van Lanen, H.A.J. (2021). Streamflow drought: implications of drought definitions and its application for drought forecasting. *Hydrology and Earth System Sciences* 25: 3991–4023. https://doi.org/10.5194/hess-25-3991-2021.

Sutanto, S.J. and Van Lanen, H.A.J. (2022). Catchment memory explains hydrological drought forecast skill. *Scientific Reports* 12: 2689. https://doi.org/10.1038/s41598-022-06553-5.

Sutanto, S.J., Van der Weert, M., Wanders, N. et al. (2019a). Moving from drought hazard to impact forecasts. *Nature Communications* 10: 4945. https://doi.org/10.1038/s41467-019-12840-z.

Sutanto, S.J., Vitolo, C., Di Napoli, C. et al. (2019b). Heatwaves, droughts, and fires: exploring compound and cascading dry hazards at the pan-European scale. *Environment International* 134: 105276. https://doi.org/10.1016/j.envint.2019.105276.

Sutanto, S.J., Van Lanen, H.A.J., Wetterhall, F., and Llort, X. (2020a). Potential of pan-European seasonal hydrometeorological drought forecasts obtained from a multihazard early warning system. *Bulletin of the American Meteorological Society* E368–E369. https://doi.org/10.1175/BAMS-D-18-0196.1.

Sutanto, S.J., Wetterhall, F., and Van Lanen, H.A.J. (2020b). Hydrological drought forecasts outperform meteorological drought forecasts. *Environmental Research Letters* 15: 084010. https://doi.org/10.1088/1748-9326/ab8b13.

Sutanto, S.J., van der Weert, M., Blauhut, V., and Van Lanen, H.A.J. (2020c). Skill of large-scale seasonal drought impact forecasts. *Natural Hazards and Earth System Sciences* 20: 1595–1608. https://doi.org/10.5194/nhess-20-1595-2020.

Tallaksen, L.M. and Van Lanen, H.A.J. (2004). Hydrological drought. Processes and estimation methods for streamflow and groundwater. In: *Developments in Water Science*, vol. 48, 579. Elsevier Science B.V.

Trambauer, P., Maskey, S., Winsemius, H. et al. (2013). A review of continental scale hydrological models and their suitability for drought forecasting in (sub-Saharan) Africa. *Physics and Chemistry of the Earth* 66: 16–26. https://doi.org/10.1016/j.pce.2013.07.003.

UNCCD (2022). Drought in numbers 2022-restoration for readiness and resilience. COP-15 Côte d'Ivoire, United Nations Convention to Combat Desertification, https://www.unccd.int/sites/default/files/2022-05/Drought%20in%20Numbers.pdf (accessed on 10 May 2022).

Van Der Knijff, J.M., Younis, J., and De Roo, A.P.J. (2010). LISFLOOD: a GIS-based distributed model for river basin scale water balance and flood simulation. *International Journal of Geographical Information Science* 24 (2): 189–212. https://doi.org/10.1080/13658810802549154.

Van Hateren, T.C., Sutanto, S.J., and Van Lanen, H.A.J. (2019). Evaluating skill and robustness of seasonal meteorological and hydrological drought forecasts at the catchment scale-case Catalonia (Spain). *Environment International* 133: 105206.

Van Lanen, H.A.J. (2006). Drought propagation through the hydrological cycle. Climate variability and change – hydrological impacts. *Proceedings of the Fifth FRIEND World Conference 122 held at Havana*, Cuba (27 November–1 December 2006), p. 308. IAHS Publ.

Van Lanen, H.A.J., Wanders, N., Tallaksen, L.M., and Van Loon, A.F. (2013). Hydrological drought across the world: impact of climate and physical catchment structure. *Hydrology and Earth System Sciences* 17: 1715–1732. https://doi.org/10.5194/hess-17-1715-2013.

Van Loon, A.F. (2015). Hydrological drought explained. *WIREs Water* https://doi.org/10.1002/wat2.1085.

Van Loon, A.F. and Laaha, G. (2015). Hydrological drought severity explained by climate and catchment characteristics. *Journal of Hydrology* 526: 3–14. https://doi.org/10.1016/j.jhydrol.2014.10.059.

Van Loon, A.F. and Van Lanen, H.A.J. (2012). A process-based typology of hydrological drought. *Hydrology and Earth System Sciences* 16: 1915–1946. https://doi.org/10.5194/hess-16-1915-.

Van Loon, A.F. and Van Lanen, H.A.J. (2013). Making the distinction between water scarcity and drought using an observation-modeling framework. *Water Resources Research* 49: 1483–1502. https://doi.org/10.1002/wrcr.20147.

Van Loon, A.F., Van Huijgevoort, M.H.J., and Van Lanen, H.A.J. (2012). Evaluation of drought propagation in an ensemble mean of large-scale hydrological models. *Hydrology and Earth System Sciences* 16: 4057–4078. https://doi.org/10.5194/hess-16-4057-2012.

Vicente-Serrano, S.M., Beguería, S., and López-Moreno, J.I. (2010). A multi-scalar drought index sensitive to global warming: the Standardized Precipitation Evapotranspiration Index-SPEI. *Journal of Climate* 23 (7): 1696–1718. https://doi.org/10.1175/2009JCLI2909.1.

Vitart, F. (2004). Monthly forecasting at ECMWF. *Monthly Weather Review* 132: 2761–2779. https://doi.org/10.1175/MWR2826.1.

Vitolo, C., Di Napoli, C., Di Giuseppe, F. et al. (2019). Mapping combined wildfire and heat stress hazards to improve evidence-based decision making. *Environment International* 127: 21–34. https://doi.org/10.1016/j.envint.2019.03.008.

Vogt, J., Sepulcre-Canto, G., Magni, D. et al. (2013). *The European Drought Observatory (EDO).* Current State and Future Directions, Geophysical Research Abstracts. EGU2013-7374.

Weisheimer, A. and Palmer, T.N. (2014). On the reliability of seasonal climate forecasts. *Journal of the Royal Society Interface* 11: 20131162.

Wetterhall, F. and Di Giuseppe, F. (2018). The benefit of seamless forecasts for hydrological prediction over Europe. *Hydrology and Earth System Sciences* 22: 3409–3420. https://doi.org/10.5194/hess-22-3409-2018.

Wilks, D.S. (2011). *Statistical Methods in the Atmospheric Sciences*, 3e, 676. Elsevier.

WMO (2012). *Standardized Precipitation Index User Guide* (ed. M. Svoboda, M. Hayes, and D. Wood). Geneva: World Meteorological Organization Report WMO-No. 1090. ISBN 978-92-63-11091-6.

Yevjevich, V. (1967). An objective approach to definition and investigations of continental hydrologic droughts, hydrology papers 23. Colorado State University, Fort Collins, USA.

Yuan, X., Wood, E.F., Chaney, N.W. et al. (2013a). Probabilistic seasonal forecasting of African drought by dynamical models. *Journal of Meteorology* 14: 1706–1720. https://doi.org/10.1175/JHM-D-13-054.1.

Yuan, X., Wood, E.F., Roundy, J.K., and Pan, M. (2013b). CFSv2-based seasonal hydroclimatic forecasts over the conterminous United States. *Journal of Climate* 26: 4828–4847. https://doi.org/10.1175/JCLI-D-12-00683.1.

3
Experiences and Lessons Learnt in Wildfire Management with PROPAGATOR, an Operational Cellular-Automata-Based Wildfire Simulator

Andrea Trucchia[1], Mirko D'Andrea[1], Francesco Baghino[1,2], Nicolò Perello[1,2], Nicola Rebora[1], and Paolo Fiorucci[1]

[1] *Cima Research Foundation, Savona, Italy*

[2] *DIBRIS – Department of Informatics, Bioengineering, Robotics and Systems Engineering, University of Genoa, Genoa, Italy*

3.1 Introduction

Mediterranean basin is notoriously prone to wildfires, a natural hazard that jeopardizes the environment, properties and human lives. Even in countries characterized by severe fire danger patterns and equipped for firefighting, there is still a lack of prevention and preparedness capacities to gather in a very reduced time vital information to cope with direct impacts on exposed people and swiftly share it among the Civil Protection Authorities (CPAs) and the first responders. The tragic wildfires that occurred in Greece, Turkey and in Portugal in the last years, which caused many fatalities (San-Miguel-Ayanz et al. 2019), the Gran Canaria summer wildfires of August 2019, where 9000 people were evacuated (Copernicus-Emergency Management Service 2019), and 2021–2022 wildfires of Algeria, with a death toll of more than 130 casualties (ReliefWeb 2021, 2022), exemplify the consequences of those shortcomings.

Responding to Extreme Weather Events, First Edition. Edited by Daniel Sempere-Torres, Anastasios Karakostas, Claudio Rossi, and Philippe Quevauviller.

The vast majority of the wild-land fires in the Mediterranean landscape are caused by human activities being them accidental or deliberate; however, natural ignitions caused by lightning are not entirely negligible and can be increased by climate change (Conedera et al. 2006). Human-caused fires result from many different reasons, ranging from negligently discarded cigarettes to campfires left unattended, mismanagement of stubble in agricultural land or intentional arson. In addition to all this, wildfires can also be ignited by infrastructures such as power lines and railways.

Most wildfire emergencies, especially in the countries of southern EU, correlate with extreme weather conditions, with persistent dry and strong winds and an extensive spread of flammable land cover (De Rigo et al. 2017). In this case, probability of a successful ignition increases. When a fire is ignited in such extreme weather conditions, its propagation is fast and difficult to contrast in most of the recent wildfire emergencies, casualties took place few hours after the fire start. It is therefore extremely urgent to support first responders and CPAs with suitable operational tools in emergency response, adopting wildfire risk maps and efficient emergency plans. The availability of cutting-edge technology and the relative training of skilled personnel results in the anticipation, discrimination and selection of the most probable scenarios and the relative best strategies. High-quality, risk-informed decisions, especially in the first phases of the event, help in minimizing the risk for the exposed population.

Prevention and preparedness, as well as firefighting activities, are managed by several operational structures and different stakeholders, which must be promptly informed coordinated in case of emergency. ICT solutions for disaster risk management (DRM) procedures and computer modelling represent a great opportunity, putting wildfire emergency managers in the position to share useful information to coordinate civil protection activities and firefighting. Effective and fast decisions in evacuation and traffic management can save lives in fast and extreme wildfire events. The dramatic events occurred in Turkey, Algeria, Greece and Portugal made evident the need for fast and reliable tools to forecast the potential behaviour of the wildfire to implement prevention and communication activities in a timely fashion to reduce the harm to affected population. This can be achieved using wildfire numerical simulators based on ad hoc mathematical models, or implementing some other kind of decision-making process, such as the fire potential polygon approach of Castellnou et al. (2019).

This chapter describes PROPAGATOR, a wildfire simulator that relies on Cellular Automata for fast scenarios developments.

In the following Subsection 3.1.1, the taxonomy of the state-of-art mathematical models for wildfire simulation is described. In Section 3.2, the history of PROPAGATOR development and implementations is outlined. In Section 3.3, the model of PROPAGATOR is described in detail. In Section 3.4, the case study of several recent wildfires in Liguria, Italy and Catalonia, Spain are described. Results are presented in Section 3.5 while conclusion are drawn in Section 3.6.

3.1.1 Mathematical Models for Wildfire Management

The burned area patterns and rate of spread of wildfires are governed by complex mechanisms (Morandini and Silvani 2010; Li and Banerjee 2021; Sahila et al. 2021). The effects of terrain orography, wind field and fuel structure, arrangement and moisture combine together to determine wildfire dynamics, who results to be a multi-scale, multi-physics and non-linear phenomenon. This makes the formulation of efficient and reliable mathematical models particularly hard, as well as their computational implementation. Nevertheless, in literature, there are many different approaches and models dedicated to this specific task (Rothermel 1972; Grishin 1997).

Detailed models do not always constitute the optimal solution to a problem as they are notoriously difficult to learn and parameterize and may require a high amount of computational resources. While for many scientific and operational tasks require advanced analytic instruments, lightweight and simple solution can outperform them for daily yet crucial civil protection routines.

Mathematical models for wildfires usually fall into one of the following categories (Sullivan 2009a, b, c; Alexandridis et al. 2011; Trucchia 2019; Trucchia et al. 2019):

1. *empirical and semi-empirical models*, which make use of statistically derived laws of fire propagation (Rothermel 1972; Fernandes 2001);
2. *macroscopic-deterministic models*, which models fire spread using a continuum, with computational fluid dynamics techniques and detailed atmospheric, heat transfer and combustion models (Albini 1985; Asensio and Ferragut 2002; Serón et al. 2005; Mell et al. 2007);
3. *stochastic lattice/grid-based models*, where wildfire is considered as a random phenomenon conceived on a grid of spatial locations, with the propagation of the fire front from a cell to cell by the use of localized probabilistic evolution rules inspired by the underlying physics at the desired level of resolution (Green et al. 1990; Berjak and Hearne 2002; Alexandridis et al. 2011; Sun et al. 2013; Ghisu et al. 2015; Freire and DaCamara 2019).

However, the distinction in categories presented above in most cases is not a strict one, since in many models different approaches are combined (Sullivan 2009c; Alexandridis et al. 2011; Trucchia et al. 2019). A specific approach may or may not be the right one for the specific task intended by the practitioner. Empirical models are known to be straightforward to implement and use, and do not require a high computational budget. They have proved to approximate, under certain restrictions due to their simplified nature, fire-spread dynamics in an acceptable way. However, empirical models are derived from laboratory experiments, and they may fail to correctly predict wildfires that spreads under different conditions and landscapes (Sullivan 2009a).

Macroscopic-deterministic models are usually on first principles. They attempt to model in detail a complex, nonlinear, stochastic, multi-scale phenomenon, and their prediction accuracy cannot be always ensured. On top of that, these models typically need high computational resources for large-scale simulations on heterogeneous areas; even if massive resources are employed, the computational time may not be comparable to the simulated time, making such model not suitable for fast response scenarios and tactical intervention.

Grid-based stochastic modelling techniques may fill the gap between the above mentioned formulations. In this case the complex and inherent stochastic underlying physics are approximated in the cell transition laws and times, grasping the mechanisms of fire spread by describing via probabilistic or deterministic methods the physics at the local scale. The front propagation at the macroscopic scale is the result of the rules operating at the detailed (local) level. They are often lightweight models, which can be integrated in the framework of existing databases with ease.

At the cost of some preliminary modelling, these models can (i) at the physical level, incorporate first principles and (semi-)empirical relations inside of the cell-to cell transition rule sets; and (ii) easily integrate spatial and temporal heterogeneity of the initial and boundary conditions of the simulation, i.e. dealing with spatial patterns of topography, land cover, fuel moisture, meteorological conditions. They can easily be integrated to Geographical Information Systems (GIS) and have (possibly real-time) meteorological data as input (Alexandridis et al. 2011). Adopting grid-based stochastic modelling techniques, also more complex fire propagation patterns, such as the fire spotting effect, can be also simulated in a rather straightforward way via ad hoc probabilistic rules that may make use of fire intensity, wind field and fuel characteristics (Perryman et al. 2013).

Cellular Automatas (CA) are the most well-known examples of the latter category of models (Von Neumann 1966; Wolfram 1984). CA models for wildfire simulation model discretize spatial interactions by adopting a square or hexagonal grid (Encinas et al. 2007). The macroscopic fire spread dynamics is simulated as an ensemble of different realization of a stochastic process. In every realization, the spreading of the fire front from burning cells to neighbouring ones is computed with probabilistic rules. Although CA models always simplify the physical processes of combustion and fire propagation, their modular nature allows the scientist to reach the wanted degree of complexity and accuracy. In Collin et al. (2011), a CA model has been integrated with existing forest physical models to achieve better accuracy. On the other hand, Ghisu et al. (2015) described an elaborate CA model that overcome grid constraints and can perform comparably to deterministic models such as FARSITE (Finney 1998), requiring, however, higher computational times. In Rui et al. (2018), a CA model has been granted improved time-accuracy via the coupling to an existing forest fire model.

3.2 Synopsis of Propagator Development: More than a Decade of Wildfire Simulations

The development of PROPAGATOR took place in more than a decade, and it is described in the following paragraphs. Ease of implementation, robustness and quick operational deployment have always been the objective kept in mind during this long process. The first implementation of PROPAGATOR originated from a specific request of the Italian Civil Protection Department, to support the organization of the G8 summit (2009). The G8 summit was originally planned in La Maddalena, Sardinia, an island frequently affected by severe forest fires in summer season. There was the need of a tool capable of evaluating the best prevention measures and support the firefighting activities in case of a wildfire emergency. This first release relied on a Google Web Toolkit-based interface, and its server application was written in MATLAB®; this implementation was running all the concurrent processes as a unique large process. The core would have been revisited for many times, but also from this seminal version it was possible to run fire propagation simulations all over Italy, from an easy-to-use web interface. PROPAGATOR was able to reproduce wildfires up to 10,000 ha in a reduced of computational time (a few minutes). At this stage, given an ignition point, it highlighted the zones more likely to be reached by wildfire, without giving insights on the temporal dynamics of wildfire spread. No fuel moisture content was implemented in the model, and the output in this version has been used as a *worst-case scenario*, discriminating the most endangered areas under the hypothesis of a totally dry fuel.

In 2011, the second release of PROPAGATOR was complete. At this stage, the model has been implemented in a 3D environment, NAZCA. The algorithm and server code were no longer mixed, and the algorithm was running as a standalone MATLAB® script, therefore being more maintainable. PROPAGATOR was shipped as a plugin, relying on the dataset of DEM and fuel cover of the whole Italy, divided into tiles. This approach still proved to be quite cumbersome and motivated further updates. In this release, timing algorithms were added, using empirical formulas of wind and topography. Probability maps at this stage were time-dependent, and isochrones (that is, contours of the fire advance at given time steps) were added as visual outputs. PROPAGATOR did not include at that time a full-fledged parametrization of the propagation speed, with no information on time scale of the wildfire spread: only different contours of the fire front, ordered from the ignition point to the final burned area, were provided. In 2014, the third release took place. The web interface underwent a redesign and PROPAGATOR was integrated into the multi-purpose MyDewetra platform (Italian Civil Protection Department and CIMA Research Foundation 2014), a web-based tool for the forecasting, monitoring and real-time management of environmental risks (http://www.mydewetra.org).

Wildfire simulations could be launched also on Lebanon thanks to the efforts of the project '*Establishment of Sustainable Natural Resources Management Platform and Early warning system*' implemented for the National Council for Scientific Research (CNRS-Lebanon), according to the initiative '*Regional Co-ordination on Improved Water Resources Management and Capacity Building*', funded by World Bank/GEF.

PROPAGATOR was one of the nowcasting algorithms to assess weather-induced fire impacts within ANYWHERE European project (*EnhANcing emergencY management and response to extreme WeatHER and climate Events*) (Habdank et al. 2019). Thanks to the active participation of end users and first responders a great effort was done to make the whole framework more robust in order to ensure the model up and running every time was requested. During ANYWHERE the fourth release saw a total translation of the source code in Python programming language, with a new server stack, with a Django REST API and database, and a Celery task dispatcher. Routines for slope and wind corrections on fire spread have been improved, and wind conditions could change over time for a given wildfire event. During ANYWHERE, an open API had been released to several developers and the landcover – DEM dataset had been extended from the sole Italian territory to Portugal, Spain (Catalonia, Cantabria, Asturias), France (Cote D'Azur and Corsica), Finland and Switzerland. Figure 3.1 highlights the regions where PROPAGATOR can deliver wildfire simulations with the Dewetra platform and-or the ANYWHERE open API, updated up to 2022. The operational experience accumulated by the first responders in the partner countries helped to identify needs and rooms for improvement of the model. One of the main common user need requirement was to include in the boundary conditions the fuel moisture in

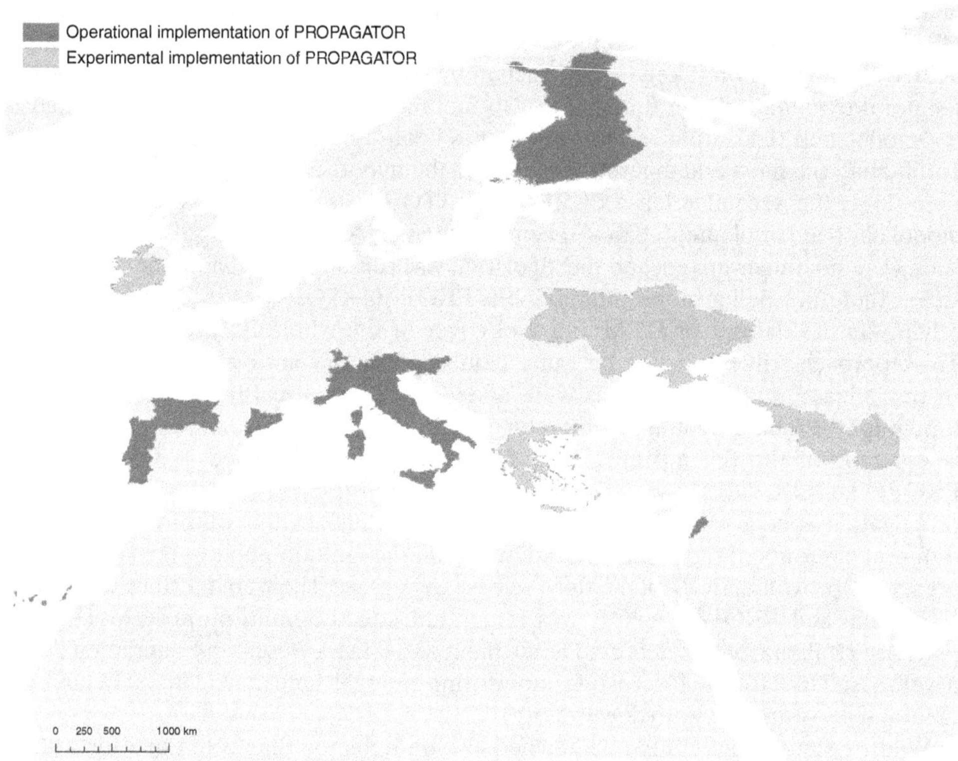

Figure 3.1 Countries where operational (dark grey, orange for online products) or experimental (light grey, yellow for online products) implementations of PROPAGATOR are in place up to 2022.

order to better represent the actual behaviour of a fire ignited behind the worst-case scenario. This request helped to transform the algorithm from a rapid risk assessment tool to a more detailed model able to discriminate extreme wildfire events from fires that can be controlled considering the firefighting capacity of the country. To make this the model was coupled with a dynamic Fine Fuel Moisture model daily available, which represents the core of the Italian Fire Danger Rating System RISICO (Fiorucci et al. 2008, 2011, 2015; Perello et al. 2022) also extended out of the Italian territory. The model was equipped with a tool for downloading boundary conditions (from NWP models) for the next hours to be simulated.

Thanks to this integration, PROPAGATOR has been still advanced to the fifth release within the INTERREG Marittime Project MEDCOOPFIRE, including the capacity to simulate the impact of firefighting activities on fire behaviour (lines and polygons where some kind of firefighting procedure is going to be put in charge) that may be prescribed by the user in a time-dependent way. Recently, already having a high Technology Readiness Level (TRL) PROPAGATOR was proposed within the SAFERS European project where the user requests was extended to produce maps of Rate of Spread (RoS) in order to give the isochrones a more realistic time parametrization and fireline intensity maps behind the isochrone of the fire perimeters. In addition, one of the main advancements of the model is related with the capacity to make run on demand in areas where static inputs are not available. In this case, the fuel map is built based on the most recent clear Sentinel 2 images that coupled with CORINE 1 and cover allow to better define the distribution of different fuel types and to identify non-burnable areas, such as roads or other fuel breaks.

Also during the fifth release, the Python3 code has been substantially rebuilt according to Object Oriented Programming procedures, enhancing its modularity. In the standalone version, there is the opportunity to give as input the geo-referenced rasters for DEM and fuel cover to launch simulations on any part of the world.

Starting from early 2021, PROPAGATOR has been tailored and adopted in the PPRDEAST3 programme for the beneficiary countries Moldova, Ukraine, Georgia, Armenia, Azerbaijan. It is used in conjunction with forest fire hazard maps and Wildfire forecast models and daily bulletins to increase Situational knowledge in all of the DRM phases.

In addition, the model represents the basis for new research activities conducted by Ph.D. students, which already proved the capacity of the model to provide reliable run at different spatial and temporal scale. The model allow to produce simulation using 10 minutes observed wind speed and direction at 5 m resolution. This allows a better validation of the model on real case studies, fostering the use of the model also to plan fuel reduction activities using prescribed burning.

3.3 Propagator Model

In this section, the model is described in detail. The PROPAGATOR model is a quasi-empirical stochastic CA model based on a raster implementation, see Trucchia et al. (2020). It discretizes the space into a grid composed of square cells of arbitrary length $\Delta x = \Delta y = L$. In this work, L has been fixed to 20 m, allowing PROPAGATOR to give high-resolution output, fundamental for reproducing the fires in presence of anthropized wild-land urban interface described in the next sections. At the centre of each cell, the elevation and vegetation type are interpolated from input rasters.

For each time step, every cell is characterized by a state. There is a finite set of possible states to choose from. Each cell can assume one out of three states:

- State 1 means that the cell is *burning* during the current simulation step;
- State 0 means that the cell is *already burned*;
- State −1 means that the cell is *unburned*, but that can burn in the following steps of the fire propagation.

An unburned cell can become a burning cell with a probability that depends on the simulation setting, or it can remain unburned; a burning cell becomes a totally burned cell after one time step with probability equal to 1; a burned cell cannot exit its state, remaining burned for the rest of the simulation.

When the fire propagates from a cell to another, the cell who is getting burned is assigned a specific time for the fire to completely take over it, described in detail in the following paragraphs. When this time is elapsed, the cell switches from unburned

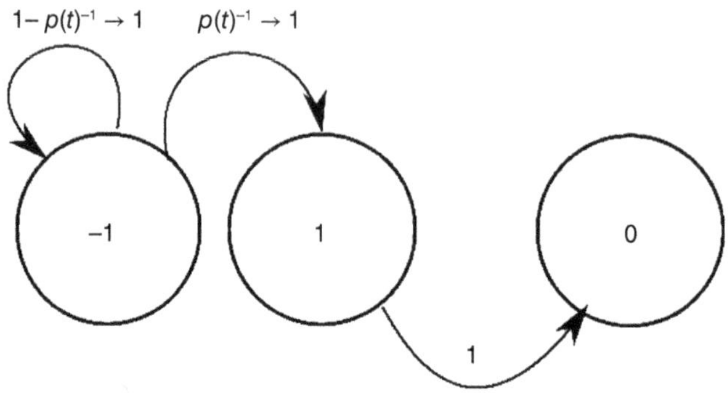

Figure 3.2 The state diagram of the CA. States −1, 1, 0 represent, respectively, an unburned cell, a burning cell and a burned cell. At a given time t, a cell still unburned has a probability $p^{-1} \rightarrow 1$ to burn. This probability comes from the position of the cell, the overall state of the wildfire propagation process, initial conditions and boundary conditions. After a single time step, every burning cell is becoming a burned, inert cell.

(state −1) to burning (state 1), and it attempts to spread fire to the adjacent unburned cells. In operational terms, the computed time step Δt for the state change is appended accordingly to a scheduler which manages the fire propagation mechanism of the cellular automaton.

The state diagram of this automata is depicted in Figure 3.2.

The model can be thought as a contamination process between adjacent cells; the probability of fire spreading from a cell to one of its neighbourhood, p_{ij}, is determined modifying a *nominal fire spread probability* (named p_n in the following). The modifiers integrate topography, wind field and fuel moisture content in the model. The evolution in time of the fire front is modelled by combining a *nominal fire spread velocity* (v_n in the following) and analogous topography-wind-moisture modifiers, with an implementation of an RoS model. For each cell, and more specifically its central point $\boldsymbol{x}_P = (x, y)$, PROPAGATOR computes the probability $u(\boldsymbol{x}_P, t)$ of being burnt at time t and space $\boldsymbol{x}$, using the fire frequency for each cell, based on a large enough ensemble of stochastic simulations and each simulation is performed for identical ignitions, static input rasters and weather conditions. As in Trucchia et al. (2020), a number $N = 100$ of realization is considered. This procedure is shown in Figure 3.3.

The CA uses a Moore neighbourhood of order one, a two-dimensional square lattice composed of a central cell, the i-cell that represents the ignited one, and the eight cells that surround it (see Figure 3.4), which can be ignited by the i-cell. The fire spreading is calculated in a probabilistic way considering the directions between the centre of the i-cell and the ones of the neighbouring cells, the slopes between the cells and the fuel moisture conditions.

Vegetation type is encoded in the domain cells trough an ad hoc fuel model. Fuel models are adopted by many fire propagation models (Burgan and Rothermel 1984; Finney 1998) to classify the physical characteristics (fuel load, arrangement, heat content, height of live and dead biomass . . .) that plays a role to determine the rate of spread, intensity and duration of a fire (Burgan and Scott 2015). Several guides and

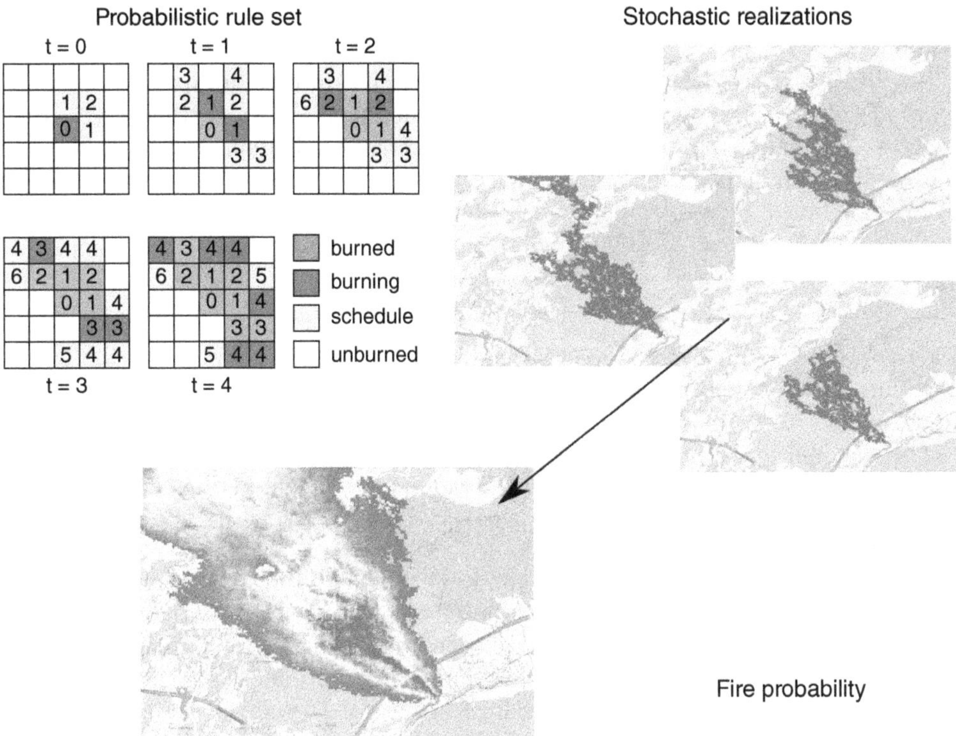

Figure 3.3 Averaging procedure used by PROPAGATOR to furnish probabilistic outputs.

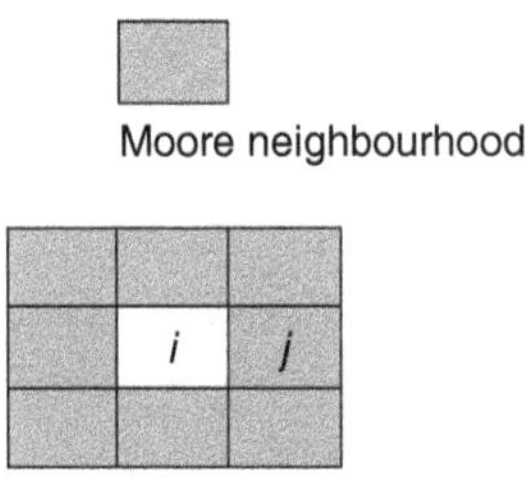

Figure 3.4 Moore neighbourhood implemented in PROPAGATOR.

collections of standard (Anderson 1982; Burgan and Scott 2015) and custom-local fuel models are available in literature, see Dimitrakopoulos (2002); Cruz and Fernandes (2008); Cai et al. (2014). PROPAGATOR adopts a simplified custom model with seven fuel types corresponding to seven types of vegetation. The available fuel types are the following: grasslands, shrubs, broadleaves, fire-prone conifers, agro-forestry areas, non-fire prone forest and non-vegetated areas.

The class called 'non-vegetated areas' includes buildings and infrastructures (streets and settlements) and natural bare soil. Fire propagation cannot take place at all in this class. Also water bodies (rivers, lakes, and seas) are considered by default as non-burnable.

The fire propagates from a cell i to the neighbour cell j with a fire spread probability p_{ij}, which depends heavily on the involved vegetation types, influenced also by the wind speed and direction, the slope between the two cells, and the fuel moisture content in the j-cell. p_{ij}, that is, the probability of a successful fire spread from an ignited i-cell to a j-cell at the time t_k is calculated using a formulation similar to the cumulative binomial probability (Hargrove et al. 2000):

$$p_{ij} = \left(1-(1-p_n)^{\alpha_{wh}}\right)e_m \tag{3.1}$$

where p_n is the nominal Fire Spread Probability, α_{wh} is the topographic and wind influence correction on the spread probability and e_m is the correction due to fine fuel moisture content.

The nominal fire spread probability p_n expresses the probability for the burning i-cell, with a certain vegetation cover, to spread fire to a neighbouring j-cell, characterized by another (or the same) vegetation cover. The values it can assume are tabulated in Table 3.1.

'Not fire-prone forest' class models the low-flammable forests: it is unlikely to get ignited from neighbouring cells, except when the burning cell class is 'Fire-prone conifers'. The 'Broadleaves' class represents tall vegetation with medium flammability, while highly flammable tall vegetation falls into 'Fire-prone conifers' class. The medium-to-low vegetation is organized in three classes: 'Agro-forestry areas' (areas with a low vegetated density, with low propagation probability); 'Shrubs' class (medium-flammable low vegetation); 'Grassland' (highly-flammable very low vegetation).

Table 3.1 The first six rows show the nominal fire spread probability p_n between all the vegetation covers. The last row shows nominal fire spread velocity v_n.

		Burning cell					
		Broadleaves	Shrubs	Grassland	Fire-prone conifers	Agro-forestry areas	Not fire-prone forest
Neighbor cells	Broadleaves	0.3	0.375	0.25	0.275	0.25	0.25
	Shrubs	0.375	0.375	0.35	0.4	0.3	0.375
	Grassland	0.45	0.475	0.475	0.475	0.375	0.475
	Fire-prone conifers	0.225	0.325	0.25	0.35	0.2	0.35
	Agro-forestry areas	0.25	0.25	0.3	0.475	0.35	0.25
	Not fire-prone forest	0.075	0.1	0.075	0.275	0.075	0.075
	Nominal Fire spread velocity (mm h^{-1})	100	140	120	200	120	60

The p_n values can influence drastically the results of fire propagation, see the parameter response analysis in Trucchia et al. (2020). They were defined by the means of a continuous calibration through all the development years, receiving both valuable inputs from experts of the involved institutions and also information deriving from fire susceptibility mapping (Tonini et al. 2020; Trucchia et al. 2022).

The slope and the wind speed and direction will eventually correct the initial value of p_n. The topography correction increases the propagation probability p_n when the slope increases in the direction of propagation (that is, the uphill case) and in turn it decreases p_n when the slope decreases in the direction of propagation (the downhill case).

The topographic correction factor, α_h, is shown in Figure 3.5.

Wind direction and speed are considered relative to the direction from the burning cell and its adjacent cells. For each time step, the values of wind magnitude and direction are considered to be homogeneous into the whole domain, but these values are perturbed in every cell. A multiplicative uniform noise $U[0.8, 1.2]$ is applied to wind magnitude, where $U[a, b]$ corresponds to the uniform distribution with support in $[a, b]$. The wind direction is perturbed using a uniform additive noise $U[-11.25°, 11.25°]$. The wind influencing factor α_w is shown in Figure 3.6, where spherical coordinates are adopted. The influence is very relevant only for quite high wind speed values, while the probability of propagation is not modified in the low-speed case, nor increasing nor decreasing. The wind direction plays a key role in the process increasing the burning probability in case of propagation direction aligned with the wind direction, while it is decreased in case of opposite directions.

The topographic influence and the wind influence on the Fire Spread Probability are then combined into a factor, named α_{wh}, that is obtained as multiplication of the two different effects:

$$\alpha_{wh} = \alpha_w \alpha_h \tag{3.2}$$

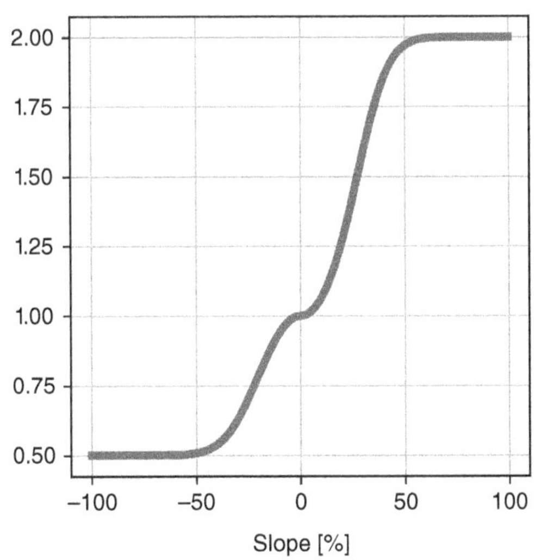

Figure 3.5 Topographic correction factor α_h.

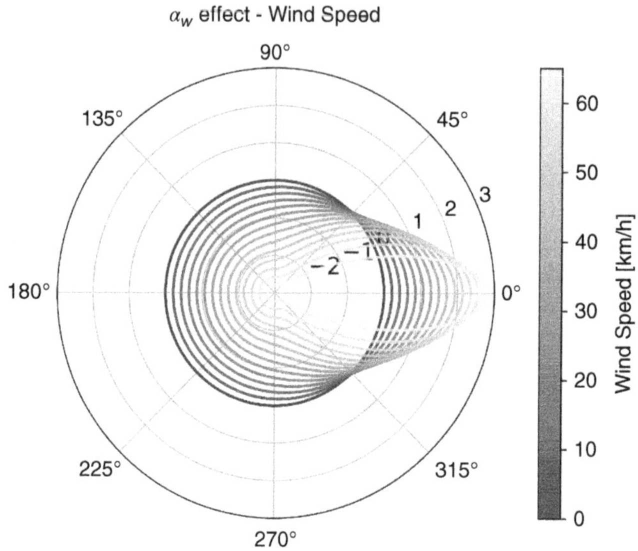

Figure 3.6 Wind influencing factor α_w.

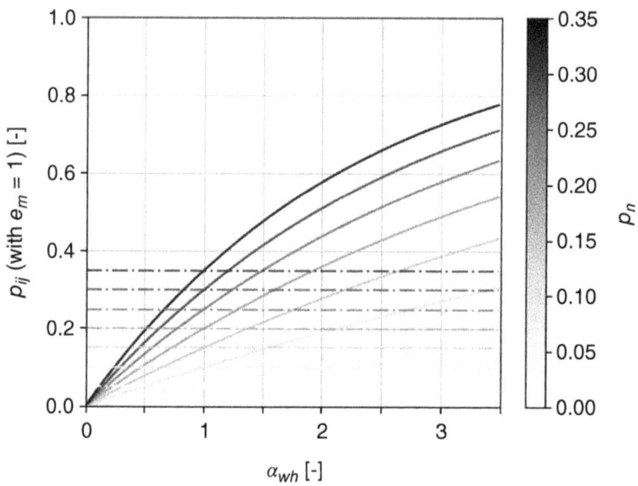

Figure 3.7 Influence of the combined slope-wind factor on the Fire Spread Probability. The plot portrays the dependence of p_{ij} of Formula (1) on the slope-wind factor α_{wh}, given a fixed $e_m = 1$ and several values for p_n (represented by the dash-dotted lines).

In Figure 3.7, the impact of α_{wh} on Fire Spread Probability p is shown: the nominal transitional probability is obtained for $\alpha_{wh} = 1$; when this factor is not unitary, the effect of the possible combinations of slope, wind speed, and direction can be evaluated. The Fire Spread Probability vary from values very close to zero to a maximum of about 0.75. Numerical experiments (Trucchia et al. 2020) show fire propagation is quite always possible for this value, see Figure 3.7.

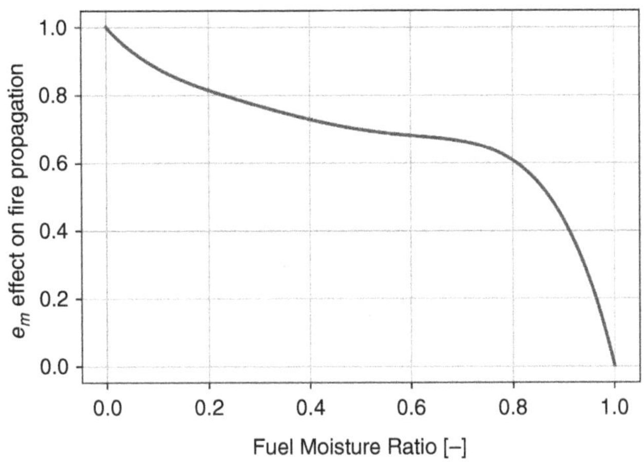

Figure 3.8 Effect of the fuel moisture ratio (dead fuel moisture over dead moisture of extinction) on the Fire Spread Probability.

The factor e_m encloses the effect of fuel moisture content on Fire Spread Probability, multiplying the nominal probability p_n. This factor has been implemented as described by Burgan and Rothermel (1984). It is portrayed in Figure 3.8. Its values ranges in the interval [0, 1] and it is a function of the fuel moisture ratio, i.e. the dead fuel moisture over the dead moisture of extinction. For simplicity, a fixed value of 0.3 has been used as moisture of extinction for all the vegetation classes considered in PROPAGATOR, applying a conservative estimate (Anderson 1982).

When a cell is ignited, the fire transition time is computed by combining the Rate of Spread v_{prop} and the fuel moisture factor f_m, with the distance d between the centre of the burning i-cell and the centre of the newly ignited j-cell. The transition time Δt is then calculated as:

$$\Delta t = \frac{d}{v_{\text{prop}} f_m} \tag{3.3}$$

There is consensus that the flammability of the vegetal fuel and then the rate of the spread of a fire depends exponentially on the fuel moisture content, e.g. see Alexandridis et al. (2011). f_m is then calculated using the formulation proposed by Marino et al. (2012):

$$f_m = e^{cM_n} \tag{3.4}$$

where c is a constant set at −0.014 and M_n is the fuel moisture ranging from 0 to 1.

The Rate of Spread v_{prop} is computed starting from the nominal v_n, which corresponds to the Fire Spread Velocity for each vegetation type without slope and wind effects; then this value is modify considering the slope and the wind effects. Values for v_n are reported in Table 3.1. Wind and slope effects have been evaluated using the formulations proposed by Sun et al. (2013), which are calculated as Equations (3.5) and (3.6). The wind speed factor K_w is computed as:

$$K_w = \exp(0.1783V) \tag{3.5}$$

where V is the wind velocity in the direction of propagation, in [m s^{-1}].

The slope factor K_ϕ is computed as:

$$K_\phi = \exp\left(3.533(\tan\phi)^{1.2}\right) \tag{3.6}$$

where ϕ is the terrain slope angle [°] in the direction of propagation.

The final Rate of Spread v_{prop} is then computed by multiplication of the nominal Fire Spread Velocity v_n and the two factors, K_w and K_ϕ. This method has been proposed by Sun et al. (2013).

PROPAGATOR involves several input factors that have an impact on the simulated burned area and fire arrival probability. A response analysis to variation in vegetation type and spatial pattern, wind speed, wind direction, slope and moisture content of fuel was performed in Trucchia et al. (2020).

In the latest releases of PROPAGATOR, firefighting actions can be selected, distinguishing operations that use water lines from other heavier operations that act on the vegetation, such as interventions made using earth-moving machines. The first actions are implemented setting the fine fuel moisture content at a prescribed value μ_{wl} in the domain cells where fighting actions are enforced, so that the fire propagation probability is strongly reduced. The μ_{wl} value depends on the actions implemented: $\mu_{wl} = 0.8$ for waterlines, $\mu_{wl} = 0.9$ for Canadair activities and $\mu_{wl} = 0.6$ for helicopters. In case of Canadair activities, the moisture is also affected to a minor degree around the area of actions; in case of helicopter actions, a random noise is introduced in order to reproduce wind effects. Heavy equipment actions are implemented changing the values of the vegetation type at 'Non-vegetated areas' where these actions are modelled, so that fire cannot propagate in those areas. This action can also be used to insert roads and non-vegetated areas not already present in the fuel map. In the proposed test cases of Section 3.4 different actions are tested.

3.4 Case Studies

PROPAGATOR has been tested intensively over hundreds of different wildfires by the hand of different end users, researchers, and CPAs (Trucchia et al. 2020). In this chapter, five recent wildfires are analysed, in which two regions constituted a pilot area for the ANYWHERE project, that is Liguria region in Italy and Catalonia (Spain). For what concerns Liguria, the three analysed wildfires were *Taggia*, *Cisano sul Neva* and *Arnasco*; in Catalonia the *Llançà* and *La Torre de l'Espanyol* cases were simulated.

3.4.1 Data Retrieval

3.4.1.1 Ignition Point, Wind Speed and Wind Direction Information on the simulated wildfires is reported in Table 3.2. 'Latitude' and 'Longitude' are reported in WGS84

Table 3.2 Description of the several case studies simulated in this work with PROPAGATOR.

Fire event	Region (Nation)	Date	Ignition latitude	Point longitude	Burnt area (ha)	Wind speed ($km\,h^{-1}$) (direction [∘])	Human activity	Duration
Cisano	Liguria (Italy)	21/09/2021	44.075052 N	8.148162 E	51.57	7.9 (162)	High	12 h
Arnasco	Liguria (Italy)	20/07/2022	44.08315 N	8.09577 E	85.79	1.5 (129)	Medium	12 h
Taggia	Liguria (Italy)	19/06/2022	43.872362 N	7.853571 E	77.52	5.6 (164)	Medium	24 h
Llançà	Catalunia (Spain)	16/07/2021	42.352879 N	3.167882 E	430.35	23.36 (319)	High	24 h
La Torre de l'Espanyol	Catalunia (Spain)	26/06/2019	41.213455 N	0.640268 E	5000	50 (150)	High	24 h

CRS; the 'Burnt area' (in hectares) is the area burnt by each wildfire; 'Wind speed and direction' represent the average speed and direction, which characterized the wildfires, while, in the simulation, time-dependent wind conditions have been considered; also the duration in hours of the simulation is reported in the table for each event.

For the Catalan wildfires, data regarding ignition points, moisture condition and wind field were retrieved from the web archives of the Ministry of Home Affairs (*Departament D'Interior*) of the Government of Catalonia http://interior.gencat.cat/ca/arees_dactuacio/bombers/foc-forestal/incendis_forestals/informes-dincendis-forestals/. This information was of paramount importance in setting the ignition point, and analysing the firefighting effort put in place. For the simulation, instead of the tabulated data in the Catalan Governmental archives, we used a Limited Area Model, in order to simulate the situation of a fully operational tactical context where no exact information on the future is furnished, but only weather forecasts can be adopted.

4.5 km resolution WRF synoptic data (Apicella et al. 2021) has been downloaded and used as the wind boundary condition. On the other hand, the data related to the synoptic wind field for the Italian events were furnished by the anemometers of the sensor network of the Italian Civil Protection Department (http://www.protezionecivile.gov.it/), located closer than 4 km from the wildfire event. The ignition points for the Italian wildfires have been estimated on the basis of information obtained during firefighting actions and after the described events.

3.4.1.2 Firefighting Actions The fighting actions for Catalan fires are formulated according to the reports produced by the Catalan Fire-Fighters ('Cos de Bombers de la Generalitat de Catalunya') in the aforementioned website of the Government of Catalonia. For the Italian cases, firefighting actions are located along secondary streets and as barriers protecting affected settlements, where major human intervention to has supposedly been deployed.

3.4.1.3 Burned Area Geometries The shapefiles of the final burned area for the Catalan fires is taken from the web archives of the Department of Agriculture, Hunting, Fishing and Food of the Catalan Government (http://agricultura.gencat.cat/ca/serveis/cartografia-sig/bases-cartografiques). The actual Italian wildfire areas are recorded on the online platform MyDewetra, while the most recent ones are obtained from satellite data by AUTOBAM, an automated algorithm for Near Real-Time Mapping of burned Areas Using Sentinel-2 data (Pulvirenti et al. 2020a, b).

3.4.1.4 Land-Cover Files In Catalan fires, the Land Cover Map of Catalonia (MCSC) has been considered (http://www.creaf.uab.es/mcsc/usa/descriptiu.htm). This resource is a high-resolution thematic cartography of the main types of the regional land cover (forests, crops, urban areas, etc.). The MCSC is developed in the Centre de Recerca Ecològica i Aplicacions Forestals (CREAF) and funded by the Generalitat de Catalunya. This map has 241 different categories, with a minimum resolution of 10 m.

For Liguria region, the 240 categories 1:25,000 maps made available from the geographical web portal of Ligurian Region https://geoportal.regione.liguria.it/catalogo/mappe.html was adopted. The aforementioned maps have been projected into a 7-category, 20-meter resolution map, see Section 3.3.

3.4.1.5 Orography Files To simulate Catalan wildfire events, a 20-m resolution DEM input file provided by the Catalan Ministry of Home Affairs has been used, while, for the Italian wildfires, the 20-meter resolution map of the Italian Institute for Environmental Protection and Research ISPRA has been used (http://www.sinanet.isprambiente.it/it/sia-ispra/download-mais/dem20/view).

3.5 Results and Discussion

In order to show the predicting capabilities of the presented model in recent wildfires, PROPAGATOR has been tested on the fire events presented in detail in Section 3.4. The simulator was launched on a 6-core (2.6 GHz) Intel® Core™ i7- 10750H CPU laptop with 16 GB of RAM. The CPU time to get the results was lower than the physical time of the simulation by several orders of smagnitude, since every simulation lasted less than 10 minutes.

Figure 3.9 portrays the comparison between the actual burnt area and the simulated fire probability distribution for the Italian fires of Arnasco (panel (a)), Cisano (panel (b)), Taggia (panel (c)) while, in Figure 3.10, the results of Llança (panel (a)) and La Torre de l'Espanyol (panel (b)) wildfire events are shown. From a qualitative point of view, the fire arrival probability distribution at the last time step of simulation (that is, the grayscale surfaces, white-to-red for online products) shows good agreement with the recorded fire shape of the wildfires, for both Ligurian and Catalan cases. The whole spatial distribution of the output probability can shed some light on different wildfire scenarios. This is particularly true for the Ligurian wildland fires, especially in the Taggia wildfire where a non-zero probability of the fire to spread uphill in the forested area north-west of the burned area is forecast.

3.5.1 Performance Indicators

To evaluate PROPAGATOR runs performance, three performance indicators (I_1, I_2, I_3) have been defined – these indicators, conceived for the probabilistic map of the simulator output, have been introduced and discussed in Trucchia et al. (2020). Additionally, other standard performance indicators have been used, namely the specificity, the sensitivity, the Sorensen coefficient and the McNemar test.

The notation used for the confusion matrix components (Agresti 2007) will be of use in this section, corresponding to the predictive capabilities of the model when compared with the observed burned area. To evaluate a performance indicator, a cell is considered to be burned by the model when the probability of fire arrival u exceeds a threshold, set to 0.5. The first entry for the confusion matrix, a, corresponds to the

(a) Arnasco wildfire - simulation and burned area.

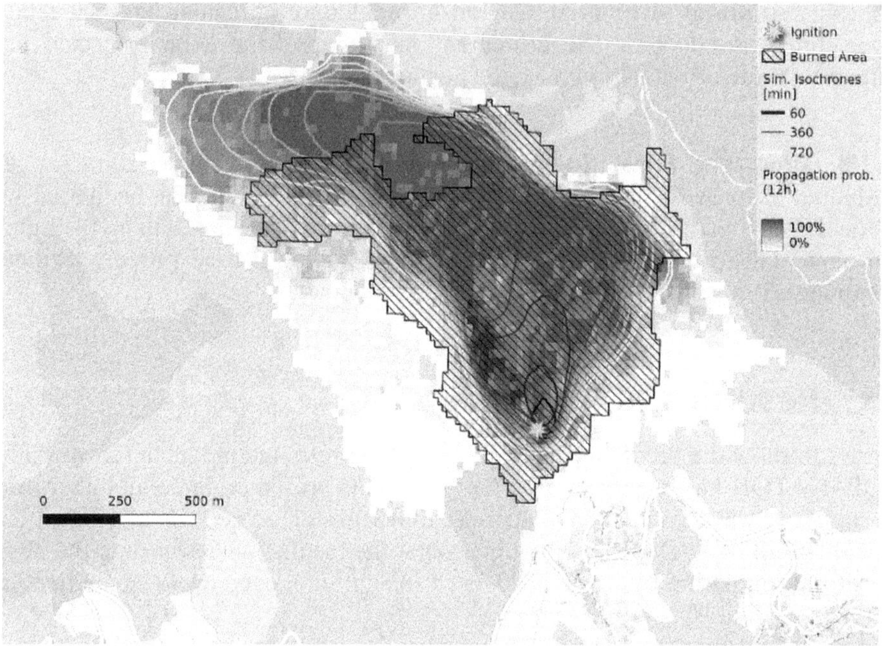

(b) Cisano wildfire - simulation and burned area.

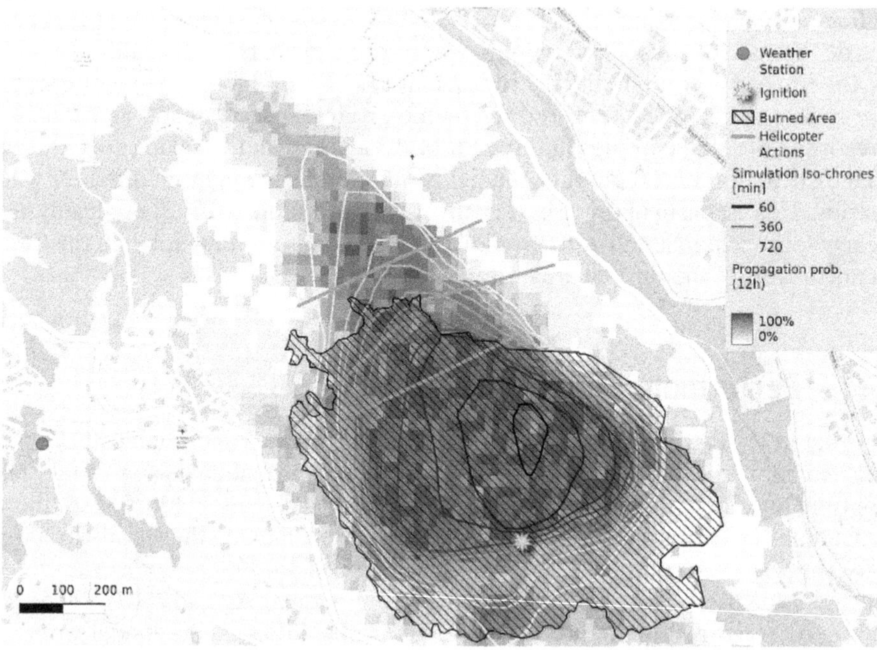

Figure 3.9 Comparison between the actual burnt areas and the simulated ones for the Ligurian events. The plots (a), (b), (c) show the actual burnt areas (represented by black hashed areas), the ignitions (stars, yellow for online products), the implemented fighting actions (solid lines, blue for online products), the simulated burnt areas (in a grayscale colourmap, white to red for online products) and the iso-chrones (grayscale solid lines, coloured solid lines for online products), which represent the contour of the 50% of arrival probability at different time steps.

(c) Taggia wildfire – simulation and burned area.

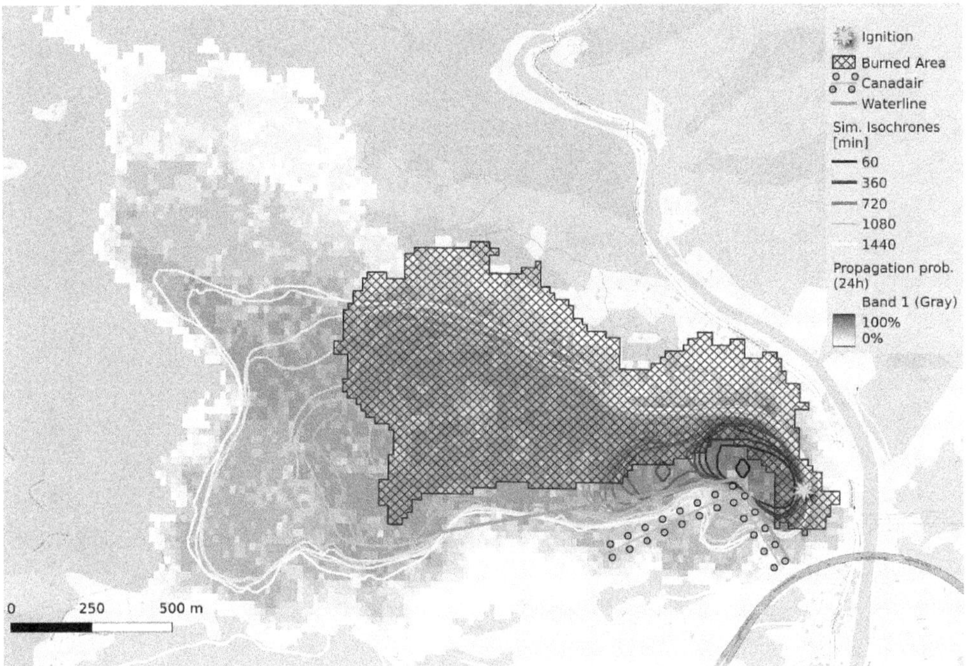

Figure 3.9 (Continued)

number of cells burned in both observed and simulated fires; b corresponds to the number of cells that did not burn in the fire but were considered burned by the model; c is the number of cells that burned in the fire but were considered as unburned by PROPAGATOR. The performance indicators I_1, I_2, I_3 are briefly described below, as well as the other standard indicators adopted. The goal value for the I_i indicators and for the p-value of McNemar test is zero, while the goal for the Sorensen coefficient, sensitivity and specificity is one.

3.5.1.1 Indicator I_1 This indicator corresponds to the Euclidean norm computed over the computational domain $\mathcal{S}$ between the indicator function $I_b(\boldsymbol{x}_\mathbf{p}, T_{\text{end}})$, which returns 1 inside of the real burnt area and 0 outside, and the probability output of PROPAGATOR, $u(\boldsymbol{x}_\mathbf{p}, T_{\text{end}})$, scaled by the actual burnt area $|\Omega_{\text{burnt}}|$:

$$I_1 = \frac{1}{|\Omega_{\text{burnt}}|}\sqrt{\int_S \left|I_b(x_P, T_{\text{end}}) - u(x, T_{\text{end}})\right|^2 dx_p} \tag{3.7}$$

3.5.1.2 Indicator I_2 This indicator corresponds to the Euclidean norm computed over the computational domain $\mathcal{S}$ between the indicator function of the actual fire

(a) Llançà wildfire – simulation and burned area.

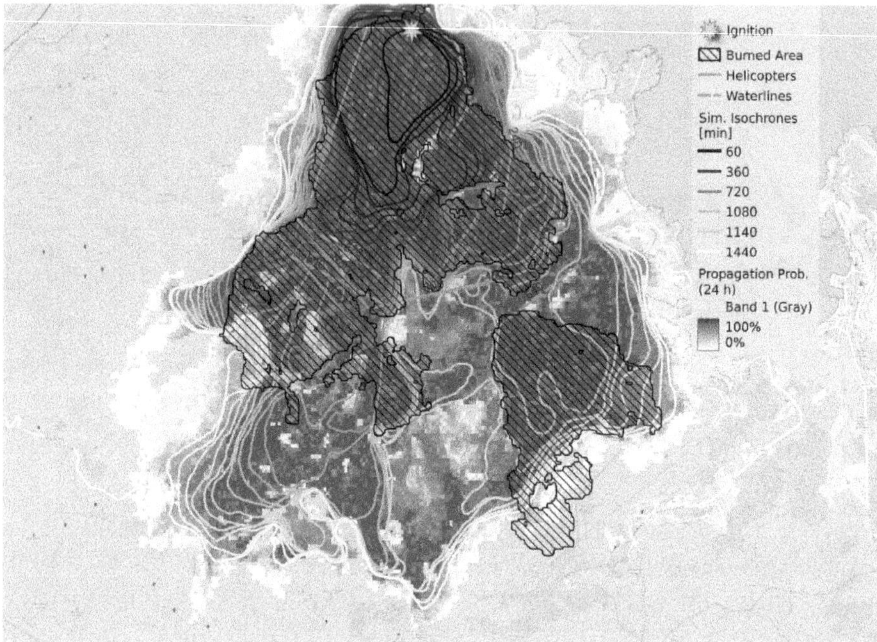

(b) La Torre de l'Espanyol wildfire – simulation and burned area.

Figure 3.10 Comparison between the actual burnt areas and the simulated ones for the Catalan events. The two plots (a) and (b) show the actual burnt areas (represented by black hashed areas), the ignitions (stars, yellow for online products), the implemented fighting actions (solid lines, blue for online products), the simulated burnt areas (grayscale colourmap, white to red for online products) and the iso-chrones (grayscale solid lines, coloured solid lines for online products), which represent the contour of the 50% of arrival probability at different time steps.

burnt area $I_b(\boldsymbol{x_P}, T_{\mathrm{end}})$ and the simulated area with burning probability u at the last computed time T_{end} higher than 0.5, $I_{>0.5}$ $(u(\boldsymbol{x_P}, T_{\mathrm{end}}))$, scaled by the actual burnt area $|\Omega_{burnt}|$:

$$I_2 = \frac{1}{|\Omega_{\mathrm{burnt}}|}\sqrt{\int_S \left|I_b\left(x_P, T_{\mathrm{end}}\right) - I_{>0.5}\left(u\left(x, T_{\mathrm{end}}\right)\right)\right|^2 dx_p} \tag{3.8}$$

Since the probability of fire arrival u ranges in [0, 1], the real burnt area may coincide or not for a given computational cell according to different probability thresholds considered. Indicator I_1 integrates for each point the squared distance between the probability of arrival and the binary information of the real burnt area (burned-unburned, represented by the indicator function I_b), in order to consider the richness of information of the probabilistic output. The indicator I_2, instead, considers the probabilistic output u as a binary indicator applying a threshold as a level of confidence.

3.5.1.3 Indicator I_3 This indicator corresponds to the area of the measured burnt area where PROPAGATOR did not forecast correctly the fire arrival according to the 0.5 threshold u $(\boldsymbol{x_P}, T_{\mathrm{end}}) < 0.5$, scaled by the actual burnt area $|\Omega_{\mathrm{burnt}}|$. I_3 can be thought as an estimator of *false negative*, and for this reason its minimization is of paramount importance in case a *worst-case scenario* is pursued:

$$I_3 = \frac{|\mathcal{S}^*|}{|\Omega_{\mathrm{burnt}}|} \tag{3.9}$$

where $\mathcal{S}^*$ is the burnt domain part defined by the relation:

$$x \in \mathcal{S}^* \Leftrightarrow I\left(x_P, T_{\mathrm{end}}\right) = 1 \text{ and } u\left(x_P, T_{\mathrm{end}}\right) < 0$$

3.5.1.4 Sorensen Coefficient S_c The Sorensen similarity index (S_c) (Sorensen 1948) is a statistical index used to compute the similarity value between two different samples. S_c values were calculated as follows:

$$S_c = \frac{2a}{2a + b + c} \tag{3.10}$$

The result is a value between 0 and 1: 1 means a perfect agreement between the two samples, while 0 means that there is no agreement.

The value of S_c is not affected by the extent of the domain $\mathcal{S}$, because areas that are unburned for both observations and simulations are not considered.

3.5.1.5 McNemar Test The McNemar test (Agresti 2007; Mallinis et al. 2014, Arca et al. 2019) is a non-parametric test based on the chi-square (χ^2) distribution used to evaluate the symmetry of rows and columns of the confusion matrix.

In the following, the corrected version of the McNemar test is adopted (Edwards 1948):

$$\chi^2 = \frac{\left(|c-b|-1\right)^2}{(b+c)} \tag{3.11}$$

The McNemar test was performed to evaluate the null hypothesis of marginal homogeneity between actual burned areas and the simulated one.

3.5.1.6 Sensitivity and Specificity The sensitivity (Montealegre et al. 2014; Arca et al. 2019) is the ratio between cells correctly classified as burned by the model and the total number of actually burned cells. Its complementary measure (1-sensitivity) is related to the probability of false negatives (error of omission). Conversely, the specificity (Montealegre et al. 2014; Arca et al. 2019) is the ratio between the number of cells correctly classified as unburned by PROPAGATOR and the total number of actually unburned cells. Its complementary measure (1-specificity) is related to the probability of false positives (error of commission).

3.5.2 Performances of Test Cases

The performance of the five test cases is available in Table 3.3. The overall results are almost always in line with the one of Trucchia et al. (2020). In general they are acceptable given that in this case, no optimization process has been performed, and after having set the inputs from model or from network recording little post process has been made, to be in line with the operational spirit of the tool. In the Catalan wildfires, the wind and fine fuel moisture conditions are, respectively, from a NWP model and an European-scale implementation of RISICO, guaranteeing that PROPAGATOR is capable to deliver fast and reliable scenarios using data that is available to the responders at the very beginning of a potentially hazardous wildfire. In the Ligurian wildfires, nearby weather stations are used and used to compute input wind conditions and fine fuel moisture (via and *ad hoc* implementation the RISICO model, which runs in weather sensor networks). Those cases can therefore be seen as reanalysis to see the possible 'what if' scenarios useful for tactical intervention trainings and wildfire management.

Table 3.3 Performance indicators computed for the case studies of PROPAGATOR.

Fire event	I_1	I_2	I_3	Sorensen coefficient	McNemar test p-value	Sensitivity	Specificity
Cisano	0.0157	0.0196	0.3622	0.7142	$<10^{-2}$	0.6378	0.8113
Arnasco	0.0145	0.0170	0.3507	0.6907	$<10^{-2}$	0.6493	0.7378
Taggia	0.0192	0.0233	0.3722	0.5178	$<10^{-2}$	0.6278	0.4405
Llançà	0.0079	0.0093	0.1013	0.6609	$<10^{-2}$	0.8987	0.5226
La Torre de l'Espanyol	0.0022	0.0027	0.1651	0.5956	$<10^{-2}$	0.8349	0.4629

3.5.3 An Example of Continuous Improvement and Operational Deployment: Implementation in Ireland

As a follow-up for ANYWHERE project, following a collaborative approach with other scientific teams and implementing partners which area active in supporting Civil Protection stakeholders, PROPAGATOR was customized to support *MetE'ireann*, Ireland's National Meteorological Service, in delivering fast scenarios of wildfires on the whole Irish territory. This could help in fostering prevention and preparedness also in view of Climate Change that would in the next years change wildfire regimes, bringing large and frequent wildfires to landscapes that are not fire prone at the moment. Climate Change require a rethinking natural hazard management in many European Countries (Stults 2017). In Figure 3.11, a fire in Loughglinn, County Roscommon, Ireland, is simulated. The PROPAGATOR client hosted by myDewetra platform is used to deliver this scenario analysis.

3.6 Conclusions

In this paper, the implementation of an operational wildfire simulator designed to respond to a broad scenario of wildfires, via an ad hoc implementation of a CA, is presented. The model is applied to a set of Spanish and Italian case studies, which range from small fires in presence of Wild-land urban interface to large fires which spanned several days and developed away from WUI. When available, information about

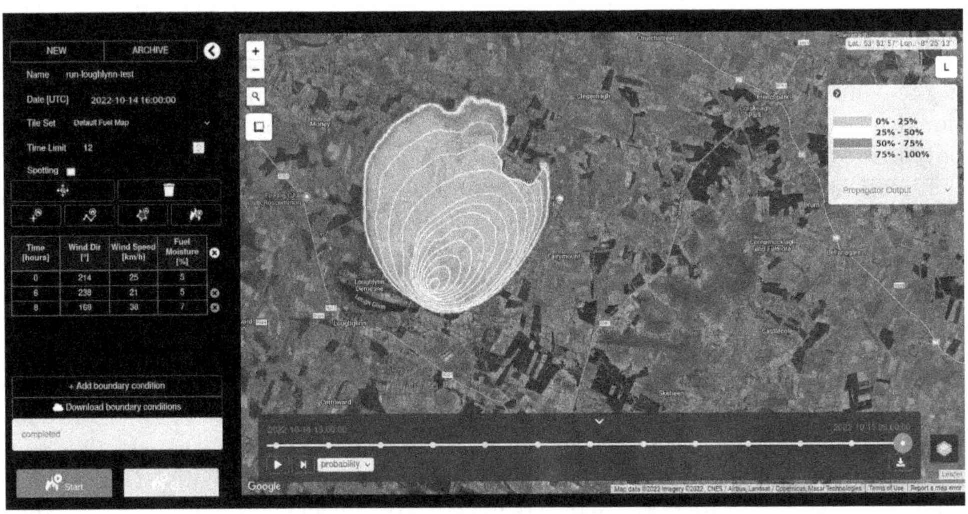

Figure 3.11 In this Figure, a wildfire in Loughglinn, County Roscommon, Ireland is simulated with PROPAGATOR using the client available in MyDewetra web platform (Italian Civil Protection Department and CIMA Research Foundation 2014). In this very dry and windy worst-case scenario, 1500 hectares of cropland are burned by a fire pushed by strong winds from the South.

firefighting procedures has been included in the simulations. The proposed test cases constituted again a good benchmark to test this CA-based model, which is designed for fast simulation of wildfires. Such CA has been equipped with a physical core that accounts for the effects of wind field, slope, vegetation distribution, fuel moisture content and eventual firefighting actions. The simulation of the test cases has been made using a probabilistic approach, taking averages of ensembles of CA simulations to derive the probability of burning of every cell for each time step. The probabilistic output of PROPAGATOR gives useful information for the fire managers, since determining in a few minutes the area where fire spread will be most probable under forecast wind and moisture conditions is of vital importance to correctly deploy resources as for tactical firefighting operations. The results have been validated via the comparison with satellite and ground-retrieved burned area polygons, and in a more quantitative way via several fitness indicators. The good overall performance confirms than a decade of continuous scientific and technical improvement of an operational tool conceived for practitioners and decision-makers.

The lessons learnt that originated from the proposed work are reported below. The first releases of the presented simulator gave good results in very heterogeneous topographic and vegetation conditions where the burned area is mainly determined by the distribution of flammable fuel and its topographic configuration. However, in practical scenarios, wind, fine fuel moisture and firefighting actions have a great impact on wildfire dynamics and its rate of spread. The constant dialogue with end users, stakeholder and implementing partners has given PROPAGATOR developers the momentum to obtain a simulator that can give reasonable results in a way wider set of configurations and scenarios. End users and first responders need to know in a timely fashion not only the potential area impacted by the wildfire, but also information on the front temporal evolution for a better coordination of firefighting actions, and a better decision making process during operational activities. This need led to the inclusion in PROPAGATOR of rate of spread of the fire front and to the implementation in the simulator of firefighting actions. Another feature that originated from operational users requests was the introduction of fuel moisture effect, to discriminate potential extreme events from easily controllable ones. In order to consider and test the latest improvements, the presented case studies regarded two wind-driven cases (the Catalan wildfires) and three other cases (the Ligurian ones) that have been characterized by a smaller total burned area. Fire simulators tend to be highly complex, time-consuming to learn, and difficult to parameterize (Schumaker et al. 2022). As a result, adopting these models in their daily routines can result to be impractical for practitioners and end users who are not computational wildfire specialists. PROPAGATOR is by design of rapid uptake by practitioners that do not need to be skilled in the physics of fire spread. PROPAGATOR has proven to be of use to such non-specialists to model wildfire spread, simulate scenarios in different weather conditions and different choices of firefighting actions. Training activities (such as the ones delivered in the framework of ANYWHERE project) and field exercises (such as the ones carried out in MEDCOOPFIRE project) benefited from the simple nature of this simulator. It should be remarked that the purpose of the presented tool is not to perform real-time high-fidelity and high-resolution simulations. The adopted CA approach relies on empirical grounds to deliver useful probabilistic scenarios of the fire spread over time, without any downscaling of the wind field. Nonetheless, such results can be helpful for Civil Protection Authorities, decision-makers and fire managers. That is precisely the spirit of the presented case study, where rapid scenarios relying on NWP forecast or sensor data are presented. PROPAGATOR is the result of a shared scientific, technical and human growth in the wildfire community, where an algorithm born from computational wildfire research has been transferred to field applications, to reach a wide audience of end users.

This would not have been possible without the many research projects, such as ANYWHERE, where the synergy between implementing partners, stakeholders and end users pushed the model way beyond a simple set of formulas and scripts.

In the process of continuous improvement, the next steps are to explore the change in resolution of the computational grid, to analyse the effect of fuel cover retrieved from Earth observations and to see the effect of different fire spotting models to be integrated in the simulator's CA framework (López-De-Castro et al. 2022).

References

Agresti, A. (2007). An introduction to categorical data analysis. In: *Wiley Series in Probability and Statistics*, 2ee, 400pp. Hoboken, NJ: Wiley ISBN: 978-0-471-22618-5.

Albini, F.A. (1985). A model for fire spread in wildland fuels by-radiation. *Combustion Science and Technology* 42: 229–258.

Alexandridis, A., Russo, L., Vakalis, D. et al. (2011). Wildland fire spread modelling using cellular automata: evolution in large-scale spatially heterogeneous environments under fire suppression tactics. *International Journal of Wildland Fire* 20: 633–647.

Anderson, H.E. (1982). Aids to determining fuel models for estimating fire behavior, *General Technical Report INT-122, Ogden, UT: U.S. Department of Agriculture, Forest Service, Intermountain Forest and Range Experiment Station* (accessed 1 March 2019).

Apicella, L., Puca, S., Lagasio, M. et al. (2021). The predictive capacity of the high resolution weather research and forecasting model: a year-long verification over Italy. *Bulletin of Atmospheric Science and Technology* 2: 3.

Arca, B., Ghisu, T., Casula, M. et al. (2019). A web- based wildfire simulator for operational applications. *International Journal of Wildland Fire* 28: 99–112.

Asensio, M.I. and Ferragut, L. (2002). On a wildland fire model with radiation. *Journal for Numerical Methods in Engineering* 54: 137–157.

Berjak, S.G. and Hearne, J.W. (2002). An improved cellular automaton model for simulating fire in a spatially heterogeneous savanna system. *Ecological Modelling* 148: 133–151.

Burgan, R.E. and Rothermel, R.C. (1984). *BEHAVE: Fire Behavior Prediction and Fuel Modeling System, Fuel Subsystem.* U.S. Dept. of Agriculture, Forest Service, Intermountain Forest and Range Experiment Station.

Burgan, R. and Scott, J. (2015). *Standard Fire Behavior Fuel Models: A Comprehensive Set for Use with Rothermel's Surface Fire Spread Model. CreateSpace Independent Publishing Platform.*

Cai, L., He, H.S., Wu, Z. et al. (2014). Development of standard fuel models in boreal forests of northeast china through calibration and validation. *PLoS One* 9: e94043.

Castellnou, M., Prat-Guitart, N., Arilla, E. et al. (2019). Empowering strategic decision-making for wildfire management: avoiding the fear trap and creating a resilient landscape. *Fire Ecology* 15: 1–7.

Collin, A., Bernardin, D., and Séro-Guillaume, O. (2011). A physical-based cellular automaton model for forest-fire propagation. *Combustion Science and Technology* 183: 347–369.

Conedera, M., Cesti, G., Pezzatti, G. et al. (2006). Lightning-induced fires in the alpine region: an increasing problem. *Forest Ecology and Management* 234: S68.

Copernicus–Emergency Management Service (2019). The copernicus emergency management service monitored the impact of forest fires in Gran Canaria, Spain. *EMS INFORMATION BULLETIN 120, Emergency Management Service of Copernicus programme*. https://emergency.copernicus.eu/mapping/ems/copernicus-ems-monitored-impact-forest-fires-gran-canaria-spain (accessed 1 April 2019).

Cruz, M. and Fernandes, P. (2008). Development of fuel models for fire behaviour prediction in maritime pine (pinus pinaster ait.) stands. *International Journal of Wildland Fire* 17: 194–204.

De Rigo, D., Liberta, G., Durrant, T., Artes Vivancos, T. and San-Miguel-Ayanz, J. (2017). Forest fire danger extremes in Europe under climate change: variability and uncertainty, EUR 28926 EN, Publications Office of the European Union, Luxembourg. ISBN 978-92-79-77046-3. doi:10.2760/13180, JRC108974.

Dimitrakopoulos, A. (2002). Mediterranean fuel models and potential fire behaviour in Greece. *International Journal of Wildland Fire* 11: 127–130.

Edwards, A.L. (1948). Note on the "correction for continuity" in testing the significance of the difference between correlated proportions. *Psychometrika* 13: 185–187.

Encinas, L.H., White, S.H., del Rey, A.M., and Sánchez, G.R. (2007). Modelling forest fire spread using hexagonal cellular automata. *Applied Mathematical Modelling* 31: 1213–1227.

Fernandes, P. (2001). Fire spread prediction in shrub fuels in Portugal. *Forest Ecology and Management* 144: 67–74.

Finney, M.A. (1998). Farsite: fire area simulator-model development and evaluation. *General Technical Report RMRS-RP-4, Ogden, UT: U.S. Department of Agriculture, Forest Service, Rocky Mountain Research Station* (accessed 1 March 2019).

Fiorucci, P., Gaetani, F., and Minciardi, R. (2008). Development and application of a system for dynamic wildfire risk assessment in Italy. *Environmental Modelling & Software* 23: 690–702.

Fiorucci, P., D'Andrea, M., Negro, D., and Severino, M. (2011). Manuale d'uso del sistema previsionale della pericolosità potenziale degli incendi boschivi *RIS.I.CO. Tech. Rep., Italian Department of Civil Protection – Presidency of the Council of Ministers, and CIMA Research Foundation.*

Fiorucci, P., D'Andrea, M., Negro, D. et al. (2015). *I°* aggiornamento del manuale d'uso del sistema previsionale della pericolosità potenziale degli incendi boschivi RIS.I.CO. –RISICO2015. *Tech. Rep., Italian Department of Civil Protection – Presidency of the Council of Ministers, and CIMA Research Foundation.*

Freire, J.G. and DaCamara, C.C. (2019). Using cellular automata to simulate wildfire propagation and to assist in fire management. *Natural Hazards and Earth System Sciences* 19: 169–179.

Ghisu, T., Arca, B., Pellizzaro, G., and Duce, P. (2015). An optimal cellular automata algorithm for simulating wildfire spread. *Environmental Modelling & Software* 71: 1–14.

Green, D.G., Tridgell, A., and Gill, A. (1990). Interactive simulation of bushfires in heterogeneous fuels. *Mathematical and Computer Modelling* 13: 57–66.

Grishin, A.M. (1997). *Mathematical Modeling of Forest Fires and New Methods of Fighting them. Publishing House of the Tomsk State University.*

Habdank, M., Schäfer, C., Scholle, P. et al. (2019). Deliverable 5.3: report on best practices and strategies for innovative self-preparedness and self-protection and the summary of lessons learned from case studies. *Tech. Rep., ANYWHERE Deliverable Report. 10.*

Hargrove, W., Gardner, R., Turner, M. et al. (2000). Simulating fire patterns in heterogeneous landscapes. *Ecological Modelling* 135: 243–263.

Italian Civil Protection Department and CIMA Research Foundation (2014). The dewetra platform: a multi-perspective architecture for risk management during emergencies. *Information*

Systems for Crisis Response and Management in Mediterranean Countries: First International Conference, ISCRAM-med 2014 (eds. C. Hanachi, F. Bénaben, and F. Charoy), pp. 165–177, Toulouse, France (15–17 October 2014). Springer International Publishing. Proceedings.

Li, S. and Banerjee, T. (2021). Spatial and temporal pattern of wildfires in California from 2000 to 2019. *Scientific Reports* 11: 1–17.

López-De-Castro, M., Trucchia, A., Fiorucci, P., and Pagnini, G. (2022). A comparison study between fire-spotting models by a wildfire simulator based on a cellular automata approach. *EGU General Assembly 2022,* https://meetingorganizer.copernicus.org/EGU22/EGU22-4600.html.

Mallinis, G., Galidaki, G., and Gitas, I. (2014). A comparative analysis of eo-1 hyperion, quickbird and landsat tm imagery for fuel type mapping of a typical mediterranean landscape. *Remote Sensing* 6: 1684–1704.

Marino, E., Dupuy, J.-L., Pimont, F. et al. (2012). Fuel bulk density and fuel moisture content effects on fire rate of spread: a comparison between firetec model predictions and experimental results in shrub fuels. *Journal of Fire Sciences* 30: 277–299.

Mell, W., Jenkins, M., Gould, J., and Cheney, P. (2007). A physics-based approach to modeling grassland fires. *Internationl Journal of Wildland Fire* 16: 1–22.

Montealegre, A.L., Lamelas, M.T., Tanase, M.A., and De la Riva, J. (2014). Forest fire severity assessment using als data in a mediterranean environment. *Remote Sensing* 6: 4240–4265.

Morandini, F. and Silvani, X. (2010). Experimental investigation of the physical mechanisms governing the spread of wildfires. *International Journal of Wildland Fire* 19: 570–582.

Perello, N., Trucchia, A., D'Andrea, M. et al. (2022). Risico, an enhanced forest fire danger rating system: validation on 2021 extreme wildfire season in southern Italy. *Environmental Sciences Proceedings* 17: 37.

Perryman, H.A., Dugaw, C.J., Varner, J.M., and Johnson, D.L. (2013). A cellular automata model to link surface fires to firebrand lift-off and dispersal. *International Journal of Wildland Fire* 22: 428–439.

Pulvirenti, L., Squicciarino, G., and Fiori, E. (2020a). A method to automatically detect changes in multitemporal spectral indices: application to natural disaster damage assessment. *Remote Sensing* 12: 2681.

Pulvirenti, L., Squicciarino, G., Fiori, E. et al. (2020b). An automatic processing chain for near real-time mapping of burned forest areas using sentinel-2 data. *Remote Sensing* 12: 674.

ReliefWeb (2021). Algeria: forest wildfires – emergency appeal no. mdrdz007 – algeria, Nov 2021. https://reliefweb.int/report/algeria/algeria-forest-wildfires-emergency-appeal-mdrdz007 (accessed 21 June 2023).

ReliefWeb (2022). Algeria: Wildfires – Aug 2022. https://reliefweb.int/disaster/fr-2022-000297-dza (accessed 21 June 2023).

Rothermel, R.C. (1972). A mathematical model for predicting fire spread in wildland fuels. Intermountain Forest & Range Experiment Station, Forest Service, U.S. Dept. of Agriculture.

Rui, X., Hui, S., Yu, X. et al. (2018). Forest fire spread simulation algorithm based on cellular automata. *Natural Hazards* 91: 309–319.

Sahila, A., Zekri, N., Clerc, J.-P. et al. (2021). Fractal analysis of wildfire pattern dynamics using a small world network model. *Physica A: Statistical Mechanics and its Applications* 583: 126300.

San-Miguel-Ayanz, J., Durrant, T., Boca, R. et al. (2019). Advance Effis report on forest fires in Europe, Middle East and North Africa 2018. *Tech. Rep., Publications Office, Luxembourg.*

Schumaker, N.H., Watkins, S.M., and Heinrichs, J.A. (2022). Hexfire: a flexible and accessible wildfire simulator. *Land* 11: 1288.

Serón, F.J., Gutiérrez, D., Magallón, J. et al. (2005). The evolution of a wildland forest fire front. *The Visual Computer* 21: 152–169.

Sorensen, T. (1948). A method of establishing groups of equal amplitudes in plant sociology based on similarity of species content and its application to analyses of the vegetation on danish commons. *Kongelige Danske Videnskabernes Selskabs. Biologiske Skrifter* 5: 1–34.

Stults, M. (2017). Integrating climate change into hazard mitigation planning: opportunities and examples in practice. *Climate Risk Management* 17: 21–34.

Sullivan, A. (2009a). Wildland surface fire spread modelling, 1990–2007. 2: empirical and quasi-empirical models. *International Journal of Wildland Fire* 18: 369–386.

Sullivan, A. (2009b). Wildland surface fire spread modelling, 1990–2007. 3: simulation and mathematical analogue models. *International Journal of Wildland Fire* 18: 387–403.

Sullivan, A. (2009c). Wildland surface fire spread modelling, 1990–2007. 1: physical and quasi-physical models. *International Journal of Wildland Fire* 18: 349–368.

Sun, T., Zhang, L., Chen, W. et al. (2013). Mountains forest fire spread simulator based on geo-cellular automaton combined with Wang Zhengfei velocity model. *IEEE Journal of Selected Topics in Applied Earth Observations and Remote Sensing* 6: 1971–1987.

Tonini, M., D'Andrea, M., Biondi, G., and degli Esposti, S., Trucchia, A. and Fiorucci, P. (2020). A machine learning-based approach for wildfire susceptibility mapping. the case study of the Liguria region in Italy. *Geosciences* 10: 105.

Trucchia, A. (2019). Front propagation in random media. PhD thesis. Facultad de Ciencia y Tecnlogìa, UPV-EHU, Leioa, Spain. https://bird.bcamath.org/handle/20.500.11824/1036.

Trucchia, A., Egorova, V., Butenko, A. et al. (2019). Randomfront 2.3: a physical parameterisation of fire spotting for operational fire spread models – implementation in WRF-Sfire and response analysis with LSFire+. *Geoscientific Model Development* 12: 69–87.

Trucchia, A., D'Andrea, M., Baghino, F. et al. (2020). Propagator: an operational cellular-automata based wildfire simulator. *Fire* 3: 26.

Trucchia, A., Meschi, G., Fiorucci, P. et al. (2022). Defining wildfire susceptibility maps in Italy for understanding seasonal wildfire regimes at the national level. *Fire* 5: 30.

Von Neumann, J. (1966). *Theory of Self-Reproducing Automata*. University of Illinois Press.

Wolfram, S. (1984). Cellular automata as models of complexity. *Nature* 311: 419–424.

4

Building an Operational Decision Support System for Multiple Weather-Induced Health Hazards

ANYWHERE Developments and Future Applications

Claudia Di Napoli

School of Agriculture, Policy and Development, University of Reading, Reading, UK

Department of Geography and Environmental Science, University of Reading, Reading, UK

4.1 Introduction

Extreme temperatures, such as those experienced during a heatwave, represent a dangerous meteorological hazard to human health. Heat disorders such as heat exhaustion and sunstroke are harmful to people of all ages and responsible for excess morbidity and mortality in affected areas. In 2003 more than 70,000 people died in western and southern Europe because of a severe and sustained episode of summer heat (Robine et al. 2008). During that event high air pollution levels were also recorded (Fiala et al. 2003). The combination of high temperatures and stable anticyclonic meteorological conditions (sunny weather; light winds) enhanced the photochemical production of harmful air pollutants and prevented their dispersion. The elevated concentrations of surface ozone O_3 and particulate matter PM_{10} (i.e. inhalable particles with diameters up to 10 μm) reached during the 2003 European heatwave have been associated with 21–38% of the total 2045 excess deaths reported in the United Kingdom (Stedman 2004).

Responding to Extreme Weather Events, First Edition. Edited by Daniel Sempere-Torres, Anastasios Karakostas, Claudio Rossi, and Philippe Quevauviller.

In the Netherlands, 400–600 deaths out of the total 1000–1400 excess deaths were related to O_3 and PM_{10} (Fischer et al. 2004). In France the excess risk of death in nine cities was found statistically significant for an O_3 increase of 10 μg m^{-3} (Filleul et al. 2006). Since 2003 Europe has witnessed multiple heatwaves, namely in 2007, 2010, 2015, 2017, and 2019 (Founda and Giannakopoulos 2009; Barriopedro et al. 2011; Dong et al. 2016; Sánchez-Benítez et al. 2018; Sulikowska and Wypych 2020). In each of these, local populations were exposed to above-average temperatures and poor air quality (Theoharatos et al. 2010; Zvyagintsev et al. 2011; Tarrasón et al. 2016; DEFRA 2018; Tarrasón et al. 2020). Most notable was the prolonged heatwave that affected eastern Europe and large parts of Russia between July and August 2010 (Barriopedro et al. 2011). A death tool of 55,000 was reported and 2000 deaths were attributed to the interaction between high temperatures and air pollution (Shaposhnikov et al. 2014).

The heatwaves affecting Europe in the past decades provide clear evidence that unusually high levels of heat and air pollution, especially when concurrent, are detrimental to human health. The ability to deliver timely, accurate, and useful early warnings for upcoming weather-induced health hazards is pivotal to maximize the impact of public health initiatives, as well as to prepare for and respond to periods of heat. Early warning systems linking forecasts of heatwaves with risk communication and risk reduction activities have been proved successful in reducing morbidity and mortality (Toloo et al. 2013). To date, 16 heat-health warning systems exist in Europe (Casanueva et al. 2019). They are implemented at single country level, forced by different numerical weather prediction (NWP) models, and triggered on the basis of various heat-sensitive meteorological parameters, namely air temperature or combinations of air temperature and relative humidity. Belgium is the only country where ozone concentration is also considered in determining the severity of a heatwave alert (IRCEL-CELINE 2022). Furthermore, existing heat-health warning systems are prevalently active in Western Europe (Casanueva et al. 2019). Recent evidence has, however, highlighted a greater increasing trend in heatwave hazards for south-eastern Europe than northern/south-western Europe during the 1998–2015 period and a higher increase in surface ozone concentrations during future heatwave days for the same areas (Morabito et al. 2017; Meehl et al. 2018). This evidence supports the creation of a pan-European decision support platform where potentially hazardous conditions of heat and air pollution are identified, predicted, and visualized seamlessly across the continent, thus fostering preparedness and coordination among first responders at any scale. Such a platform would answer the call by the World Meteorological Organisation to incorporate the forecasts of meteorological hazards detrimental to human health, such as heatwaves and air pollution events, into the suite of products at the disposal of national meteorological services (Kusch 2004).

This chapter introduces the ANYWHERE system as the first implementation of a real-time decision support platform for two weather-induced health hazards: heatwaves and air quality (Figure 4.1). The system is based on state-of-the-art forecasting algorithms that deliver eight 24/7 heatwave and air quality products generated by a common NWP model, i.e. the integrated forecasting system from the European Centre for Medium-Range Weather Forecasts (IFS-ECMWF). The products are integrated with more than 300 products tailored to the forecast of drought, forest fires, flash floods, landslides, and storm surges to provide the first pan-European multi-hazard early

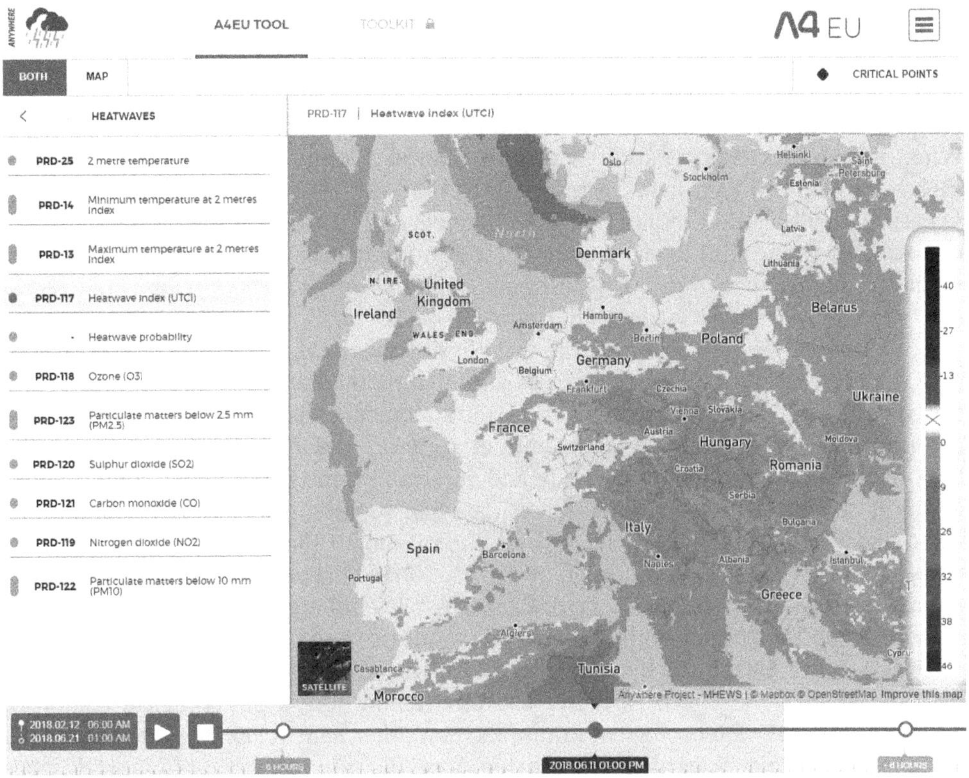

Figure 4.1 Products of weather-induced health hazards in the ANYWHERE multi-hazard early warning system (MH-EWS) platform. The map displays the forecast of the heatwave index product for the 11 June 2018. It shows, in orange to dark red colours, the areas where heat stress, as identified by the Universal Thermal Climate Index (UTCI), is expected to reach levels detrimental to human health.

warning system (MH-EWS). The chapter provides a detailed description of the ANYWHERE heatwave and air quality products, and their implementation in seven pilot sites. The usability and effectiveness of the products in predicting the 2017 European heatwave is presented alongside an evaluation from end users. The chapter concludes with a discussion on future developments and applications for ANYWHERE heatwave and air quality products.

4.2 Heatwave Prediction in ANYWHERE

The ANYWHERE MH-EWS includes two products for the prevention and management of health hazards caused by heatwaves. These are a heatwave index and a heatwave probability. Both are based on the UTCI, a state-of-the-art metrics representing the physiological stress experienced by an individual in a hot environment.

4.2.1 The Universal Thermal Climate Index

Thermoregulation is a process that allows the human body to maintain its core internal temperature within certain boundaries so that body's comfort, performance, and health are guaranteed even when the surrounding ambient temperature is very different. When the ambient temperature is greater than skin temperature, thermoregulation causes the body to gain heat. In response to that, the body attempts to lose heat by sweating. Anything that reduces the effectiveness of sweating, such as the lack of wind breeze, tight-fitting clothes, or an impaired thermoregulation system, will lead the internal body temperature to rise. This response, called *body overheating*, causes the excess morbidity and mortality recorded during heatwaves.

Body overheating and, in general, the physiological consequences of exposure to heat (and its extremes) can be assessed by modelling the human body's thermoregulation. The UTCI-Fiala model allows these physiological consequences to be simulated and predicted (Fiala et al. 2012; Havenith et al. 2012). It defines the *heat stress*, i.e. the heat-related factors acting on the body, and determines the corresponding body's physiological responses by using (i) ambient data (air temperature, humidity, ventilation, solar, and thermal radiation); (ii) a physiological model that includes heat and mass transfer within the body, thermoregulatory reactions of the central nervous system, and perceptual responses; (iii) an adaptive Western-style clothing insulation model for outdoor climates. The output of the UTCI-Fiala model is the UTCI.

The choice of UTCI as a heatwave index for the ANYWHERE MH-EWS is based on the ability of the UTCI to better express human body's physiological responses under a variety of meteorological situations than other heat-related indices developed in the past (Błażejczyk et al. 2013). Furthermore, the rate of increase in mortality has been shown to depend on heat stress levels, as represented by the UTCI, in 17 European countries (Di Napoli et al. 2018).

4.2.2 Forecasting Algorithms

As described in the former section, the UTCI-Fiala model calculates the UTCI from ambient data. For the purpose of the ANYWHERE MH-EWS, the UTCI-Fiala model has been revised by means of an *operational* procedure. The procedure is computationally more efficient than the original model (Bröde et al. 2012) and therefore suitable to represent the algorithm for the computation of UTCI forecasts.

The algorithm consists of four steps. First, it retrieves the forecast fields of 2 m air temperature, 2 m dew point temperature, wind speed, and multiple radiation components as computed by the ensemble forecast system (ENS) of the IFS-ECMWF. The choice to use ENS outputs rather than single deterministic forecasts is based on their superior forecasting skill (Pappenberger et al. 2015). Second, the algorithm computes relative humidity from air temperature and dew point temperature via the Clausius–Clapeyron relation, and the mean radiant temperature from solar and thermal radiation fields (Alduchov and Eskridge 1996; Di Napoli et al. 2020). Third, the UTCI is calculated via the *operational* procedure from each of the 50 members (i.e., perturbed simulations) output by the ENS for air temperature, relative humidity, wind speed, and mean radiant temperature. From the 50 UTCI ensemble members, the algorithm computes the ensemble mean UTCI and the ensemble probability to exceed strong heat stress, with the latter defined as the proportion of members exceeding 32 °C. This threshold has been chosen because of its relation to an increase in mortality (Di Napoli et al. 2018). The algorithm outputs are in GRIB (GRIdded Binary) format, a WMO standard for the storage and distribution of gridded data.

4.2.3 Heatwave Products

In the ANYWHERE MH-EWS, the heatwave index is the grid cell-by-grid cell mean UTCI of the 50 UTCI ensemble forecasts and the heatwave probability is the grid cell-by-grid cell percentage of the 50 UTCI ensemble forecasts predicting UTCI values above 32 °C. Both products consist of a sequence of pan-European maps with a spatial resolution of approximately 18 × 18 km and have a 10-day lead time at 6 hourly steps. They are updated at every IFS-ENS release, i.e. at 00UTC and 12UTC (Coordinated Universal Time).

4.2.4 Integration in the MH-EWS

As the algorithm is executed on ECMWF servers, the heatwave index and the heatwave probability products are pushed from MARS (ECMWF Meteorological Archival and Retrieval System), where they are stored, to the ANYWHERE MH-EWS via SFTP protocol. The transfer occurs twice a day when products from the latest IFS-ENS release are made available.

After being connected, the heatwave products are included in the 'Heatwave' category of the ANYWHERE MH-EWS. An interactive interface allows the end user to visualize manually or automatically forecasted maps at each time step up to 10 days ahead. For the heatwave index product, maps are colour-coded according to UTCI category scale, where each category of heat stress corresponds to specific physiological responses and protective measures (Table 4.1). As higher UTCI corresponds to higher heat stress, areas for which moderate to extreme heat stress is forecasted to occur are indicated with an orange to dark red palette. This allows ANYWHERE MH-EWS users to efficiently visualize and promptly identify areas with forthcoming detrimental heat-health hazard (Figure 4.1). The heatwave probability product is shown in a 0–100% scale with a white-to-red palette, where the redder the colour the higher the confidence in the 50 UTCI ensemble members to predict UTCI values above 32 °C.

4.2.5 Temperature Products

To complement the information provided by the heatwave index and the heatwave probability products, the ANYWHERE MH-EWS includes four other products related to and meaningful for heatwave forecasting. These are 2 m temperature, daily minimum temperature at 2 m, and daily maximum temperature at 2 m. The choice to consider these products is based on the fact that temperature is widely used in the definition and forecasting of heatwaves (Casanueva et al. 2019). Furthermore, daily minimum and

Table 4.1 Category scales of the UTCI and corresponding physiological responses.

Range (°C)	Category	Physiological responses	Protection measures
Above 46	Extreme heat stress	Increase in rectal temperature time gradient Steep decrease in total net heat loss Averaged sweat rate > 650 g h^{-1}, steep increase	Temporary body cooling. Drinking > 0.5 l h^{-1} necessary. No physical activity.
38–46	Very strong heat stress	Low core to skin temperature gradient Increase in rectal temperature at 30 minutes	Temporary use of air condition Shaded places necessary Drinking > 0.5 l h^{-1} Reduce physical activity
32–38	Strong heat stress	Averaged sweat rate > 200 g h^{-1}. Increase in rectal temperature at 120 minutes Instantaneous change in skin temperature	Use shaded places. Drinking > 0.25 l h^{-1} Temporary reduce physical activity
26–32	Moderate heat stress	Change of slopes in sweat rate, rectal and skin (mean, face, hand) temperature Occurrence of sweating at 30 minutes. Steep increase in skin wettedness	Drinking > 0.25 l h^{-1}.
9–26	No thermal stress	Averaged sweat rate > 100 g h^{-1} Plateau in rectal temperature time gradient	Physiological thermoregulation sufficient to keep thermal comfort
0–9	Slight cold stress	Local minimum of hand skin temperature	Use gloves and hat
−13–0	Moderate cold stress	Vasoconstriction. Face skin temperature at 30 minutes < 15 °C (pain)	Intensify activity and protect face and extremities against cooling
−13–27	Strong cold stress	Numbness. Increase in core to skin temperature gradient	Intensify activity and protect face and extremities against cooling Use warmer clothing
−27–−40	Very strong cold stress	Frostbite, numbness, shivering. Steeper decrease in rectal temperature	Intensify activity and protect face and extremities against cooling Use warmer clothing Reduce outdoor exposure time
Below −40	Extreme cold stress	Frostbite. Decrease in rectal temperature time gradient	Stay indoors. If outdoor exposure is necessary, use heavy and windproof clothing

Source: Adapted from Blazejczyk et al. (2013).

maximum temperatures are of relevance to human health as they are representative of the night-time and day-time conditions experienced by an individual (McGregor et al. 2015). The first may be associated with heat disorders such as heatstroke; the latter reflect the potential deterioration in sleep quality during heatwaves and the associated lack of recovering from daytime heat exposure.

The 2 m temperature product corresponds to the temperature of air at 2 m above the surface of land, sea or in-land waters output from the IFS-ECMWF. ECMWF forecasts based on this variable have been demonstrated to have a good predictability skills for European heatwaves (De Perez et al. 2018; Lavaysse et al. 2019). The daily minimum temperature at 2 m and the daily maximum temperature at 2 m products correspond to the Extreme Forecast Index (EFI) for the minimum and maximum temperature forecasted by the IFS-ECMWF, respectively. The EFI compares the ENS forecast probability distribution of an event to the NWP-climate distribution to establish the abnormality of that event (Lalaurette 2003). The EFI ranges from −1 to +1 with 0 indicating that the forecasted probability distribution agrees with the climatology and positive (resp. negative) EFI indicating that most values are above (resp. below) the climatology. Experience suggests that EFI magnitudes of 0.5–0.8 (irrespective of sign) can be generally regarded as 'unusual' weather being likely whilst magnitudes above 0.8 usually signify that 'very unusual' weather is likely. Although proven effective in identifying heat extremes across Europe, the EFI is used for this specific scope in Switzerland only (Casanueva et al. 2019, Magnusson 2019). With the ANYWHERE system, this limitation is overcome and EFI is made available as an operational supporting tool for heatwave forecasting at the continental level.

Twice a day, an SFTP protocol pushes the products from ECMWF MARS, where they are stored after being generated by the IFS, to the ANYWHERE MH-EWS. In the platform, the 2 m temperature product is visualized as a pan-European map at 9 × 9 km spatial resolution and can be explored with a timestep of one hour for the first six days and a timestep of six hours for the rest of the horizon. The daily minimum temperature at 2 m and the daily maximum temperature at 2 m products are visualised as pan-European maps at 18 × 18 km spatial resolution with a 10-day lead time and daily resolution.

4.3 Air Pollution Prediction in ANYWHERE

The ANYWHERE MH-EWS includes six products for the prevention and management of health hazards caused by detrimental levels of air pollution. These are ozone (O_3), nitrogen dioxide (NO_2), sulphur dioxide (SO_2), carbon monoxide (CO), particulate matter below 10 μm (PM_{10}), and particulate matter below 2.5 μm ($PM_{2.5}$).

4.3.1 Air Quality

The above-listed air pollutants have been selected based on their documented relevance for human health during events of extreme heat. This is summarized in Table 4.2. During heatwaves ozone levels increase following the interactions of multiple pollutants with sunlight. The limited air circulation promotes the build-up and stagnation of air pollutants, such as carbon monoxide and fine particulate matter, that are produced by events (wildfires) that often concur with heatwaves (Shaposhnikov et al. 2014; Kollanus et al. 2017; dos Santos et al. 2018).

Table 4.2 Air pollutants included in the ANYWHERE MH-EWS, their impacts and corresponding references in the literature.

Air pollutant	Sources	Health impacts	References
Ozone, O_3	Formed by the photochemical reaction of pollutants, such as nitrogen oxides (NOx) from vehicle and industry emissions and volatile organic compounds (VOCs) emitted by vehicles, solvents, and industry. Highest levels of ozone pollution occur during periods of sunny weather, e.g. heatwaves, as a consequence	Breathing problems, asthma, lung diseases, reduced lung function. Larger mortality on heat-wave days with high concentration levels of ground-level ozone	WHO (2021), Analitis et al. (2014)
Nitrogen dioxide, NO_2	Combustion processes, e.g. heating, power generation, engines in vehicles, wildfires. Main source of nitrate aerosols, which form an important fraction of $PM_{2.5}$ and, in the presence of ultraviolet light, of O_3	Asthma, bronchial symptoms, lung inflammation, reduced lung function	WHO (2021), Vieno et al. (2010)
Sulphur dioxide, SO_2	Burning of sulphur-containing fossil fuels (coal and oil) for domestic heating, power generation (e.g. for air conditioning during heatwaves), motor vehicles	Inflammation of the respiratory system, lung function impairment, eye irritation	WHO (2021), NHS (2011)
Carbon monoxide, CO	Incomplete-combustion of carbon-containing fuels. Major contributions from road transport and industrial fuel burning. Produced by home gas appliances. Component of wildfires smoke	Dizziness, confusion, unconsciousness, death. During the 2010 Russian heatwave, levels of carbon monoxide were several times the maximum allowable concentration across the Moscow region	EPA (2022), Golitsyn et al. (2011), Konovalov et al. (2011)
Particulate matters below 10 microns, PM_{10}, and below 2.5 µg, $PM_{2.5}$ (fine particles)	Complex mixture of solid and liquid particles of organic and inorganic substances suspended in the air, such as sulphate, nitrates, ammonia, sodium chloride, black carbon, mineral dust and water. Components of wildfires smoke	Cardiovascular diseases, respiratory diseases, lung cancer. Larger mortality on heat-wave days with high concentration levels of PM_{10}	WHO (2021), Analitis et al. (2014)

4.3.2 Forecasting Algorithms

ANYWHERE air quality products are predicted via the regional air quality (RAQ) forecasting system (Marécal et al. 2015). The RAQ system is developed by the Copernicus Atmosphere Monitoring Service (CAMS) through a series of research activities, amongst which the MACC (Monitoring Atmospheric Composition and climate) project, that aims to develop tools for air quality monitoring in Europe.

The RAQ system is based on the combined used of an ensemble of seven state-of-the-art numerical air quality models that are developed and run in Europe and have been extensively evaluated: CHIMERE from INERIS, France; EMEP (European Monitoring and Evaluation Programme), from MET Norway, Norway; EURAD-IM (European Air pollution Dispersion Inverse Model), from University of Cologne, Germany; LOTOS-EUROS (Long Term Ozone Simulation – European Ozone Simulation) from KNMI and TNO, Netherlands; MATCH (Multi-scale Atmospheric Transport and Chemistry) from SMHI, Sweden; MOCAGE (Model Of atmospheric Chemistry At larGE scale) from METEO-FRANCE, France; and SILAM (System for Integrated modelling of Atmospheric coMposition) from FMI, Finland. All seven models share the same meteorological parameter settings (IFS-ECMWF), the same boundary conditions for chemical species (from the CAMS IFS-MOZART global production), and the same emission surface fluxes as compiled by the CAMS emission inventories for anthropogenic emissions over Europe and for biomass burning.

The general objective of the RAQ system is to provide air quality forecasts at the pan-European scale (25 °W–45 °E; 30 °N–70 °N), which aligns with the objectives of the ANYWHERE MH-EWS. The forecasts consist in the concentrations ($\mu g\,m^{-3}$) of air pollutants at the surface level and are provided in netCDF (Network Common Data Form) format.

4.3.3 Air Quality Products

In the ANYWHERE MH-EWS, the ozone, nitrogen dioxide, sulphur dioxide, carbon monoxide, particulate matter below 10 μm, and particulate matter below 2.5 μm products are the mean concentration of the corresponding pollutant as computed by the multi-model RAQ ensemble. Each product is updated daily and consists of a sequence of pan-European maps with a spatial resolution of approximately 0.1° × 0.1° (approx. 11 × 11 km), a four-day lead time, and one-hour steps.

4.3.4 Integration in the MH-EWS

RAQ ensemble outputs are retrieved through a WSDL (Web Services Description Language) protocol from the Copernicus Atmosphere Monitoring European Air Quality Service and pushed to the ANYWHERE MH-EWS. The transfer occurs once a day when products from the latest model release are made available.

After being connected, the air quality products are included in the 'Heatwave' category of the ANYWHERE MH-EWS and ready to be visualized. Maps are colour-coded according to the Air Quality Index (AQI) by the European Environment Agency

Table 4.3 Air quality index classification and corresponding protection measures as defined by the European environment agency for the air pollutants included in the ANYWHERE MH-EWS (European Environment Agency 2017, European Environment Agency 2021). For carbon monoxide, a concentration threshold equal to 10 mg m^{-3} is considered based on the guidelines by the European Commission (2022) to indicate moderate or poorer air quality.

Pollutant	Air quality index (based on hourly concentration in μg m^{-3})				
	Good	Fair	Moderate	Poor	Very poor
Ozone, O_3	0–80	80–120	120–180	180–240	240–600
Nitrogen dioxide, NO_2	0–40	40–100	100–200	200–400	400–1000
Sulphur dioxide, SO_2	0–100	100–200	200–350	350–500	500–1250
Particles less than 2.5 μm, $PM_{2.5}$	0–10	10–20	20–25	25–50	50–800
Particles less than 10 μm, PM_{10}	0–20	20–35	35–50	50–100	100–1200
Protection measures	Enjoy your usual outdoor activities		Consider reducing intense activities outdoors if you experience symptoms such as sore eyes, cough or sore throat		

(European Environment Agency 2017). The AQI defines different levels of air quality based on specific air pollutants concentrations and their adverse effects to human health, providing protection measures correspondingly (Table 4.3). In the platform, concentrations corresponding to moderate, poor, and very poor air quality are displayed with a yellow-orange-red colour palette. This allows forecast conditions of air quality potentially detrimental to human health to be effectively identified in space and in time by platform users.

4.4 ANYWHERE MH-EWS in Action: The European 2017 Heatwave

The heatwave and air pollution that affected Europe in June 2017 represent a significant example of weather-related health hazards that the ANYWHERE MH-EWS is able to predict.

From 17 to 22 June, a heatwave remarkable for its precocity and intensity hit the continent (Sánchez-Benítez et al. 2018). A high-pressure system, anchored over most of Western Europe, favoured the maintenance of a dome of hot air extending from North Africa up to the United Kingdom. High temperatures broke local records and triggered heat action plans in France, Italy, and the United Kingdom (Le Parisien 2017; La Repubblica 2017; PHE 2017a). In Belgium air pollution alerts were also issued, particularly for ozone (VRT 2017).

A number of deaths higher than average was recorded in Belgium, France, Portugal, Spain, and the United Kingdom (235, 580, 296, 780 and 598 excess deaths, respectively). Excess deaths differed by age groups. In Spain, Belgium, and the United Kingdom, the elderly paid the largest toll (MSSSI 2017; Sciensano 2017; PHE 2017b). On the contrary, in France, the 15–64 age group was the most affected, with 7 deaths reported among workers. This has been explained by precocity of the episode, which took place outside the summer holiday period (Santé publique France 2017).

The products integrated into the ANYWHERE MH-EWS captured the start, duration, and end of the June 2017 heatwave (Figure 4.2). The heatwave index product issued on 17 June (00UTC release) predicted an increase in heat stress level in Western Europe from that day up to 19 June and the permanence of moderate to very strong stress conditions until 24 June (panel b). The heatwave probability product issued on the same day forecasted a very high percentage (90% or greater) of UTCI ensemble forecasts above 32 °C across the region and during the same period (panel c). Comparison to UTCI data from the ERA5-HEAT reanalysis data set (Di Napoli et al. 2021), here used as a proxy for observations, confirms that ANYWHERE heatwave products were able to predict the intensity of the heatwave with moderate, strong, and very strong levels (between 26 °C and 46 °C UTCI) endured over most of Western Europe. It also highlights the products' capability to forecast the end of the June 2017 heatwave and its duration one week in advance (panel a). This result agrees with previous findings indicating four to six days as the maximum lead time at which the anomaly correlation coefficient (ACC)[1] reaches 60%. Furthermore, this case study highlights the importance of ensemble forecasting for heatwave events. For 23 June, the heatwave index product predicted moderate and strong heat stress in Germany, western Poland, and Czech Republic. According to the heatwave probability product, however, less than 10% UTCI ensemble forecasts predicted heat stress above 32 °C on that day. Reanalysis data confirmed for the area UTCI values mostly associated with no thermal stress and not exceeding moderate heat stress levels.

During the June 2017 heatwave, ozone concentrations overpassed the EU ozone threshold of $180\,\mu g\,m^{-3}$ in Belgium, northern France and Italy, and southern UK (DEFRA 2017; IRCEL-CELINE 2017; Lombardia 2017; Prev'Air 2017). CAMS reanalysis data reveal the spatial distribution of ozone concentration across Western Europe and its relation to the heatwave. Highest concentrations were achieved after 19 June, once the heatwave had become established (Figure 4.3a). Forecast maps of the ANYWHERE ozone product for 21 June, the day when ozone levels reached the highest concentrations, show that critical levels could be predicted up to four days in advance, especially in congested areas like the Po Valley in Italy (Figure 4.3b). This result is in agreement with the evaluation of RAQ forecasts in other extreme ozone events (Marécal et al. 2015).

This case study confirms the ability of the heatwave and air pollution products implemented in the ANYWHERE MH-EWS to foresee and locate ahead of time a large-scale event with documented impacts on human health and demonstrates the potential of the platform as an operational decision support tool for multiple weather-induced health hazards at the pan-European level.

4.5 Implementation at Pilot Sites

Alongside the pan-European ANYWHERE platform, heatwave and air pollution products were integrated into MH-EWS prototypes aimed to support the decision-making process of first emergency responders at the local level. The prototypes were

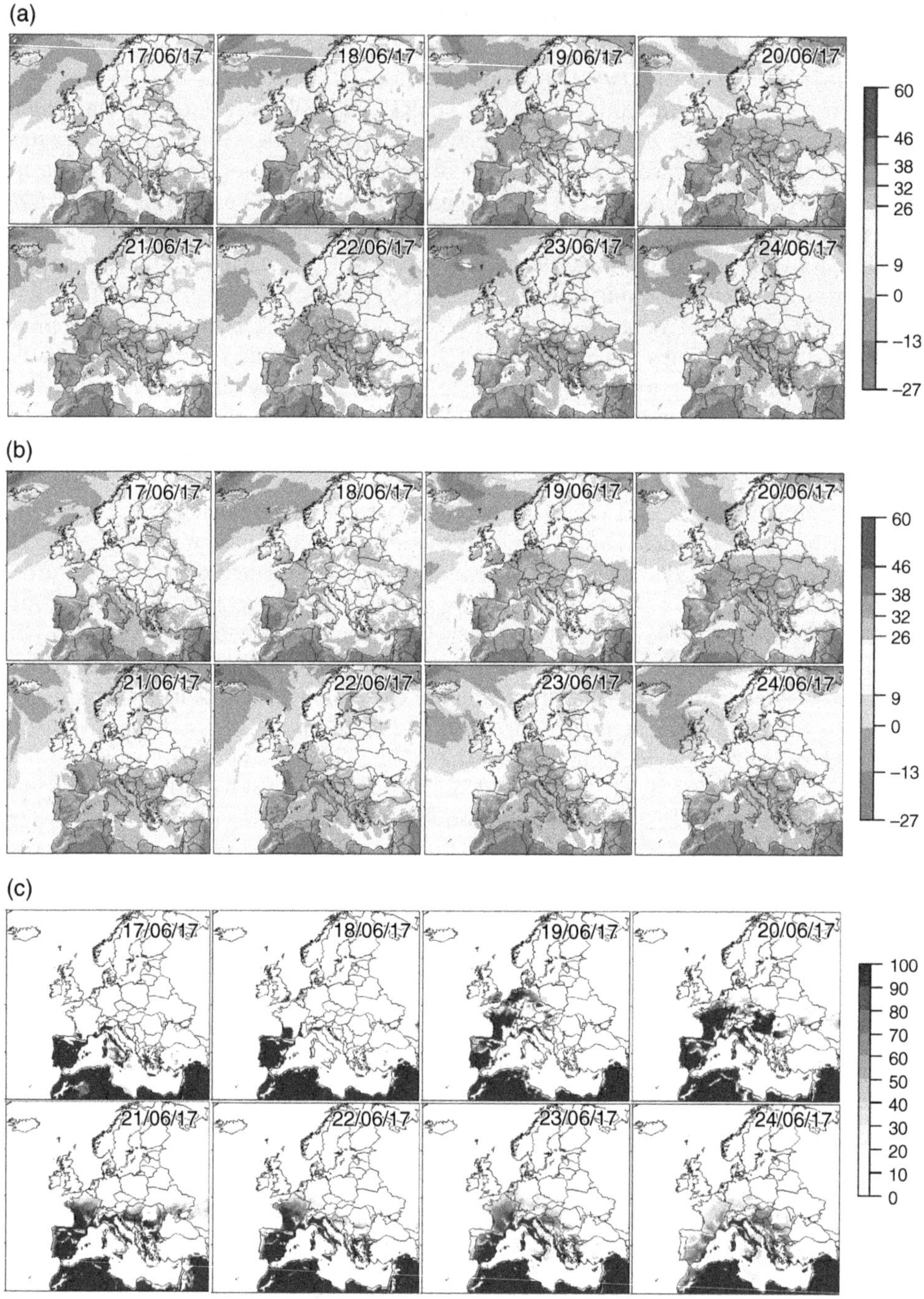

Figure 4.2 Heat stress as expressed by the UTCI (12UTC) during the June 2017 heatwave. (a) UTCI values (in °C) between 17 and 24 June from the ERA-HEAT reanalysis dataset. (b) UTCI forecasts as issued on 17 June (heatwave index product). (c) Forecasted percentage of ensemble forecasts predicting UTCI values above 32 °C as issued on 17 June (heatwave probability product).

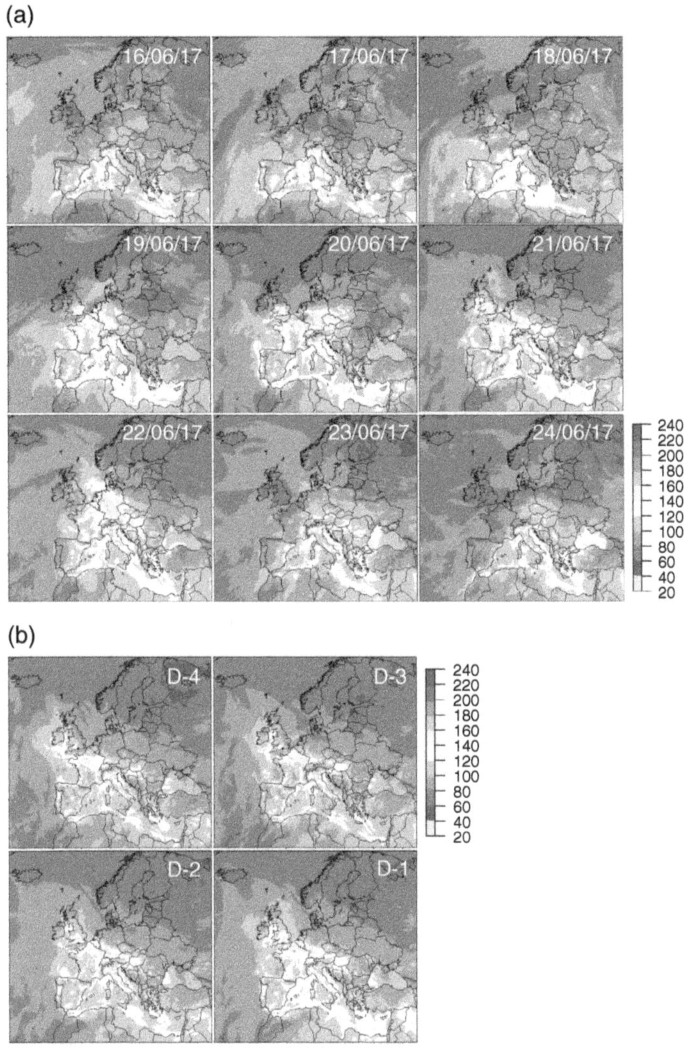

Figure 4.3 Ozone concentrations (μg m^{-3}) during the June 2017 heatwave. (a) CAMS reanalysis data from 16 to 24 June. (b) Ozone forecasts for 21 June as incorporated in the ANYWHERE platform once issued by the RAQ service 24 (D-4), 48 (D-3), 72 (D-2), 96 (D-1) hours before.

implemented, used, and evaluated in seven pilot sites. These include six regional pilot sites spread across as many countries and each characterized by specific climates, population densities (e.g. cities and rural areas), and high-impact weather phenomena: Liguria, Italy (A4Lig); Catalonia, Spain (A4Cat); Rogaland, Norway (A4Nor); South Savo, Finland (A4Finn); Canton of Bern, Switzerland (A4Alps); Corsica, France (A4Cor). One pilot site (A4CENEM) had national coverage and was utilized by the National Control Centre for Emergency Management (CENEM, *Centro Nacional de Seguimiento y Coordinación de Emergencias*), which belongs to the Directorate-General for Civil Protection and Emergencies (DGPCE, *Dirección General de Protección Civil y Emergencias*) and is responsible for the management, coordination, and dissemination of weather alerts in Spain.

4.5.1 Integration of Local Heatwave and Air Pollution Products

The ANYWHERE prototype A4CENEM is shown in Figure 4.4 and is here described. For the prediction and management of heatwaves-induced health hazards, A4CENEM embeds four pan-European products named as 'Universal Thermal Climate Index (UTCI)', 'Heatwave probability (UTCI > 32 °C)', 'Temperature forecasts (IFS-ECMWF)', and 'Maximum temperature Index Forecasts (IFS-ECMWF)'. In addition, it includes three locally customized products:

1. 'Thermometers (AEMET)' is the temperature recorded at the meteorological stations of the Spanish National Meteorological Agency (AEMET, *Agencia Estatal de Meteorología*) and is displayed as a map of icons (Figure 4.4a). The product allows end users to visualize the last 24 hours of data in 1-hour intervals at the national level. Clicking on an icon selects one station and displays the recent evolution of temperature over time.
2. 'Rel. humidity forecasts (IFS-ECMWF)' corresponds to the forecast of relative humidity and is derived from the forecasts of 2 m temperature and 2 m dew point temperature as issued from the IFS-ECMWF. The product, which appears as maps, has a lead time of 10 days with a timestep of 1 hour for the first 5 days, and 3 and 6 hours afterwards. Its inclusion in A4CENEM reflects the evidence that in Spain low relative humidity is associated to an increase in heat-related mortality and it reinforces the detrimental effects of air pollutants, ozone in particular, on human health (Díaz et al. 2002a, b; Montero et al. 2012; Linares et al. 2015).
3. 'Heat Warnings (AEMET)' shows the warnings for maximum temperatures as issued by AEMET when region-specific thresholds are exceeded (AEMET 2018b). The thresholds are defined based on climatological criteria, which correspond to the concept of 'rare or very infrequent' event and of adversity, with the latter depending on the threat they may pose to the population (AEMET 2018a). A navigation menu allows the user to explore up to 72 hours of forecasted warnings. Clicking on a region where a heat warning is issued brings up pop-up windows showing the name of the region, the level of the warning, and its evolution in time (Figure 4.4b).

AEMET warnings are integrated into A4CENEM also for dust, together with the six pan-European air quality products. Sahara dust intrusions have been shown to produce an increment of particulate matter concentrations in all regions of Spain, and an increase in daily mortality in some of these (Díaz et al. 2017). The 'Dust Warnings (AEMET)' product shows warnings of dust content in the atmosphere as issued by AEMET, thus providing evidence-based support for the prevention of health hazards associated with detrimental levels of air pollution. The product has a 72-hour lead time

Figure 4.4 A4CENEM as an example of ANYWHERE prototype at a pilot site. (a) The 'Thermometers (AEMET)' product showing the temperatures recorded at the AEMET meteorological stations on 19 May 2022. (b) The 'Heat Warnings (AEMET)' for 20 May 2022 as issued the day before for the area Campiña Sevillana.

and can be accessed by users via the same navigation tools as the 'Heat Warnings (AEMET)' product.

The four location-specific products – 'Thermometers (AEMET)', 'Rel. humidity forecasts (IFS-ECMWF)', 'Heat Warnings (AEMET)', 'Dust Warnings (AEMET)' – are integrated also into A4Cat, the MH-EWS prototype geographically tailored to the autonomous community of Catalonia and utilized by the local Civil Protection Agency and its Operations Coordination Centre (CECAT, *Centre de Coordinació Operativa de Catalunya*) as well as the Catalan Water Agency (ACA, *Agència Catalana de l'Aigua*). A4Cat includes three additional locally customized heatwave and air pollution products. These are temperature and air quality measurements from the thermometers and pollution sensors of the Meteorological Service of Catalonia (SMC, *Servei Meteorològic de Catalunya*), and heat warnings issued by SMC. The 'SMP Heat Warnings (SMC)' product is defined by local forecasters on the predicted probability of maximum daily temperature to exceed a threshold for three consecutive days or more (SMC 2022). The threshold corresponds to the 98th percentile of maximum daily temperatures recorded in June, July, and August between 2011 and 2020, and is specific to single municipalities (SMC 2021). The product has thus a higher spatial resolution than 'Heat Warnings (AEMET)' and complements the latter with very local information and expertise. Should a heat warning be issued by SMC, this is promptly communicated to CECAT who activates a heat response plan in affected municipalities.

Alongside heatwave products driven by the IFS-ECMWF, three MH-EWS prototypes – A4Finn, A4Lig, A4Nor – include the high-resolution 'FMI Temperature p100' product from MEPS HarmonEPS. HarmonEPS is a limited-area, short-range, convection-permitting ensemble prediction system; MPES is the operational implementation of HarmonEPS in Finland, Norway, and Sweden (Frogner et al. 2019). The 'FMI Temperature p100' product corresponds to the highest temperature forecast by the 10 MPES members at 4 hourly steps with 54 hour lead time and a spatial resolution of approximately 18 × 18 km.

4.5.2 Evaluation at Pilot Sites

Pan-European and locally customised heatwave and air pollution products were evaluated at each pilot site by end users. They consulted the MH-EWS prototypes as part of their daily operational activities and used the products as decision support tools in their emergency management operations during a demonstration period (1 October 2018–30 September 2019). After the demonstration period, end users were interviewed, and their feedback collected about the usability and effectiveness of the products.

In A4Cor, the MH-EWS prototype tailored on Corsica and used by the local Fire and Rescue Service (SIS2B, *Service d'Incendie et de Secours de la Haute Corse*), heatwave products were the most used among all the available products. A4CENEM users found both national and pan-European products useful in their operations. In July 2019, a heatwave affected Spain with record high temperatures (AEMET 2019). During that event, A4CENEM users consulted the heatwave products and found the prototype easy to consult, as well as helpful in their decision-making process. Particularly, they recognized the benefit of the animation play/bar which allowed them to visualize forecasts at multiple time steps in the future and thus identify in advance how dangerous the heatwave could have been in medium (7–10 days) and short (1–5 days) lead time.

Heatwave products were at the focus also of A4Lig, the MH-EWS prototype tested and validated by the municipality of Genoa (CDG, *Comune di Genova*) and the hydrometeorological centre for civil protection in the Liguria region (ARPAL, *Agenzia regionale per la protezione dell'ambiente – Liguria*). A4Lig stores historical data in a seamless way and offers the possibility to access them alongside newly forecasted products. Thanks to this, end users who could compare a run with previous releases and check for any inconsistency, thus evaluating the reliability of the forecasts. Furthermore, in A4Lig, forecast maps of heatwave and air pollution products can be overlayed in real time to show where and when multiple weather-induced health hazards are predicted to simultaneously happen. These implementations, specific to A4Lig, were evaluated favourably and have been requested by end users to be included also in the other MH-EWS prototypes.

Common to all MH-EWS prototypes, end users highlighted the importance of having a single platform with all the information (forecasts, warnings, observations), thus favouring the use of different products. They found that interfaces are simple and easy to use in their daily operations, which is key when a high-impact weather-related event demands for a prompt, effective emergency response. End users also reported a change in the way they prepare and approach to emergencies (Vicente et al. 2019). A4Lig users 'have increased their response capacity [. . .] through the information available and through the tools developed in ANYWHERE'. A4Cat users 'have managed to move from a reactive action (before A4Cat) to a pro-active action (after A4Cat)'.

4.6 Future Applications

Multiple are the future implementations where ANYWHERE heatwave and air pollution products can provide a more comprehensive picture of weather-induced health hazards and thus contribute to better forecasting and response management. Five thematic areas are identified and are here described.

4.6.1 Impact-Based Warnings

Impact-based forecasting combines weather forecasts with an assessment of possible impacts with the aim to minimize human and economic cost of extreme weather events. For heatwaves, the development of impact-based forecasting relies on finding relationships between the predictors (temperature or heat stress anomalies) and the impact occurrence and severity (human mortality) (Merz et al. 2020). According to a study based on climate and mortality records spanning 24 years, UTCI values above the 95th climatological percentile are associated with periods of excess mortality and the threshold can therefore be implemented as an 'action trigger' for heat-health warnings (Di Napoli et al. 2019). This hypothesis was successfully validated during European 2017 heatwave. For the Catalonia region, the ANYWHERE heatwave index product forecasted UTCI values above the 95th climatological percentile on 14, 15, 17, 22 and

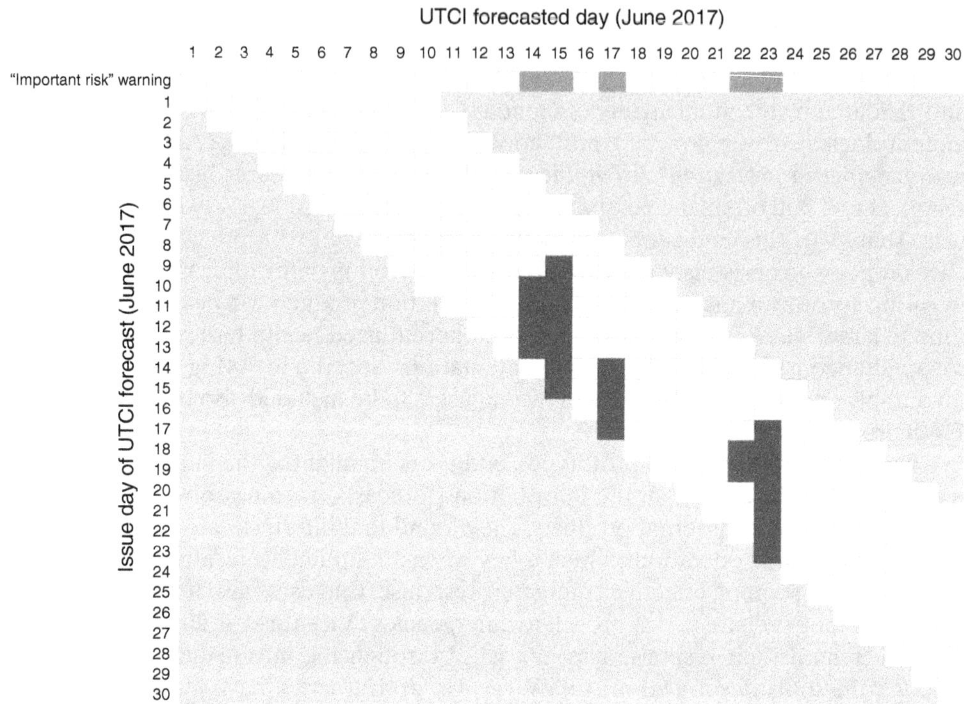

Figure 4.5 Predictability of the ANYWHERE heatwave index product for Catalonia during the European 2017 heatwave. Green cells indicate days when 12UTC forecasts, averaged over the A4Cat pilot site, exceeded the 95th climatological percentile of the UTCI. Yellow cells indicate days when an 'important risk' warning from AEMET was in place.

23 June and up to six days in advance (Figure 4.5). On those days, AEMET warnings for 'important risk' were in place. Following this, a new ANYWHERE product could show where and when UTCI values are predicted by the IFS-ENS model to exceed the 95th climatological percentile threshold. In A4CENEM, the product could complement the current 'Heat Warnings (AEMET)' product, which notifies on adverse temperature extremes up to a maximum period of 72 hours, by providing information on detrimental heat stress with a 10-day lead time. Furthermore, the product could map out the corresponding exceedance probability as predicted by the 50 UTCI ensemble forecasts. A cost-benefits analysis showed that many European areas would have benefited from medium-range forecasts during the 2006 heatwave if public health interventions had been triggered at an exceedance probability of 30%, suggesting the adoption of this value in an operational setting (Koppe 2007). The product would represent an improvement from the heatwave probability product because, being based on climatology, it would forecast any area where local populations are exposed to a heat stress higher than the one they are adapted to.

As far as air pollution is concerned, the potential of combining air quality estimates with population density information was recognized by the World Health Organisation as a powerful methodology to map the impacts on the health of the population already in 1999 (WHO 1999). Following this, the possibility to forecast the impact of air quality on European populations was explored in ANYWHERE using the MH-EWS air quality products and ENACT population density maps (Batista e Silva et al. 2020). Specifically, the forecasts of pollutants' concentrations were classified in five air quality index levels based on the AQI classification (Table 4.3). At each grid cell, the overall air quality was determined by the poorest index level of any of the five pollutants and population density values were classified into three population exposure classes (low for 10–999 people km^{-2}, medium for 1000–4999 people km^{-2}, high for 5000 and more people km^{-2}). Population exposure classes were arbitrarily chosen and manually adjusted by testing a variety of threshold values and analysing the resulting population exposure maps in various locations across Europe. The combination of AQI and population density was done in each grid cell by means of the impact matrix, resulting in three impact levels: low (yellow), medium (orange) and high (red) (Figure 4.6, panel b). It was assumed that only moderate, poor and very poor air quality situations cause relevant impacts; in locations with good or fair air quality, no significant impacts are expected. Figure 4.6 shows an example of impact forecasting map for an event of poor air quality affecting the south-west Europe on 11 July 2019 (panels a and c). For each NUTS region,[2] AQI forecast maps and the population density map were overlayed, and the total population affected by moderate (N_{moderate}), poor (N_{poor}) and very poor (N_{verypoor}) air quality determined. An impact index I_{NUT} was defined as

$$I_{\text{NUT}} = \frac{\left(1 \cdot N_{\text{moderate}}\right) + \left(2 \cdot N_{\text{poor}}\right) + \left(3 \cdot N_{\text{verypoor}}\right)}{\text{Area}_{\text{NUT}}}$$

where Area_{NUT} is the surface area of the NUT region; higher weight was allocated to the population affected by poorer air quality. To determine an impact level (low, medium, high) for the region, I_{NUT} was compared to thresholds based on Mdn_{dens}, the median of population densities of all NUTS regions (Mdn_{dens}= 82.9 people km^{-2}, Table 4.4). An example of the resulting air quality impact map is shown in Figure 4.6 (panel d). The Peloponnese shows a low impact level because it has a comparably low population density (33.8 people km^{-2}) and only a small share of it was affected by moderate or poor air quality, resulting in a low impact index of I_{NUT}= 18.6 people km^{-2}. The more densely populated island of Crete (84.0 people km^{-2}) was instead fully affected by poor air quality, yielding an impact index of I_{NUT}= 168.0 people km^{-2} and thus a high impact level. This approach provides a coherent pan-European perspective on air pollution-related health impacts, which are thus made comparable across borders. Impact-based forecast maps are currently integrated as a product into A4CENEM and are being assessed as an operational forecasting tool for rising preparedness to acute harmful air pollution events.

4.6.2 Multi-Hazard Forecasting

Every year, millions of people worldwide are exposed to two or more hazards with consequences on personal health as well as preparedness responses and recovery processes that go beyond what is seen from single disasters (Dilley 2005; Leppold et al. 2022).

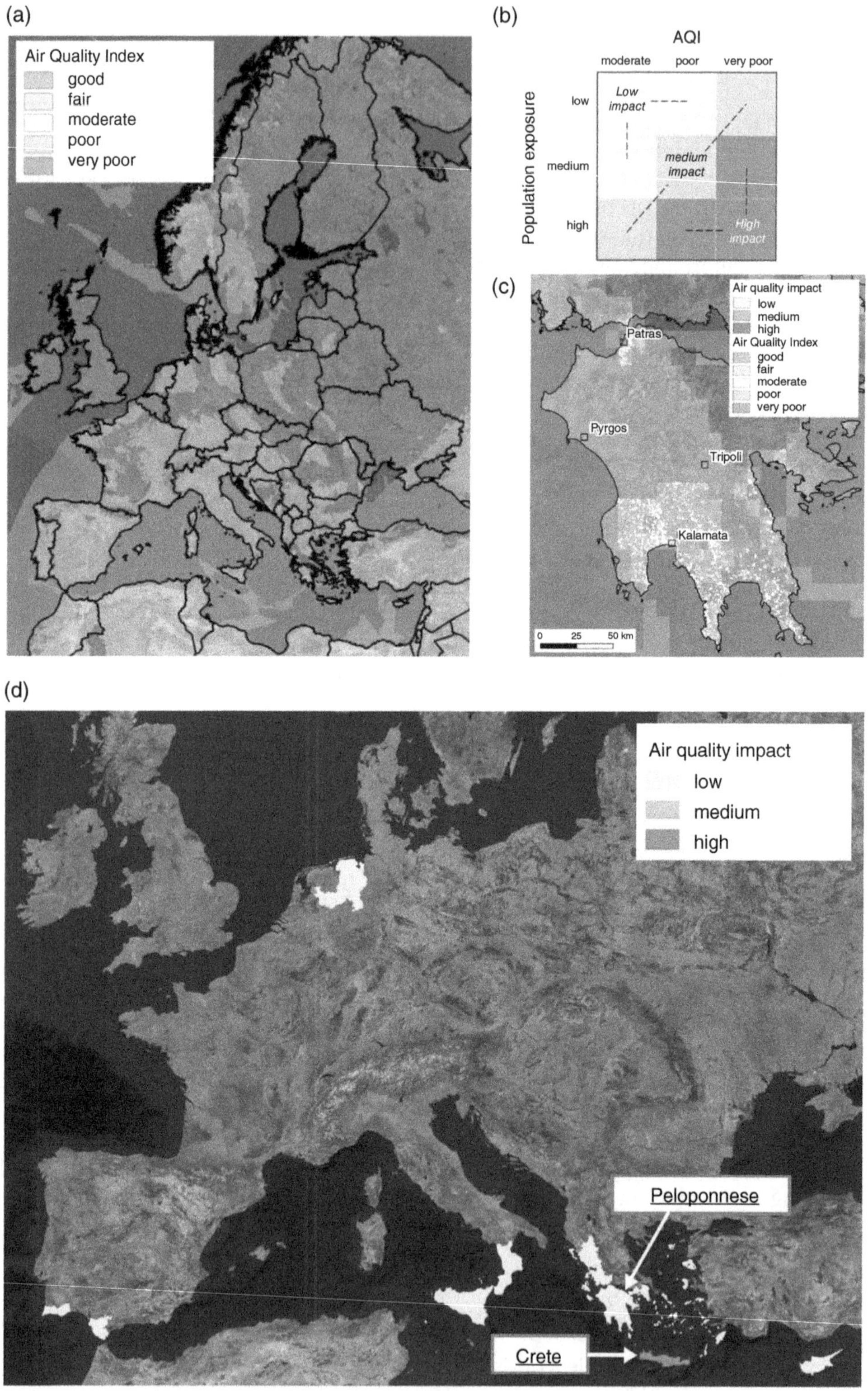

Figure 4.6 A framework for impact-based forecasting of air pollution health hazards. (a) Air Quality Index (AQI) forecast for 11 July 2019 at 10 UTC. (b) Matrix used to estimate the air pollution impact on population by combining the AQI forecast level with the population exposure class in each grid cell. (c) Resulting air quality impact map for the Peloponnese, Greece. (d) Pan-European air quality impact-based forecast by NUTS regions.

Table 4.4 Threshold definitions to determine the impact levels in the NUTS regions, based on the impact index (I_{NUT}) and the median of population densities of all NUTS regions (Mdn_{dens}).

Impact level	Threshold	Threshold values
None	$I_{NUT} = 0$	$I_{NUT} = 0$
Low	$0 < I_{NUT} \leq Mdn_{dens}$	$0 < I_{NUT} \leq 82.9$ people km^{-2}
Medium	$Mdn_{dens} < I_{NUT} \leq 2 \cdot Mdn_{dens}$	82.9 people km^{-2} $< I_{NUT} \leq 165.8$ people km^{-2}
High	$I_{NUT} > 2 \cdot Mdn_{dens}$	$I_{NUT} > 165.8$ people km^{-2}

Creating an early warning, which is able to warn of concurring hazards, can help local responders coordinate, prioritize and deploy resources on the ground more efficiently (WMO 2018). Vitolo et al. (2019) demonstrated that ANYWHERE heatwave and wildfire products can successfully provide such a system. In June 2017, a heatwave and a wildfire affected simultaneously southern Europe with reported casualties and damages. ANYWHERE products could predict the location and time of the combined heatwave-wildfire danger 10 days before it occurred. As stakeholders need a five-day (or more) notice to allow sufficient time for planning and allocating resources, they evaluated this multi-hazard forecasting information as useful during the event. With a similar approach, future applications could consider forecasts of compound heatwave-pollution events and the inclusion of other health-relevant hazards, such as pollen (Jacobs et al. 2014).

4.6.3 Cold Spells as a Health Hazard

As heatwaves, periods of unusually low temperature (*cold spells*) can be detrimental to human health and are associated with increased mortality rates in affected populations (Weilnhammer et al. 2021). Ten European countries have operational cold weather alert systems in place, each based on the forecasts of different meteorological variables at different lead times (Laaidi et al. 2013). Forecasts of minimum temperatures from the IFS-ECMWF and MEPS HarmonEPS are integrated as products into A4Cat, A4CENEM, A4Finn, A4Lig, and A4Nor prototypes. A4Cat, A4CENEM, and A4Finn also include cold warnings issued for Catalonia, Spain, and Finland by corresponding meteorological services. Thanks to its thermo-physiological relevance, the UTCI can be applied to predict cold stress at the pan-European level and complement existing cold weather alert systems. Where these are missing, it can provide valuable information for decision-making. Days when UTCI < 9 °C (i.e. at least cold stress), for instance, witness the most frequent medical consultations and emergency interventions for hypothermia, hypertension and respiratory tract infections (Krzyżewska et al. 2017; Romaszko et al. 2019; Skutecki et al. 2019). A cold spell product could use that same threshold for impact-based cold hazard forecasting.

4.6.4 Social Sensing

There is increasing evidence that social media can be used for predicting the occurrence of heat extremes and their impacts on human health, thus allowing real-time management of related risks (Cecinati et al. 2019; Young et al. 2021). Acting as a distributed network of mobile sensors that react to heat events by exchanging messages, Twitter has been found to experience an increase in daily tweets as temperatures increase. This suggests that the volume of tweets in certain locations can be used to estimate thresholds of local discomfort conditions (Grasso et al. 2017). A positive correlation has also been found between the number of air-pollution-related tweets and pollution concentrations (Sammarco et al. 2017; Gurajala et al. 2019; Hswen et al. 2019). The ANYWHERE system already includes a crowdsourcing product that analyses the content of tweets and determines a flood risk probability according to their semantic content [Ref to Milan's chapter]. Building from this experience, a similar product targeting heat- and air-pollution-related tweets at the pan-European level could be deployed and integrated.

4.6.5 Protecting the Vulnerable

Heat and pollution hazards do not affect everybody in the same way. Among the most vulnerable are the elderlies, children, outdoor workers, athletes, and those with pre-existing medical conditions (Mayrhuber et al. 2018; Dominski et al. 2021). In future applications, ANYWHERE heatwave products could be targeted for vulnerable groups by implementing thresholds that, once exceeded, define harmful heat conditions in the workplace, e.g. in the agriculture sector, or for recreational activities, such as racewalking and marathons (Hollander et al. 2021; Romaszko et al. 2022). Further, the products could be complemented with the forecasts of multiple thermal indices, such as the wet bulb globe temperature (WBGT), which is an international ISO standard for the estimation of heat stress on workers (ISO 2017). The forecasts could be seamlessly calculated from the same NWP model as the current ANYWHERE heatwave products via opensource software (Brimicombe et al. 2022).

4.7 Conclusions

Heatwaves and air pollutions are well-known hazards to human health. Predicting when and where they are expected to hit is essential for a timely and effective preparedness to their impacts. The ANYWHERE system is the first implementation of a real-time decision support platform providing the forecasts of detrimental heat and pollution extremes at the pan-European level. The forecasts, known as products, are generated seamlessly from a common NWP model (IFS-ECMWF) and consist of continental-wide maps displaying levels of human heat stress and pollutants' concentrations for up to 10 days into the future. Innovatively in ANYWHERE, the products include probabilistic forecasting of heatwaves and adopt a 'human-centric' approach where extremes are defined according to thresholds associated with demonstrated effects on population health. The products are delivered via a user-friendly interface in seven pilot sites. At each pilot site, they are customized and proactively targeted to the needs and requirements of local emergency management operation services, with inclusion of real-time measurements from weather stations and location-specific warnings. Following a 12-month trial period, the usability and effectiveness of ANYWHERE heatwave and air pollution products were evaluated positively by end users. The products could predict the high levels of heat stress and ozone achieved during the 2017 European heatwave at least 4 days in advance.

The products were of value also when used with other ANYWHERE products; combined with the wildfire product, they successfully predicted the compound heatwave-wildfire event of June 2017. By increasing the availability of and access to early warnings for multiple hazards, the ANYWHERE system demonstrates the usefulness of a MH-EWS to safeguard human health, thus translating into practice the Sendai Framework for Disaster Risk Reduction (Aitsi-Selmi et al. 2015). Future implementations of the ANYWHERE system could include warnings based on health or social media data, targeted also to the vulnerable and other hazards such as cold spells.

Funding

This work was funded by the ANYWHERE project (EnhANcing emergencY management and response to extreme WeatHER and climate Events, Project ID: 700099) within the European Commission's HORIZON2020 programme. At the time of writing, Claudia Di Napoli is supported by a Wellcome Trust grant (209734/Z/17/Z).

Acknowledgements

The author wishes to thank Tuomo Bergman (FMI), Eurico de Carvalho Filho (PREDICT Services), Ilona Láng-Ritter (FMI), Josias Láng-Ritter (CRAHI-UPC) and Xavi Llort (HYDS) for providing informational material on the MH-EWS prototypes at the pilot sites, and Samuel Sutanto (Wageningen University and Research) for providing feedback on the original draft.

Notes

1. ACC represents a measure of how well forecast anomalies correspond to observed anomalies. ACC values lie between +1 and −1. The lead time when the ACC value falls below 0.6 is considered the time when the positioning of synoptic scale features ceases to have value for forecasting purposes. Source: ECMWF (2022). Anomaly correlation coefficient. https://confluence.ecmwf.int/display/FUG/Anomaly+Correlation+Coefficient (accessed May 2022).
2. Nomenclature of Territorial Units for Statistics (NUTS, *Nomenclature des unités territoriales statistiques*) is a geocode standard for referencing the subdivisions of EU member states for statistical purposes.

References

AEMET (2018a). Interpretation: warnings [Interpretación: Avisos]. http://www.aemet.es/es/eltiempo/prediccion/avisos/ayuda (accessed May 2022).

AEMET (2018b). Thresholds and warning levels [Umbrales y niveles de aviso]. http://www.aemet.es/documentos/es/eltiempo/prediccion/avisos/plan_meteoalerta/umbrales_niveles_aviso.pdf (accessed June 2022).

AEMET (2019). July 2019 was very hot in the whole of Spain [Julio de 2019 fue muy cálido en el conjunto de España]. https://www.aemet.es/es/noticias/2019/Avance_clima_jul19 (accessed May 2022).

Aitsi-Selmi, A., Egawa, S., Sasaki, H. et al. (2015). The Sendai framework for disaster risk reduction: renewing the global commitment to people's resilience, health, and well-being. *International Journal of Disaster Risk Science* 6: 164–176.

Alduchov, O.A. and Eskridge, R.E. (1996). Improved magnus form approximation of saturation vapor pressure. *Journal of Applied Meteorology and Climatology* 35: 601–609.

Analitis, A., Michelozzi, P., D'Ippoliti, D. et al. (2014). Effects of heat waves on mortality. Effect modification and confounding by air. *Epidemiology* 25: 15–22.

Barriopedro, D., Fischer, E.M., Luterbacher, J. et al. (2011). The hot summer of 2010: redrawing the temperature record map of Europe. *Science* 332: 220–224.

Batista e Silva, F., Freire, S., Schiavina, M. et al. (2020). Uncovering temporal changes in Europe's population density patterns using a data fusion approach. *Nature Communications* 11: 1–11.

Błażejczyk, K., Jendritzky, G., Bröde, P. et al. (2013). An introduction to the universal thermal climate index. *Geographia Polonica* 86: 5–10.

Brimicombe, C., Di Napoli, C., Quintino, T. et al. (2022). Thermofeel: a python thermal comfort indices library. *SoftwareX* 18: 101005.

Bröde, P., Fiala, D., Błażejczyk, K. et al. (2012). Deriving the operational procedure for the universal thermal climate index (UTCI). *International Journal of Biometeorology* 56: 481–494.

Casanueva, A., Burgstall, A., Kotlarski, S. et al. (2019). Overview of existing heat-health warning systems in Europe. *International Journal of Environmental Research and Public Health* 16: 2657.

Cecinati, F., Matthews, T., Natarajan, S. et al. (2019). Mining social media to identify heat waves. *International Journal of Environmental Research and Public Health* 16: 762.

De Perez, E.C., Van Aalst, M., Bischiniotis, K. et al. (2018). Global predictability of temperature extremes. *Environmental Research Letters* 13: 054017.

DEFRA (2017). Air quality information bulletin – 21st June 2017. https://uk-air.defra.gov.uk/news?view=234 (accessed March 2022).

DEFRA (2018). *Air Pollution in the UK 2017*. Department for Environment, Food & Rural Affairs. https://uk-air.defra.gov.uk/assets/documents/annualreport/air_pollution_uk_2017_issue_1.pdf (accessed June 2022).

Di Napoli, C., Pappenberger, F., and Cloke, H.L. (2018). Assessing heat-related health risk in Europe via the universal thermal climate index (UTCI). *International Journal of Biometeorology* 62: 1155–1165.

Di Napoli, C., Pappenberger, F., and Cloke, H.L. (2019). Verification of heat stress thresholds for a health-based heat-wave definition. *Journal of Applied Meteorology and Climatology* 58: 1177–1194.

Di Napoli, C., Hogan, R.J., and Pappenberger, F. (2020). Mean radiant temperature from global-scale numerical weather prediction models. *International Journal of Biometeorology* 64: 1233–1245.

Di Napoli, C., Barnard, C., Prudhomme, C. et al. (2021). ERA5-HEAT: a global gridded historical dataset of human thermal comfort indices from climate reanalysis. *Geoscience Data Journal* 8: 2–10.

Díaz, J., Garcia, R., Velázquez de Castro, F. et al. (2002a). Effects of extremely hot days on people older than 65 years in Seville (Spain) from 1986 to 1997. *International Journal of Biometeorology* 46: 145–149.

Díaz, J., Jordán, A., Garcia, R. et al. (2002b). Heat waves in Madrid 1986–1997: effects on the health of the elderly. *International Archives of Occupational and Environmental Health* 75: 163–170.

Díaz, J., Linares, C., Carmona, R. et al. (2017). Saharan dust intrusions in Spain: health impacts and associated synoptic conditions. *Environmental Research* 156: 455–467.

Dilley, M. (2005). *Natural Disaster Hotspots: A Global Risk Analysis*. World Bank Publications.

Dominski, F.H., Branco, J.H.L., Buonanno, G. et al. (2021). Effects of air pollution on health: a mapping review of systematic reviews and meta-analyses. *Environmental Research* 201: 111487.

Dong, B., Sutton, R., Shaffrey, L., and Wilcox, L. (2016). The 2015 European heat wave. *Bulletin of the American Meteorological Society* 97: S57–S62.

Dos Santos, L.R., Alves-Correia, M., Câmara, M. et al. (2018). Multiple victims of carbon monoxide poisoning in the aftermath of a wildfire: a case series. *Acta Médica Portuguesa* 31: 146–151.

ECMWF (2022). Anomaly correlation coefficient. https://confluence.ecmwf.int/display/FUG/Anomaly+Correlation+Coefficient (accessed May 2022).

EPA (2022). Basic information about carbon monoxide (CO) outdoor air pollution. https://www.epa.gov/co-pollution/basic-information-about-carbon-monoxide-co-outdoor-air-pollution (accessed March 2022).

European Commission (2022). Air quality standards. https://ec.europa.eu/environment/air/quality/standards.htm (accessed March 2022).

European Environment Agency (2017). Air quality index. https://www.eea.europa.eu/themes/air/air-quality-index (accessed May 2018).

European Environment Agency (2021). European air quality index. https://www.eea.europa.eu/themes/air/air-quality-index (accessed March 2022).

Fiala, J., Cernikovsky, L., De Leeuw, F. and Kurfuest, P. (2003). Air pollution by ozone in Europe in summer 2003. Overview of exceedances of EC ozone threshold values during the summer season April–August. Report to the European Commission by the European Environment Agency, p. 33. https://www.eea.europa.eu/publications/topic_report_2003_3/Topic_No_3_2003_web.pdf

Fiala, D., Havenith, G., Bröde, P., and Kampmann, B. (2012). UTCI-Fiala multi-node model of human heat transfer and temperature regulation. *International Journal of Biometeorology* 56: 429–441.

Filleul, L., Cassadou, S., Médina, S. et al. (2006). The relation between temperature, ozone, and mortality in nine French cities during the heat wave of 2003. *Environmental Health Perspectives* 114: 1344–1347.

Fischer, P.H., Brunekreef, B., and Lebret, E. (2004). Air pollution related deaths during the 2003 heat wave in the Netherlands. *Atmospheric Environment* 38: 1083–1085.

Founda, D. and Giannakopoulos, C. (2009). The exceptionally hot summer of 2007 in Athens, Greece – a typical summer in the future climate? *Global and Planetary Change* 67: 227–236.

Frogner, I.-L., Andrae, U., Bojarova, J. et al. (2019). HarmonEPS – the HARMONIE ensemble prediction system. *Weather and Forecasting* 34: 1909–1937.

Golitsyn, G.S., Gorchakov, G.I., Grechko, E.I. et al. (2011). Extreme carbon monoxide pollution of the atmospheric boundary layer in Moscow region in the summer of 2010. *Doklady Earth Sciences* 441: 1666–1672.

Grasso, V., Crisci, A., Morabito, M. et al. (2017). Public crowdsensing of heat waves by social media data. *Advances in Science and Research* 14: 217–226.

Gurajala, S., Dhaniyala, S., and Matthewss, J.N. (2019). Understanding public response to air quality using tweet analysis. *Social Media+ Society* 5: 2056305119867656.

Havenith, G., Fiala, D., Błażejczyk, K. et al. (2012). The UTCI-clothing model. *International Journal of Biometeorology* 56: 461–470.

Hollander, K., Klöwer, M., Richardson, A. et al. (2021). Apparent temperature and heat-related illnesses during international athletic championships: a prospective cohort study. *Scandinavian Journal of Medicine & Science in Sports* 31: 2092–2102.

Hswen, Y., Qin, Q., Brownstein, J.S., and Hawkins, J.B. (2019). Feasibility of using social media to monitor outdoor air pollution in London, England. *Preventive Medicine* 121: 86–93.

IRCEL-CELINE (2017). Extreme heat and ozone peaks plan. Bulletin of June 18, 2017 [Plan fortes chaleurs et pics d'ozone. Bulletin du 18 juin 2017]. https://www.irceline.be/fr/smog/ozone_heat/2017/forte-chaleur-ozone-18-06-2017 (accessed June 2022).

IRCEL-CELINE (2022). What is the 'Ozone and heat plan' in Belgium and what are the different phases? https://www.irceline.be/en/documentation/faq/what-is-the-ozone-and-heat-plan-in-belgium-and-what-are-the-different-phases (accessed March 2022).

ISO (2017). 7243: ergonomics of the thermal environment – assessment of heat stress using the wbgt (wet bulb globe temperature) index. In: *ISO/TC 159/SC 5 Ergonomics of the Physical Environment*, 3e, 18. Geneva: International Organization for Standardization.

Jacobs, S.J., Pezza, A.B., Barras, V., and Bye, J. (2014). A new 'bio-comfort'perspective for Melbourne based on heat stress, air pollution and pollen. *International Journal of Biometeorology* 58: 263–275.

Kollanus, V., Prank, M., Gens, A. et al. (2017). Mortality due to vegetation fire–originated PM2. 5 exposure in Europe – assessment for the years 2005 and 2008. *Environmental Health Perspectives* 125: 30–37.

Konovalov, I.B., Beekmann, M., Kuznetsova, I.N. et al. (2011). Atmospheric impacts of the 2010 Russian wildfires: integrating modelling and measurements of an extreme air pollution episode in the Moscow region. *Atmospheric Chemistry and Physics* 11: 10031–10056.

Koppe, C. (2007). EuroHEAT. Climate information decision support tool for heat in Europe. Scientific background. http://www.euroheat-project.org/dwd/flyer_background.pdf (accessed June 2022).

Krzyżewska, A., Dobek, M., Domżał-Drzewicka, R., and Rząca, M. (2017). Emergency interventions due to weather-related hypothermia. *Weather* 72: 8–12.

Kusch, W. (2004). *Guidelines on Biometeorology and Air Quality Forecasts*. WMO.

Laaidi, K., Economopoulou, A., Wagner, V. et al. (2013). Cold spells and health: prevention and warning. *Public Health* 127: 492–499.

Lalaurette, F. (2003). Early detection of abnormal weather conditions using a probabilistic extreme forecast index. *Quarterly Journal of the Royal Meteorological Society: A Journal of the Atmospheric Sciences, Applied Meteorology and Physical Oceanography* 129: 3037–3057.

Lavaysse, C., Naumann, G., Alfieri, L. et al. (2019). Predictability of the European heat and cold waves. *Climate Dynamics* 52: 2481–2495.

Leppold, C., Gibbs, L., Block, K. et al. (2022). *Public Health Implications of Multiple Disaster Exposures*. The Lancet Public Health.

Linares, C., Díaz, J., Tobías, A. et al. (2015). Impact of heat and cold waves on circulatory-cause and respiratory-cause mortality in Spain: 1975–2008. *Stochastic Environmental Research and Risk Assessment* 29: 2037–2046.

Lombardia, A. (2017). Ozone: levels still high throughout Lombardy [Ozono: Livelli ancora alti in tutta la Lombardia]. https://www.arpalombardia.it/Pages/Ozono-livelli-ancora-alti-in-tutta-la-Lombardia.aspx (accessed March 2022).

Magnusson, L. (2019). The 2019 western European heatwaves. *ECMWF Newsletter* 161. https://www.ecmwf.int/en/newsletter/161/news/2019-western-european-heatwaves (accessed March 2022).

Marécal, V., Peuch, V.-H., Andersson, C. et al. (2015). A regional air quality forecasting system over Europe: the MACC-II daily ensemble production. *Geoscientific Model Development* 8: 2777–2813.

Mayrhuver, E.A.-S., Dückers, M.L., Wallner, P. et al. (2018). Vulnerability to heatwaves and implications for public health interventions–a scoping review. *Environmental Research* 166: 42–54.

McGregor, G.R., Bessmoulin, P., Ebi, K., and Menne, B. (2015). *Heatwaves and Health: Guidance on Warning-System Development*. WMOP.

Meehl, G.A., Tebaldi, C., Tilmes, S. et al. (2018). Future heat waves and surface ozone. *Environmental Research Letters* 13: 064004.

Merz, B., Kuhlicke, C., Kunz, M. et al. (2020). Impact forecasting to support emergency management of natural hazards. *Reviews of Geophysics* 58: e2020RG000704.

Montero, J., Mirón, I., Criado-Álvarez, J. et al. (2012). Influence of local factors in the relationship between mortality and heat waves: Castile-La Mancha (1975–2003). *Science of the Total Environment* 414: 73–80.

Morabito, M., Crisci, A., Messeri, A. et al. (2017). Increasing heatwave hazards in the southeastern European Union capitals. *Atmosphere* 8: 115.

MSSSI (2017). National plan for preventive actions to the effects of excess temperatures on health [Plan Nacional de actuaciones preventivas de los efectos de los excesos de temperaturas sobre la salud]. https://www.sanidad.gob.es/ciudadanos/saludAmbLaboral/planAltasTemp/2021/docs/Balance_Plan_Calor_2017.pdf (accessed June 2022).

NHS (2011). Heatwave plan for England. Protecting health and reducing harm from extreme heat and heatwaves. https://assets.publishing.service.gov.uk/government/uploads/system/uploads/attachment_data/file/216193/dh_127235.pdf (accessed June 2022).

Pappenberger, F., Jendritzky, G., Staiger, H. et al. (2015). Global forecasting of thermal health hazards: the skill of probabilistic predictions of the universal thermal climate index (UTCI). *International Journal of Biometeorology* 59: 311–323.

Parisien, L. (2017). Heatwave alert: orange-level vigilance extended to 66 departments [Alerte canicule: la vigilance orange étendue à 66 départements]. https://www.leparisien.fr/societe/canicule-les-demandes-de-soins-augmentent-dans-51-departements-20-06-2017-7070244.php (accessed March 2022).

PHE (2017a). Heat health warning: 2017. https://www.gov.uk/government/news/heat-health-warning-2017 (accessed March 2022).

PHE (2017b). PHE heatwave mortality monitoring. Summer 2017. https://assets.publishing.service.gov.uk/government/uploads/system/uploads/attachment_data/file/942651/PHE_heatwave_mortality_monitoring_report_2017.pdf (accessed June 2022).

PREV'AIR (2017). Ozone pollution episode: situation and evolution – June 21, 2017 [Episode de pollution à l'ozone: situation et évolution – 21 juin 2017]. http://www2.prevair.org/actualites/episode-de-pollution-lozone-situation-et-evolution-21-juin-2017 (accessed March 2022).

Repubblica, L. (2017). Heat, the ministry of health announces: "Heatwave in 10 cities between saturday and Sunday" [Caldo, il ministero della Salute: "Tra sabato e domenica ondata di calore in 10 città"]. https://www.repubblica.it/cronaca/2017/06/23/news/afa_ministero_salute_ondate_calore_in_dieci_citta_-168893050/ (accessed March 2022).

Robine, J.-M., Cheung, S.L.K., Le Roy, S. et al. (2008). Death toll exceeded 70,000 in Europe during the summer of 2003. *Comptes Rendus Biologies* 331: 171–178.

Romaszko, J., Skutecki, R., Bocheński, M. et al. (2019). Applicability of the universal thermal climate index for predicting the outbreaks of respiratory tract infections: a mathematical modeling approach. *International Journal of Biometeorology* 63: 1231–1241.

Romaszko, J., Dragańska, E., Jalali, R. et al. (2022). Universal climate thermal index as a prognostic tool in medical science in the context of climate change: a systematic review. *Science of the Total Environment* 9: 154492.

Sammarco, M., Tse, R., Pau, G., and Marfia, G. (2017). Using geosocial search for urban air pollution monitoring. *Pervasive and Mobile Computing* 35: 15–31.

Sánchez-Benítez, A., García-Herrera, R., Barriopedro, D. et al. (2018). June 2017: the earliest European summer mega-heatwave of Reanalysis period. *Geophysical Research Letters* 45: 1955–1962.

Santé Publique France (2017). Heatwave from 17 to 25 June 2017. National assessment of the impact on health as of 07/26/2017 [Canicule du 17 au 25 juin 2017. Bilan national de l'impact sur la santé au 26/07/2017].

SCIENSANO (2017). Moderate excess mortality during heat wave in June [Matige oversterfte tijdens hittegolf in juni]. https://www.sciensano.be/nl/pershoek/matige-oversterfte-tijdens-hittegolf-juni (accessed March 2022).

Shaposhnikov, D., Revich, B., Bellander, T. et al. (2014). Mortality related to air pollution with the Moscow heat wave and wildfire of 2010. *Epidemiology (Cambridge, Mass.)* 25: 359.

Skutecki, R., Jalali, R., Dragańska, E. et al. (2019). UTCI as a bio-meteorological tool in the assessment of cold-induced stress as a risk factor for hypertension. *Science of the Total Environment* 688: 970–975.

SMC (2021). Territorial distribution of thresholds [Distribució territorial de llindars]. https://www.meteo.cat/wpweb/divulgacio/la-prediccio-meteorologica/situacio-meteorologica-de-perill/distribucio-territorial-de-llindars/ (accessed May 2022).

SMC (2022). Definition of Hazardous Weather Situation (SMP) [Definició de Situació Meteorològica de Perill (SMP)]. https://www.meteo.cat/wpweb/divulgacio/la-prediccio-meteorologica/situacio-meteorologica-de-perill/ (accessed May 2022).

Stedman, J.R. (2004). The predicted number of air pollution related deaths in the UK during the August 2003 heatwave. *Atmospheric Environment* 38: 1087–1090.

Sulikowska, A. and Wypych, A. (2020). How unusual were June 2019 temperatures in the context of European climatology? *Atmosphere* 11: 697.

Tarrasón, L., Hamer, P., Guerreiro, C. et al. (2016). Interim annual assessment report for 2015 - European air quality in 2015. CAMS (Copernicus Atmosphere Monitoring Service). https://policy.atmosphere.copernicus.eu/reports/CAMS71_2016SC1_D71.1.1.2_201609_final.pdf (accessed June 2022).

Tarrasón, L., Hamer, P., Meleux, F. et al. (2020). Interim annual assessment report for 2019 - European air quality in 2019. CAMS (Copernicus Atmosphere Monitoring Service). https://policy.atmosphere.copernicus.eu/reports/CAMS71_IAR_2019_final.pdf.

Theoharatos, G., Pantavou, K., Mavrakis, A. et al. (2010). Heat waves observed in 2007 in Athens, Greece: synoptic conditions, bioclimatological assessment, air quality levels and health effects. *Environmental Research* 110: 152–161.

Toloo, G.S., Fitzgerald, G., Aitken, P. et al. (2013). Are heat warning systems effective? *Environmental Health* 12: 1–4.

Vicente, C., Rebora, N., Botey, P. et al. (2019). ANYWHERE deliverable 6.4: report compiling the conclusions and the recommendations of the users in the use of the A4EU systems implemented in each pilot site. http://www.anywhere-h2020.eu/wp-content/uploads/docs/D6.4_submitted.pdf (accessed June 2022).

Vieno, M., Dore, A.J., Stevenson, D.S. et al. (2010). Modelling surface ozone during the 2003 heat-wave in the UK. *Atmospheric Chemistry and Physics* 10: 7963–7978.

Vitolo, C., Di Napoli, C., Di Giuseppe, F. et al. (2019). Mapping combined wildfire and heat stress hazards to improve evidence-based decision making. *Environment International* 127: 21–34.

VRT (2017). Waarschuwingsfase ozon- en hitteplan geactiveerd. https://www.vrt.be/vrtnws/nl/2017/06/18/waarschuwingsfaseozon-enhitteplangeactiveerd-1-3004818/ (accessed March 2022).

Weilnhammer, V., Schmid, J., Mittermeier, I. et al. (2021). Extreme weather events in Europe and their health consequences–a systematic review. *International Journal of Hygiene and Environmental Health* 233: 113688.

WHO (1999). *Environmental Health Indicators: Framework and Methodologies* (ed. D. Briggs), 119. World Health Organization.

WHO (2021). Ambient (outdoor) air pollution. https://www.who.int/en/news-room/fact-sheets/detail/ambient-(outdoor)-air-quality-and-health (accessed March 2022).

WMO (2018). Multi-hazard early warning systems: a checklist. Outcome of the first Multi-hazard Early Warning Conference.

Young, J.C., Arthur, R., Spruce, M., and Williams, H.T. (2021). Social sensing of heatwaves. *Sensors* 21: 3717.

Zvyagintsev, A., Blum, O., Glazkova, A. et al. (2011). Air pollution over European Russia and Ukraine under the hot summer conditions of 2010. *Izvestiya, Atmospheric and Oceanic Physics* 47: 699–707.

5

The EUMETNET OPERA Radar Network – European-Wide Precipitation Composites Supporting Rainfall-Induced Flash Flood Emergency Management

Shinju Park[1], Marc Berenguer[1], Daniel Sempere-Torres[1], and Annakaisa Von Lerber[2]
[1] *Center of Applied Research in Hydrometeorology, Universitat Politècnica de Catalunya (UPC), Barcelona, Spain*
[2] *Finnish Meteorological Institute, Helsinki, Finland*

5.1 Introduction

Precipitation is one of the essential climate variables (ECVs), and its changes in frequency and magnitude over the past years in Europe had considerable impacts on our society (EEA 2017). For example, hydrological hazards such as river floods, flash floods, pluvial floods or debris flows triggered by heavy rainfalls result in over one-third of the total economic losses caused by weather and climate-related events (about EUR 500 billion in the 32 EEA Member States over the period 1980–2020, EEA Briefing 2021). In particular, flash floods have been one of the challenges of emergency management because of their short onset times, from a few minutes to a few hours (Borga et al. 2011; Cloke et al. 2017; Harrison et al. 2017). Flash floods are often triggered by local heavy rainfall over small to medium catchments frequently ungauged (Sempere-Torres et al. 2010; Alfieri et al. 2019; Corral et al. 2019).

Reliable and accurate observations of causative precipitation with access to their historic archives are fundamental to raising preparedness to flash floods in the context

Responding to Extreme Weather Events, First Edition. Edited by Daniel Sempere-Torres, Anastasios Karakostas, Claudio Rossi, and Philippe Quevauviller.

of risk reduction and development of climate change adaptation plans at national, regional and local levels (Casajus Valles et al. 2020). For the past years, a need for climatological products of coherent high-resolution precipitation observations providing coverage beyond the borders of a single country has been identified for improved cross-border risk assessment (UNDRR 2019).

Currently, several precipitation products based on European-wide gauge precipitation records exist. These include the datasets collected from the national and regional hydrometeorological institutes (e.g., Meteorological Data Collection Center of Copernicus Emergency Management Service, Rehfeldt et al. 2021), the WMO Global SYNOP data (public access), or the datasets from citizen voluntary networks together with an official network of automatic surface stations such as the WOW product (Kirk et al. 2021). These point measurements are highly inhomogeneous in space and time (Kidd et al. 2017). Some gridded gauge products also are available only with coarse spatial grids or for the past years (not as a real-time product): e.g., Copernicus E-OBS daily data (Haylock et al. 2008; Cornes et al. 2018 at spatial resolutions of 0.1° and 0.25°) and EMO-5 with 5-km resolution and from 1990 to 2019 (Thiemig et al. 2022).

Satellite observations or NWP models have also produced European-wide precipitation products, e.g., half-hourly IMERG accumulations, combining the information from the GPM satellite constellation at 0.1° grid with a 30-minute update (e.g., O et al. 2017), or NWP-based hourly accumulations obtained with the ECMWF IFS model over a 7 km grid with 6-hour update (e.g., Lavers et al. 2021). Nevertheless, their coarse spatial resolution and additional sampling uncertainty still limit the stand-alone applicability of these products to resolve intense local evolution of storms for emergency monitoring and forecasting.

Alternatively, over Europe, more than 200 radars are operated by the National Meteorological and Hydrological Services (NMHSs). The EUMETNET project OPERA (https://www.eumetnet.eu/opera), established in 1999 (Saltikoff et al. 2018), has coordinated the exchange of expertise in running operational weather radars among the 30 members and operationally produce the European-wide radar-based precipitation composites. Thanks to its continental coverage and real-time operation at high spatial and temporal resolutions (2 km and 15 minutes), the OPERA composites can capture the evolution of intense rainfall, thus providing critical information to assess and anticipate rainfall-induced hazards and impacts.

This chapter presents the characteristics of radar-based OPERA precipitation composites and the results of systematic monitoring of their quality, and their use for high-resolution flash flood hazard nowcasting at the European scale with the ERICHA system. In the framework of the DG-ECHO project ERICHA ('European Rainfall-InduCed Hazard Assessment' 2016–2017), the system was integrated as the module for flash flood nowcasting of the European Flood Awareness System (EFAS).

5.2 The EUMETNET OPERA Radar Precipitation Composites

OPERA has operationally produced radar precipitation composites and volumetric data collected from more than 150 radars since 2012 (figure 1; see Huuskonen et al. 2014). The near-ground precipitation 2-D composites include three products updated every 15 minutes: (i) Instantaneous Surface Rain Rate obtained from the lowest level reflectivity composite applying the reflectivity-rain rate relation of Marshall and Palmer (1948), (ii) 1-Hour Rainfall Accumulation (simple sum of previous four 15-minute rain-rate products), and (iii) Instantaneous Maximum Reflectivity (the maximum of all the observations located in the vertical of each composite pixel location). These composites are produced over a Lambert Equal Area 2-km grid and are free to access to the members of OPERA and currently available to third parties upon license agreement.

Due to the heterogeneity in hardware, processing algorithms, and measuring conditions among the member states, the quality control of the composite data has been a challenge since the beginning of the operational composite production (Huuskonen et al. 2014). Through continuous discussions among the OPERA community (Saltikoff et al. 2017) implemented some techniques including the use of the satellite filtering of erroneous radar data and the development of a quality index (Saltikoff et al. 2019). The common data processing strategy has also been made upon the agreement of the member states in the different phases of the OPERA project. In recent years, the need for quality-ensured homogeneous precipitation data (European-wide, high-resolution and sub-daily) has increased among groups working on different applications such as NWP data assimilation and verification, climate studies or hydrological applications (Saltikoff et al. 2019). Therefore, the current phase of OPERA (phase 5, from 2019 to 2023) is re-organizing the Odyssey production flow (shown in Figure 5.1) into three production lines focusing on specific needs: namely (i) producing five-minute maximum reflectivity composite for current weather, severe weather alerts, and aviation applications; (ii) serving the incoming data back to the members for their own processing (e.g., regional composites or NWP assimilation) through a real-time data hub (Stratus in Figure 5.1) and (iii) producing centrally quality-controlled volume data and products

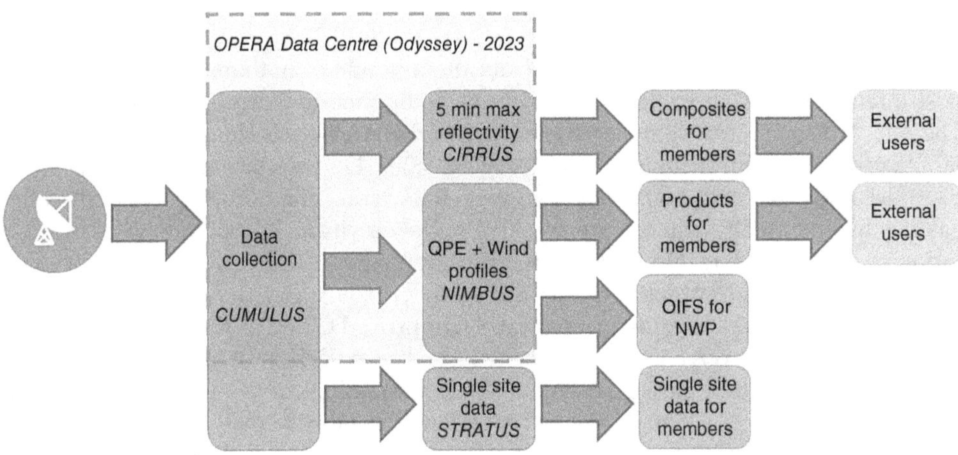

Figure 5.1 Current status of the OPERA data production chain plotted together with the previous OPERA Data Centre (Odyssey). The Odyssey generates and archives the quality-controlled precipitation composites and volume data for Numerical Weather Prediction (NWP) modellers via the OPERA Internet File Server (OIFS) from single-site raw radar data. *Source:* Courtesy of OPERA.

(e.g., wind profiles, composites of instantaneous surface rain rate and one-hour rainfall accumulation) for applications such as precipitation climatology or hydrology.

5.3 Monitoring the Quality of the Opera Precipitation Composites

The real-time applications of OPERA precipitation composites require systematic monitoring of their quality. The analyses done in the framework of some applications showed that OPERA precipitation composites underestimated precipitation amounts (Park et al. 2016; Delobbe and Lukach 2017; see also, Saltikoff et al. 2019). In particular, the quality of the OPERA rainfall rates has been systematically monitored by CRAHI-UPC in the context of six consecutive European projects (namely HAREN, EDHIT, ERICHA, ANYWHERE, SMUFF and TAMIR) focusing on the development of algorithms for rainfall-induced flood impact forecasting to support emergency managers and first responders. Park et al. (2019) presented the long-term evaluation of OPERA rainfall estimates and illustrated the bias reduction obtained by applying a simple real-time gauge-adjustment method.

The daily monitoring of the OPERA rainfall accumulations has been done from the comparisons between the valid Radar (R)-Gauge (G) pairs before and after applying the adjustment method (e.g., in Figure 5.2, an R/G value around 1 indicates small biases in rainfall estimates).

After applying the gauge-adjustment method, the corrected accumulations clearly show much smaller biases (as also indicated by the R/G ratio median values of Table 5.1). The results of Figure 5.2 and the Table 5.1 also show the changes in the non-adjusted OPERA data (overall improvement indicated by the increase of the R/G ratio median values). The results shown in Table 5.1 (see also Figure 5.3) show a significant change of the quality of the non-adjusted OPERA estimates in 2017. This coincides with the change in the processing strategy of the OPERA composites since late 2017 (Saltikoff et al. 2019 reported that in 2017 the OPERA processing was updated, including better hardware monitoring from the upgrade of national radars, the use of best-quality data from some countries and the application of some on/off quality). The effect of such improvements in the recent OPERA data is clearer during summer months (e.g. June, July, August in 2021, Figure 5.2b). During the winter months, however, the biases are still high, showing lower R/G ratios compared to those after the gauge adjustment. This can be explained by the lower vertical development of precipitation systems (which results in beam overshooting) and by the snowfall events in some areas of the continent, which is not identified by OPERA and, thus, the Z-R transformation is not adapted for snow (Lopez 2014). Improved QPE for winter would benefit from improved radar observational capacity (e.g., dual-polarization radar measurements providing hydrometeor classification and local climatological snow characteristics on crystal type, degree of riming and aggregation and density over the conterminous United States by Zhang et al. 2020), which are not available in all the OPERA member states and are not yet applied in the processing of OPERA composites.

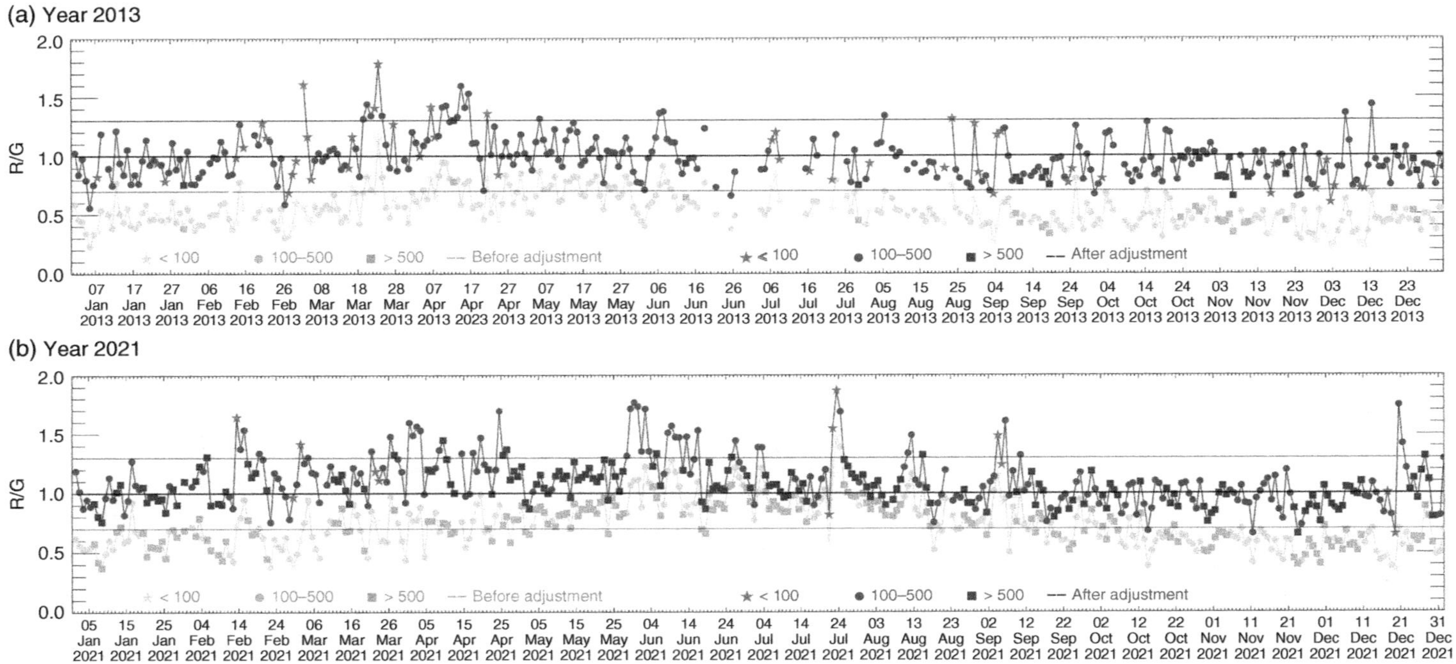

Figure 5.2 Time series of daily R/G in (a) 2013 and (b) 2021, based on the comparisons between OPERA radar precipitation estimates both before (light grey) and after (dark grey) applying the gauge-adjustment method. R/G values close to 1 indicate a good agreement with gauge observations. The total number of gauges used in the analysis varied from 2339 to 3412 within the OPERA coverage, counting GLOBAL SYNOP records at Continental scale and additional Spanish AEMET stations (in 2021). The collocated R-G daily accumulations estimated from the valid pairs (the different colours indicate the number of pairs) are selected after filtering out those in clutter contaminated areas or erroneous records from both measurements (see Park et al. 2019 for further details).

Table 5.1 A summary table of the evaluation of OPERA daily rainfall accumulations without and with the gauge adjustment in terms of R/G ratio at the valid daily R-G pairs (1 indicates there is no bias between radar and gauge estimates).

Year	Skill score quantile (%) – R/G					
	Non-adjusted			Gauge-adjusted		
	15	50	85	15	50	85
2013	0.28	0.47	0.60	0.67	0.90	1.06
2014	0.35	0.46	0.54	0.75	0.92	1.08
2015	0.34	0.45	0.57	0.78	0.94	1.12
2016	0.38	0.47	0.58	0.86	1.01	1.15
2017	0.46	0.58	0.71	0.89	1.0	1.22
2018	0.55	0.71	0.94	0.88	1.05	1.25
2019	0.52	0.67	0.87	0.87	1.01	1.23
2020	0.5	0.66	0.88	0.86	1.02	1.24
2021	0.52	0.69	0.91	0.88	1.05	1.27

Three quantiles (15, 50, and 85 percentiles) are computed over the period of Jan–Dec for each year from 2013 to 2021.

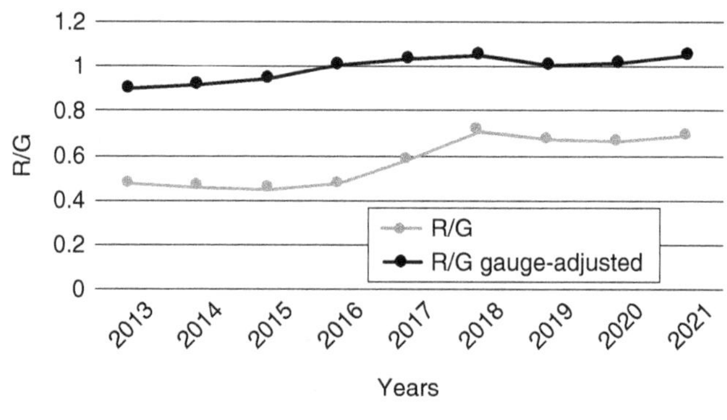

Figure 5.3 Median values of R/G ratio obtained from daily OPERA accumulations without and with the gauge adjustment (respectively, light grey and black lines) from 2013 to 2021 (50 percentile of Table 5.1).

On average, the gauge-adjusted estimates in winter mitigate the biases shown by OPERA estimates in winter, but its application for effective real-time flash flood hazard nowcasting would require careful interpretation, and the use of additional information (e.g., probability of snow from NWP shown in Figure 5.4 adopted from von Lerber et al. 2021) to identify the phase of observed precipitation.

5.4 Application of Opera Precipitation Composites for Flash Flood Hazard Nowcasting

Flash flood (FF) hazard is many times characterized in terms of the probability of occurrence (or return period) that measures how rare the flood event is. The FF hazard assessment is often carried over the catchment areas by analysing the potential of

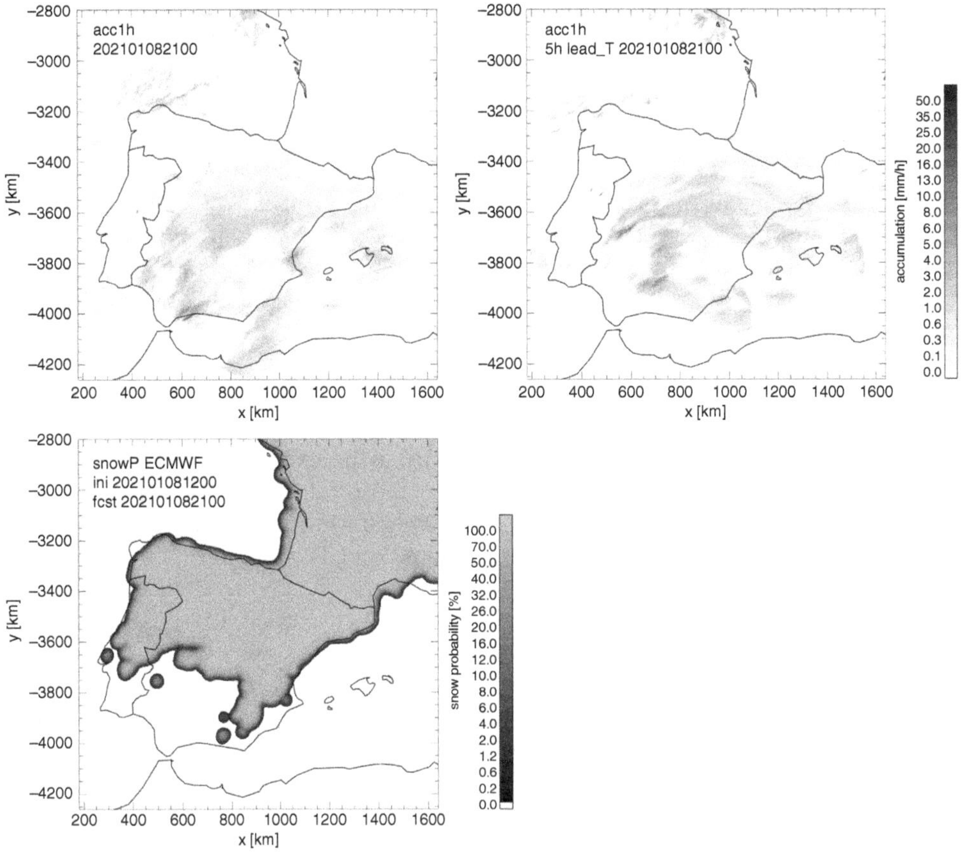

Figure 5.4 Gauge-adjusted OPERA hourly accumulation observed on 8 Jan 2021 at 2100 UTC over the Iberian Peninsula (upper left), nowcasts with lead-time five hours (upper right), an example of FMI's snow probability fields derived from the ECMWF precipitation-type products predicted when actually snowed over the area of Madrid located at the centre of Spain (bottom).

storms (e.g., ingredient-based by Doswell et al. 1996), based on the basin-aggregated rainfall amounts, or the runoff simulated with a rainfall-runoff model (e.g., Hapuarachchi et al. 2011; Corral et al. 2019). The availability of OPERA precipitation composites has provided a unique opportunity to develop a high-resolution FF hazard assessment and nowcasting system at a pan-European scale using coherent high-resolution transnational information (Park et al. 2019; Ritter et al. 2021).

As a simple approach to characterizing FF hazard, the basin-aggregated rainfall (i.e., the rainfall averaged in the basin upstream of each point of the drainage network) can be used assuming that the catchment response will be strongly dominated by this catchment-aggregated rainfall (Guillot and Duband 1967; Alfieri and Thielen 2015; Alfieri et al. 2019). Although this approach neglects some of the factors that affect the hydrological response of the upstream basin (e.g., initial moisture state of the catchment, groundwater deficit, the presence of accumulated snow, flow routing, or the presence of vegetation in the flood plain), it is well adapted to high-resolution FF hazard assessment over large domains in real-time, using operational precipitation inputs (e.g., real-time radar rainfall composites or NWP precipitation forecasts). Additionally, it does not require model calibration in the catchments of interest (typically small and medium areas from 10 to 1000 km^2 and often ungauged).

The ERICHA FF hazard nowcasting system developed by CRAHI-UPC uses European-wide precipitation observations and nowcasts obtained from the radar composites generated by OPERA (Figure 5.5). Precipitation nowcasts are generated by applying a Lagrangian persistence technique (Berenguer et al. 2005, 2011) with the gauge-adjusted OPERA precipitation composites. It estimates the motion field from consecutive radar composites and generates precipitation nowcasts by advecting the most recent radar composite for a lead time of up to six hours with a backward semi-Lagrangian scheme (Germann and Zawadzki 2002) and neglecting the evolution of precipitation intensities. The obtained precipitation observations and nowcasts are used to calculate the basin aggregated rainfall over the European drainage network with a spatial resolution of 1 km.

The FF hazard level is estimated by comparing the calculated basin-aggregated rainfall values for a duration corresponding to the catchment's concentration time (using accumulation windows of 1, 3, 6, 12, and 24 hours) with climatological reference thresholds for the same variable and over the same grid. If a long-term rainfall dataset exists in the application domain, the climatological reference thresholds can be estimated by extreme value analysis of the basin-aggregated rainfall building Intensity-Duration-Frequency (IDF) curves for each point of the analysis domain (e.g., Corral et al. 2019 over Liguria, Italy; Alfieri and Thielen 2015 at the European scale for NWP precipitation forecasts). These datasets for high-resolution gridded basin-aggregated rainfall over Europe are limited. The original version of the ERICHA system adapted

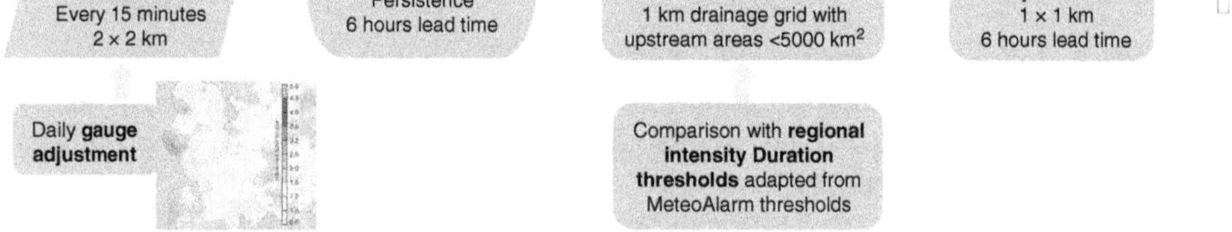

Figure 5.5 Processing chain of the ERICHA FF hazard nowcasting system implemented in the European Flood Awareness System producing European-wide precipitation and flash flood hazard nowcasts. The gauge-adjusted OPERA QPE composites are used to generate precipitation nowcasts in the range of zero to six hours. Together, precipitation observations and nowcasts are the inputs to the flash flood hazard module, which generates real-time flash flood hazard nowcasts over the European drainage network with 1-km and 15 minutes resolutions by comparing the calculated basin-aggregated rainfall with reference thresholds to estimate the hazard level (the ERICHA flash flood indicator). *Source*: Adapted from Park et al. (2019).

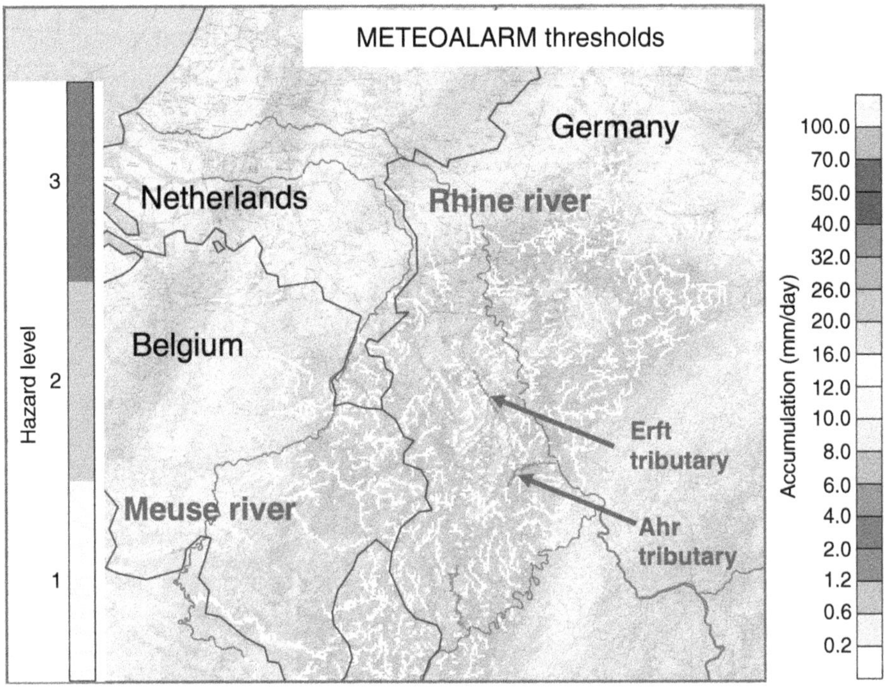

Figure 5.6 Daily rainfall accumulation and flash flood hazard summary obtained with the ERICHA system for the case of 14 July 2021 that affected Central Europe. The streams where the ERICHA system issued the high hazard level (in the darkest color along the Erft and Ahr tributaries) coincide with the areas where the most devastating floods were reported by Kreienkamp et al. (2021).

the regional Intensity-Duration (ID) thresholds for intense rainfall warnings over Europe used in the EUMETNET system METEOALARM (www.meteoalarm.eu). The ERICHA system has shown its skill to identify the most significant flash floods that have affected Europe since 2017, thus benefiting of the high-resolution OPERA precipitation composites (e.g., see Figure 5.6 the FF hazard levels identified for the case on 14 July 2021).

The ERICHA system has been running in real time in the European Flood Awareness System (EFAS) using the gauge-adjusted OPERA composites (www.efas.eu, Figure 5.7). Its capacity can complement the ERIC product for medium-range FF hazard forecasts (based on precipitation derived from COSMO-LEPS, e.g. Alfieri and Thielen 2015; Raynaud et al. 2015; Smith et al. 2016).

5.5 Conclusions and Outlook

Thanks to its coverage and real-time operation at high spatial and temporal resolutions (2 km, 15 minutes), the OPERA precipitation composites are a fundamental input to the European-wide flash flood hazard and impact forecasting that can complement existing regional, national and European products.

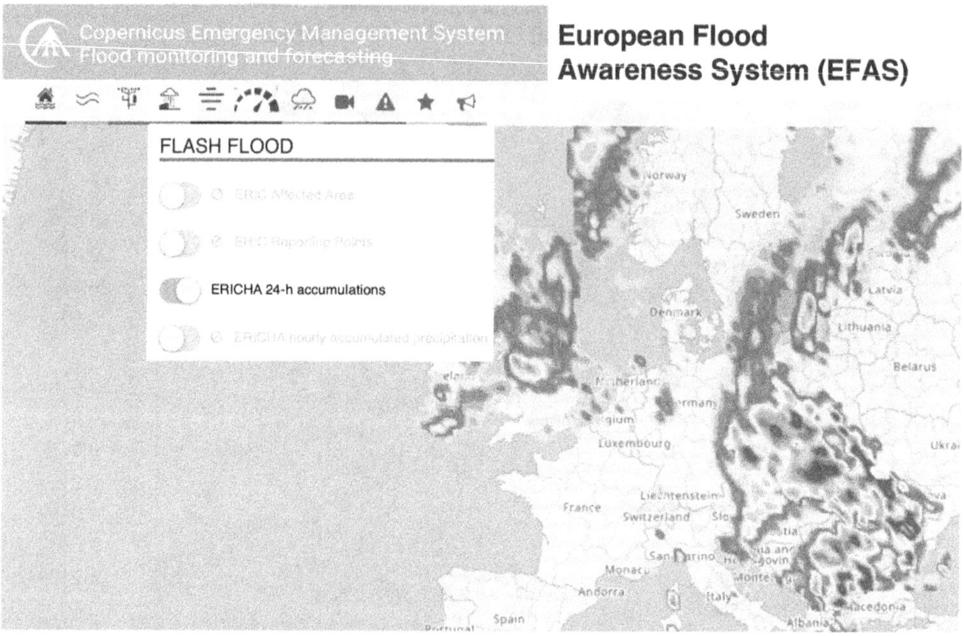

Figure 5.7 ERICHA 24-hour precipitation accumulation computed from the gauge-adjusted OPERA radar observations shown in EFAS (the map viewer as of 2022, www.efas.eu). This product is available over the OPERA grid (with a spatial resolution of 2 km), and is calculated twice a day, at 00 and 12 UTC.

The biases identified in OPERA-based Quantitative Precipitation Estimates are compensated with the gauge-adjustment method of Park et al. (2019) in real time. However, further work to improve their quality for hydrological applications is still necessary (e.g., better ground and sea clutter filtering, beam blockage mitigation or identification of precipitation phase) and is continuous work in the framework of the current and next phases of OPERA projects (Saltikoff et al. 2019).

The ERICHA flash flood hazard nowcasting system uses the gauge-adjusted OPERA composites for real-time monitoring of flood events and increased situational awareness to prepare for emergencies. Some of the functionalities of the original ERICHA system have been recently enhanced for better flash flood hazard and impact nowcasting to support emergency management at both European and regional scales. These include:

- Improved flash flood hazard assessment with the retrieval of reference rainfall thresholds over Europe from the OPERA archive. As of 2022, the availability of almost 10 years of gauge-adjusted OPERA precipitation composites has enabled the possibility to derive a set of radar-based reference thresholds of basin-aggregated rainfall coherent with the observations used for FF hazard assessment in real time: from the 1-hour accumulations of gauge-adjusted composites from 2013 to 2022 the thresholds for return periods of 2, 5 and 20 years have been obtained (Figure 5.8). Although the values for the higher return periods are subject to large uncertainties,

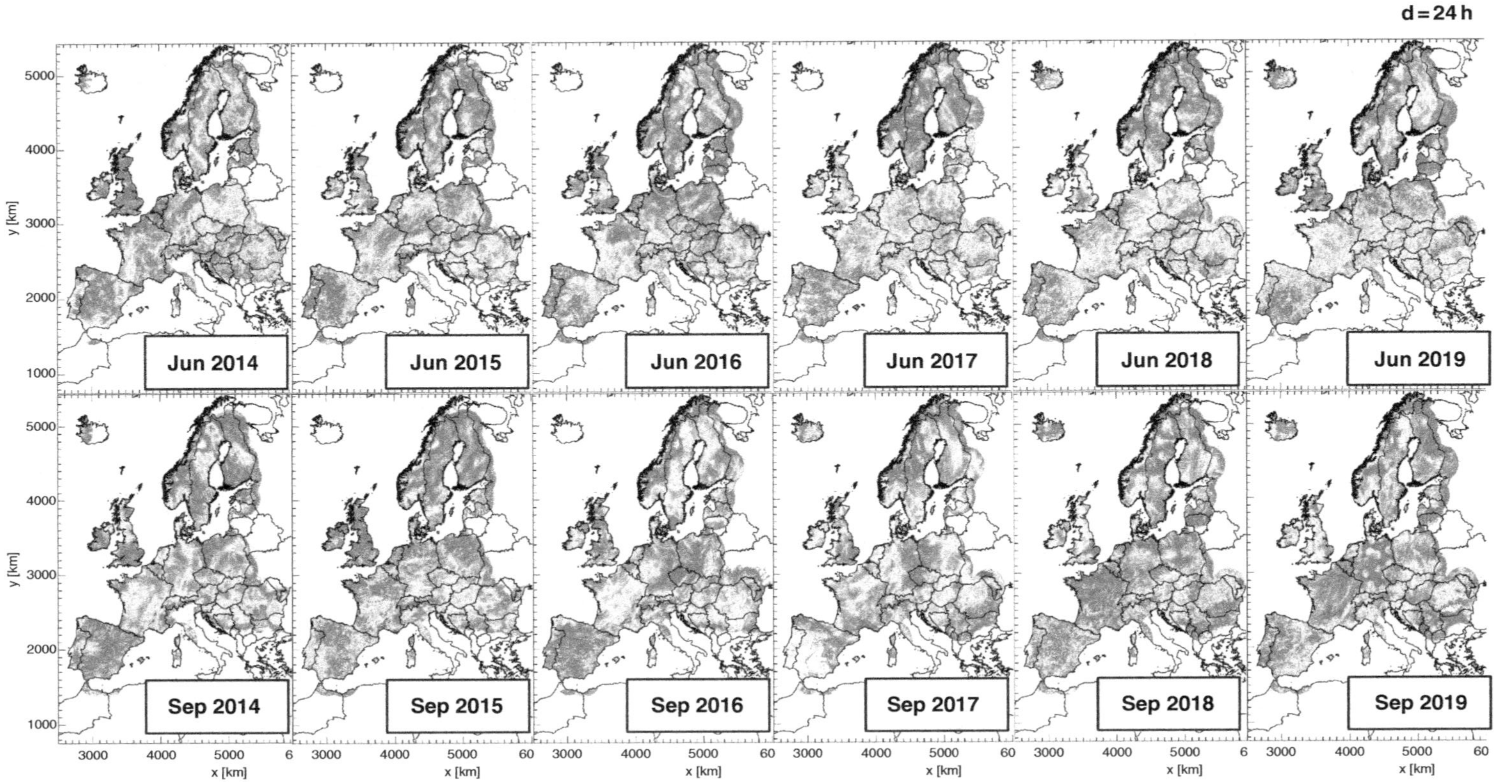

Figure 5.8 Examples of monthly maximum basin-aggregated hourly accumulations obtained over the European drainage network (1 km grid) for the months of June (top) and September (bottom).

they still guide high values and can be updated as more OPERA composites are collected in the future.

- Uncertainty assessment to support the decision-making during emergency management with the generation of probabilistic hazard forecasts. The system has been adapted to account for the uncertainty in precipitation inputs using ensemble precipitation forecasts obtained with (i) the radar-based nowcasting technique SBMcast (Berenguer et al. 2011), or (ii) the system for seamless blending of SBMcast nowcasts and NWP ensembles forecasts developed in the SMUFF project.

Flash flood impact assessment at the regional, national or continental scale (Ritter et al. 2020, 2021) developed in the framework of the ANYWHERE Multi-Hazard Early Warning System (www.anywhere-h2020.eu; van Lanen et al. 2019). Impact assessment is based on the combination of the ERICHA flash flood hazard with the hazard (regional), vulnerability and exposure datasets developed by the Member States in the framework of the EU Floods Directive.

References

Alfieri, L. and Thielen, J. (2015). A European precipitation index for extreme rain-storm and flash flood early warning. *Meteorological Applications* 22 (1): 3–13. https://doi.org/10.1002/met.1328.

Alfieri, L., Berenguer, M., Knechtl, V. et al. (2019). Flash flood forecasting based on rainfall thresholds. In: *Handbook of Hydrometeorological Ensemble Forecasting* (ed. Q. Duan, F. Pappenberger, J. Thielen, et al.), 1–38. Springer Berlin Heidelberg.

Berenguer, M., Corral, C., Sanchez-Diezma, R., and Sempere-Torres, D. (2005). Hydrological validation of a radar-based nowcasting technique. *Journal of Hydrometeorology* 6 (4): 532–549.

Berenguer, M., Sempere-Torres, D., and Pegram, G.G.S. (2011). SBMcast – an ensemble nowcasting technique to assess the uncertainty in rainfall forecasts by Lagrangian extrapolation. *Journal of Hydrology* 404: 226–240.

Borga, M.E.N., Anagnostou, G., and Blöschl, J.D.C. (2011). Flash flood forecasting, warning and risk management: the HYDRATE project. *Environmental Science & Policy* 14: 833–844.

Casajus Valles, A., Marin Ferrer, M., Poljanšek, K., and Clark, I. (ed.) (2020). *Science for Disaster Risk Management 2020: Acting Today, Protecting Tomorrow*'. EUR 30183 EN. Luxembourg: Publications Office of the European Union. ISBN: 978-92-76-18181-1 https://doi.org/10.2760/438998. JRC114026.

Cloke, H., Di Baldassarre, G., Landeg, O. et al. (2017). Hydrological risk: floods. In: *Science for Disaster Risk Management 2017: Knowing Better and Losing Less*. EUR 28034 EN (ed. K. Poljanšek, M. Marín Ferrer, T. De Groeve, and I. Clark). Luxembourg: Publications Office of the European Union. Chapter 3.4, https://doi.org/10.2788/688605.

Cornes, R., van der Schrier, G., van den Besselaar, E.J.M., and Jones, P.D. (2018). An ensemble version of the E-OBS temperature and precipitation datasets. *Journal of Geophysical Research: Atmospheres* 123: https://doi.org/10.1029/2017JD028200.

Corral, C., Berenguer, M., Sempere-Torres, D. et al. (2019). Comparison of two systems for regional flash flood hazard assessment. *Journal of Hydrology* 572: 603–619. https://doi.org/10.1016/j.jhydrol.2019.03.026.

Delobbe, L. and Lukach, M. (2017). Monitoring rainfall accumulations. Final report, OPERA Development Task OD11b. https://www.eumetnet.eu/wp-content/uploads/2019/01/Monitoring_rainrates_OPERA.pdf (accessed 18 September 2017).

Doswell, C.A., Brooks, H.E., and Maddox, R.A. (1996). Flash flood forecasting: an ingredients-based methodology. *Weather and Forecasting* 11: 560–581.

EEA Reports (2017). *Climate Change Adaptation and Disaster Risk Reduction in Europe Enhancing Coherence of the Knowledge Base, Policies and Practices*. Luxembourg: European Environment Agency, Publications Office of the European Union ISSN 1977-8449, https://doi.org/10.2800/938195.

Germann, U. and Zawadzki, I. (2002). Scale-dependence of the predictability of precipitation from continental radar images. Part I: description of the methodology. *Monthly Weather Review* 130 (12): 2859–2873.

EEA, Briefing no. 21/2021 (2021). Economic losses and fatalities from weather- and climate-related events in Europe. http//doi: 10.2800/530599.

Guillot, P. and Duband, D. (1967). La méthode du Gradex pour le calcul de la probabilité des crues à partir les pluies. In: *Proceedings Fort Collins Symposium*, vol. 1, 506–515. Fort Collins, CO: Colorado State University.

Hapuarachchi, H.A.P., Wang, Q.J., and Pagano, T.C. (2011). A review of advances in flash flood forecasting. *Hydrological Processes* 25: 2771–2784.

Harrison, G., Hewson, T., and Roberts, N. (2017). Meteorological risk: extra-tropical cyclones, tropical cyclones and convective storms. In: *Science for Disaster Risk Management 2017: Knowing Better and Losing Less*. EUR 28034 (ed. K. Poljanšek, M. Marín Ferrer, T. De Groeve, and I. Clark). Luxembourg: EN, Publications Office of the European Union. Chapter 3.7, https://doi.org/10.2788/688605.

Haylock, M.R., Hofstra, N., Klein Tank, A.M.G. et al. (2008). A European daily high-resolution gridded data set of surface temperature and precipitation for 1950-2006. *Journal of Geophysical Research: Atmospheres* https://doi.org/10.1029/2008JD010201.

Huuskonen, A., Saltikoff, E., and Holleman, I. (2014). The operational weather radar network in Europe. *Bulletin of the American Meteorological Society* 95: 897–907.

Kidd, C., Huffman, G.J., Becker, A. et al. (2017). So, how much of the Earth's surface is covered by rain gauges? *Bulletin of the American Meteorological Society* 98: 69–78. https://doi.org/10.1175/BAMS-D-14-00283.1.

Kirk, P.J., Clark, M.R., and Creed, E. (2021). Weather observations website. *Weather* 76: 47–49.

Kreienkamp, F., Philip, S.Y., Tradowsky, J.S. et al. (2021). *Rapid Attribution of Heavy Rainfall Events Leading to the Severe Flooding in Western Europe During July 2021*. World Weather Attribution (WWA). http://hdl.handle.net/1854/LU-8732135 (accessed 23 August 2021).

van Lanen, H.A.J. S. M. Schauwecker, M. Stoffel et al. (2019). Final report describing: (i) Uncertainty of forecasted weather- related hazards, (ii) robustness under worst-case climate change projections and/or coinciding natural hazards, (iii) operationalizability, and (iv) impact forecasting. ANYWHERE Deliverable 2.5. http://www.anywhere-h2020.eu/wp-content/uploads/docs/D2.5_submitted.pdf (accessed 31 December 2019).

Lavers, D.A., Harrigan, S., and Prudhomme, C. (2021). Precipitation biases in the ECMWF integrated forecasting system. *Journal of Hydrometeorology* 22: 1187–1198.

von Lerber, A., Park, S., Berenguer, M. et al. (2021). Precipitation-type nowcasting and evaluation of precipitation products. *TAMIR Deliverable* 2 (1): 33.

Lopez, P. (2014). Comparison of ODYSSEY precipitation composites to SYNOP rain gauges and ECMWF model. *ECMWF Technical Memorandum* 717: 1–17.

Marshall, J.S. and Palmer, W.M.K. (1948). The distribution of raindrops with size. *Journal of Meteorology* 5 (4): 165–166.

O, S., Foelsche, U., Kirchengast, G. et al. (2017). Evaluation of GPM IMERG early, late, and final rainfall estimates using WegenerNet gauge data in southeastern Austria. *Hydrology and Earth System Sciences* 21: 6559–6572. https://doi.org/10.5194/hess-21-6559-2017.

Park, S., Berenguer, M., Sempere-Torres, D. et al. (2016). The quality evaluation of OPERA quantitative precipitation estimates. *ERICHA Deliverable* A.1: 26pp.

Park, S., Berenguer, M., and Sempere-Torres, D. (2019). Long-term analysis of gauge-adjusted radar rainfall accumulations at European scale. *Journal of Hydrology* 573: 768–777. https://doi.org/10.1016/j.jhydrol.2019.03.093.

Raynaud, D., Thielen, J., Salamon, P. et al. (2015). A dynamic runoff co-efficient to improve flash flood early warning in Europe: evaluation on the 2013 central European floods in Germany. *Meteorological Applications* 22: 410–418.

Rehfeldt, K., Schirmeister, Z., Pichon, D. et al. (2021). The CEMS meteorological data collection centre – annual report 2020. European Commission, JRC126414.

Ritter, J., Berenuger, M., Corral, C. et al. (2020). ReAFFIRM: real-time assessment of flash flood impacts – a regional high- resolution method. *Environment International* 136: https://doi.org/10.1016/j.envint.2019.105375.

Ritter, J., Berenguer, M., Park, S., and Sempere-Torres, D. (2021). Real-time assessment of flash flood impacts at pan-European scale: The ReAFFINE method. *Journal of Hydrology* 603: 127022.

Saltikoff, E., Kurri, M., Leijnse, H. et al. (2017). Maintenance keeps radars running. *Bulletin of the American Meteorological Society* 98: 1833–1840. https://doi.org/10.1175/BAMS-D-16-0095.1.

Saltikoff, E., Haase, G., Leijnse, H. et al. (2018). OPERA past, present and future. *Proc. 10th European Conf. on Radar in Meteorology and Hydrology*, 025, Ede-Wageningen, The Netherlands (1–6 July 2018). https://www.researchgate.net/profile/E_Saltikoff/publication/326552217_OPERA_-_ past_present_a.

Saltikoff, E., Haase, G., Delobbe, L. et al. (2019). OPERA the Radar Project. *Atmosphere* 10: 320. https://doi.org/10.3390/atmos10060320.

Sempere-Torres, D., Velasco, D., Berenguer, M. et al. (2010). *Improving Risk Management for Flash Floods and Debris Flow Events. A Book Chapter of The Water Framework Directive: Action Programmes and Adaptation to Climate Change* (ed. P. Quevauviller, U. Borchers, K.C. Thompson, and T. Simonart), 139–169. The Royal Society of Chemistry, ch. 4.2. http://pubs.rsc.org/en/content/ebook/978-1-84973-053-2.

Smith, P., Pappenberger, F., Wetterhall, F. et al. (2016). Chapter 1255 11 – on the operational implementation of the european flood awareness system (EFAS). In: *Flood Forecasting* (ed. T.E. Adams and T.C. Pagano), 313–348. Boston, https://doi.org/10.1016/B978-0-12-801884-2.00011-6: Academic Press.

Thiemig, V., Gomes, G.N., Skøien, J.O. et al. (2022). EMO-5: A high-resolution multi-variable gridded meteorological data set for Europe (1990-2019). *Earth System Science Data* 14: 3249–3272.

UNDRR (2019). *Global Assessment Report on Disaster Risk Reduction*. Geneva, Switzerland: United Nations Office for Disaster Risk Reduction (UNDRR) eISBN: 978-92-1-004180-5.

Zhang, J., Tang, L., Cocks, S. et al. (2020). A dual-polarization radar synthetic QPE for operations. *Journal of Hydrometeorology* 21: 2507–2521.

6

Towards Impact-Based Communication During Climate Emergencies

A Community-Based Approach to Improve Flood Early Warning Systems

Erika Meléndez-Landaverde, Daniel Sempere-Torres, and Shinju Park

Center of Applied Research in Hydrometeorology, Universitat Politècnica de Catalunya (UPC), Barcelona, Spain

6.1 Introduction

In recent years, record-breaking rainfall events have caused catastrophic consequences worldwide (Cornwall 2021). In Germany and Belgium, severe flooding resulted in hundreds of fatalities, evacuations, power outages, and extensive damage to infrastructure (The Lancet Regional Health 2021). Despite timely rainfall and flood warnings issued by national meteorological and hydrological services (NMHSs) days before the events, they did not guarantee appropriate preparation and protection actions from at-risk areas (Fekete and Sandholz 2021). The alarming number of reports about communities misunderstanding or diminishing the local risk and thus being caught unprepared to effectively respond exposes a critical breach in the local emergency management and communication chain (Netzel et al. 2021). Moreover, heavy precipitation events are expected to increase in frequency and intensity based on climate change projections (IPCC 2021), rapidly transforming extreme rainfall events into the 'new normal'. In this context, early warning systems (EWSs) can be key to enhancing the local preparedness and building resilience for effective community response during emergencies

Responding to Extreme Weather Events, First Edition. Edited by Daniel Sempere-Torres, Anastasios Karakostas, Claudio Rossi, and Philippe Quevauviller.

(UNISDR 2015). Significant advances have been achieved in the capacity and accuracy to forecast intense rainfall episodes and their associated impacts (Kaltenberger et al. 2020). The multi-hazard impact-based EWS (MH-EWS) tools developed and tested during the 18-month pilot demonstrations of the ANYWHERE-h2020 have proven capable of triggering high-resolution rainfall and flood warnings with extended lead time to help enable earlier preparedness (Corral et al. 2019; Ritter et al. 2020). Nevertheless, recent events have shown that regardless of the availability of advanced technology and impact forecasting systems (Cornwall 2021), communities and authorities need urgent support to understand their local risks and vulnerabilities if they are expected to take the necessary actions to ensure their safety.

The shift towards impact-based early warning system (IB-EWS) services can be a promising cost-effective approach to increase the likelihood that recipients will understand their risk and take appropriate protective actions (German Red Cross 2018). The combination of hazard forecast information with vulnerability and exposure data can enable IBEWS to trigger actions that can save lives and reduce damages during emergencies (Schroeter et al. 2021). However, EWSs are traditionally developed in a top-down and hazard-centred style, focusing on the technological developments for risk detection and monitoring and only approaching communities at the end of the design process (Tarchiani et al. 2020). As identified in previous studies, failure to integrate end users and their knowledge into the design and implementation of EWSs can lead to poor reactions and disastrous impacts on vulnerable and exposed communities (Baudoin et al. 2016). Thus, to design a community-based system that can appropriately support the protection actions of citizens and sectors and reduce their risk, communities need clear guidance on how to integrate key aspects of IB-EWS within their local capacities realistically. Elements that must be considered within a community perspective are, but not limited to, the development of partnerships with all relevant stakeholders; identification and understanding of local needs; the collection of local knowledge, vulnerability, and impact data; and the evaluation and verification of IBEWS. As seen in the ANYWHERE-h2020 project, the technological capabilities for running advanced rainfall and flood forecasts are available at a national and regional scale and have been successful (Corral et al. 2019). Therefore, the next step is to downscale these tools to a community level to enhance the local response capabilities to extreme weather-induced events.

This chapter presents the innovative concept of IB-EWS services to trigger actionable decisions at a community scale. Using the rainfall and flood warning algorithms of the ANYWHERE project as a foundation, a framework for adopting them to a local level, named the Site-Specific EWS (SS-EWS), has been developed to co-design, implement and co-evaluate a community-based IB-EWS (Melendez-Landaverde and Sempere-Torres 2022). The SS-EWS framework guides communities and local experts on using hydrometeorological, local vulnerability, impact and exposure information to derive impact-based rainfall thresholds for triggering site-specific warnings (SSWs). These SSWs aim to communicate impacts that community sectors (i.e. transportation) or specific vulnerable locations could experience due to extreme rainfall events alongside appropriate protection actions from official or pre-approved community emergency protocols to reduce them. The SS-EWS has been implemented in two vulnerable communities in the Catalonia region to analyse its performance in real-time and the influence of the SSWs in the protective decision-making process of end-users during emergencies.

The chapter is structured as follows: Section two introduces the general concept of IB-EWS and key elements for their successful design and implementation. Section three presents the SS-EWS framework for rainfall events and its 'community-based' work scheme for implementing a co-design and co-evaluation environment. Section four deals with the results and lessons learned from the SS-EWS framework implementation in two vulnerable communities across the Catalonia region, Terrassa and Blanes. Finally, some concluding remarks and an outlook for the future of community-based IB-EWS are presented in section five.

6.2 Impact-Based Early Warning Systems (IB-EWS) for Actionable Decisions: Key Aspects

In recent years, societies and international organizations have recognized the need for EWS that actively integrate the requirements and local knowledge of sectors at risk (Puzyreva et al. 2022) on the basis that a good and accurate meteorological forecast does not guarantee appropriate protection responses will be taken by exposed communities (The Lancet Regional Health 2021). In this context, IB-EWSs have been promoted as the next step on the EWSs chain to help overcome the disconnections between forecasts and the emergency management chain to enable early actions (Weyrich et al. 2018). In other words, IB-EWSs are services that combine hazard information with vulnerability (and sometimes exposure) data to translate what the upcoming meteorological conditions could mean for areas or vulnerable sectors. Thus, by moving from 'what the weather will be' to 'what the weather will do', IB-EWSs aim to communicate the possible impacts that could be experienced and provide guidance on what actions can be performed to help ensure the safety of recipients (see example in Table 6.1).

Recent research on IB-EWS suggests that warnings coming from these services can influence the decision-making process of end users and that appropriate early action is more likely to be taken when information of potential impacts (Casteel 2016; Potter

Table 6.1 Comparison between traditional hazard-orientated and impact-based warning messages for urban floods.

Hazard-orientated warning	Impact-based warning
A heavy rainfall episode is forecasted for the city of Terrassa. In the next hour, accumulations higher than 35 mm in 30 minutes are expected. Follow the advice of the emergency services and local authorities	A highly dangerous rainfall is forecasted for the city of Terrassa. **Extreme weather conditions can cause significant surface water accumulations and result in traffic delays.** Take the indicated self-protection actions immediately and follow the advice of the emergency services and local authorities For more information on protective actions, check the official communication channels

et al. 2018; Meléndez-Landaverde et al. 2020) and local knowledge is included (Bucherie et al. 2022). In work done by Kox et al. (2018), responders and emergency managers from Germany expressed that they would be better prepared for upcoming weather-induced emergencies if they received timely impact-based information. Furthermore, as presented in the study by Kaltenberger et al. (2020), many European NMHSs are currently working on transitioning from only-hazard-focused EWs to warnings triggered by impact criteria.

Although there are visible benefits to moving towards IB-EWSs, their development and implementation can be challenging. Due to the social element incorporated into the IB-EWS, additional resources must be considered and integrated beyond the conventional hydrometeorological information. For instance, the combination of historical and future impact data (e.g. evacuations, damages, disruptions) with spatial–temporal vulnerability and exposure information must be adequately collected and assessed to develop an effective IB-EWS. Additionally, the warning content of the IB-EWS must be constantly evaluated to determine if they influence the risk perception and understanding of users and thus, support their decision-making process during emergencies.

The following sections introduce key aspects of IB-EWS that should be considered during their design, implementation, and evaluation.

6.2.1 Partnerships for an Effective Co-Design IB-EWS

IB-EWSs aim to support the protection actions of vulnerable societies and their different sectors, including local authorities, citizens, emergency responders, and private and public sectors. In this sense, NMHs have established the necessary expertise to develop and implement advanced systems for detecting and monitoring the hazard parameters (i.e. intensities) of upcoming rainfall events (Alfieri et al. 2019; Schroeter et al. 2021). However, other critical elements for developing IB-EWSs, such as understanding local vulnerabilities, impact data, response capacities and local disaster risk measures, are generally outside the work sphere of these agencies (WMO 2015). Therefore, partnerships and co-creation frameworks are also considered as one of the key aspects of the IB-EWSs design and developing chain. These systems must be built on a foundation of strong collaboration with a wide range of key institutions, agencies and community sectors to (i) incorporate detailed vulnerability and exposure knowledge, (ii) identify communication requirements and needs and (iii) foster trust and ownership of the system (Sufri et al. 2020). All the elements of the IB-EWS mentioned above, especially the ones involving direct interaction with the public (e.g. warning information and behavioural recommendations), can benefit significantly from the combined experience and knowledge that key community partners can incorporate during the development process. For example, disaster reduction and civil protection agencies (DRCPAs) can take the leading role in identifying local risks and their impacts as they usually have information related to community action protocols, vulnerability maps, and past emergency measures for effective disaster response. The NMHs can provide hydrometeorological data on the frequency and observed physical parameters of weather-induced events in the area and indicate the available technological resources for monitoring and forecasting. Similarly, stakeholders from the public and private sectors can express their requirements in terms of emergency communications and provide detailed local impact data and feedback for warning validation. In this context, establishing and maintaining long-term partnerships requires a strong commitment from all key stakeholders.

Elements that must be considered for developing successful partnerships are, but are not limited to, the definition of roles and responsibilities, formalized by official agreements (e.g. Memorandums of Understanding); the number of resources (e.g., time, software and personnel) needed to foster a shared work environment; establishment of official communication between partners; and verification and quality strategies of produced forecasts and triggered warnings (WMO 2012).

6.2.2 End Users: Identifying Needs for Emergency Response

DRR strategies have been negatively affected when local perception was overlooked, as authorities (i.e. DRCPAs) and citizens can understand and view risks differently (Bradford et al. 2012). To develop an IB-EWS that can effectively trigger appropriate self-protection responses and emergency actions from citizens and authorities, it is critical to understand how they will use the warnings and if this information meets their specific needs. As evidenced by past events (Fekete and Sandholz 2021; Netzel et al. 2021), one type of general warning does not fit all community groups since different vulnerability levels, and response capacities affect their ability to react and decrease their overall risk during emergencies. For instance, the transport department might be interested in how an upcoming rainfall event can affect the timetables of public services. At the same time, local authorities might focus on coordinating the emergency actions that might be needed at vulnerable points (e.g. recreational areas around water bodies). Thus, specific impact information and concrete behavioural recommendations within impact-based warnings (IBWs) must be customized for the target audience and sector to enable early planning and action. Moreover, the criteria to establish the appropriate impact-based threshold used for triggering IBWs can also vary due to the risk tolerance level (*What is considered a high and unacceptable impact for this sector?*) not being homogenous across the community (IFRC 2020). To align with the people-centred approach promoted by the Sendai Framework of Disaster Risk Reduction (SFDRR), IB-EWSs must be developed to target a local (even an individual) level as much as possible (Potter et al. 2021).

To collect this detailed understanding of end-user needs for IB-EWS, communities and relevant stakeholders must be involved from the early stages of the development. The local risk perception and expectations of the target audience regarding emergency management and response have to be assessed to design a tailored IB-EWS that can be useful for them. In this sense, the knowledge and expertise of social and behavioural sciences have a primary role in assessing end users' perceived risks and how they comprehend, utilize, and respond to warning information (Rogers et al. 2019). Moreover, their insights and data collection methodologies (e.g. surveys and interviews) can be valuable for identifying needs and developing a targeted and helpful system including people-centred approaches.

6.2.3 Risk Identification and Impact Data Collection

To design and verify IB-EWSs is critical to identify and gather local data regarding the hazards, vulnerabilities, exposure and specific impacts and area or site might experience. This information is essential to building an IB-EWS better aligned with the risk profile and response capabilities of exposed communities. However, their poor systematic collection and limited availability have been identified as the main challenges in implementing and evaluating IB-EWS (Kaltenberger et al. 2020). Nevertheless, a combination of different approaches and data sources is suggested by previous studies to collect current and historical information (Harrison et al. 2022; Lalancette and Charles 2022).

For the local hazard identification in the context of rainfall episodes, NMHs and research institutes generally have databases of the historical events and their observed hydrometeorological parameters for specific regions (Tschoegl et al. 2006; Kaltenberger et al. 2020). In real-time, forecasts from numerical weather prediction models and radar-based nowcasting can help monitor weather phenomena and localized rainfall-induced events, especially in areas with poor or insufficient monitoring networks (Corral et al. 2019; Park et al. 2019).

Information sources of impacts triggered by extreme rainfall events can be divided between official and unofficial. In this sense, official impact data refers to the information collected and organized by the recognized authorities or emergency response groups of a community involved in the disaster reduction activities, such as civil protection, police, and fire departments. Conversely, unofficial data can come from media coverage networks (e.g. local news reports), social media, and reports coming from citizens and volunteers (Barriendos et al. 2014; Taylor et al. 2015). Due to the challenges of verifying in situ impact information, the combination of both data sources is valuable for creating local and regularly updated impact databases. These usually include detailed descriptions of previously observed impacts such as flood extent areas, number of evacuations and 112 calls, injuries, loss of life, damage to buildings and disruption of public services. Therefore, impact databases can provide a general picture of the recurrent impacts in the area, their triggers and hotspots during emergencies. In the context of IB-EWS for extreme rainfall events, impact databases are critical for defining impact-based rainfall thresholds to use as warning triggers.

Vulnerability and exposure are situation-specific, interacting with the hazard to dynamically increase (or decrease) the risk of communities (WMO 2021). Therefore, it is essential to assess them within the local context of communities to understand how susceptible and resilient they are to face the impacts of extreme rainfall events. Vulnerability indicators can be developed using public census (e.g. land-use, infrastructure), demographic statistics and economic and ecologic datasets. These are usually collected and stored by public agencies, research institutes and risk and insurance specialists (Aroca-Jimenez et al. 2017; Mason et al. 2021). Exposure is typically defined by overlaying the target area, community or site assets with hazard information such as flood extends or sea-level rise. However, the dynamic nature of exposure and vulnerability can make their up-to-date assessment challenging. Non-traditional data sources to assess exposure in real time are being studied, with their implementation now limited to increasing the situational awareness of authorities (Harrison et al. 2021).

6.2.4 Evaluation of IB-EWSs

The evaluation of any type of EWS traditionally involves the statistical analysis between forecasts and observations to measure specific components of the EWS, such as the timeliness and accuracy of the system (WMO 2013). However, since IB-EWSs aim to improve the local risk understanding of users and trigger appropriate actions, additional indicators must be incorporated into their evaluation. Therefore, beyond collecting quantitative observed impact data (e.g. number of houses flooded), qualitative evaluation methods such as interviews, focus groups and surveys must be implemented to obtain feedback from end users and stakeholders. As suggested by international guidelines (IFRC 2020; WMO 2021), IB-EWS can make use of social sciences to evaluate the clarity of the IBWs and the understanding of users (e.g. *Was the warning easily understood? Were there any misinterpretations?*); usefulness (e.g. *Were the warnings helpful during the emergency? Why?*); and the effectiveness of the warning to trigger responses (e.g. *Was protective action initiated? If there were actions taken, were they appropriate?*). Some NMHs, such as the Met Office and the Australian Bureau of Meteorology, are currently implementing evaluation forms to score the effectiveness of the IBWs triggered during specific events from the point of view of stakeholders (IFRC 2020).

6.3 The Next Step for Community-Based EWS: The Site-Specific EWS Framework (SS-EWS)

The results from the 18-month pilot demonstrations of the ANYWHERE-h2020 project have shown that we currently have the technological platforms and advanced rainfall and flood forecasting algorithms available to enhance the response capabilities of national and regional DRCPAs (Corral et al. 2019; Ritter et al. 2020). Recent studies (Amores et al. 2020; Bosseler et al. 2021) demonstrated that accurate and timely warnings do not guarantee an appropriate response from at-risk citizens (Casteel 2016) despite the advancement of recently developed rainfall-based flood warning tools, algorithms, and forecast models. Hence, EWSs cannot be seen exclusively as a technological solution for emergency management but as a system that interacts with the communities and authorities across a social context to support their collective (and individual) decision-making processes (Sufri et al. 2020). In this context, rainfall and flood warning services until recently have been developed with a strong focus on improving the tools, algorithms and forecast models with little to no engagement of targeted end users (Tarchiani et al. 2020). Consequently, these systems traditionally have a solid technological foundation but can be weak in their risk communication strategies and in integrating appropriate emergency response and preparedness measures (Baudoin et al. 2016).

As promoted by the Hyogo and Sendai international frameworks (UNISDR 2015), communities must be active participants in the design and implementation of DRRs measures such as IB-EWSs to incorporate their local knowledge and specific emergency requirements and thus increase the probability that appropriate actions will be taken to reduce or avoid negative impacts.

6.3.1 The Site-Specific Early Warning System Framework (SS-EWS)

As seen in Figure 6.1, the next innovative phase in the EWS chain is to take the available rainfall and flood warning services and rethink their design from a community-based perspective by focusing on the specific vulnerabilities of exposed communities. These efforts can enable IBEWSs systems to interpret the upcoming forecasts within a community context and translate them into concrete and actionable decisions that can be undertaken to decrease their risk during emergencies. Melendez-Landaverde and Sempere-Torres (2022) have recently developed a community-based Site-Specific Early Warning System (SS-EWS) as a step forward in the line of IB-EWSs. The SS-EWS aimed to guide communities and their local representatives (e.g. authorities, decision-makers, vulnerable groups) in how to co-design, implement and co-evaluate a rainfall-based IB-EWS for vulnerable points (e.g. a school in a flood zone area) or community sectors (e.g. transport).

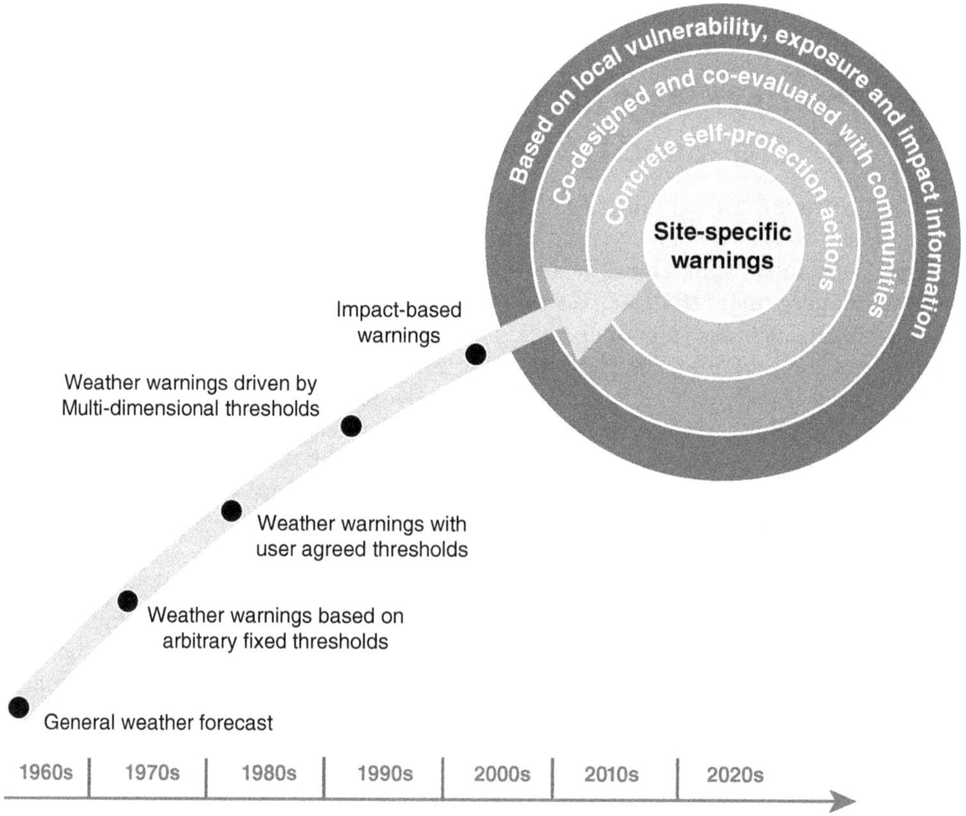

Figure 6.1 The next step in the evolution of early warning systems: the site-specific warnings. *Source*: Adapted from Meiyan et al. 2015.

To achieve this, the framework SS-EWS proposes steps to utilize available meteorological data from radar-based nowcasting (up to 2 hours lead-time) and numerical weather prediction models (forecasts up to 48 hours) with local vulnerability and impact information (see Section 6.2.3) to derive specific impact-based rainfall thresholds. Thus, the system works under the assumption that beyond these defined hydrometeorological thresholds, previously seen impacts in the area might be possible. For these thresholds, the framework guides communities in developing hazard-impact databases containing the observed parameters (e.g. intensity or return periods) of past rainfall events and the triggered local impacts at specific sectors or points. Categories within these impact databases can vary according to the area and specific community sector exposed (e.g., transport, the general public), but they commonly include descriptive aspects of previous events such as the number of evacuations, injuries, loss of life and 112 calls; damages to buildings; and disruption of public services. These databases are subsequently co-evaluated with community representatives on an impact severity three-level scale (i.e. low, medium and high) to establish the critical warning thresholds based on impacts, coupled with the frequency (return periods) and observed rainfall accumulations. Moreover, advisory tables are co-developed to identify the specific protection actions from pre-approved emergency protocols that should be performed at each warning level. This assessment allows the SS-EWS to integrate the risk tolerance and knowledge from communities and, thus, understand the upcoming rainfall events in a local context to trigger appropriate warning levels and emergency actions for specific vulnerable groups (see more in Melendez-Landaverde and Sempere-Torres 2022). The final step of the SS-EW framework is to communicate this emergency information alongside the concrete actions from established community protocols that citizens and local authorities have co-designed through all the previous steps. For this, site-specific warnings (SSWs) are developed and disseminated across local available platforms (e.g. mobile apps or social media).

To carry out the co-design, implementation and co-evaluation of the warning system, the SS-EWS framework adopts a community-based approach consisting of quantitative and qualitative data collected from the community sessions composed of work meetings, focus groups, interviews, and online surveys with selected local stakeholders, hereafter community representatives. These sessions are constructed to actively engage the communities and guide them in co-designing the SS-EWS components (i.e. hazard-impact databases, impact-based rainfall thresholds, advisory tables SSWs). In this context, enabling communities to generate and establish their own criteria provides insight into what the users and vulnerable groups are hoping to achieve from adopting a community-based SS-EWS (Maskrey et al. 2019). Moreover, although through the co-design process the SS-EWS components are being constantly discussed, a designated number of community sessions are exclusively dedicated to evaluating the real-time performance of the warning system with community representatives. In this regard, the amount and frequency of the co-evaluation sessions can follow the annual rainy season, after mid-to-high impact rainfall events criteria, or if changes in the local requirements have been made (IFRC 2020).

It is challenging in an impact-based approach that these evaluations are often beyond whether recipients approve or 'like' the impact-based warnings (Potter et al. 2018), requiring specific criteria. For example, the online evaluation surveys of the SS-EWS framework are dedicated to understanding if these warnings positively influence the protective decision-making process and local risk understanding of users during emergencies compared to the current alerting system in the area. Finally, the dissemination channels of the SSWs are also evaluated through online experimental demonstrations.

6.4 The SS-EWS in Catalonia, NE Spain: Experiences and Lessons Learned

The SS-EWS framework has been applied to two representative cities of Catalonia, Terrassa and Blanes. Both cities have experienced extreme floods (due to enhanced convective rainfall events located over the mountainous coastal basins) that have resulted in numerous casualties and significant socio-economic damages (Martín-Vide and Llasat 2018; Navarro 2020). For instance, the record-breaking storm Gloria in 2020 caused the Tordera river next to Blanes to overflow along its banks while its high discharge (600 $m^3\ s^{-1}$) destroyed the two bridges connecting the city by car and train with its neighbour municipality, Malgrat de Mar.[1] Based on climate projections (IPCC 2021), these extreme rainfall events and their negative socio-economic consequences are only expected to increase in the region.

Therefore, due to their high vulnerability, Catalonia has started to implement climate adaptation strategies to increase the resilience of its most vulnerable areas. In 2015, The Generalitat of Catalonia passed a new law requiring all activities within risk areas to develop and implement a self-protection plan to be activated by a local EWS in case of emergencies (Departamento de Interior 2015). In this context, the Department of Interior of Catalonia (INT) and the cities of Terrassa and Blanes actively participated in the EU H-2020 Innovation Action ANYWHERE project focused on improving the anticipation and response of extreme weather-induced events. Within the project, local authorities of these three regions were able to implement real-time tools from the MH-EWS to support their decision-making process during emergencies. Taking advantage of these initiatives established over Catalonia, the cities of Terrassa and Blanes participated as local pilot sites in the different stages of the SS-EWS implementations. The implementation includes connecting the EWS platforms to the appropriate self-protection actions needed by vulnerable communities, which can be as a reference point for the best practices towards climate resilience.

Terrassa city served as the first local evaluation for the real-time implementation of the SS-EWS system and the mobile application developed to disseminate the SSWs warnings. As active stakeholders in the H2020 ANYWHERE project, the municipality implemented a city-scale version front-end platform of the ANYWHERE MH-EWS, ARGOS-City, for proactive weather-induced emergency management. In parallel, the SS-EWS framework was implemented at three vulnerable points, one public school located in a flood zone area and two ford crossings. As a result of this work synergy, the current ARGOS system in Terrassa operates in real time with the co-designed impact-based rainfall thresholds (linked to specific self-protection actions) and the warning levels coming from the SS-EWS co-design process.

Similarly, Blanes implemented a location-specific prototype of the MH-EWS products for 14 camping sites in the delta of the Tordera river basin[2] called ANYWHERE A4Campsite. This platform was able to provide meteorological information in real-time alongside protection actions. However, unlike the ARGOS-City in Terrassa, the platform and warning levels were not driven by local impacts nor co-produced with the community-based SS-EWS framework. Hence, designing the SS-EWS at Blanes also included additional steps of developing rainfall and flood impact-based warning messages (SSWs). The content SSWs were evaluated alongside their influence on the risk perception and decision-making process of recipients in contrast to the official warnings issued by the official meteorological agency in the region.

The experiences and lessons learned in Terrassa and Blanes are presented in the following sections.

6.4.1 Community-Based Sessions in Terrassa: The Co-Design Process and Experiences

Understanding the potential users and their emergency needs is critical to developing a warning system that delivers useful warnings for triggering actionable decisions. In Terrassa, a group of end-user stakeholders representing different community sectors (i.e. community representatives) were invited to participate in the co-design and co-evaluation process of the local SS-EWS system. This group included the local civil protection authorities, members of the police force, fire-fighters, road and transport operators, and members of the local public school system alongside experts in community resilience and EWS developments. In this sense, the task of recruiting the relevant community representatives was possible due to the involvement of the civil protection authorities of Terrassa. Their leadership, support, risk knowledge and pre-established partnerships with the different community sectors were essential to creating an active interdisciplinary group that could represent a community of Terrassa and the vulnerable locations for the co-design process.

From mid-2020 to late 2021, seven general meetings and five focus groups were held to co-design the elements of the SS-EWS framework mentioned in Section 6.3.1. The meetings with the group were dedicated to understanding and exposing the current warning needs and gaps, presenting the SS-EWS framework and work plan, the progress made regarding the co-design stages. Most importantly, such personal interactive sessions were essential to coordinate the collection of local inputs as well as locally adapted and tailored warning outputs. Challenges regarding the seven general meetings mainly focused on scheduling aspects to gather all representatives as these reunions were more informative in nature.

On the other hand, the focus groups served to discuss and make collective decisions regarding the local risk evaluation and the direction of the SS-EWS for the community and sites. These dialogues were fundamental in helping to build a joint concept with representatives for managing emergencies in Terrassa and exposing any gaps or current critical challenges in emergency response. The first focus group session concentrated on guiding the community to identify the vulnerable points or hotspots (based on local experience); i.e. a public school located in a flood-prone area;

two ford crossings at one of the main rivers in the city (Palau), where a citizen lost his life while driving during an intense rainfall episode in 2018; and the public of Terrassa as a community sector.

The second focus group session helped organized previous rainfall events and the local impacts experienced in the area. However, to better guide the discussion and facilitate the collection of local knowledge data, a preliminary online mix-method survey was conducted beforehand to the community representatives (see more Melendez-Landaverde and Sempere-Torres 2022). The survey included questions and Likert-scale statements to understand their perception regarding the current EWS at Terrassa, its advantages, disadvantages and areas of improvement (Bradford et al. 2012; Lalancette and Charles 2022). Additionally, a section of the survey was dedicated to classifying (from low to high) previously seen local impacts and identifying examples of past triggering events (i.e. dates). The local information gathered from the online survey and sessions one and two helped create hazard-impact databases for Terrassa and the identified specific sites. These databases were complemented by information coming from official historical rainfall data, frequency analysis, official flood maps, regional and local news outlets, social media posts, and insurance data claims from the Spanish Insurance Compensation Consortium. A visible challenge in this co-design stage was guiding the discussions for establishing the impact-severity criteria for the selected locations and Terrassa. As expected, community representatives had different concerns and views based on their background, personal experiences, risk tolerance and work responsibilities (Uccellini and ten Hoeve 2019). As a previously identified challenge of IB-EWSs (WMO 2021), not all the requirements of stakeholders can be integrated, and a balance must be reached between the diverse communication needs and actions during emergencies and the available resources. In this sense, instead of developing a general EWS where essential requirements and vulnerabilities could be overseen, the SS-EWS framework aims to help communities identify vulnerable points to create specific and localized IB-EWSs adapted to their needs. In this context, the discussions for classifying the severity of impacts focused on the previous rainfall experiences of the selected locations while prioritizing the opinions and requests from the representatives linked to these sites. This same approach was followed in the third and fourth focus group sessions devoted to defining the impact-based rainfall thresholds, with the community representatives from the vulnerable sites (e.g. the school) having the leading role in defining these aspects for triggering their SSW levels. However, when moving from the individual vulnerable locations to the co-design of the SSWs for Terrassa city, a compromise had to be reached in the initial implementation sector. During emergencies, this segmentation (dividing users into subgroups based on their characteristics and needs) between sectors and vulnerable locations can result in a dynamic system with SSWs issued simultaneously across the community to support the response actions of specific groups. For Terrassa, the developed SSW levels were designed to target the general public sector and the impacts they could experience in an urban setting (difficulties in mobility or danger to life caused by rainfall-induced urban floods). Beyond providing behavioural recommendations for citizens, sectors such as the civil protection in Terrassa could link these impact-based rainfall thresholds to specific emergency actions needed (e.g. rescue activities) to be better prepared once a SSW is activated for the area of Terrassa. The SSW functionality based on SS-EWS allows the integration of emergency actions from the civil protection unit to support the public in a manner that local communities need to enable rapid reactions (Morss et al. 2016). The fifth and last session was dedicated to designing the content for the SSW messages. For the design of the SSW messages, an effort was made to include local knowledge, concepts or words within the warnings to help increase the personalization and to exclude technical jargons (Rogers et al. 2019). Table 6.2 presents the impact-based rainfall thresholds, the final warning message, potential impacts and specific protection actions for the SSW level 2 (i.e., alert) developed for the school in Terrassa.

Table 6.2 The impact-based rainfall thresholds, potential impacts, the warning message and the specific protection actions for the site-specific warning (SSW) level 2: alert developed for the school in Terrassa, Spain.

SSW Level 2: Alert			
Impact-based rainfall thresholds	Specific impacts	Warning message	Protection actions
24 hours mm^{-1} > = 60 mm 12 hours mm^{-1} > = 40 mm 30 minutes mm^{-1} > = 25 mm	Possible localized water accumulations at the school grounds, affecting the classrooms located in the east modules	A **dangerous heavy rainfall episode** is forecasted for the school area. Weather conditions **can cause localized water accumulations in the school playground and low points** Additionally, the heavy rainfall may cause **urban flooding in the vicinity of the school.** Take the recommended protective actions and follow the advice of the emergency services and local authorities	1. Take the water levels at the identified low point of the school. If the height of the water is greater than or equal to 10 cm: • Call civil protection to report the current situation and authorize the evacuation of the modules in the lower part of the school • In case of evacuation, mobilize the students to the public library • Cancel all outdoor activities • Prepare the emergency kit

6.4.2 Community-Based Emergency Response: SS-EWS Real-Time Application in Terrassa

Since late 2020, the ARGOS platform (where the H2020 ANYWHERE MH-EWS runs in real-time) has operated in Terrassa with the SSW levels, providing the warnings for urban, river, and flash flooding. After the initial co-design process of SS-EWS in-line with the ARGOS warnings, an iterative loop of co-evaluation began in the area to evaluate and improve the inputs and outputs for the real-time SS-EWS demonstration. Specifically, the co-evaluations focused on (i) revising the early actions from pre-defined self-protection plans triggered by the SSWs (*Did the system trigger appropriate specific self-protection actions for each event?*), and (ii) adjusting the impact-based rainfall thresholds based on the forecasted hazard and the observed impacts.

For the evaluation period of 2021, the SSWs notified about upcoming hydrometeorological events with enough time, with most reported impacts at the sites and Terrassa falling within the pre-defined impact severity categories. Moreover, in the co-evaluation sessions held during this period, community representatives expressed their satisfaction with the activated warnings and, more importantly, the specific actions triggered by the system during rainfall events. However, adjustments (i.e. increase of rainfall threshold based on the historical rainfall records at specified sites) had to be made to the SSW levels associated with low impacts to reduce the overload of warning notifications for minor events and thus reduce the false alarms. Furthermore, in connection to the local civil protection emergency management tools such as the site of the ford crossings with newly installed automatic barriers, additional rainfall thresholds obtained from the physical sensors (e.g. river gauges nearby) were also applied. Currently, the system is expected to be reviewed after every rainfall season or any extreme event in the area based on changes to the local vulnerability or user requirements.

6.4.3 The Site-Specific Warnings (SSWs): Their Influence on the Risk Perception and Understanding of Users in Blanes

By including local impact information and self-protection actions in warnings, the SSWs recipients are assumed to improve their local risk perception and understanding of meteorological events and thus perform appropriate responses (Ripberger et al. 2015; Casteel 2016; Meléndez-Landaverde et al. 2020). Similarly done in the previous studies on IBWs (Potter et al. 2018; Weyrich et al. 2018; Taylor et al. 2015), the evaluation in Blanes focused on identifying the influence of the SSW messages on the perceptions, understanding and confidence of participants while making self- and community-protective decisions in a real-case flood scenario (the 23–24th October 2011 event) within an online experimental survey (see Lucena 2021). In the survey, participants (mostly the local civil protection and first responders) took the role of the principal at the public school in Blanes where a prototype of the SS-EWS was developed. While taking this role, they had to decide whether to perform or not the official specific protection actions linked to the school based on the warning information they had just

Table 6.3 Statements to evaluate the variables influencing the decision-making process and protective behavioural intention of participants during emergencies.

Variables	Statement	Response options
Credibility	S1: I believe the message to be credible	1 = strongly disagree 2 = disagree 3 = neither agree nor disagree 4 = agree 5 = strongly agree
Concern	S2: the warning makes me concerned for my safety	
Overall understanding	S3: the warning is easy to understand	
Impact understanding	S4: I find it easy to understand the possible impacts of the weather event	
	S5: based on the warning, I understand the threats to my safety	
Self-protection action understanding	S6: based on the warning, it is clear how I should modify my behaviour	
	S7: the warning gives me enough information to take actions	
Usefulness	S8: I believe the message to be useful to my decision-making process	
Confidence	S9: I feel confident about the decisions taken	

received (see Table 6.3). Two types of warnings were evaluated in this survey: the newly implemented impact-based SSWs and the existing hazard-driven (i.e., no local impact information provided) official warnings from the Meteorological Service of Catalonia (SMC). After being presented with the rainfall scenario and warnings, participants were asked questions to explore the variables identified in previous research to influence the decision-making process and protective behavioural intention of end users during emergencies (see Table 6.3). These variables were *assessed through positively framed statements and a five-point Likert scale from 'I completely* disagree' to 'I completely agree'.

The results obtained from the survey (In total, 47 participants) were analysed using a paired-samples t-test by assuming the Likert scale to be linear (Potter et al. 2018). The type of warning (SSWs vs SMC) was chosen as an independent variable to study its effects on the different selected attributes of Table 6.3. The p-values < 0.05 were considered in this test to be statistically significant and the effect size is presented by the Cohen's D. As seen in Table 6.4, the SSWs scored higher than the official SMC warnings in the nine selected variables regarding warning credibility, level of concern, overall understanding, impact and self-protection understanding, warning usefulness and confidence. In this sense, the SMC official warnings generated in participants a neutral standpoint (3 or 'I neither agree nor disagree') across the nine statements, while the SSWs pushed this stance on average 1.4 towards higher levels of agreement (4 or 'I agree'). The results indicate that participants perceived the information coming from SSWs to be more credible and easier to understand in terms of potential local impacts and what self-protection actions they should perform in contrast to the SMC warnings. Furthermore, recipients were more concerned about their safety and more confident in the decisions to be made during the emergency scenario with SSWs. Notably, the most significant increase in mean scores between SSWs and SMC (1.47) appears with the

Table 6.4 Results from the influence of the site-specific (SSWs) and the official SMC warnings on the variables in Table 6.4.

Statement	SSWs		SMC		*p*-value	Cohen's D
	Mean	SD*	Mean	SD*		
S1	4.45	0.74	3.73	0.86	<0.01	0.89
S2	4.35	0.69	3.28	1.26	<0.01	1.2
S3	4.41	0.67	3.43	0.96	<0.01	0.8
S4	4.45	0.74	3.31	1.14	<0.01	1.04
S5	4.43	0.5	3.53	0.84	<0.01	0.92
S6	4.31	0.74	3.2	1.09	<0.01	1.16
S7	4.2	0.71	2.73	1.11	<0.01	1.17
S8	4.43	0.61	2.96	0.93	<0.01	1.08
S9	4.02	0.59	3.25	0.88	<0.01	0.89

SD: Standard deviation*

statements representing understanding of what self-protection action should be performed and warning usefulness (S7 and S8, respectively).

A recurrent criticism in the context of EWSs is their difficulty in effectively delivering understandable risk information to trigger appropriate protection actions from communities (Ching et al. 2015; Uccellini and Hoeve 2019). As reported in recent events, this has contributed to widespread damage, many lost lives and numerous questions about what could have been done differently (Fekete and Sandholz 2021). As suggested by the results in Blanes, impact-based warnings such as the SSWs can be a promising communication tool to empower vulnerable communities to understand their risks and identify the appropriate self-protection actions they should perform during a flood emergency. However, co-designing SSWs within a community-based framework requires significant time and effort to integrate the specific local risk knowledge (Bucherie et al. 2022) and to foster trust and active collaboration among community members. Although they involve additional resources for their development, the difference in mean scores between the current official SMC warnings and the SSWs supports their implementation to help at-risk communities and local authorities during rainfall emergencies in Catalonia. In this regard, future work on SSWs and emergency communication will focus on improving the warning information and the dissemination channels for effectively delivering them to the most vulnerable communities across Catalonia and Europe.

6.4.4 A4alerts: Mobile Application for Emergency Communication

In recent years, there have been initiatives to integrate information and communication technology (ICT) during emergencies (Palen and Anderson 2016). In 2021, around 6.3 billion people were using smartphones worldwide, a number that is only expected to increase (Statista 2022). In this aspect, crisis apps have come forward as a promising tool to support both first (and second) responders and citizens in (i) providing short-term and real-time risk information, (ii) increasing their situational awareness, and (iii) informing what actions must be taken to decrease their personal and local risk. Location-specific warnings such as the SSWs can benefit from these opportunities crisis apps can offer to reach as many vulnerable points as possible, especially when the response time window is limited. Some official crisis apps are currently in operation (e.g. Katwarn in Germany and FEMA app in the United States) with different levels of success (Kaufhold et al. 2018). As with any technological implementation, there are limitations regarding crisis apps in terms of the level of engagement from users (e.g. installation and continuous use intention), malfunctions and the appropriate integration of the needs of vulnerable population groups (Tan et al. 2020). Therefore, it is crucial that when developing crisis apps, a community-centred approach is established to increase the likelihood of appropriately addressing the limitations from the perspective of the needs and requirements of end users (Dallo and Marti 2021).

As part of the SS-EWS framework, a mobile app has been co-designed to disseminate the local SSWs interactively. The 'A4alerts' app (see Figure 6.2) can quickly communicate the triggered local warnings and inform users about the approved official actions according to the current forecast in the area. Following good practices and previous results on usability guidelines for crisis apps (Tan et al. 2020), functionalities to increase risk perception and situational awareness have been implemented; e.g., users can send geotagged photos of impacts to be later shown within the app after screening and official approval.

The A4alerts app is expected to be extensively evaluated in the future (refer here the working paper) with communities across Catalonia and serve as a point of reference in integrating technological informatics in the field of community crisis communication.

6.5 An Outlook on Future Community and Impact-Based Communication Tools for Floods

The SS-EWS guides communities, local authorities and NMHs on the co-design and co-evaluation of local IBEWS for identified vulnerable sites or community sectors. The SS-EWS framework has been successfully implemented and evaluated in two vulnerable cities in Spain, Terrassa and Blanes. The co-design and co-evaluation processes enabled the communities to develop impact-based rainfall thresholds to trigger SSWs and appropriate self-protection actions for specific vulnerable locations and sectors in real time. Although some visible challenges were experienced during co-design and co-evaluation processes, the implemented community-based framework provided a pathway tool adapted for participants to discuss from their diverse requirements to derive the final messages of SSWs to be disseminated across the two communities. Moreover, when the SSWs were evaluated against the current official warnings, the co-designed warnings from the community-based framework significantly increased the risk perception and understanding of recipients and improved their intended behavioural response during emergencies.

Currently, the SS-EWS and its SSWs have been developed and tested for rainfall-induced urban or river floods. Future research and implementation work will focus on

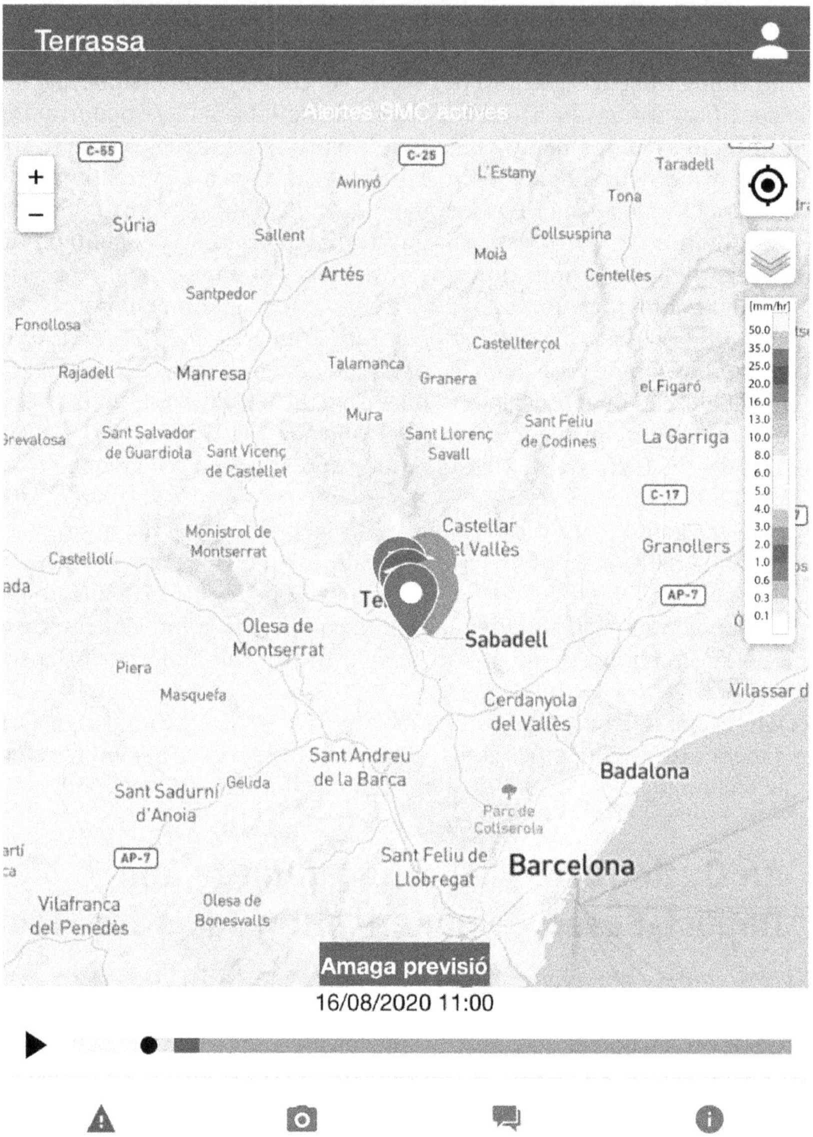

Figure 6.2 The A4alerts mobile app home screen.

integrating other types of hazard warnings (e.g. wildfire, winds) developed from the ANYWHERE MH-EWS for a complete community-based approach during emergencies. The 'A4alerts' app will also be upgraded with its functionalities to enable bi-directional communication between authorities and citizens to (i) increase situational awareness, (ii) impact verification and (iii) improve the self-protection responses of citizens or designated volunteers.

In the current climate emergency we are facing, regional and national EWS have limitations on effectively disseminating risk information at a community scale to appropriately reduce the local impacts. Implementing the community-based SS-EWS can be a valuable complement to the overall emergency management across the region.

Notes

1. Source: https://tinyurl.com/366vjx67.
2. Source: https://tinyurl.com/yeywhxzr.

References

Alfieri, L., Berenguer, M., Knechtl, V. et al. (2019). Flash flood forecasting based on rainfall thresholds. In: *Handbook of Hydrometeorological Ensemble Forecasting* (ed. Q. Duan, F. Pappenberger, A. Wood, et al.), 1223–1260. Berlin, Heidelberg: Springer https://doi.org/10.1007/978-3-642-39925-1_49.

Amores, A., Marcos, M., Carrió, D.S., and Gomez-Pujol, L. (2020). Coastal impacts of Storm Gloria (January 2020) over the north-western Mediterranean. *Natural Hazards and Earth System Sciences* 20 (7): 1955–1968. https://doi.org/10.5194/NHESS-20-1955-2020.

Aroca-Jimenez, E., Bodoque, J.M., Antonio Garcia, J., and Diez-Herrero, A. (2017). Construction of an integrated social vulnerability index in urban areas prone to flash flooding. *Natural Hazards and Earth System Sciences* 17 (9): 1541–1557. https://doi.org/10.5194/nhess-17-1541-2017.

Barriendos, M., Ruiz-Bellet, J.L., Tuset, J. et al. (2014). The Prediflood database of historical floods in Catalonia (NE Iberian Peninsula) AD 1035–2013, and its potential applications in flood analysis. *Hydrology and Earth System Sciences* 18 (12): 4807–4823. https://doi.org/10.5194/HESS-18-4807-2014.

Baudoin, M.A., Henly-Shepard, S., Fernando, N. et al. (2016). From top-down to "community-centric" approaches to early warning systems: exploring pathways to improve disaster risk reduction through community participation. *International Journal of Disaster Risk Science* 7 (2): https://doi.org/10.1007/s13753-016-0085-6.

Bosseler, B., Salomon, M., Schlüter, M., and Rubinato, M. (2021). Living with urban flooding: a continuous learning process for local municipalities and lessons learnt from the 2021 events in Germany. *Water* 13 (19): 2769. https://doi.org/10.3390/w13192769.

Bradford, R.A., O'Sullivan, J.J., van der Craats, I.M. et al. (2012). Risk perception – issues for flood management in Europe. *Natural Hazards and Earth System Sciences* 12 (7): 2299–2309. https://doi.org/10.5194/nhess-12-2299-2012.

Bucherie, A., Werner, M., van den Homberg, M., and Tembo, S. (2022). Flash flood warnings in context: combining local knowledge and large-scale hydro-meteorological patterns. *Natural Hazards and Earth System Sciences Discussions* 22: 461–480. https://doi.org/10.5194/nhess-22-461-2022.

Casteel, M.A. (2016). Communicating increased risk: an empirical investigation of the national weather service's impact-based warnings. *Weather, Climate, and Society* 8 (3): 219–232. https://doi.org/10.1175/WCAS-D-15-0044.1.

Ching, P., Reyes, V., Sucaldito, M., and Tayag, E. (2015). An assessment of disaster-related mortality post-Haiyan in Tacloban City|Western Pacific surveillance and response. https://ojs.wpro.who.int/ojs/index.php/wpsar/article/view/344 (accessed 14 June 2023).

Cornwall, W. (2021). Europe's deadly floods leave scientists stunned. *Science* 373 (6553): https://doi.org/10.1126/science.373.6553.372.

Corral, C., Berenguer, M., Sempere-Torres, D. et al. (2019). Comparison of two early warning systems for regional flash flood hazard forecasting. *Journal of Hydrology* 572: 603–619. https://doi.org/10.1016/j.jhydrol.2019.03.026.

Dallo, I. and Marti, M. (2021). Why should I use a multi-hazard app? Assessing the public's information needs and app feature preferences in a participatory process. *International Journal of Disaster Risk Reduction* 57: 102197. https://doi.org/10.1016/J.IJDRR.2021.102197.

Departamento de Interior (2015). Disposiciones Departamento De La Presidencia. *Diari Oficial de la Generalitat de Catalunya* 5). Article 8 (6824). https://portaljuridic.gencat.cat/ca/document-del-pjur/?documentId=686710&traceability=01&language=ca.

Fekete, A. and Sandholz, S. (2021). Here comes the flood, but not failure? Lessons to learn after the heavy rain and pluvial floods in germany 2021. *Water (Switzerland)* 13 (21): https://doi.org/10.3390/W13213016.

German Red Cross (2018). Report: 6th global dialogue platform on forecast-based financing (Issue 04). https://www.forecast-based-financing.org/wp-content/uploads/2019/03/Report_DP18Berlin.pdf (accessed 14 June 2023).

Harrison, S.E., Potter, S.H., Prasanna, R. et al. (2021). 'Where oh where is the data?': identifying data sources for hydrometeorological impact forecasts and warnings in Aotearoa New Zealand. *International Journal of Disaster Risk Reduction* 66: 102619. https://doi.org/10.1016/j.ijdrr.2021.102619.

Harrison, S.E., Potter, S.H., Prasanna, R. et al. (2022). Identifying the impact-related data uses and gaps for hydrometeorological impact forecasts and warnings. *Weather, Climate, and Society* 14 (1): 155–176. https://doi.org/10.1175/WCAS-D-21-0093.1.

International Federation of Red Cross (IFRC) (2020). The future of forecasts: impact-based forecasting for early action. https://www.forecast-based-financing.org/wp-content/uploads/2020/09/Impact-based-forecasting-guide-2020.pdf (accessed 14 June 2023).

Kaltenberger, R., Schaffhauser, A., and Staudinger, M. (2020). "What the weather will do" – results of a survey on impact-oriented and impact-based warnings in European NMHSs. *Advances in Science and Research* 17: 29–38. https://doi.org/10.5194/asr-17-29-2020.

Kaufhold, M. A., Rupp, N., Reuter, C. et al. (2018). 112.Social: design and evaluation of a mobile crisis app for bidirectional communication between emergency services and citizens. *26th European Conference on Information Systems: Beyond Digitization – Facets of Socio-Technical Change, ECIS 2018*, Portsmouth, UK (23–28 June 2018). https://aisel.aisnet.org/ecis2018_rp/81.

Kox, T., Lüder, C., and Gerhold, L. (2018). Anticipation and response: emergency services in severe weather situations in Germany. *International Journal of Disaster Risk Science* 9 (1): 116–128. https://doi.org/10.1007/S13753-018-0163-Z/FIGURES/2.

Lalancette, A. and Charles, A. (2022). Factors influencing hazard management by municipalities: the case of coastal communities. *Global Environmental Change* 73: https://doi.org/10.1016/J.GLOENVCHA.2021.102451.

Lucena, E. (2021). *Design and Evaluation of a Sitespecific Flood Early Warning System: A Case Study in Blanes, Spain*. Universitat Politécnica de Catalunya. https://upcommons.upc.edu/handle/2117/360715.

Martín-Vide, J.P. and Llasat, M.C. (2018). The 1962 flash flood in the Rubí stream (Barcelona, Spain). *Journal of Hydrology* 566: 441–454. https://doi.org/10.1016/J.JHYDROL.2018.09.028.

Maskrey, S.A., Priest, S., and Mount, N.J.J. (2019). Towards evaluation criteria in participatory flood risk management. *Journal of Flood Risk Management* 12 (2): e12462. https://doi.org/10.1111/jfr3.12462.

Mason, K., Lindberg, K., Haenfling, C. et al. (2021). Social vulnerability indicators for flooding in aotearoa New Zealand. *International Journal of Environmental Research and Public Health* 18 (8): 3952. https://doi.org/10.3390/ijerph18083952.

Meiyan, J., Lianchun, S., Tong, J. et al. (2015). China's implementation of impact and risk-based early warning. *World Meteorological Organization Bulletin* 64 (2): 9–12. https://library.wmo.int/doc_num.php?explnum_id=6985.

Meléndez-Landaverde, E.R. and Sempere-Torres, D. (2022). Design and evaluation of a community and impact-based site-specific early warning system: the SS-EWS framework. *Journal of Flood Risk Management* https://doi.org/10.1111/jfr3.12860.

Meléndez-Landaverde, E.R., Werner, M., and Verkade, J. (2020). Exploring protective decision-making in the context of impact-based flood warnings. *Journal of Flood Risk Management* 13 (1): e12587. https://doi.org/10.1111/jfr3.12587.

Morss, R.E., Mulder, K.J., Lazo, J.K., and Demuth, J.L. (2016). How do people perceive, understand, and anticipate responding to flash flood risks and warnings? Results from a public survey in Boulder, Colorado, USA. *Journal of Hydrology* 541: 649–664. https://doi.org/10.1016/j.jhydrol.2015.11.047.

Navarro, S. (2020). Gestió del ric d'inundació al Delta i la Baixa Tordera. Agència Catalana de l'Aigua. http://isacc.creaf.cat/wp-content/uploads/2020/02/PR_2020_02_19_INUNDABILITAT_Taula_Tordera_compressed.pdf (accessed 14 June 2023).

Netzel, L.M., Heldt, S., Engler, S., and Denecke, M. (2021). The importance of public risk perception for the effective management of pluvial floods in urban areas: a case study from Germany. *Journal of Flood Risk Management* 14 (2): https://doi.org/10.1111/JFR3.12688.

Palen, L. and Anderson, K.M. (2016). Crisis informatics-new data for extraordinary times: focus on behaviors, not on fetishizing social media tools. *Science* 353 (6296): 224–225. https://doi.org/10.1126/SCIENCE.AAG2579/ASSET/17FA270B-AE2C-41A0-9609-C01517DA9A0C/ASSETS/GRAPHIC/353_224_F1.JPEG.

Park, S., Berenguer, M., and Sempere-Torres, D. (2019). Long-term analysis of gauge-adjusted radar rainfall accumulations at European scale. *Journal of Hydrology* 573: 768–777. https://doi.org/10.1016/j.jhydrol.2019.03.093.

Potter, S.H., Kreft, P.V., Milojev, P. et al. (2018). The influence of impact-based severe weather warnings on risk perceptions and intended protective actions. *International Journal of Disaster Risk Reduction* 30: 34–43. https://doi.org/10.1016/J.IJDRR.2018.03.031.

Potter, S., Harrison, S., and Kreft, P. (2021). The benefits and challenges of implementing impact-based severe weather warning systems: perspectives of weather, flood, and emergency management personnel. *Weather, Climate, and Society* 13 (2): 303–314. https://doi.org/10.1175/wcas-d-20-0110.1.

Puzyreva, K., Henning, Z., Schelwald, R. et al. (2022). Professionalization of community engagement in flood risk management: insights from four European countries. *International Journal of Disaster Risk Reduction 71*: https://doi.org/10.1016/j.ijdrr.2022.102811.

Ripberger, J., Silva, C., Jenkins-Smith, H., & James, M. (2015). The influence of consequence-based messages on public responses to tornado warnings. *Bulletin of the American Meteorological Society*, **96**(4), 577– 590. https://doi.org/10.1175/BAMS-D-13-00213.1

Ritter, J., Berenguer, M., Corral, C. et al. (2020). ReAFFIRM: real-time assessment of flash flood impacts – a regional high-resolution method. *Environment International* 136: 105375. https://doi.org/10.1016/j.envint.2019.105375.

Rogers, D.P., Anderson-Berry, L., Bogdanova, A.-M. et al. (2019). Learning from multi-hazard early warning systems to respond to pandemics. http://documents.worldbank.org/curated/en/429511592591445701/Learning-from-Multi-Hazard-Early-Warning-Systems-to-Respond-to-Pandemics (accessed 14 June 2023).

Schroeter, S., Richter, H., Arthur, C. et al. (2021). Forecasting the impacts of severe weather. *Australian Journal of Emergency Management* 36 (1): 76–83. https://doi.org/10.47389/36.1.76.

Statista (2022). Number of smartphone subscriptions worldwide from 2016 to 2027. https://www.statista.com/statistics/330695/number-of-smartphone-users-worldwide/ (accessed 14 June 2023).

Sufri, S., Dwirahmadi, F., Phung, D., and Rutherford, S. (2020). A systematic review of community engagement (CE) in disaster early warning systems (EWSs). In: *Progress in Disaster Science*, vol. 5, 100058. Elsevier. https://doi.org/10.1016/j.pdisas.2019.100058.

Tan, M.L., Prasanna, R., Stock, K. et al. (2020). Understanding end-users' perspectives: towards developing usability guidelines for disaster apps. *Progress in Disaster Science* 7: 100118. https://doi.org/10.1016/j.pdisas.2020.100118.

Tarchiani, V., Massazza, G., Rosso, M. et al. (2020). Community and impact based early warning system for flood risk preparedness: the experience of the Sirba river in Niger. *Sustainability* 12 (5): 1802. https://doi.org/10.3390/SU12051802.

Taylor, F.E., Malamud, B.D., Freeborough, K., and Demeritt, D. (2015). Enriching great britain's national landslide database by searching newspaper archives. *Geomorphology* 249: 52–68. https://doi.org/10.1016/j.geomorph.2015.05.019.

The Intergovernmental Panel on Climate Change (IPCC) (2021). *Climate Change 2021: The Physical Science Basis. Contribution of Working Group I to the Sixth Assessment Report of the Intergovernmental Panel on Climate Change*. Cambridge University Press. https://www.ipcc.ch/report/ar6/wg1/downloads/report/IPCC_AR6_WGI_SPM_final.pdf (accessed 14 June 2023).

The Lancet Regional Health (2021). Germany floods – a warning for future extreme weather events. *The Lancet Regional Health – Europe* 7: 100194. https://doi.org/10.1016/j.lanepe.2021.100194.

Tschoegl, L., Below, R., and Guha-Sapir, D. (2006). *An Analytical Review of Selected Data Sets on Natural Disasters and Impacts*. Centre for Research on the Epidemiology of Disasters.

Uccellini, L.W. and ten Hoeve, J.E. (2019). Evolving the national weather service to build a weather-ready nation. *Bulletin of the American Meteorological Society* 100 (10): 1923–1942. https://doi.org/10.1175/BAMS-D-18-0159.1.

UNISDR (2015). Sendai framework for disaster risk reduction 2015-2030. *United Nations Office for Disaster Risk Reduction Publications* 30 (3): 23–24. https://www.undrr.org/publication/sendai-framework-disaster-risk-reduction-2015-2030.

Weyrich, P., Scolobig, A., Bresch, D.N., and Patt, A. (2018). Effects of impact-based warnings and behavioral recommendations for extreme weather events. *Weather, Climate, and Society* 10 (4): 781–796. https://doi.org/10.1175/WCAS-D-18-0038.1.

WMO (2015). WMO guidelines on multi-hazard impact-based forecast and warning services (Issue 1150). https://library.wmo.int/doc_num.php?explnum_id=7901 (accessed 14 June 2023).

World Meteorological Organization (2012). Guidelines for creating a memorandum of understanding and a standard operating procedure between a national meteorological or hydrometeorological service and a partner agency. http://www.wmo.int/pages/prog/amp/pwsp/publicationsguidelines_en.htm (accessed 14 June 2023).

World Meteorological Organization (2021). WMO guidelines on multi-hazard impact-based forecast and warning services – part II: putting multi-hazard IBFWS into practice. https://library.wmo.int/doc_num.php?explnum_id=10965 (accessed 14 June 2023).

World Meteorological Organization (WMO) (2013). Flood forecasting and early warning. *Integrated Flood Management Tools Series* 19: 21–22. https://library.wmo.int/index.php?lvl=notice_display&id=16351.

7

Challenges for a Better Use of Crowdsourcing Information in Climate Emergency Situational Awareness and Early Warning Systems

Milan Kalas[1], Joy Ommer[1,2], Amin Shakya[3], Saša Vraníc[1], Denys Kolokol[1], and Tommaso Sabattini[1]

[1] *KAJO s.r.o., Bytca, Slovakia*

[2] *Department of Geography and Environmental Science, University of Reading, Reading, UK*

[3] *Faculty of Geo-information and Earth Observation, University of Twente, WRS, Enschede, Netherlands*

7.1 Introduction

Available information to civil protection in emergency situations and early warning is commonly based on data provided by authorities such as meteorological institutions. Throughout the last decade, advances are highlighting several advantages of enriching these 'traditional' emergency information sources with non-conventional data sources such as remote sensing, drones, and citizen science. Among them, citizen-generated content presents its own unique set of opportunities but also challenges in disaster risk management (DRM).

Citizen-generated content is produced through citizen science activities. In other words, it is a process where citizens are involved in research such as data collection, analysis or validation. Citizen science is a broad concept where citizens can be actively (e.g. crowdsourcing and crowdsensing) or passively (e.g. through social media posts) involved: crowdsourcing refers to data collection, data analysis or even design of a program framework outsourced to a large group of people – the crowd (ANYWHERE 2019).

Responding to Extreme Weather Events, First Edition. Edited by Daniel Sempere-Torres, Anastasios Karakostas, Claudio Rossi, and Philippe Quevauviller.

An example of crowdsourcing can be taken from phenology where citizens are monitoring plant and tree developments throughout the year. Similarly, crowdsensing builds on the input of a crowd but can be more passive than crowdsourcing since the data is commonly collected through (low-cost) sensors. Crowdsensing is often applied in environmental monitoring (e.g. monitoring of the ambient temperature, noise or air quality) to collect a high amount of data (European Commission 2021). Contrarily, using social media tools, citizens are (unintentionally) providing input by posting on social media channels such as Twitter, which is used for analyses in various fields.

The participation of citizens in science offers a wide spectrum of benefits for the research but also for the participants themselves. For instance, citizen science can enable the integration of under-represented groups and, therefore, diversify the participant group (Albert et al. 2021; Kloetzer et al. 2021; Pelacho et al. 2021). Citizens ideally benefit from this approach with increased knowledge through the project and awareness on certain topics. This raised awareness and knowledge can introduce changes in behaviour, for instance, towards more environmentally friendly behaviours. As such, citizen science as a practice not only affects the knowledge generation but also it can shift the power and agency dynamics in the same social hierarchy (empowering the crowd). In a way, citizen science is the democratization of science in academia and in practice. Besides advantages for science and participants, governments and the policy sector can benefit from engaged citizens with more legitimate evidence, political advocacy, or even higher interests of citizens in local actions and involvement (Toomey and Domroese 2013; van Noordwijk et al. 2021).

As juicy as it appears to be, major challenges are still in the way of a seamless integration of citizen-generated content. Commonly discussed challenges are related to data quality as well as the reliability of the data. In particular, data quality is linked to data accuracy, metadata and appropriateness (European Commission 2021). Balázs et al. (2021) summarizes further challenges around data that are reflected in the credibility of the data, its potential for reuse and the ownership. Depending on the context, potential obstacles may also be the access of the data or their interoperability.

The ANYWHERE project aimed at integrating different tools of citizen science into various phases of the disaster management cycle, especially, at improving emergency situational awareness and early warning systems (EWS). In particular,

- crowd mapping was applied to improve real-time risk knowledge,
- social media was scraped to gain real-time emergency information, and
- a crowdsourcing application was developed to gather information on the magnitude of disaster events as well as their impact.

The aim of this chapter is not to provide an extensive literature review of the challenges of citizen science in DRM but to introduce how crowd-generated content was integrated into emergency management in ANYWHERE and the lessons learnt from these activities to increase the understanding of the potential, remaining challenges but also identified best practices. To pave the way to the experiences of the ANYWHERE project, the following Section 7.2 will briefly introduce possible ways for the adoption and inclusion of citizen-generated content in the domain of emergency management, commonly used tools, and challenges identified from literature. Section 7.3 will briefly present the aforementioned crowd solutions developed and tested within the project while chiefly focusing on the lessons learnt from these activities to indicate challenges but also solutions of crowd-generated content in DRM.

7.2 Crowd-Generated Content to Support Emergency Management and Early Warning

Large-scale disasters, such as hurricanes, storms, floods and landslides, are difficult to manage, despite rapid advancements in technology. In emergency situations of large-scale disasters, it is crucial to generate a timely reaction to their impacts (Ernst et al. 2017). This requires quick, reliable intelligence, and high-resolution information for emergency management. Citizen science offers ways to gather such reliable information and collaborate with citizens to collect and share information during emergencies. The timely reaction granted by such an approach leads to more flexibility and network effects to support improvisation (Xu et al. 2017).

7.2.1 Examples of the Citizen Science in Disaster Risk Management

In DRM, commonly practised methods are partly complemented with citizen science methods. For instance, participatory risk assessment is more and more practised as participatory risk mapping where citizens can add risk information to a map-based application (Klonner et al. 2021). Another important example is the crowd mapping by the Humanitarian OpenStreetMap Team (HOT) that offers volunteers to help identify and map building damages of disasters on satellite images (Danylo et al. 2018; Herfort et al. 2021). In 2020, HOT funded a community-led microgrant for mapping in order to support response to the Bolivian wildfires (HOT 2020). The aim of this activity was to build a community-generated mapping system where citizens can map damages to their property (e.g. crop and infrastructure), map new fires and to map evacuation facilities. This activity highlights the effective use of citizen science and citizen engagement in various phases of the DRM and the increase in resilience and self-protection of the communities but also provides a solid base for sustainable mechanisms to engage citizens in affected communities in adaptation and build-better strategies in DRM.

An example of crowdsourcing is the *I-REACT* app (Bielski et al. 2017), which allows citizens to engage in real-time in-field wildfire reporting by entering information on the location, smoke, flames and wind – as they observe it. In exchange, the app provides early warning, as well as wildfire education features. App users can be further engaged in the validation process of early warning messages and social media entries.

To increase real-time emergency management by local authorities and citizens themselves, citizen science is increasingly adopted to improve emergency situational awareness and EWS. For this purpose, weather observations are enriched with data from citizens' own (low-cost) weather sensors to densify the coverage of observation stations (Alfonso, Chacón and Peña-Castellanos 2015; de Vos et al. 2017; Mandement and Caumont 2020). For instance, in the *FloodCitiSense* project aimed at improving the preparedness and response of cities but also of citizens themselves, crowdsensing was integrated to complement existing data for flood early warning services (Verbeiren et al. 2019). Crowdsensing of meteorological variables is a common practice as a large number of citizens possess personal weather stations that automatically upload the data to platforms such as WeatherCloud (2012). Similarly, Huang (2020) developed a methodology to use social media messages as a proxy for rainfall estimates.

Moving aside from crowd generated response during an ongoing crisis event, Santoni and Rufat (2021) report on cases where social media posts are analysed retrospectively – after an event – to increase the understanding of citizens' behaviours and actions during a disaster event. Negri et al. (2021) retrospectively analysed public images to measure adherence of the public to social distancing and mask mandates during the COVID crisis.

In the context of emergency response, the *Facebook Data For Good Annual Report 2020* (Facebook Company 2020) presents how Facebook crisis apps supported governmental agencies with facilitating the targeted alerting and providing public safety information. The Facebook population data helped state agencies to distribute material to affected communities and were used to reveal the evacuation patterns for communities affected by bushfires in Australia in January 2020.

Combining several types of crowd-generated content, Gaitan et al. (2014) developed a smart response platform to increase local emergency awareness of the city for effective response. The platform builds on a heterogeneous set of input data including crowd-sensed real-time data but also Twitter posts, and complaint reports from citizens entered through a municipal system.

7.2.2 Tools

As introduced above, different citizen science tools can be applied in DRM. This section aims to briefly introduce these tools in the context of DRM and the ANYWHERE project.

7.2.2.1 Social Media Sensing Social media proves to be a very powerful tool when it comes to gathering and sharing information and collaborating with citizens. The potential that lies herein has been recognized and is analysed in several studies. These studies have found that Twitter can be used for event detection purposes, events ranging from harmless traffic jams to earthquakes and tropical storms (Reuter et al. 2016). In other cases, Twitter has been mined and monitored for information about a specific event, e.g. information about floods in the United Kingdom (Reuter and Spielhofer 2017) and other disasters (Spielhofer et al. 2016). During incidents, social media and crowdsourcing can play a vital role, for instance, to raise situational awareness. While several emergency agencies in countries as the United States already use social media to monitor and share information, most German agencies avoid social media, or have a profile in a social network but neglect to use its full potential, although studies found that emergency services' staff already use social media or increasingly want to use social media (Reuter et al. 2016). Citizens are already actively using social media in emergencies, as activity bursts show but also expect emergency services and managers to use social media and react to information originating here (Reuter and Spielhofer 2017). Large-scale disasters lead to spikes of activity on social networks. This enables indirect crowdsourcing approaches, where people do not actively and immediately participate in given tasks. These generated spikes, or 'bursts', can be analysed to identify and track disasters as they appear, potentially leading to a gain value when reacting to emergencies (Yin et al. 2012; Spielhofer et al. 2016; Ernst et al. 2017; Xu et al. 2017). This can lead to more direct approaches as well, when emergency managers decide to actively involve the public in crowdsourced tasks, such as active information sharing, direct communication, or promoting preparedness (Yin et al. 2012; Reuter and Spielhofer 2017).

7.2.2.2 Crowdsourcing and Crowdsensing Crowd-generated content from crowdsourcing, crowdsensing, and crowd mapping are creating new possibilities in the domain of DRM and community responses since these tools are empowering communities to self-organize and respond to crises in real time. Crowdsourcing enables timely gathering and processing of reliable information to be generated and dissipated in a timely fashion for disaster response. Such an approach increases the flexibility of the disaster response system and further induces a network effect enabling the improvisation by communities to respond to disaster. In contrast to social media tools, they are striving for citizens' engagement, awareness and education while completing a specific task. This active form of citizen science certainly requires more engagement and activation of citizens to reach a sufficient number of volunteers. In crowdsourcing activities, data is collected through online reporting or mapping tools such as a stand-alone software (e.g. web-based, or mobile phone applications) where participants can fill in a predefined form (including elements such as text, location or photographs). Considering the large number of participants, this form is structured in an easy-to-understand framework. After filling in the information, the data can be aggregated to produce a combined dataset. Similarly, crowdsensing is collecting sensed data shared by citizens using devices such as mobile phones, transportable sensors or personal weather stations. Sensing can take place either in a specific location or on-the-fly, e.g. while walking, cycling or driving with so-called wearable sensors. Citizen science is not only defined by the degree of participation but also by the frequency: long-term and short-term data collection. Long-term can be associated with continuous monitoring for preparedness, recovery or prevention actions while short-term refers to rush modes for emergency preparedness and response.

7.2.3 Challenges in the Integration and Application of Citizen-Generated Content in DRM

Even though citizen science enables a large number of possibilities and entails several benefits, its application is limited as challenges are unravelling in different parts of citizen science ranging from the participation of citizens to the accuracy of collected data to the adoption of data or results by practitioners and policymakers.

Active modes of citizen science require direct engagement of citizens in, for instance, data collection. A great challenge is already the recruitment of volunteers considering their willingness to participate. Citizens' motivation for participation is often dependent on the project settings including the theme or topic, types of activities and technical aspects, but it may also be determined by the (co-) benefits they receive from their participation (Asingizwe et al. 2020; Neset et al. 2021; Pateman et al. 2021). Technical issues can be the usability of applications, or generally, internet-based surveys referring to the digital divide. From a citizen's perspective, one restricting factor can be difficulties of identifying projects where one could get involved. Furthermore, the increased adoption of citizen science comes along with an overload of applications as commonly each project has its own application. Nonetheless, these challenges are not subject for social media tools since they usually do not require active engagement of volunteers.

Barriers for participation and, therefore, data collection are one side of the story. A long list of barriers for the adoption of collected data into disaster management is another side of the story. Challenges arise constantly around the quality and the acceptance of crowd-generated data, internal barriers such as the need for technical and structural changes in management of local authorities or legal concerns regarding the adoption.

The acceptance and integration of crowd-generated content or more generally of open-source data is unfortunately not naturally given. The main reason for this is the lack of trust in data collected or produced by non-experts. Data may be lacking in quality due to own perceptions brought into the data collection. Data may further be biased considering the fact that participants could be only active citizens in the specific theme but does not include other less active citizens (World Meteorological Organization 2017; Basiri et al. 2019; See 2019). Another issue related to disaster management is the possibility that online data collection may be interrupted during disaster events because of communication and reception disruptions (Haworth 2016). The data collection system may sometimes be sub-optimally suited for the purpose or may result in poor user interface or ambiguity in design hindering the best use of the available citizen-scientists to serve the project.

Institutional barriers to the adoption and integration may be the lack of in-house ICT expertise that entails the need for investments in training, time and money which may not be supported by the institution. Overall, the integration of new technologies comes along with additional costs. Legal aspects such as the conformity with policies, the liability and credibility of data were mentioned as additional obstacles. One major issue to be raised at this point is the security of citizens' privacy – sharing of personal or sensitive information (Haworth 2016; Harrison and Johnson 2019).

7.3 ANYWHERE Applications and Their Lessons Learnt

The unique composition of the ANYWHERE project – having experts covering all different aspects of DRM – allowed the activation of crowd mapping (Section 7.3.1), designing a social media tool (Section 7.3.2), and a crowdsourcing solution (Section 7.3.3) to fully exploit the potential of crowd-generated content in DRM. The following sections will present these solutions and related activities within the ANYWHERE project while focusing on the lessons learnt from these activities.

7.3.1 Crowd Mapping to Support Real-Time Risk Assessment

ANYWHERE created complementary knowledge on exposure and vulnerability necessary to perform a proper risk assessment. For this purpose, ANYWHERE established contacts with HOT and proposed a semi-automated mechanism for triggering the HOT mapping community (via the HOT Tasking Manager) based on a signal from the ANYWHERE Multi-hazard Early Warning System (MH-EWS). An example of the application of HOT data is the crowd mapping of sensitive infrastructure in the ANYWHERE impact forecast described in Láng-Ritter et al. (2022). Early activation of the HOT community – before the disaster onset – could speed up the mapping task and, hence, expose more complete and representative exposure data for the real-time risk assessment. This crowd task can be repeated for areas with predicted high risk and low HOT or OSM data coverage. Early and targeted crowd activation to perform various tasks to assess and mitigate the risk is a successful example of so-called no regret actions.

7.3.2 Social Media Streaming to Increase Emergency Situational Awareness

Social media sensing was adopted in order to increase emergency situational awareness. The proposed system is called Social Media Analytical Toolkit (SMAT), which streams Twitter posts to detect emergencies such as flooding based on preselected DRM keywords (e.g. flood). Figure 7.1 shows the detection of the flooding in Pakistan and India in summer 2022.

SMAT was launched in 2017 and since then is collecting on average 15 K relevant and geolocated tweets per day (before early 2023 changes in Twitter access policy). SMAT was tested during the project execution in actual emergency events occurring in the pilot sites in collaboration with the local stakeholders (e.g. flooding in Valencia in October 2018). The main users of the system are the ANYWHERE project partners, providing feedback on system design and performance.

7.3.2.1 Technical Background

Machine learning: tweets were collected based on a set of static DRM keywords and, additionally, on keywords for different hazards provided by experts from the ANYWHERE pilot sites. Initially, the keywords were used to collect a set of tweets for training the machine learning classifier. Collected tweets were manually annotated based on their relevance and mixed with some openly available training datasets (Alam et al. 2021). In total about 4000 annotated tweets per language were used in training of the ML. All posts are being processed through a pipeline while each message is assigned a likelihood to be relevant (based on this trained machine learning classifier).

Localization: the possible location of a tweet is derived from – using both – the original location of the message (mobile GPS if available) and a location derived from the

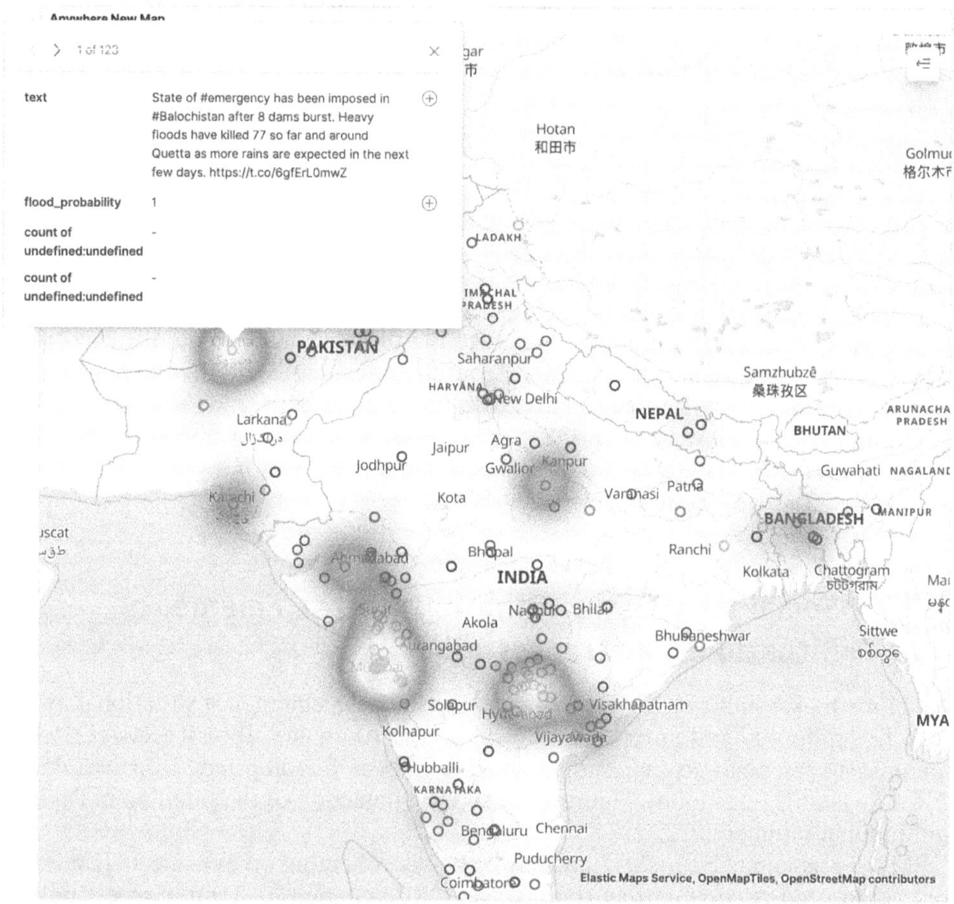

Figure 7.1 Example of emergency detection with the ANYWHERE social media tool: detected flood hot spots in Asia in June and July 2022.

text. Relevant posts are then displayed in a map view that can indicate arising emergencies. For instance, an area with a larger number of posts popping up indicates an event like the floods in Pakistan and India in 2022 depicted in Figure 7.1.

Aggregation: in order to assist users with the interpretation of the social media stream and explore its potential to improve the situational awareness, a new functionality was designed to aggregate tweets based on their geographic position. Tweets can be aggregated on a user-defined time interval (e.g. last 30 minutes). Hence, each geolocated message is assigned an information of an administrative region and the country to allow fast aggregations at the regional and national level. Average counts can highlight the locations where more tweets mention the hazardous phenomenon. The sudden increase in normalized tweet counts is expected during the event onset; thus, it can serve users to identify locations where new emerging events may require their attention.

Content Analysis & Languages: in addition, well-curated, text-processing algorithms are used to extract the meaning holding word chunks, which can be used to provide more insights in the magnitude and impact of the event purely based on the language used (e.g. disastrous flood, wide-spread damage). Overall, the tool supports the analysis of tweets in different languages (depending on the hazard):

- floods (IT, ENG, FR, DE, ES, CA)
- wind (IT, DE)
- heatwave (ENG)
- storm surge (IT)
- forest fires (IT, ENG)

7.3.2.2 SMAT Integration in ANYWHERE Applications During the project execution, the SMAT data stream was integrated into all ANYWHERE platforms. Integration example showing the integration in ANYWHERE A4EU platform is shown in Figure 7.2. After the project, the real-time SMAT data feed is integrated in the Predict Services (ANYWHERE partner) platform and is used by Predict teams, bringing a complementary information from the field, including information about event impact which helps in decision-making and for communication towards the territories. Example of the Predict integration is in Figure 7.3.

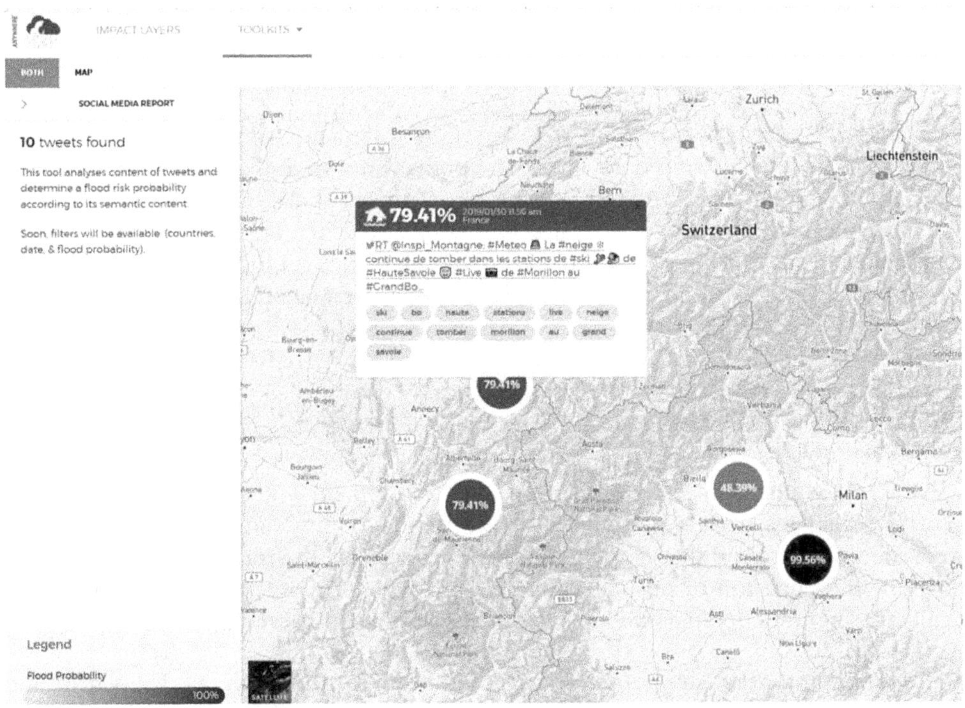

Figure 7.2 Example of the SMAT integration in the ANYWHERE A4EU Pan-European system.

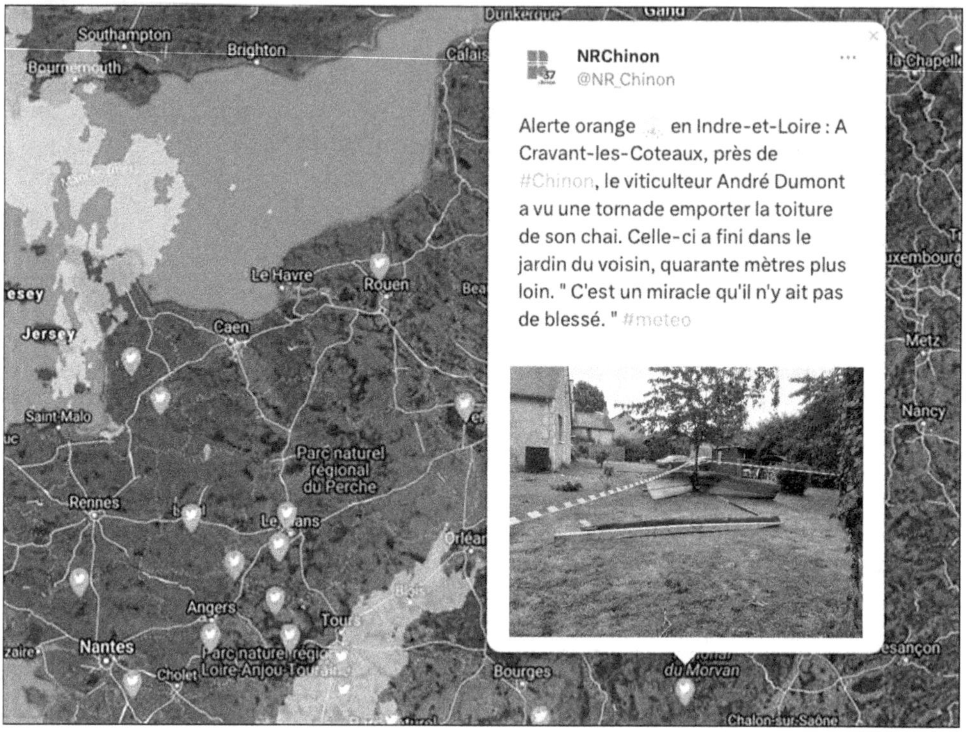

Figure 7.3 Example of integration of SMAT multi-hazard stream in the predict platform.

SMAT data are also used to monitor flood events in India within the INDRA Reporter of the Indian Institute of Technology (Saharia et al. 2020).

The processed data stream is available via API interface upon a request. The API allows advanced filtering of data (e.g. by date, relevance, geolocation, language, etc.), which is delivered as standardized GeoJSON. Data stream can be easily integrated in any desktop or web GIS system.

The social media stream was integrated in all ANYWHERE platforms. Since the system is based on ElasticSearch (www.elastic.co), the richest analytical access to the SMAT data is via Kibana dashboard, which is available upon a request.

7.3.2.3 Lessons Learnt Throughout the development and testing of the SMAT, several challenges arose which (initially) lowered the quality of the results. Challenges were identified around the scraping of social media feeds, the relevance analysis of the posts, or the inclusion of posts in different languages. This section will explain these challenges in more detail as well as identified solutions to overcome them.

- *Limited real-time streaming*: social media platforms such as LinkedIn and Facebook offer a programmatic access to the posts via API, but they do not offer access to real-time posts, which is essential for the situational awareness in DRM applications (Monaco and Adnaudo 2020). Therefore, the SMAT is solely based on Twitter post (before early 2023 changes in Twitter access policy).

- *Retweets*: during real emergency situations, ANYWHERE project partners from the civil protection complained that high number of tweets, correctly classified as relevant, were retweets of the news articles. Civil protection authorities in Europe did not consider those new articles very informative during the ongoing crisis, since the news articles are typically based on the official communication form the authority itself. The retweets of the news articles present a very high percentage of the overall crisis-related communication. This was also confirmed by other studies, Wang et al. (2021) reported more than 70% of the tweets being related to the news and comments on the crisis. News articles remain an extremely valuable source of information for situational awareness in less developed countries and for international emergency coordination and humanitarian organizations. In Europe, the main challenge for social media analytics lies at identifying tweets from eyewitnesses.
- *Spatial tweeting differences*: regional differences in the usage of Twitter were uncovered in the uptake of the Twitter use in crisis management by civil protection authorities (Reuter et al. 2016), and in the quality of the language processing tools and trained classifiers. This can affect the proportional coverage of different languages during the crisis. For instance, tweets in German are much less frequent as in other languages (English, Spanish or Italian). This also accounts for events occurring in German-speaking regions. In this context, Figure 7.4 shows the distribution of languages in flood related tweets collected from 1 July to 31 August 2022. This period covers the disastrous flooding in Germany, Belgium and Netherlands in mid-July 2021. It becomes obvious that only a small fraction of tweets is in German

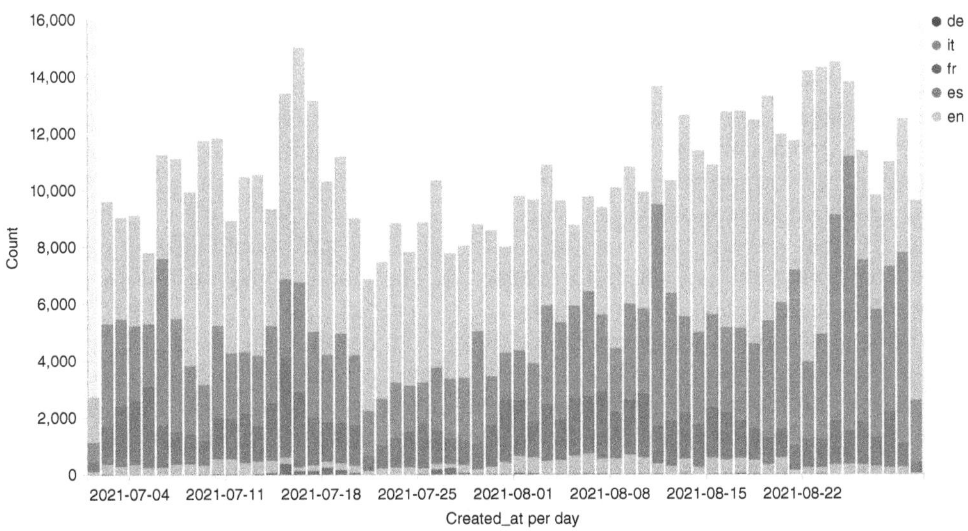

Figure 7.4 Language distribution of flood related tweets between 1 July 2022 and 31 August 2022.

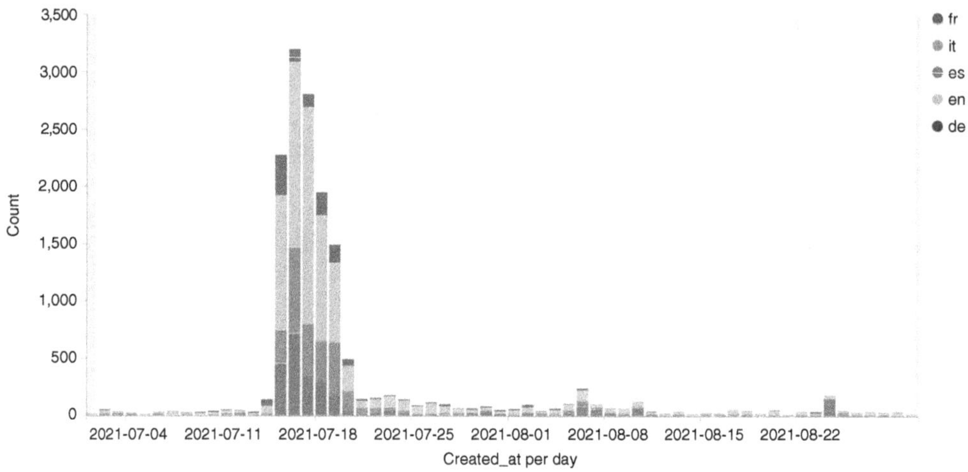

Figure 7.5 Language distribution of flood-related tweets geolocated in Germany between 1 July 2022 and 31 August 2022.

language (dark grey bars in the Figure 7.4) and the same trend can be observed in entire tweet collection (search keywords are similarly distributed in all languages). Figure 7.5 shows tweets from the same period, which were geolocated in Germany.

Ideally, this low coverage of German tweets will improve with better trained language models and classifiers in SMAT in the future. An important aspect is to encourage and assist German authorities to improve their presence in social media and to moderate the crisis related communications to become more structured and informative.

- *Tweet annotations*: labelling of tweets was easily done in English or Spanish as tasks could be completed in one or two days at most. However, the recruiting of workers for labelling tweets in German, French or Italian was more challenging as it took much longer (even up to a week) to complete a batch of messages for annotations. In order to maximize the information gain during the crisis events, a new feedback mechanism was implemented. This allows users to annotate each tweet as relevant or irrelevant and tag selected tweets with the categories which the tweet is related to. For example, the tweet can be related to floods and should also be related to extreme wind and impact on infrastructure. These annotations will allow users to validate the information provided by the automated classifier, as well as collecting the annotations during the event and using them in retraining the classifier.
- *Anticipated emergency peaks due to early warning*: a number of obstacles unveiled in context of the analysis of the post content. One typical issue was the application of the event detection in the areas with well-established EWS and good communication campaigns on social media; thus, all social media traffic related to pre-event information posted by governmental institutions can contaminate the signal. In other words, these warnings can create anticipated peaks in social media traffic which are correctly classified as relevant. However, the peak does not indicate the

actual emergency situation related to the extreme event. Integrating the ANYWHERE MH-EWS, the proposed solution aimed at running an event-specific social media monitoring (triggered by a signal from the MH-EWS) in addition to the continuously monitoring the social media at the global scale. This event-based search used information about timing, location of the event, and the names of potentially affected infrastructure. Those were then used as additional keywords for the SM streaming. To increase the interoperability of the proposed solution, the MH-EWS trigger was using the Common Alerting Protocol (OASIS 2010) as standardized format for alerting. The social media signal was employed to confirm signals derived from the MH-EWS; thus, only a few messages are needed to approve the forecast. This process speeded up the event detection part compared to the social media standalone solution and reduced the uncertainty of EWS.

- *Improving the messages for better monitoring*: to improve the information gain from the social media, it is important to communicate and train citizens to post crisis-related content on the social media in structured way – ideally applying a set of easy to remember, intuitive but standardized hashtags. An extensive list of hashtags to be used in crisis or humanitarian related activities is provided by the Humanitarian Exchange Language (HXL).[1] HXL provides a very extensive vocabulary targeted to speed up processing and data exchange in humanitarian sector but could serve as a good starting point for defining a basic set of hashtags for disaster-related social media content (Keßler and Hendrix 2015). An important task remains at the side of the civil protection authorities to make an effort in continuously communicating the best practices to the citizens. Despite a constant progress, there is still a big gap in effective and systematic use of social media by civil protection authorities especially during the emergencies. Significant effort in creating an emergency 'tweeting culture' is being done by Virtual Operations Support Teams (VOSTs). They bridge the gap between the regular volunteers and civil protection authorities. Having a clear organization with the national and regional structures VOST groups are well accepted by the official authorities. VOST is an excellent example of how crisis communication on social media should be done. Promoting the official channels, amplifying the official information and combatting the fake news. Another task, where VOST could assist civil protection to gather information from social media is the training of the machine learning algorithms used in different tasks of the social media filtering and knowledge making (annotating, geo-referencing). Trained volunteers could step in during the ongoing crisis and fine-tune pre-trained models. This so-called adaptive training should ideally be performed by trained volunteers during the ongoing crisis and together with event-based real-time human geo-tagging of wrongly located tweets can significantly improve the quality and informativeness of the SMAT data stream, as well as improve the acceptability of the SM data by official authorities.

7.3.3 A Crowdsourcing Solution for Collecting Information on the Magnitude and Impact of Disasters

In order to collect more complex information from the citizens, ANYWHERE proposed a dedicated crowdsourcing solution. The desired platform was supposed to be open source, easy to configure and deploy, having an existing community of users and being available on both desktop and mobile devices. Following the initial requirements assessment and expert consultations, the USHAHIDI platform was selected as a vehicle to drive the ANYWHERE campaigns (Figure 7.6). The primary use of the crowdsourcing platform was to collect structured information about ongoing events and their impact. The campaigns were designed in all pilot languages (English, Spanish, German, Catalan, Italian and French) providing a possibility to send geo-referenced images and videos, with a set of metadata about event magnitudes and impacts. The taxonomy used in the campaigns was on the one hand aligned to fit the risk indicators defined in main

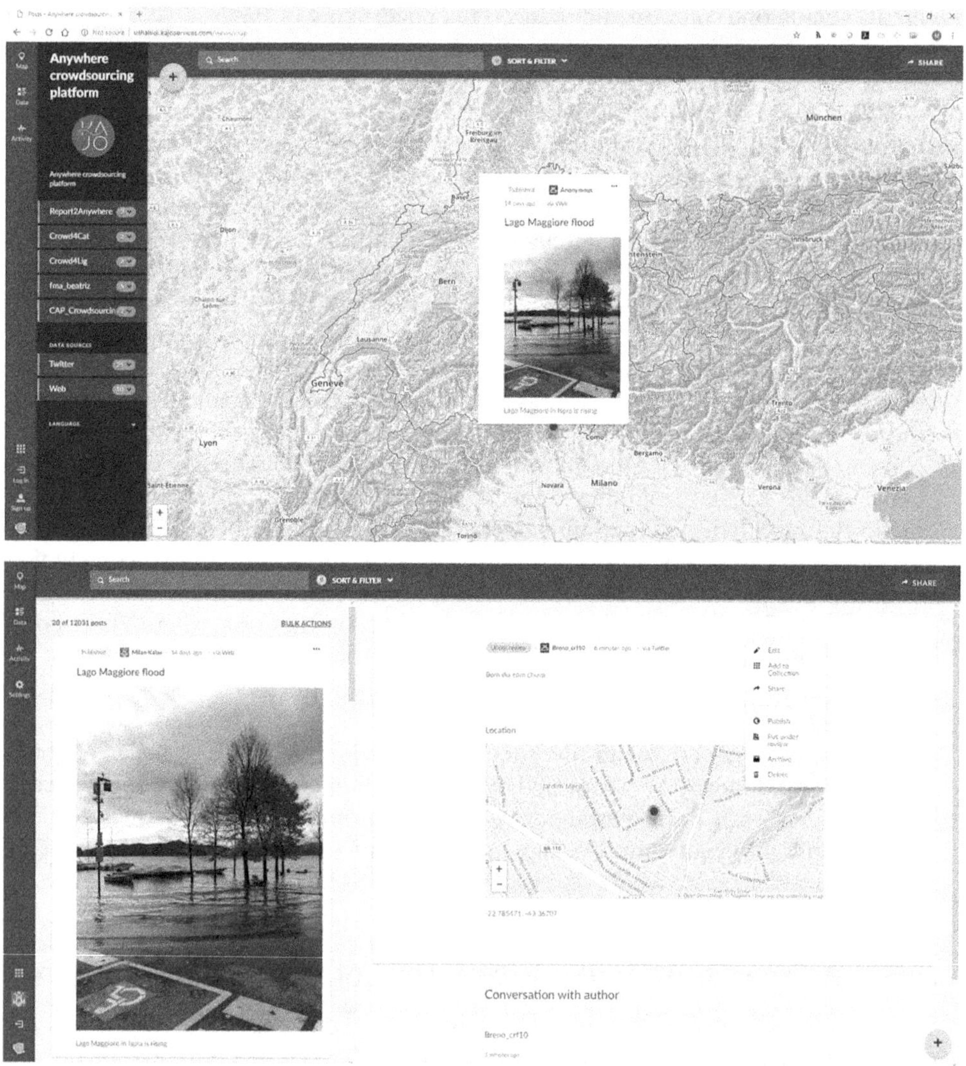

Figure 7.6 Examples of ANYWHERE crowdsourcing platform.

DRR Frameworks (e.g. the Sendai Framework (UNISDR 2015)) and loss and damage reporting best practices. On the other hand, it was made user-friendly and easy to complete by using tick boxes, drop-down menus. The platform was primarily targeting to collect real-time information from eyewitnesses and people in the field (e.g. second responders) to improve situational awareness during the crisis. Information collected was also planned to fill the database of extreme events, which would serve for damage reporting, plausibility checks of the risk models, scenario building and as a living memory of the past extreme events. Another important aspect of the crowdsourcing platform (and crowdsourcing as such) was to engage citizens in DRM and improve their understanding of the good practices for self-preparedness and self-protection.

7.3.3.1 Lessons Learnt After discussions with the leading organizations in crowdsourcing (Standby Task Force and VOST Europe), the critical element of a successful crowdsourcing campaign is the crowd activation. EWS can provide a crucial role in activating the crowd before the event starts. This requires certain considerations and ideally should be done with a well-trained and institutionalized group of volunteers to perform the crowd tasking management. This is mainly to avoid spread of panic and misinformation. Based on the feedback from practitioners, it also increases the trust and acceptability of the crowdsourcing by the civil protection authorities. Furthermore, having information about possible impact of a predicted event helps in creating a story to attract volunteers to participate in the task and helps to effectively shape the crowd tasks.

7.3.4 Towards a Holistic System

Building on the ANYWHERE concept of the MH-EWS, which was intended to provide cutting-edge predictions for the main climate-driven extreme events and their potential impacts, the main idea of the SMAT and the crowdsourcing platform was to employ it to support risk reduction in all phases of the disaster management cycle. The key ingredient of the Anywhere solution was to maximize the possible interplay between the EWS, crowdsourcing, and social media analytics to speed-up the information flow, fill the information gaps and uncertainty of the single systems used in crisis response (EWS, EO, SM and Crowd) as well as to stimulate the participatory knowledge gathering, citizen's self-protection through improved understanding of the driving factors and the mechanism of the disaster.

It was not feasible to develop all components and bring the triggering and information exchange between different components into operational service. Therefore, based on the research and experience gathered in the ANYWHERE project, we propose a holistic approach how social media analytics can be combined with EWS and integrated the disaster risk management workflows. Similar approach can be proposed to complete the solution with other relevant sources of information (e.g. EO, in situ). There is work to be done in establishing standardized protocols on the triggering mechanism and data exchange between the impact-based modelling, e.g. humanitarian mapping applications. Same issues should be tackled between the social media analytics and EWS. Standardized trigger from EWS can be reused in many other applications (pre-tasking of rapid mapping, crowd pre-tasking, forecast based financing, etc.).

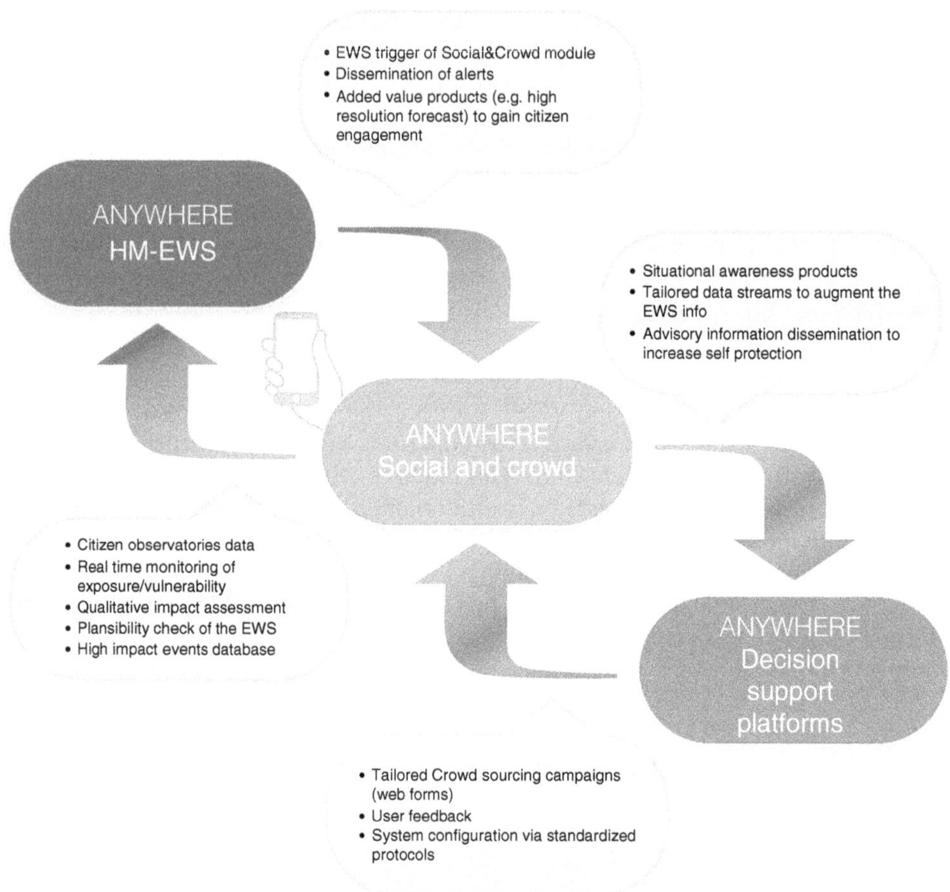

Figure 7.7 Schematic view of the ANYWHERE use of CS information.

In Figure 7.7 we illustrate the ideal future solution, building on the parts developed under ANYWHERE and focusing on the integration of SM analytics and crowdsourcing integration with EWSs.

In the following section, we describe the main components of the system. Starting from an EWS alert about possible extreme event in a standardized Common Alerting Protocol format is issued. This alert is used as a trigger to of the SMAT (Section 7.3.1) and Crowdsourcing (Section 7.3.2). Event-based, social media search is launched using filtering criteria from EWS (geographical constraint, tweets only in local language, names of possibly affected infrastructure and/or dominant points of interests). Information derived from EWS is also used in pre-configuring the crowdsourcing campaigns, which can be launched by responsible authority from the decision support platform. Campaign should be targeting local communities and potential eyewitnesses.

The system should also allow making better use of social media, civil protection authorities can launch a pre-configured information campaign, providing basic information about the event, information about potential impact (ideally tailored to crucial types of respondents), basic self-protection instruction and directives on how to provide feedback about the event. This information should be provided in parts, with the approaching start of the event. In order to further engage citizens in this information exchange, added value services (e.g. local impact forecasts, activity based forecast) and gamification approach can be used.

Information collected from SMAT and Crowdsourcing will be fed into the decision support system. Since all data feeds are provided as a standardized service (OGC, CAP, etc.), their integration in ANYWHERE or any other existing platform is almost seamless. Data derived from SMAT and Crowdsourcing can also be used in the EWS system as a plausibility check (before any other data become available, e.g. EO) and to augment the situational awareness products (e.g. impact information reported on social media). Information from SMAT can also help civil protection authorities in moderating the advisory SM campaigns depending on the feedback and live risk perception information derived from SM stream.

Another component building on the SMAT and Crowdsourced data stream is a citizen observatory of extreme events. This can serve as a living memory of the extremes but can also be complementing the existing databases of the extreme events. An important aspect to consider is to find the best balance between the best practices on loss and damage reporting and the feasibility of gathering this information via citizen science. Some simplifications should be introduced in the complex hazard and risk taxonomies, but the full data interoperability should be preserved. Impact Information derived from both sources can support disaster damage reporting obligation, evaluating the performance of the impact-based forecasting, and fine-tuning of critical thresholds used in early warning.

Another underexploited potential of using open volunteer geographic information is the monitoring of rapid changes in exposure and vulnerability. The location of vulnerable population and sensitive infrastructure is the key essence of the impact-based forecasting, but we mostly rely on the official information which is often missing or outdated. The EWS information about timing, magnitude and location of the extreme event can be used to trigger a volunteer mapping campaigns to complement available data on exposure and vulnerability as well as their temporal development during the crisis. This can be seamlessly integrated in the risk assessment modelling to continuously improve impact-based forecasts. Volunteer mapping serves as a very effective citizen engagement and educational activity. Improving citizen's understanding of the risk leads usually stronger commitment to participate. Practical applicability of data in all phases of the DRM cycle and other applications (e.g. navigation, urbanization) creates a side story is another highly motivating factor.

7.3.5 Facilitating Communication Between Actors in Emergency Management

It is important to highlight that one of the very substantial outcomes of the ANYWHERE project is the contribution to the general acceptance of crowd-generated content in the crisis management. Another primary outcome is the creation of a sustainable community of institutionalized and non-institutionalized volunteers connected to the ANYWHERE project who are ready to provide assistance in response to extreme events. This community actively participated in the development of ANYWHERE tools (partly introduced in this book) that were tailored to their specific needs. The network of ANYWHERE volunteers is being built around the main European and Global volunteer crisis response initiatives like the Virtual Operations Support Teams (VOST), Stand By Task Force, and HOT.

The ANYWHERE project organized three workshops where renowned scientists and representatives of the leading virtual volunteering organizations came together with all stakeholders involved in the civil protections and emergency management from the ANYWHERE pilot sites and beyond. This included national and regional civil protection officers, ministries of the interior, hydro-meteorological services, fire-brigades, civic organization and school managers. At these workshops different aspects of possible use of social media, crowdsourcing and citizen science in general were discussed with the participants. These ANYWHERE workshops created a unique ecosystem stimulating the practical application of these alternative data sources in effective crisis management. A constructive dialog between the experts and practitioners identified many problematic issues and barriers in the use of crowd-generated content limiting the use in its full potential. It also helped in shaping collaboration between volunteer organizations and civil protection. Overall, the workshops increased openness and acceptability of the citizen-generated content in DRM. At later stage of the project, the positive attitude of the practitioners resulted in a live co-design session where all innovative ways of combining EWS and social media monitoring as well as the use of crowd-generated data in impact forecasting were tested and tuned in actual emergency events (e.g. floods in Valencia in October 2018, or the storm surge event in Liguria in October 2018).

7.4 Conclusion

This chapter focused on the challenges and solutions identified in ANYWHERE for a better use of crowd-generated content in climate emergency situational awareness and early warning systems. In this regard, the chapter briefly introduced the concept of citizen science and its application in disaster risk management. The second half of the chapter presented the integration of crowd-generated content into DRM, which was approached in the project along the main lessons learnt that can be derived from these activities.

Several challenges are associated with the adoption and inclusion of citizen-generated content in DRM. The activities of ANYWHERE highlighted the challenges that are commonly acknowledged within literature such as those around the activation and engagement of the crowd in data collection. Whilst arisen challenges of the social media streaming were more context relate identified solutions can be adopted for different projects. Here below, the authors provide a summary of the most relevant best practices derived from the project that may support others the development of other projects.

First, an important lesson that was learnt is the adoption of a common language that is easy-to-understand for anyone engaged in citizen science activities but also that is integrable into existing (governmental) structures. This refers to the adoption of, e.g. EU taxonomies, best practices for loss reporting or standardized hashtags defined in the humanitarian exchange language developed by the United Nations. Furthermore, to increase citizens engagement, it emerged that it is important to create training campaigns for interested citizens during quite periods to provide feedback and to become the earliest responders during the crises. It is also important to provide easy to deploy and configurable crowdsourcing platform ideally based on open-source technologies, which is available on desktop and mobile devices without need of downloading and installation.

Second, using social media for event detection can be supported with tweet aggregation but foremost, it is beneficial to connect the social media streaming needs to EWSs to extract actionable content as authorities tend to use social media as a warning dissemination channel.

Lastly, the workshops organized for different DRM actors turned out to be successful in the sense that they enabled a dialogue between experts and practitioners. Initial resistance towards the adoption of crowd-generated content was turned into acceptance by many participants. Various national VOST teams became officially recognized by civil protection authorities. VOST Spain received a Civil protection medal in 2017 as a recognition of their work. In 2022 VOST Portugal received European Digital Volunteering Award within 112 Awards 2022.

Ideas born from the ANYWHERE project were translated into pilot innovative solutions within the Copernicus Hackathon Barcelona 2019. Application of exposure information derived from HOT data used in the impact-based forecast developed under ANYWHERE project was illustrated in Láng-Ritter et al. (2022).

Note

1. https://hxlstandard.org/.

References

Alam, F., Qazi, U., Imran, M., and Ofli, F. (2021). HumAID: human-annotated disaster incidents data from Twitter. *15th International Conference on Web and Social Media (ICWSM),* Virtual Venue (7–10 June 2021).

Albert, A., Balázs, B., Butkevičienė, E. et al. (2021). Citizen social science: new and established approaches to participation in social research. In: *The Science of Citizen Science* (ed. A. Albert, B. Balázs, E. Butkevičienė, et al.), 119–138. Cham: Springer International Publishing https://doi.org/10.1007/978-3-030-58278-4_7.

Alfonso, L., Chacón, J.C., and Peña-Castellanos, G. (2015). Allowing citizens to effortlessly become rainfall sensors. *E-proceedings of the 36th IAHR world Congress* (28 June–3 July 2015). The Hague.

ANYWHERE (2019). Deliverable 4.4: report describing the Crowdsourcing and social media communication report tools, including the recommendations for improvement provided by the users during the 5 first months of demonstrations.

Asingizwe, D., Poortvliet, P.M., Koenraadt, C.J. et al. (2020). Why (not) participate in citizen science? Motivational factors and barriers to participate in a citizen science program for malaria control in Rwanda. *PLoS One*. Edited by. A. Soundy 15 (8): e0237396. https://doi.org/10.1371/journal.pone.0237396.

Balázs, B., Mooney, P., Nováková, E. et al. (2021). Data quality in citizen science. In: *The Science of Citizen Science* (ed. B. Balázs, P. Mooney, E. Nováková, et al.), 139–157. Cham: Springer International Publishing https://doi.org/10.1007/978-3-030-58278-4_8.

Basiri, A., Haklay, M., Foody, G. et al. (2019). Crowdsourced geospatial data quality: challenges and future directions. *International Journal of Geographical Information Science* 33 (8): 1588–1593. https://doi.org/10.1080/13658816.2019.1593422.

Bielski, C. O'Brien, V., Whitmore, C. et al. (2017). Coupling early warning services, crowdsourcing, and modelling for improved decision support and wildfire emergency management. *2017 IEEE International Conference on Big Data (Big Data)* (ed. J.Y. Nie et al.), Boston, MA (11–14 December 2017), pp. 3705–3712. EOXPLORE UG, Weil Am Rhein, Germany, WE-Conference Proceedings Citation In. IEEE.

Danylo, O., Moorthy, I., Sturn, T. et al. (2018). The picture pile tool for rapid image assessment: a demonstration using hurricane Matthew. *ISPRS Annals of the Photogrammetry, Remote Sensing and Spatial Information Sciences* 4: 27–32. https://doi.org/10.5194/isprs-annals-IV-4-27-2018.

Ernst, C., Mladenow, A., and Strauss, C. (2017). Collaboration and crowdsourcing in emergency management. *International Journal of Pervasive Computing and Communications* 13 (2): 176–193. https://doi.org/10.1108/IJPCC-03-2017-0026.

European Commission (2021). *Evaluating the Impact of Nature-Based Solutions: A Handbook for Practitioners*. Luxembourg: European Commission https://doi.org/10.2777/2498.

Facebook Company (2020). Facebook data for good annual report 2020. https://scontent-arn2-1.xx.fbcdn.net/v/t39.8562-6/239409865_949264702522493_1292335916807550455_n.pdf?_nc_cat=106&ccb= 1-7&_nc_sid=ae5e01&_nc_ohc=GF1ZXJeJBOMAX9Z4O2u&_nc_ht=scontent-arn2-1.xx&oh=00_AT8tZuhz2FFG6uz7Rhdp6z1fCyEr3ZLCxrziaBS2kKQZMQ&oe=62B 0D74B (accessed 17 June 2022).

Gaitan, S., Calderoni, L., Palmieri, P. et al. (2014). From sensing to action: quick and reliable access to information in cities vulnerable to heavy rain. *IEEE Sensors Journal* 14 (12): 4175–4184. https://doi.org/10.1109/JSEN.2014.2354980.

Harrison, S. and Johnson, P. (2019). Challenges in the adoption of crisis crowdsourcing and social media in Canadian emergency management. *Government Information Quarterly* 36 (3): 501–509. https://doi.org/10.1016/j.giq.2019.04.002.

Haworth, B. (2016). Emergency management perspectives on volunteered geographic information: opportunities, challenges and change. *Computers, Environment and Urban Systems* 57: 189–198. https://doi.org/10.1016/j.compenvurbsys.2016.02.009.

Herfort, B., Lautenbach, S., Porto de Albuquerque, J. et al. (2021). The evolution of humanitarian mapping within the OpenStreetMap community. *Scientific Reports* 11 (1): 3037. https://doi.org/10.1038/s41598-021-82404-z.

Huang, G. (2020). Flood impact mitigation and resilience enhancement. IntechOpen. https://books.google.it/books?id=J60tEAAAQBAJ (accessed 17 June 2022).

Humanitarian OpenStreetMap Team (2020). Practical action in Bolivia: mapping to support ongoing response to the Bolivian wildfires. https://www.hotosm.org/projects/practical-action-in-bolivia-mapping-to-support-ongoing-response-to-the-bolivian-wildfires/ (accessed 17 June 2022).

Keßler, C. and Hendrix, C. (2015). The humanitarian eXchange language: coordinating disaster response with semantic web technologies. *Semantic Web* 6 (1): 5–21. https://doi.org/10.3233/SW-130130.

Kloetzer, L., Lorke, J., Roche, J. et al. (2021). Learning in citizen science. In: *The Science of Citizen Science* (ed. L. Kloetzer, J. Lorke, J. Roche, et al.), 283–308. Cham: Springer International Publishing https://doi.org/10.1007/978-3-030-58278-4_15.

Klonner, C., Usón, T.J., Aeschbach, N. et al. (2021). Participatory mapping and visualization of local knowledge: an example from Eberbach, Germany. *International Journal of Disaster Risk Science* 12 (1): 56–71. https://doi.org/10.1007/s13753-020-00312-8.

Láng-Ritter, J., Berenguer, M., Dottori, F. et al. (2022). Compound flood impact forecasting: integrating fluvial and flash flood impact assessments into a unified system. *Hydrology and Earth System Sciences* 26 (3): 689–709. https://doi.org/10.5194/hess-26-689-2022.

Mandement, M. and Caumont, O. (2020). Contribution of personal weather stations to the observation of deep-convection features near the ground. *Natural Hazards and Earth System Sciences* 20 (1): 299–322. https://doi.org/10.5194/nhess-20-299-2020.

Monaco, N. and Adnaudo, D. (2020). *Data Analytics for Social Media Monitoring: Guidance on Social Media Monitoring and Analysis Techniques, Tools and Methodologies*. National Democratic Institute for International Affairs (NDI). https://www.ndi.org/sites/default/files/NDI_Social%20Media%20Monitoring%20Guide%20ADJUSTED%20COVER.pdf (accessed 21 June 2023).

Negri, V., Scuratti, D., Agresti, S. et al. (2021). Image-based social sensing: combining AI and the crowd to mine policy-adherence indicators from Twitter. *2021 IEEE/ACM 43rd International Conference on Software Engineering: Software Engineering in Society (ICSE-SEIS)* Madrid, ES (25–28 May 2021). IEEE. https://doi.org/10.1109/ICSE-SEIS52602.2021.00019.

Neset, T.-S., Wilk, J., Cruz, S. et al. (2021). Co-designing a citizen science climate service. *Climate Services* 24: 100273. https://doi.org/10.1016/j.cliser.2021.100273.

van Noordwijk, T.C.G.E., Bishop, I., Staunton-Lamb, S. et al. (2021). Creating positive environmental impact through citizen science. In: *The Science of Citizen Science* (ed. T.C.G.E. van Noordwijk, I. Bishop, S. Staunton-Lamb, et al.), 373–395. Cham: Springer International Publishing https://doi.org/10.1007/978-3-030-58278-4_19.

OASIS (2010). Common alerting protocol V1.2. http://docs.oasis-open.org/emergency/cap/v1.2/CAP-v1.2-os.pdf (accessed 17 June 2022).

Pateman, R., Dyke, A., and West, S. (2021). The diversity of participants in environmental citizen science, citizen science. *Theory Into Practice* 6 (1): 9. https://doi.org/10.5334/cstp.369.

Pelacho, M., Rodríguez, H., Broncano, F. et al. (2021). Science as a commons: improving the governance of knowledge through citizen science. In: *The Science of Citizen Science* (ed. M. Pelacho, H. Rodríguez, F. Broncano, et al.), 57–78. Cham: Springer International Publishing https://doi.org/10.1007/978-3-030-58278-4_4.

Reuter, C. and Spielhofer, T. (2017). Towards social resilience: a quantitative and qualitative survey on citizens' perception of social media in emergencies in Europe. *Technological Forecasting and Social Change* 121: 168–180. https://doi.org/10.1016/j.techfore.2016.07.038.

Reuter, C., Ludwig, T., Kaufhold, M.A. et al. (2016). Emergency services' attitudes towards social media: a quantitative and qualitative survey across Europe. *International Journal of Human-Computer Studies* 95: 96–111. https://doi.org/10.1016/j.ijhcs.2016.03.005.

Saharia, D., Singhal, S., Jain, A. et al. (2020). INDRA reporter: crowdsourcing quantitative flooding data with impact classifications to develop real-time flood risk dashboards. American Geophysical Union, Fall Meeting 2020, abstract #SY011-0009.

Santoni, V. and Rufat, S. (2021). How fast is fast enough? Twitter usability during emergencies. *Geoforum* 124: 20–35. https://doi.org/10.1016/j.geoforum.2021.05.007.

See, L. (2019). A review of citizen science and crowdsourcing in applications of pluvial flooding. *Frontiers in Earth Science* 7: https://doi.org/10.3389/feart.2019.00044.

Spielhofer, T., Greenlaw, R., Markham, D. et al. (2016). Data mining Twitter during the UK floods: investigating the potential use of social media in emergency management. *2016 3rd International Conference on Information and Communication Technologies for Disaster Management (ICT-DM)*, Vienna (13–15 December 2016), pp. 1–6. IEEE. https://doi.org/10.1109/ICT-DM.2016.7857213.

Toomey, A.H. and Domroese, M.C. (2013). Can citizen science lead to positive conservation attitudes and behaviors? *Human Ecology Review Society for Human Ecology* 20 (1): 50–62. http://www.jstor.org/stable/24707571.

UNISDR (2015). Sendai framework for disaster risk reduction 2015-2030. Geneva. https://www.undrr.org/publication/sendai-framework-disaster-risk-reduction-2015-2030 (accessed 17 June 2022).

Verbeiren, B., Seyoum, S.D., Lubbad, I. et al. (2019). FloodCitiSense: early warning service for urban pluvial floods for and by citizens and city authorities. In: *New Trends in Urban Drainage Modelling: UDM 2018 11* (ed. B. Verbeiren, S.D. Seyoum, I. Lubbad, et al.), 660–664. Springer International Publishing https://doi.org/10.1007/978-3-319-99867-1_114.

de Vos, L., Leijnse, H., Overeem, A. et al. (2017). The potential of urban rainfall monitoring with crowdsourced automatic weather stations in Amsterdam. *Hydrology and Earth System Sciences* 21 (2): 765–777. https://doi.org/10.5194/hess-21-765-2017.

Wang, B., Liu, B., and Zhang, Q. (2021). An empirical study on Twitter's use and crisis retweeting dynamics amid Covid-19. *Natural Hazards* 107 (3): 2319–2336. https://doi.org/10.1007/s11069-020-04497-5.

WeatherCloud (2012). Real-time weather monitoring online platform. https://weathercloud.online (accessed 17 June 2022).

World Meteorological Organization (2017). Crisis mapping and crowdsourcing in flood management. https://www.floodmanagement.info/publications/tools/APFM_Tool_26_e.pdf (accessed 17 June 2022)

Xu, Z., Liu, Y., Xuan, J. et al. (2017). Crowdsourcing based social media data analysis of urban emergency events. *Multimedia Tools and Applications* 76 (9): 11567–11584. https://doi.org/10.1007/s11042-015-2731-1.

Yin, J., Lampert, A., Cameron, M. et al. (2012). Using social media to enhance emergency situation awareness. *IEEE Intelligent Systems* 27 (6): 52–59. https://doi.org/10.1109/MIS.2012.6.

8

Co-Evaluation: How to Measure Achievements in Complex Co-Production Projects? ANYWHERE's Contribution to Enhance Emergency Management of Weather and Climate Events

Oliver Gebhardt[1] and Christian Kuhlicke[2,3]

[1] *Department of Economics, Helmholtz Center for Environmental Research GmbH – UFZ, Leipzig, Germany*

[2] *Department of Urban & Environmental Sociology, Helmholtz Center for Environmental Research GmbH – UFZ, Leipzig, Germany*

[3] *Institute for Environmental Sciences and Geography, University of Potsdam, Potsdam, Germany*

8.1 Introduction

Early warning is a complex socio-technical process that is based on an understanding of relevant (i) hazardous processes as well as physical, social and economic vulnerabilities, (ii) the establishment and maintenance of forecasting, sensoring, monitoring and warning devices as well as (iii) channels and networks for disseminating and communicating information and is accompanied by (iv) capacities on various levels to understand incoming information, pass it on and act upon it in a resilient manner (Basher 2006). As a consequence, any scientific attempt to establish, improve or transform an early

Responding to Extreme Weather Events, First Edition. Edited by Daniel Sempere-Torres, Anastasios Karakostas, Claudio Rossi, and Philippe Quevauviller.

warning systems needs to take into account various nodes and linkages between different entities and should have a clear understanding of needs, expectations, working routines, capacities and roles of relevant organizations as well as the distribution of responsibilities among and between them, which implies the necessity to engage with representatives from weather and hydrological services, disaster risk management, infrastructure operators (including critical infrastructures), volunteers and members of the wider public (Drobot and Parker 2007; Handmer 2000).

However, current research foci both in the natural and social sciences (as well as more interdisciplinary perspectives) hardly reflect the comprehensiveness of the process. Research is usually conducted in a segmented manner and explores single aspects of the early warning process. While there is an emerging debate arguing that decision-makers' needs and expectations should be reflected and considered more thoroughly, the need analysis is usually restricted to a narrow large-scale national or international consultation process by means of standardized quantitative survey, which hardly reflects the great variety of operational and institutional contexts of operators and users of early warning systems (Kox et al. 2015).

A fundamental transformation of the relationship between science, policymaking and practice has been demanded that aims at overcoming the linear model of tool development (Basher 2006) suggesting that basic research translates into applied research in order to generate results, which in turn are more or less automatically taken up by policymakers and operators. Yet, conducting research on pressing societal issues, such as increasing the warning capacity, does not automatically result in research outputs that help to address or even overcome this problem (Sarewitz and Pielke Jr 2007). Therefore, some time ago Morss et al. have called for an 'end-to-end' research approach by linking 'all stages of the research and development process, from research (end 1) through to delivering an application to the end user (end 2)' in order to link operators' and users' needs in the research process from the very beginning (Morss et al. 2005). The consideration of practical knowledge for research and development projects is increasingly valued (Oliver et al. 2019).

There are various forms of collaborative research and development practices and recent research examines meaningful categorizations of distinct approaches well as the delineation of different phases of co-production (Chathoth et al. 2013; Etgar 2008; Lusch and Vargo 2006). The motivations discussed in the literature for initiating co-production processes are manifold. They can be (i) substantive, i.e. improving the quality of results by enabling a more holistic understanding of contexts, problems or solutions; (ii) instrumental, i.e. aiming for an effective utilization of findings; (iii) normative, i.e. assuming an intrinsic value of co-production; or (iv) political, i.e. stressing the importance of empowerment and legitimacy of such processes (Oliver et al. 2019). Furthermore, co-production-oriented research has focussed specifically on factors influencing co-production processes (Etgar 2008; Flatscher and Riel 2016; Voorberg et al. 2015). In view of the at least implicit claim of co-production being a superior mode of developing new technologies, processes or services, surprisingly, only few evaluation studies are available to date and, interestingly, the majority of these studies focusses on evaluating the process rather than the outcome of co-production. (Voorberg et al. 2015). By presenting the ANYWHERE co-evaluation framework and reflecting on its application, we would like to contribute to filling this gap.

8.2 Application of the ANYWHERE Co-Evaluation Framework

The development of the ANYWHERE platform was designed as a collaborative process, taking the expectations and needs of the users not only as a starting point but also their assessment as central to the evaluation of the result. Hence, the evaluation of the development process as well as that of the platform prototype were also designed in a very collaborative way and, therefore, termed *co-evaluation*. The involvement of those contributing to the development of the system, especially of the users at the pilot sites, ran through almost all phases of the co-evaluation process.

Two types of co-evaluation were carried out in the development of the ANYWHERE platform. Since the majority of evaluation studies that can be found in the literature focusses on the process level, we will refrain from going into more detail on the co-evaluations of the iterative co-production process here. Instead, the comparative assessment of the outcomes of co-production will be presented.

Still, in a nutshell, it can be said that using a broad array of participatory face-to-face and online approaches, an overview of needs, expectations and available expertises was compiled, synthesized and shared among all parties. A short list of the most relevant anticipated functionalities was detailed for each pilot site. On a regular basis, it was assessed to what extends needs and expectations with regard to inputs and co-production processes were met.

The co-evaluation of the pilot site-specific configurations of the ANYWHERE platform from a user-perspective needs to be distinguished from the validation of algorithms or the verification and validation of the multi-hazard early warning system and disaster risk management platform, which are described elsewhere. The conjoint evaluation of the ANYWHERE platform aims at investigating its added value for users through comparative evaluation of their existing operational systems, i.e. legacy systems, and site-specific set-ups of their respective emergency management systems including the functionalities of the ANYWHERE platform. The co-evaluation builds upon the understanding the existing working routines, settings and demands for an improved emergency management system. The operation of the legacy systems as well as the systems including the ANYWHERE platform are conceptualized and described in detail as baseline scenario (legacy system) and ANYWHERE scenario (system including ANYWHERE functionalities).

Co-evaluations were not only aimed at supporting the market outreach of the ANYWHERE platform by providing evidence on the changes the use of the ANYWHERE platform can make in real-world emergency management operations or facilitating judgements on the transferability of these results to other context conditions but also to assist the sustainable implementation of the platform at the pilot sites. The comparative assessment was meant to enhance the visibility and, thereby, the perception of relative advantage of the ANYWHERE platform by individual operators as well as by the respective organizational unit and potential new customers. The basic structure of the ANYWHERE co-evaluation methodology can be described by the following six steps:

- Step 1: Context analysis
- Step 2: Description of baseline scenario and ANYWHERE scenario

- Step 3: Selection of suitable and feasible criteria
- Step 4: Selection of appropriate co-evaluation method
- Step 5: Data collection
- Step 6: Data aggregation and analysis

8.2.1 Step 1: Context Analysis

The context analysis was based on a comprehensive investigation of the working and decision-making context at the pilot sites during field visits and partner consultations. Operators' and coordinators' expectations as well as their process- and output-related needs were discussed and synthesized. Roles and responsibilities of different actors as well as the data sources used, interactions and information flows during the emergency management processes were mapped and described in detail. In this way, the temporal structures of the different warning phases and decision-making processes were unvealed.

8.2.2 Step 2: Description of Baseline Scenario and ANYWHERE Scenario

The context-analysis laid the foundation of the description of the baseline scenario, i.e. the emergency management processes at the pilot sites using the respective legacy systems. The description of the ANYWHERE scenarios, i.e. the specific set-up of the ANYWHERE systems at the pilot sites, was provided by partners by means of a dedicated report describing the site-specific platform prototypes, i.e. A4LIG (Genoa/Liguria region), A4CAT (Catalonia), A4NOR (Rogaland), A4FINN (South Savo), A4ALPS (Bern Canton-Alps) and A4COR (Corsica).

8.2.3 Step 3: Selection of Suitable and Feasible Criteria

The performance of the ANYWHERE platform and the respective legacy systems was planned to be assessed along the lines of the four impact areas 'economic', 'social', 'technological' and 'environmental'. For these impact areas corresponding 'effects', 'criteria', 'sub-criteria' (if applicable) and 'indicators' were specified. The comprehensive list of co-evaluation criteria was based on inputs from initial enquiry at the beginning of the development process and pilot site visits (bottom-up) as well as the so-called SEQUOIA methodology (Cucco et al. 2016; Monacciani et al. 2011)[1] and ISO/IEC norms 25010, 25022, 25023 for evaluating software quality (top-down).

Pilot site partners rated the criteria listed with regard to their suitability and feasibility. Suitability represented the ability of criteria to capture relevant performance aspects of the emergency management systems.[2] Feasibility primarily addressed the data collection process, i.e. required reflection on data availability and opportunities for data collection in the context of ANYWHERE demonstration activities.[3]

So, the determination of the final set of co-evaluation criteria was performed in an iterative manner. First, a pre-selection of criteria was conducted by the pilot site partners and, second, based on bilateral consultations with the co-evaluation experts a site-specific set of criteria was determined. This final set formed the basis for the co-evaluation questionnaire, which was comprised of criteria which were assessed to be relevant or highly relevant and for which data collection would be feasible with high or moderated effort. This questionnaire was used for the midterm co-evaluations at the pre-test sites in spring 2019. Third, the actual (and not the expected) ability to collect data for each criterion and corresponding indicator determined the final set of criteria to be used at each pilot site. As a consequence of this iterative process the number of criteria and share of impact areas covered by the criteria varied across sites.

Through the filtering process described above, i.e. local partners' preferences and data availability restrictions, depending on the pilot site, between 31 (A4NOR in Rogaland) and 50 (A4COR in Corsica) criteria belonging to the categories of social and technological impacts were finally considered in the co-evaluation.[4] About $^3/_4$ of all criteria considered, addressed social impacts while the rest focused on technological impacts. For most of the sites within the category of social impacts the shares of working routine-related and knowledge production and sharing-related criteria were quite equal.

8.2.4 Step 4: Selection of Appropriate Co-Evaluation Method

A comparative appreciation of different potential evaluation methods, i.e. Cost-Benefit Analysis (CBA),[5] Cost-Effectiveness Analysis (CEA)[6] and Multi-Criteria Analysis (MCA),[7] let to the conclusion that weighting the pros and cons of the approaches, acknowledging the properties of the performance data to be dealt with (scales, metrics, level of uncertainty) and reflecting the pre-conditions of the application of the various MCA approaches at hand,[8]we decided for the use of a probabilistic multi-criteria outranking approach to comparatively assess the performance of the legacy system and the respective system including ANYWHERE functionalities.[9]

8.2.5 Step 5: Data Collection

Once the legacy systems and ANYWHERE systems were specified, the performance criteria and indicators were selected and evaluation methods chosen the data collection process was initiated. A specific data collection strategy was determined for each pilot site. For each indicator capturing single performance aspects of the ANYWHERE system and the legacy system data was collected. Due to the lack of statistical performance data substantial shares of the information used for the co-evaluation was sourced through questionnaire-based expert interviews with operators and coordinators who were experienced in using the ANYWHERE platform.

The co-evaluation questionnaire was pre-tested with personal interviews with operators and coordinators in April and May 2019 at the two pilot sites South Savo and Catalonia, which represented the varying degrees of maturity of the prototypes at the different ANYWHERE sites. Performance data was collected through personal interviews at the end of the demonstration period, i.e. in September and October 2019.

The relative importance of the criteria was determined through an interactive exercise based on the revised SIMOS weighting technique.[10]

The team of the Helmholtz Centre for Environmental Research (UFZ) conducted the interviews for A4ALPS, A4CAT, A4FINN in person on-site and for A4NOR via a four hours Skype conference. The UFZ team trained ANYWHERE partners of the European Centre for Medium-Range Weather Forecasts – ECMWF and Consorzio Futuro in Ricerca – CFR to conduct the interviews for the co-evaluation of A4COR and A4LIG. These partners also took part in the interviews for the assessment of A4CAT to ensure that the same interview routine was to be followed. The UFZ team assisted the interviews in Liguria via Skype conference.

All completed questionnaires, documentations of the criteria weighting and other interview data were digitalized for further use. A database was compiled, which formed the basis for the statistical analyses as well as MCA conducted.

8.2.6 Step 6: Data Aggregation and Analysis

An interactive software for Probabilistic Multi-Attribute Evaluation PRIMATE[11] was used to assist data aggregation and the comparative assessment of the two systems at each pilot site. The MCA module of PRIMATE is based on the outranking concept PROMETHEE (Preference Ranking Organisation Method for Enrichment Evaluations), which performs a pairwise comparison of the alternative systems across all evaluation criteria.

The results of the MCA were presented in PRIMATE in various ways. This included not only the overall performance of the alternative systems but also their strengths and weaknesses with regard to each performance aspect considered. As a sensitivity analysis separate MCAs were conducted using the criteria weights elicited by local coordinators and operators individually.

8.3 Discussion of Co-Evaluation Results

Co-evaluations were conducted to reveal the relative advantages and disadvantages of the ANYWHERE system compared to the legacy system at pilot site level. The following site-specific results were available.

Comprehensive analysis of the user experience: This included (i) general impressions of the ANYWHERE platform and its integration into existing emergency management processes, (ii) main advantages and disadvantages the new system compared to legacy system and (iii) an assessment of usefulness of the what local coordinators and operators considered to be the most relevant features of their emergency management systems.

Figure 8.1 provides an example of such an assessment at the Corsican pilot site. The usefulness of various features was assessed for the categories *data inputs for decision-making, data processing support* and *facilitation of action* using a seven-point Likert scale. The dark line reflects the result for the legacy system and the bright line that of the new management system. The larger the enclosed area, the more useful the specific features were considered to be. Figure 8.1 shows at first glance the relative advantage of the new system in this respect from the perspective of the operators and coordinators at the Corsican site.

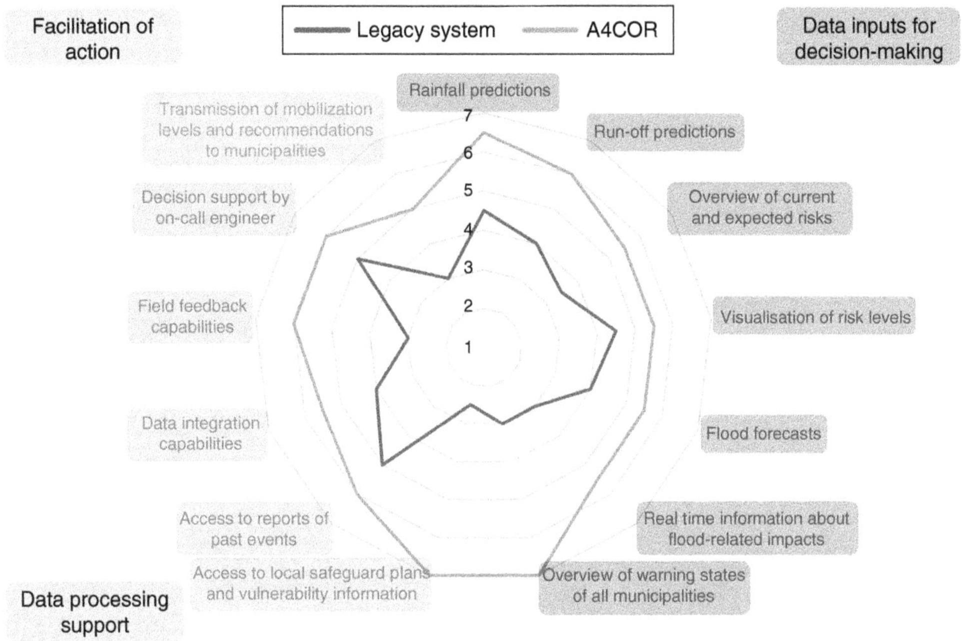

Figure 8.1 Comparative assessment of usefulness of relevant features of the emergency management systems at the Corsican pilot site.

Using such descriptive analysis, the performance of the two systems could also be compared in terms of support for work routines and various aspects of knowledge production as well as more technical aspects. Figures 8.2–8.4 illustrate the results of this comparison as mean values of the operators' and coordinators' assessments at the pilot site in Corsica.

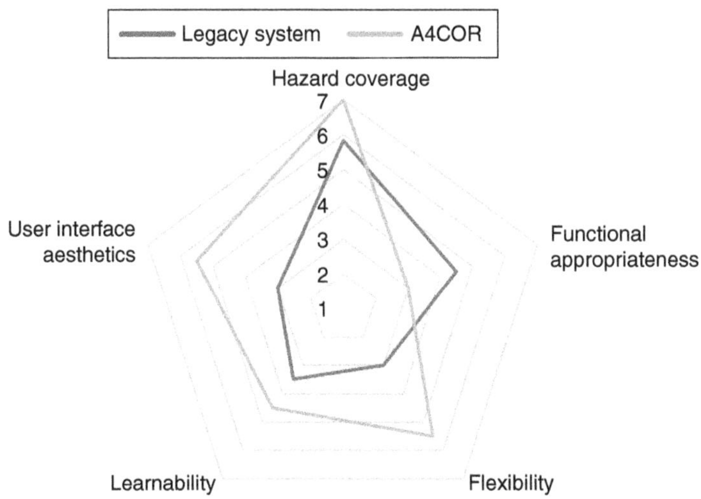

Figure 8.2 A4COR – comparison of emergency management systems based on knowledge production-related criteria in Corsica.

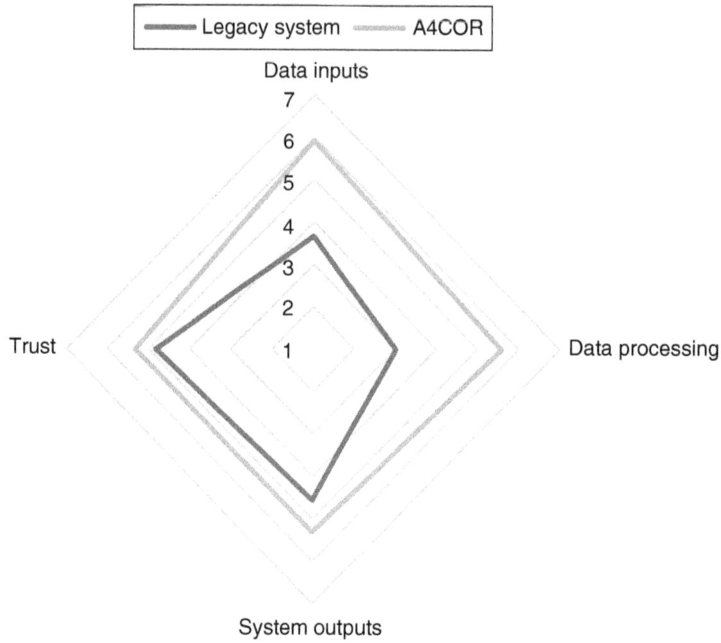

Figure 8.3 Comparison of emergency management systems based on working routine-related criteria in Corsica.

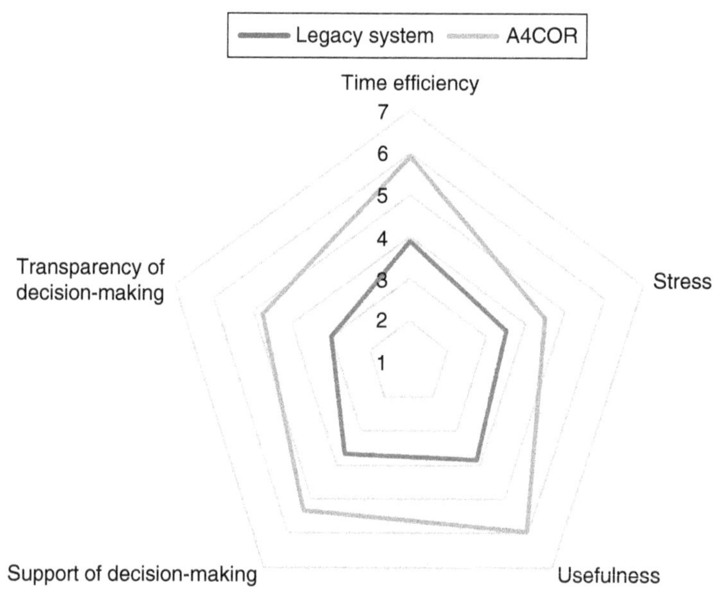

Figure 8.4 Comparison of emergency management systems based on technical criteria in Corsica.

As it was necessary to weight the individual performance aspects used as criteria for the co-evaluation in order to conduct the MCA, an overview of the relative importance of these performance aspects could be created for each pilot site. Figure 8.5 gives such an overview from the perspective of the operators as well as the coordinators in Corsica.

These descriptive analyses were supplemented by the results of the MCA, which were available at different levels of aggregation. Due to the use of a probabilistic variant of MCA, at the highest aggregation level this included a probability value with which the emergency management system, supplemented with ANYWHERE functionalities, outperformed the legacy system. At all pilot sites, the probability of the new system outperforming the legacy system was higher, but differences existed. The co-evaluation results in Catalonia, South Savo, Rogaland and Corsica (all 100%) and with some limitations in Genoa/Liguria region (80%) were very clearly in favour of the ANYWHERE systems. A neck-and-neck race took place in Bern Canton-Alps (Switzerland), where A4ALPS outperformed the legacy system with a probability of 55%. In addition, MCA results could be presented for individual criteria (see Figure 8.6) and separately for the user types operator and coordinator.

In Figure 8.5, the height of the bars indicates the relative importance of each criterion in measuring the performance of the emergency management systems. The higher the dark bar (operator's perspective) or the bright bar (coordinator's perspective), the more important the respective performance aspect was considered to be. In the Corsican case, we can see that the variation was much greater for the operators than for the coordinators.

In Figure 8.6, the dark bars represent the new system and the bright bars the legacy system. The bars can indicate negative or positive preference values. The higher/lower the respective bar, the stronger/lower the preference of the users for the respective system with regard to this aspect compared to the reference system. In this particular example, it can be seen that the ANYWHERE system was preferred especially in terms of data management and processing.

Cross-site analysis was not trivial as the main focus of co-evaluation was on the pilot site and, hence, site-specific criteria sets were used. Figure 8.7 shows a comparison based on criteria used at least at four of the six pilot sites. It reveals that the emergency management systems including the ANYWHERE products outperformed the legacy systems for $^{2}/_{3}$ of the criteria. Systems performed very similar for four criteria and the legacy systems outperformed the ANYWHERE systems with regard to only three criteria.

Figure 8.8 provides an overview of how the ANYWHERE systems performed compared to the respective legacy systems across all pilot sites by presenting the mean criteria value differences. Strong advantages of the AYNWHERE systems were identified with regard to the following performance aspects: Time efficiency (+2.07 on a 7-point scale), Usefulness (+1.85), Ease of data processing (+1.60), Quality of data input

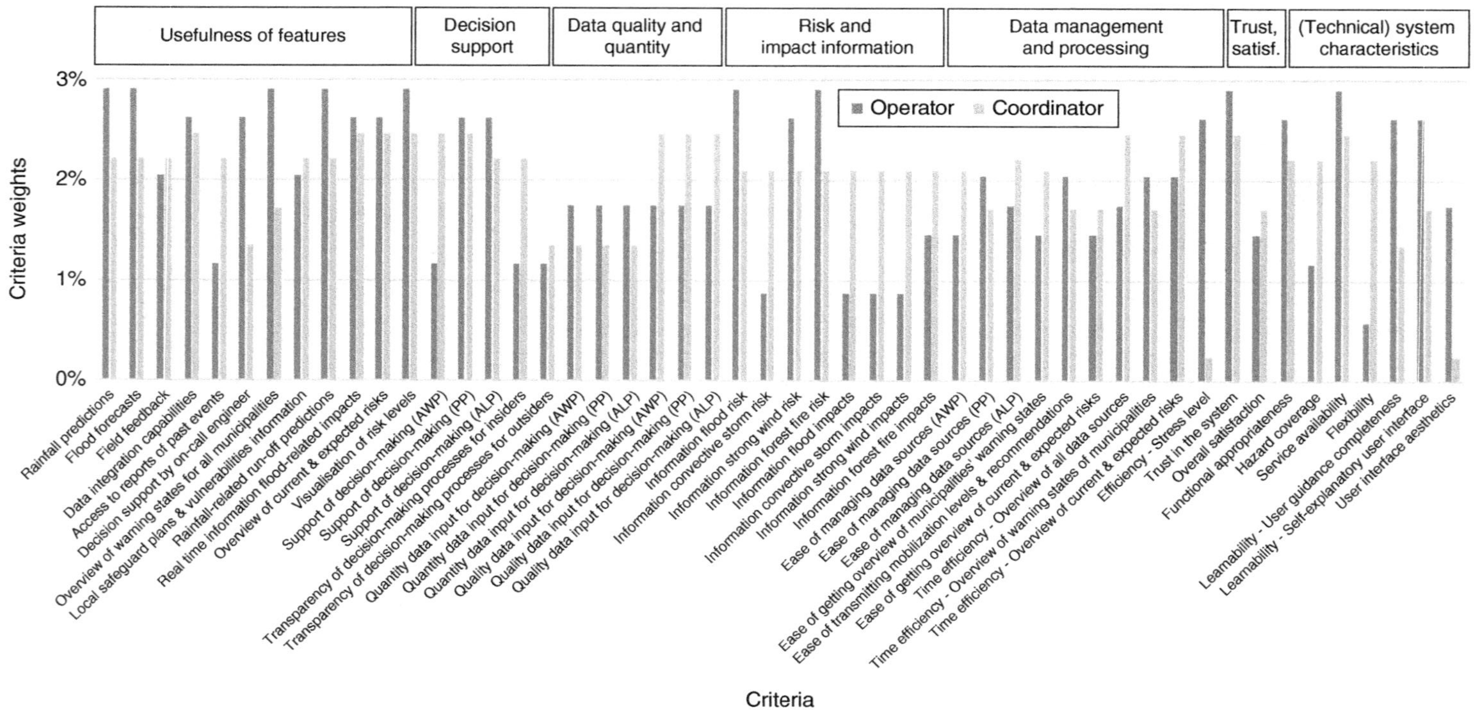

Figure 8.5 Relative importance of all co-evaluation criteria at the pilot site Corsica as specified by operators and coordinators.

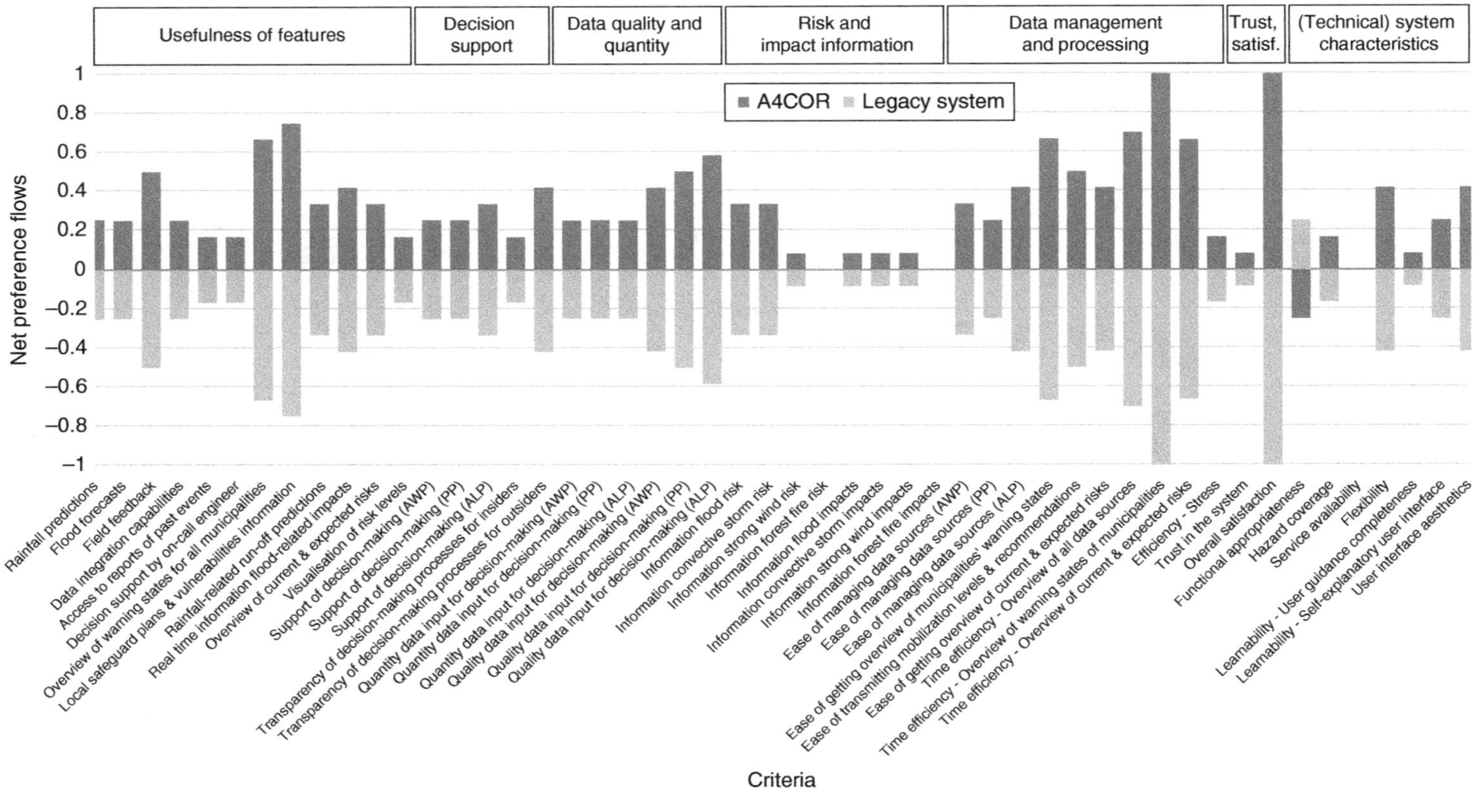

Figure 8.6 Performance comparison of A4COR and legacy system by criterion at the pilot site Corsica.

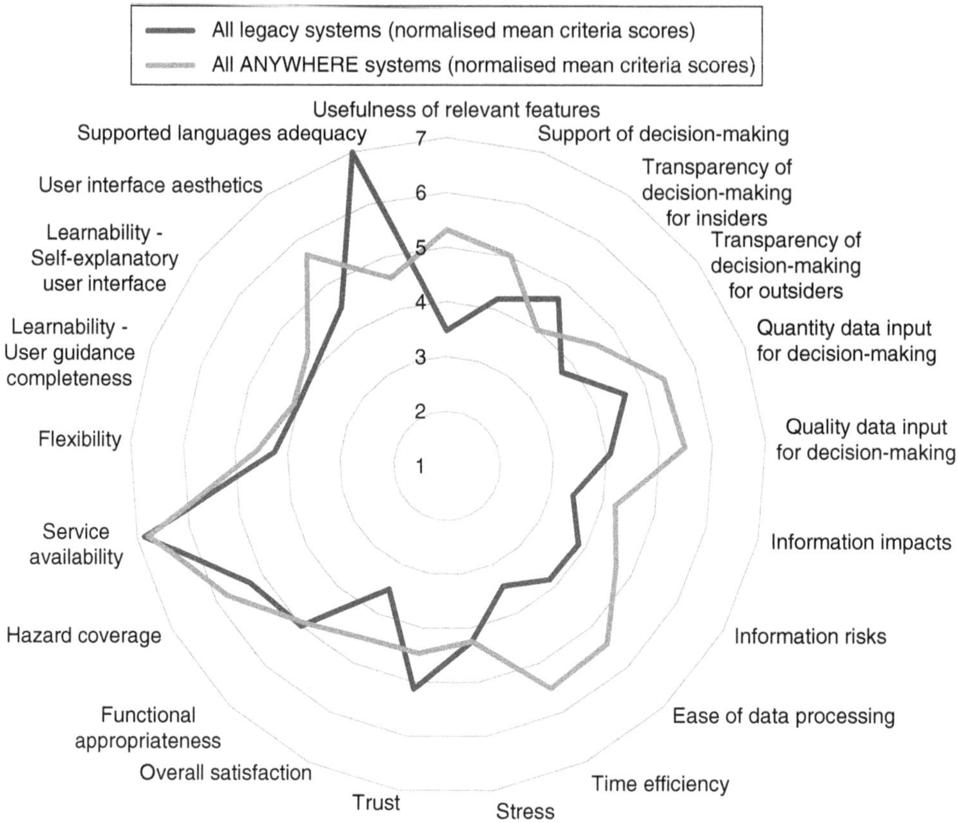

Figure 8.7 Performance comparison of ANYWHERE systems and legacy systems: all criteria.

for decision-making (+1.42), User interface aesthetics (+1.17), Overall satisfaction (+1.0). Advancements were also recorded with respect to Support of decision-making (+0.83), Knowledge about relevant impacts (+0.82), Transparency of decision-making for outsiders (+0.81), Knowledge about relevant risks (+0.81) and Quantity of data input for decision-making (+0.78). The legacy systems performed better with respect to Trust (–0.66), Transparency of decision-making for insiders (–0.70) and substantially better regarding Supported languages adequacy (–2.4). It is to be expected that this underperformance might only by temporal. A plausible reason for this assumption is the still low degree of maturity of the system at the time of data collection, i.e. the end of the demonstration period. The final version of the ANYWHERE platform will feature user interfaces in all required languages and it can be expected that trust and understanding of particular functionalities only develop over time as users become familiar with the new systems.

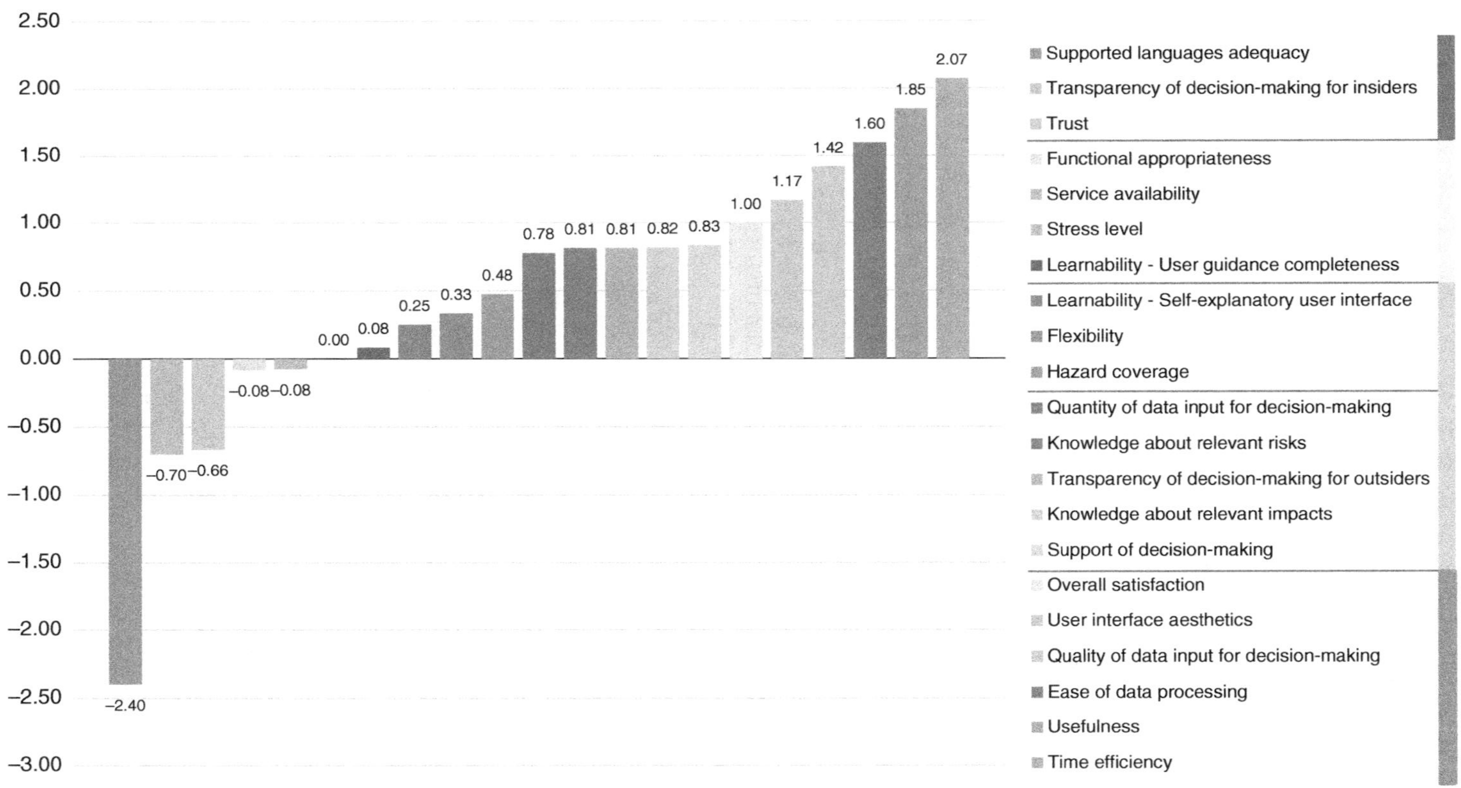

Figure 8.8 Performance comparison of ANYWHERE systems and legacy systems: mean value differences of criteria.

8.4 Discussion

When conducting such a co-evaluation the definition of the objects to be compared forms the first important step of the procedure, in this particular case defining distinct emergency management systems. Even though there are difference in the degree of substitution for most pilot sites the respective ANYWHERE systems rather complemented than substituted the legacy systems. Hence, we decided to rather compare scenarios, i.e. the use of the legacy system in a baseline scenario and the ANYWHERE scenario including the new functionalities in the emergency management process. Still, for comprehensibility reasons we chose to use the term 'system' rather than 'scenario' in this paper. From a theoretical perspective this co-existence made the comparison fuzzier, but after having clarified at the beginning of the data collection interviews what is to be compared interviewees were very capable of thinking in these two scenarios and answered accordingly.

Another key challenge was that each system installed was unique and that each pilot site was uniquely defined by a specific set of context conditions. Therefore, we took great effort to map relevant context conditions and to understand how different stakeholders were interacting and relying on early warning platform. In addition, the ANYWHERE systems were evolving during the demonstration period as the systems were constantly updated and adapted according to the requirements of users. Taking this flux into account, we also needed to constantly update our data collection strategy.

We decided to primarily rely on qualitative data for the co-evaluation. This decision was grounded both in a lack of statistical, more quantitative data and in our ambition to understand the advantages and disadvantages of the ANYWHERE system from the perspective those actually operating and steering the emergency management process on-site. Therefore, we focused on their perception and assessments. We took great effort to identify the right persons to obtain data from by means of semi-structured interviews containing both closed and a set of open questions. It was required that these persons had to be engaged in the co-development process of the new platform, they had to be knowledgeable, i.e. used both systems, and had to be able and willing to share their experiences. The interviews were conducted in the actual working environments, hence questions and answers could be contextualized easily.

MCA was used for the co-evaluation as it allowed for using qualitative data sourced the users, the consideration of their vision of what performance aspects have to be taken into consideration and what weight those aspects should have in the final aggregation of the preference scores. This method enabled us, thus, to systematically capture and compare different performance aspects to cater for different perspectives and present the results in a comprehensible way.

An additional benefit of the co-production and co-evaluation process stimulated have been the considerable learning effects as the development of the ANYWHERE system was progressing. This included, above all, peer-to-peer learning across sites. Many pilot site partners were actively exchanging with their counterparts at other sites both on the usability of the ANYWHERE platform but also on advancements in emergency management in more general terms. This included also assessments of the usefulness of existing extreme weather-related products. The iterative needs refinement became an important basis for future further developments of the emergency management systems.

8.5 Conclusion

We presented the results of testing a generic framework developed for a user-centric co-evaluation of the collaboratively produced ANYWHERE platform. The developed framework proved to be flexible enough to be adapted to heterogeneous environments, context-specific needs, preferences expectations and starting conditions as well as providing a solid basis for rigorously comparing co-evaluation outcomes between different working environments.

In this paper, we also shed some light not only on the overall performance of the ANYWHERE systems compared to the legacy systems but more importantly on their specific advantages and disadvantages, on the achievements as well as on options for improvement. This formed a highly appreciated basis for further integration of ANYWHERE products into the existing emergency management systems and their further development at the respective sites.

Our cross-site analysis revealed that the ANYWHERE systems outperformed the legacy systems for $^2/_3$ of performance criteria selected by the users. It is to be expected that with regard to the aspects for which such an advantage did not (yet) exist at the time of data collection, i.e. the end of the demonstration period, it is expected to be achieved timely through the addition of requested supported languages and increasing experience through more intensive use of the systems.

The presented methodological framework has proven to be suitable for the co-evaluation of innovative systems in the field of early warning and emergency management.

Notes

1. This approach was originally developed for project evaluation in the context of the EU FP7 research project 'Socio-Economic Impact Assessment for Research Projects (SEQUOIA)'. Since then, it was adapted and applied several times in very different fields, e.g. for a Pan-European disaster inventory (SecInCoRe), for Digital Cultural Heritage projects (Maxiculture), for Digital Social Innovation projects (iA4Si), for e-Infrastructure projects (ERINA+) and for automated service-oriented architecture testing infrastructures (MIDAS).
2. For each criterion it was assessed whether it addresses an (i) irrelevant performance aspect, (ii) moderately relevant performance aspect, (iii) relevant performance aspect, (iv) highly relevant performance aspect.
3. For each criterion it was assessed whether it is (i) very unlikely that information can be collected, (ii) information can be collected with high effort, (iii) information can be collected with moderate effort through consultation of actors outside the ANYWHERE consortium or (iv) information can be collected with moderate effort through consultation of ANYWHERE partners.
4. The following number of criteria were used for the co-evaluations of the individual systems: A4LIG (Genoa/Liguria region) 34 criteria, A4CAT (Catalonia) 45 criteria, A4FINN (South Savo) 34 criteria, A4ALPS (Bern Canton-Alps) 33 criteria.
5. A CBA's main objective is to find the most efficient, i.e. optimal option. Its basic rationale is to relate the aggregated benefits of an alternative to its costs to determine its net benefits compared to a reference 'baseline' option. If different options are compared the alternative with the highest net benefit is selected. For more information on CBA see Hanley and Spash (1993), Hansjürgens (2004), Young (2005). For applications in the context of natural hazard

management see MAFF (1999), Brouwer and Kind (2005), Pearce and Smale (2005), Turner et al. (2007), Thöni et al. (2009), Meyer et al. (2012, 2014).

6. A CEA relates the (monetary) costs of an alternative to its effects, which are measured using a non-monetary target indicator. Depending on the purpose of the evaluation different decision rules can be applied, e.g. cost-effectiveness ratio, costs to achieve defined target, effectiveness at given cost level. For more information on CEA see Messner (2006), Rheinsberger and Weck-Hannemann (2007).
7. MCA is an evaluation approach that judges the performance of alternatives against a number of objectives or evaluation criteria (Belton and Stewart 2002). It can be applied for assessments either if several objectives have to be taken into consideration or if several criteria are considered covering different performance of the alternatives to be compared. The latter option is particularly relevant if indicators measuring these aspects cannot (easily) be merged into one non-monetary (target) indicator or expressed in monetary terms. So, in contrast to the CBA and CEA the availability of monetary or other quantitative data is not an application prerequisite as performance aspects can also be qualitatively assessed, e.g. through expert judgements For more information on different MCA approaches see Bana e Costa (1990), Zimmermann and Gutsche (1991), Munda (1995), Vincke (1992), Belton and Stewart (2002). For applications in the context of natural hazards management see Bana et al. (2004), Brouwer and van Ek (2004), Akter and Simonovic (2005), Kenyon (2007), Meyer (2007).
8. An example being the need for unveiling decision makers' complete preference structure for MAUT-based approaches.
9. This is also in line with experiences from other EU projects such as SEQUOIA and ERINA+, which show that whenever impacts of actions are difficult to monetise and/or only become measurable in the medium or long-term MCAs are best suited for such types of evaluation.
10. Criteria cards were ordered from the most to the least important. On the one hand side blank cards could be used to increase the distance between criteria and on the other hand side criteria cards could be put next to each other, if criteria were considered to be equally important. Then importance scores were specified for each criteria row. In general, the score 100 was elicited to the row with the most important criteria. Then interviewees indicated the relative importance of the least important criterion compared to the most important one. If it was considered to be half as important, the score 50 was elicited. Then the scores for all row in between the most and the least important criterion were specified. For more details see Zardari et al. 2015.
11. For a guided tour offering comprehensive information on the methodical foundations of PRIMATE as well as practical advice for its use see Drechsler et al. 2009.

References

Akter, T. and Simonovic, S.P. (2005). Aggregation of fuzzy views of a large number of stakeholders for multi-objective flood management decision-making. *Journal of Environmental Management* 77 (2): 133–143.

Bana, E. and Costa, C.A. (1990). *Reading in Multiple Criteria Decision Aid*. Berlin, Heidelberg, New York, Tokyo: Springer Verlag.

Bana, E., Costa, C.A., Da Silva, P.A., and Correia, F.N. (2004). Multicriteria evaluation of flood control measures: the case of ribeira do livramento. *Water Resources Management* 18 (21): 263–283.

Basher, R. (2006). Global early warning systems for natural hazards: systematic and people-centred. *Philosophical Transactions of the Royal Society A - Mathematical Physical and Engineering Sciences* 364: 2167–2182.

Belton, V. and Stewart, T.J. (2002). *Multiple Criteria Decision Analysis – An Integrated Approach.* Boston, Dordrecht, London: Kluwer.

Brouwer, R. and Kind, J.M. (2005). Cost-benefit analysis and flood control policy in the Netherlands. Cost-benefit analysis and water resources management. In: *Cost-Benefit Analysis and Water Resources Management* (ed. R. Brouwer and D. Pearce), 71–91. Cheltenham: Edward Elgar.

Brouwer, R. and van Ek, R. (2004). Integrated ecological, economic and social impact assessment of alternative flood control policies in the Netherlands. *Ecological Economics* 50 (1–2): 1–21.

Chathoth, P., Altinay, L., Harrington, R.J. et al. (2013). Co-production versus co-creation: a process based continuum in the hotel service context. *International Journal of Hospitality Management* 32: 11–20.

Cucco, I., De Rosa, S., Passani, A. et al. (2016). SecInCoRe Deliverable 5.3. Validation strategy and first functional evaluation model of communication system concept.

Drechsler, M., Lange, M., and Meyer, V. (2009). *PRIMATE – An Interactive Software for Probabilistic Multi-Attribute Evaluation, Software Handbook*. Leipzig: UFZ.

Drobot, S. and Parker, D.J. (2007). Advances and challenges in flash flood warnings. *Environmental Hazards* 7: 173–178.

Etgar, M. (2008). A descriptive model of the consumer co-production process. *Journal of the Academy of Marketing Science* 36 (1): 97–108.

Flatscher, M. and Riel, A. (2016). Stakeholder integration for the successful product-process co-design for next-generation manufacturing technologies. *Cirp Annals-Manufacturing Technology* 65 (1): 181–184.

Handmer, J. (2000). Are flood warning futile? Risk communication in emergencies. *The Australian Journal of Disaster and Trauma Studies* 2: https://www.massey.ac.nz/~trauma/issues/2000-2/handmer.htm.

Hanley, N. and Spash, C. (1993). *Cost-Benefit Analysis and the Environment.* Cheltenham: Edward Elgar Publishing.

Hansjürgens, B. (2004). Economic valuation through cost-benefit analysis – possibilities and limitations. *Toxicology* 205 (3): 241–252.

Kenyon, W. (2007). Evaluating flood risk management options in Scotland: a participant-led multi-criteria approach. *Ecological Economics* 64: 70–81.

Kox, T., Gerhold, L., and Ulbrich, U. (2015). Perception and use of uncertainty in severe weather warnings by emergency services in Germany. *Atmospheric Research* 158: 292–301.

Lusch, R.F. and Vargo, S.L. (2006). Service-dominant logic: reactions, reflections and refinements. *Marketing Theory* 6 (3): 281–288.

Messner, F. (2006). Applying participatory multicriteria methods to river basin management: improving the implementation of the water framework directive. *Environment and Planning. C, Government & Policy* 24 (2): 159–167.

Meyer, V. (2007). GIS-based multicriteria analysis as decision support in flood risk management. *UFZ Discussion Papers, 6/2007*, Leipzig: UFZ.

Meyer, V., Priest, S., and Kuhlicke, C. (2012). Economic evaluation of structural and non-structural flood risk management measures: examples from the Mulde River. *Natural Hazards* 62: 301–324.

Meyer, V., Schwarze, R., Becker, N. et al. (2014). Assessing the costs of natural hazards – state of the art and the way forward. In: *Hydrometeorological Hazards: Interfacing Science and Policy* (ed. P. Quevauviller), 225–290. Chichester: Wiley.

Ministry of Agriculture Fisheries and Food (MAFF) (1999). *Flood and Coastal Defence Project Appraisal Guidance: Economic Appraisal.* London: UK Government Department Ministry of Agriculture Fisheries and Food.

Monacciani, F., Navarra, M., Passani, A., and Bellini, F. (2011). D3.3a – SEQUOIA final self-assessment methodology.

Morss, R.E., Wilhelmi, O.V., Downton, M.W., and Gruntfest, E. (2005). Flood risk, uncertainty, and scientific information for decision making: lessons from an interdisciplinary project. *Bulletin of the American Meteorological Society* 86: 1593–1601.

Munda, G. (1995). *Multicriteria Evaluation in a Fuzzy Environment – Theory an Applications in Ecological Economics*. Heidelberg: Physica Verlag.

Oliver, K., Kothari, A., and Mays, N. (2019). The dark side of coproduction: do the costs outweigh the benefits for health research? *Health Research Policy and Systems* 17 (1): 33.

Pearce, D.J. and Smale, R. (2005). Appraising flood control investments in the UK. Cost-benefit analysis and water resorces management. In: *Cost-Benefit Analysis and Water Resources Management* (ed. R. Brouwer and D.J. Pearce), 71–92. Cheltenham: Edward Elgar.

Rheinsberger, C. and Weck-Hannemann, H. (2007). *Economic Approaches to Risk Evaluation, Working Paper for the EU-MEDIN Book Project*, 1–15. Zurich.

Sarewitz, D. and Pielke, R.A. Jr. (2007). The neglected heart of science policy: reconciling supply of and demand for science. *Environmental Science & Policy* 10: 5–16.

Thöni, M., Leiter, A., and Weck-Hannemann, H. (2009). Protective measures against natural hazards – are they worth their costs? In: *Sustainable Natural Hazard Management in Alpine Environments* (ed. E. Veulliet, J. Stötter, and H. Weck-Hannemann), 201–227. Springer.

Turner, R.K., Burgess, D., Hadley, D. et al. (2007). A cost–benefit appraisal of coastal managed realignment policy. *Global Environmental Change* 17: 397–407.

Vincke, P. (1992). *Multicriteria Decision-Aid*. Chichester, New York, Brisbane, Toronto, Singapore: Wiley.

Voorberg, W.H., Bekkers, V.J.J.M., and Tummers, L.G. (2015). A systematic review of co-creation and co-production: embarking on the social innovation journey. *Public Management Review* 17 (9): 1333–1357.

Young, R.A. (2005). Economic criteria for water allocation and valuation. Cost-benefit analysis and water resources management. In: *Cost-Benefit Analysis and Water Resources Management* (ed. R. Brouwer and D.J. Pearce), 13–45. Cheltenham: Edward Elgar.

Zardari, N.H., Ahmed, K., Moniruzzaman Shirazi, S., and Yusop, Z. (2015). *Weighting Methods and their Effects on Multi-Criteria Decision Making Model Outcomes in Water Resources Management*. Springer Briefs in Water Science and Technology, Springer.

Zimmermann, H.J. and Gutsche, L. (1991). *Multi-Criteria Analyse – Einführung in die Theorie der Entscheidungen bei Mehrfachzielsetzungen*. Berlin, Heidelberg, New York, Tokyo: Springer.

9

Using Artificial Intelligence to Manage Extreme Weather Events

The Impact of the beAWARE Solution

Anastasios Karakostas[1], Stefanos Vrochidis[2], and Ioannis Kompatsiaris[2]

[1] *DRAXIS Environmental S.A., Thessaloniki, Greece*

[2] *Centre for Research and Technology Hellas (CERTH), Information Technologies Institute (ITI), Thessaloniki, Greece*

9.1 Introduction

After three years of progress, the beAWARE project has achieved a significant impact in the fields in which it is involved. To achieve this, the platform and the mobile application had to be tested against real-life conditions, which had been demonstrated through the three pilots, heatwave, flood, and fire and evaluated based on the interaction with the technology that the end-users experienced during each pilot differently.

This procedure allowed the Consortium to gather precise information, data, and feedback from different target groups. All the end-users' perspective was taken into account and helped the continuous evolution of the platform and the mobile application.

One of the general main criteria expressed was the simplicity of the system in order to make it as user-friendly as possible and to give trustworthy data to the operator/user. Moreover, the continuous feedback from end users to the technical partners gave the ability to adjust in order to fulfil their requirements based on the three scenarios. To add more, the results of each evaluation were considered in order to support further development of the system and to take corrective actions wherever required.

Responding to Extreme Weather Events, First Edition. Edited by Daniel Sempere-Torres, Anastasios Karakostas, Claudio Rossi, and Philippe Quevauviller.

Requirements and needs of the end users were recorded and re-examined throughout the development process to maximize the impact of the system to end users.

The overall impact of the project has shown the ability of the system to support authorities and first responders in the management of events related to extreme weather conditions, both as a stand-alone solution and as complimentary to legacy tools, demonstrating an overall value for further exploitation. All evaluation results of the pilots of the project that were derived from the feedback of the participants, supported the continuous evolution of the system, which in turn increased the system's impact in the management of events related to extreme weather conditions, showing and exploiting potential of the system.

9.2 Overall Objectives of the Project

The beAWARE project ended on 31 December 2019 and has met all its milestones. The project made a clear contribution to its main target: improving situational awareness and command and control (C2) systems. As a whole, beAWARE integrated platform encompasses high-end technologies and machine learning capabilities for obtaining information from text messages, social media and voice calls; a classification mechanism to process weather and other multimodal data and to generate early warnings and real-time alerts; deep learning techniques to detect crisis events in visual content; automatic drone routing and piloting for receiving valuable information from aerial imagery; case-based reasoning and decision support algorithms for crisis management; and automatic generation of multi-lingual reports to transform all the above into linguistic information to the authorities creating a usable system for performing in real case scenarios (Figure 9.1).

beAWARE project proposed a novel approach for disaster management, which manages the different events and information retrieved all along the disaster management process. beAWARE main objectives are enumerated below.

Obj.1 – Perform a research study on the requirements for emergency services given the current digital landscape (i.e. end user in emergency need, PSAP operator, first responder) (Antzoulatos et al. 2021; Hertweck et al. 2018).

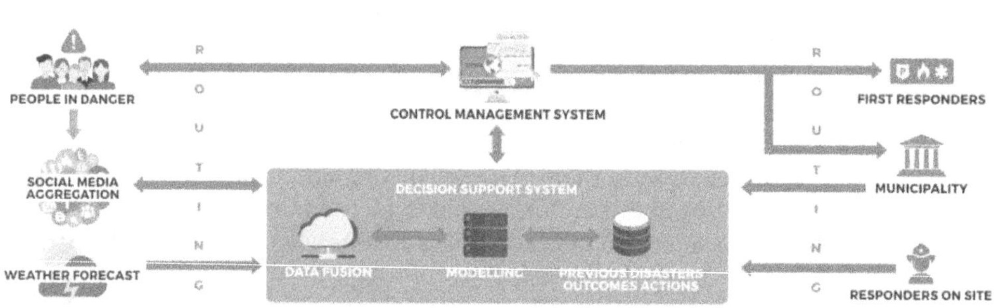

Figure 9.1 The beAWARE concept.

The aim of this objective was to identify the needs for emergency services within the current accessible digital landscape. The research tried to find the gaps in the present warning technologies and propose improvements with the aim of filling the gaps. In order to achieve this objective, a spiral process (requirements elicitation process) was established throughout the entire project, where every iteration was based on the outcomes of the previous. The members of the consortium with weekly teleconferences, plenary meetings and Workshops managed to organize and formulate a list of approximately 100 user requirements, which corresponds to the three relevant use cases of the project: Fire, Flood Heatwave.

Obj.2 – Multilingual speech and written communication analysis in emergency calls (Shvets et al. 2018)

The capability to grasp the content of any message transmitted by a citizen in danger during an emergency was very crucial in the context of beAWARE. To fulfil this objective, beAWARE designed and developed the following components: Social media analysis tool, which continuously collects tweets from the Twitter and classify them as relevant or not. Automatic speech recognition module, which is used to transcribe voice messages in four supported languages. In order to showcase the ability of the platform to integrate with existing tools, the component was adjusted in order to communicate and transcribe calls coming either from the beAWARE mobile application or from existing call centres. Transcribed and written communication analysis: In order to extract concepts and conceptual relations from textual material (texts, social media postings, and transcribed spoken language), a multimedia text analysis module was developed. beAWARE extended the knowledge extraction techniques to detect from multilingual text a wide range of incidents and impacted objects, as well as locations and states associated with incidents (Figure 9.2).

Obj.3 – Aggregate multimodal information from sensor networks, meteorological stations, etc. and social media for decision support and validation purposes and issue early warnings (Hertweck et al. 2018; Kontopoulos et al. 2018).

beAWARE has designed and developed the procedures and the infrastructure to support the aggregation of multimodal information. The system collects information from the following sources:

- Meteorological data (forecasting weather data and real-time weather observations)
- Hydrological data (river water level) forecasting and real-time observations from sensors located in the river
- Flood risk maps from and fire hazard maps from EFFIS and EFAS
- Videos and Images (from the beAWARE mobile application, static cameras and Drones)
- Audio messages (from the beAWARE mobile application and call centres)
- Social media data relevant to a specific crisis (from Twitter)

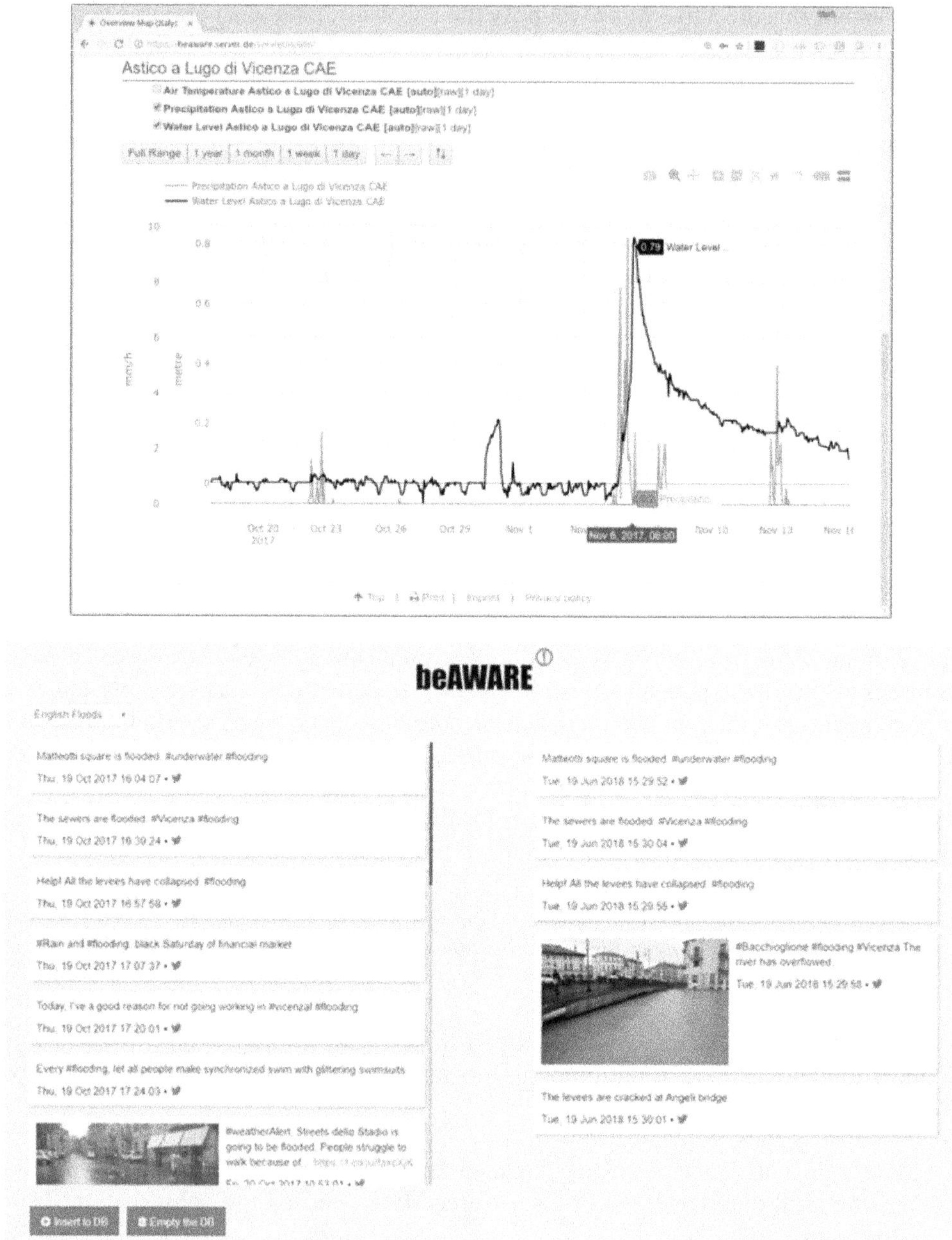

Figure 9.2 Sensor and social media data.

The sensing data (forecasts and observations for river water level and weather parameters) are stored to the SensorThings Server API (FROST), a standard-based sensor data store. The main module developed to process this data is the Crisis Classification (CRCL) module which consumes all stored data from the different sources and estimates the crisis level of an event. CRCL consists of two components, the Early Warning to provide provisional alerts and the Real-Time Monitoring component responsible for monitoring the evolvement of the crisis.

Obj.4 – Visual context analysis during emergency calls (Moumtzidou et al. 2018).

This objective aims at understanding the visual context of a person calling in an emergency or the reporting from a first responder, where the term 'call', in a wider sense, refers to the reporting of a person for an emergency situation by using all the available multimedia context that beAWARE mobile application provides.

To enable the above, beAWARE designed and developed and finalized the image and video analysis components to extract high-level information such as type of the crisis (flood, fire), detect impacted people, animals, cars, indoor/outdoor, city landscape/deserted area by extracting low-level features from visual data and translate them into high-level concepts based on supervised machine learning techniques. The extracted visual concepts lead primarily to real-time detection and assessment of occurring events and their type, as well as a global understanding of the situation in the affected area (people impacted, traffic conditions etc.).

Moreover, crisis event detection in multimedia collections has been also explored following the temporal dimension of a flood or fire dynamic texture in. In addition beAWARE has proposed novel methods for combining multi-modal information in early and late fusion schemes that deploy multi-modal stacked autoencoders along with dimensionality reduction algorithms (Figure 9.3).

Obj.5 – Semantic integration of multimodal information from the emergency calls, M2M/IoT platforms and social media for decision support and generation of early warnings (Hertweck et al. 2018; Kontopoulos et al. 2018).

Figure 9.3 Visual context analysis in emergency events. *Source*: (a) Sgt. Jose Diaz-Ramos/U.S. Department of Defense. (b) Associated Press / Alamy Stock Photo.

The key direction of this objective was to research and develop technologies for semantic integration of the diverse multimodal content to enable reasoning for decision support and for the generation of early warnings.

There is a number of tools developed within the project to enable the reasoning support. The backbone of the reasoning mechanisms consists in the use of a beAWARE ontology for the mapping of the input. Based on this ontological framework, beAWARE is able to estimate the severity level of an incident, to cluster relevant incidents, to link the heterogeneous digital information and reveal semantic hidden relations.

Moreover, a two-layer validation mechanism, both at the component level and at the Knowledge Base level, was integrated to validate the incoming data and filter conflicting or irrelevant information.

Obj.6 – Multilingual report generation from aggregated emergency data (Shvets et al. 2018).

The report generation aims in the development of techniques for the generation of multilingual written information that derives from the previous ontological representations.

The output of the analysis from all the beAWARE analysis components is used as input to the report generation component that processes the information and automatically determines which contents need to be forwarded to linguistic generation so as to provide written reports to the authorities.

Two types of multilingual reports can be issued by the system, situational and summary reports covering the occurrence of incidents in chronological order. Entities and concepts beyond those modelled by the ontology can be verbalized, thus facilitating the portability of the developed components to new emergency scenarios.

Similarly to the Text analysis component, the Report Generator can generate reports in all four supported languages (English, Italian, Spanish and Greek).

Obj.7 – Research and development of Main Public Safety Answering Point (PSAP) for emergency multimedia-enriched calls Develop a PSAP (Mordecai & Kantsepolsky 2018).

In the context of beAWARE, a main public safety answering point (MAIN-PSAP) was designed and developed in order to provide a platform for integration of multiple unconnected security applications and devices and control them through one unified user interface. Special effort was made to improve user experience with emphasis on how user receives snapshot at any given moment for decision-makers to enhance the situational awareness. Especially at the final version of the platform, the emphasis was given to the improvement of the visual design of the UI and the interoperability of the platform, by refining the appearance and functionality of the task management, making the dashboard more comprehensive, with improved and interactive charts and by enhancing the appearance of the map, with more layers and filtering options (Figure 9.4).

Obj.8–Design and execute 3 large-scale pilots.

To demonstrate and evaluate beAWARE solution, there were selected three demonstration scenarios for flood, fire, heatwave to be implemented respectively in Vicenza,

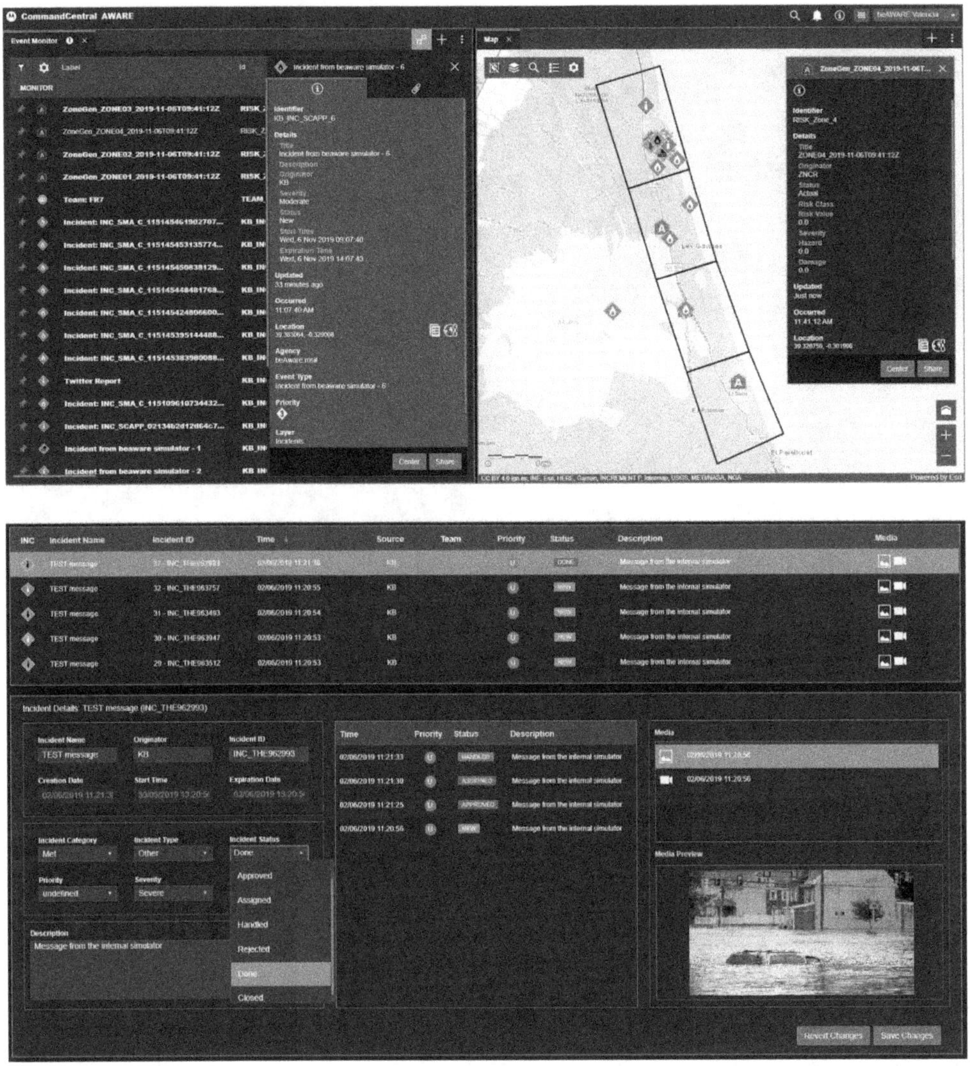

Figure 9.4 Main public safety answering point of beAWARE.

Valencia and Thessaloniki, respectively. The locations were selected based on the frequency of the occurrence of those event. Overall, three large-scale pilots successfully carried out involving in total more than 600 participants.

- 1st Pilot – Heatwave in Thessaloniki. The first beAWARE pilot was executed on 20 November 2018 in Thessaloniki, Greece, with the aim to test the first prototype along with the heatwave scenario.
- 2nd Pilot – Flood in Vicenza. The second beAWARE pilot took place on 7 March in Vicenza, Italy, from 8 : 00 CET to 14 : 00 CET at the presence of more than 90 participants (Figure 9.5).

Figure 9.5 Participants performing tasks during the Vicenza pilot.

- 3rd Pilot – Fire in Valencia. The third beAWARE pilot was executed on 14 November in Valencia, Spain, with the aim to test the final version of the beAWARE system applied to the fire scenario. This pilot involved more than 500 participants (Figure 9.6).

9.3 The Impact of beAWARE

9.3.1 Scientific and Innovation Impact

The development of a new system that combines various technologies that answer different requests from end-users' perspective resulted in a noted scientific and innovation impact of the beAWARE technology, which is presented in various scientific conferences and research papers. Additionally, innovation elements were highlighted and noted after the pilots upon discussion with the participants.

Moreover, beAWARE had achieved to give effective and faster emergency responses to extreme weather and climate events, with faster analysis of risks. In the three pilots, beAWARE decision support services for crisis management had proven to act with great efficiency answering all different tasks for the three scenarios. In addition, beAWARE system successfully improved the coordination of emergency reactions in the field to address extreme weather events thanks to the combination of

Figure 9.6 Command centre in Vicenza and in Valencia pilot.

reliable and updated meteorological data, provided in our case by FMI as also AAWA's AMICO model and the DSS that was developed by the technical partners. The trigger mechanism that beAWARE developed for the three extreme weather events tested in different periods of time for each one (before, during and after the event) resulting in codified messages from the platform to the mobile application, upon Authority decision and request proves the abilities of the platform to assess the validity of information. This combination of technologies and reliable data was proven during the pilots and mentioned at the evaluation sheets and during the discussions (hot debriefs) with the pilots' participants.

In combination with the above ones, beAWARE addresses the expected impact of shorter reaction time and higher efficiency on reaction due to extreme weather and climate events. This was achieved first, thanks to beAWARE KB that estimates the emergency level of the situation in order PSAP operators and first responders to act accordingly. Second, due to the fact that beAWARE can continuously receive and analyse data, such as meteorological, sensor (water), video, image, audio, social media (Twitter), it gives the ability to decision-makers, PSAP operators and first responders to have higher efficiency on their decisions and calls as also their actions on the field depending on the situation. It is worth to mention that UPF developed an extended text analysis module that can handle not only inputs from its own mobile application but also social media (Twitter) inputs related to the three scenarios, in English, Italian, Spanish and Greek. To add more, CERTH's automatic support recognition system, image, video, drone and traffic analysis helped in the overall creation of the beAWARE system. Supporting the above functionalities, IOSB provided a mobile application and a SensorThings API, which were filtered and notified by the M2M/IoT platform through open and standardized interfaces resulting in the overall formation of the system.

To add more, from the perspective of technology providers and industries as mentioned above, a large positive impact of the beAWARE took place from participation in scientific conferences such as ISCRAM 2019, SmartResilience, Sayso_H2020 EnviroInfo 2018, TRECVID 2018, COWM 2018, European Meteorological Society: Annual Meeting (EMS) 2018, SafeKozani 2018, SemEval-2017, The Tel Aviv Coastal Forum workshops such as CAP Workshop 2019, IEEE Smart World NVIDIA AI City Challenge, The First Multilingual Surface Realisation Shared Task (SR'18), CAP Workshop 2019, Infoday Regional H2020 and many more, but also 43 publications giving an overall boost to science in this field. This large and through the years positive impact that was given for the beAWARE system from different fields of science helped the overall impact as also the dissemination strategy of the project.

Furthermore, beAWARE with its provided geolocation system not only allows for PSAP to view the exact location of the event but also inform first responders or citizens to head to a specific location with coordinates and of course citizens to give exact location with its coordinates of an event, without being at the spot, if their life is in danger. In all this effort MSIL integrated all the above data, creating the PSAP platform, supported by IBM's communication bus, cloud-based messaging service, based on standards and upon cooperation with the rest of the consortium.

Finally, cooperation between end users and technical partners throughout the whole process for the development of the system, allowed end users to support the development of an innovative solution through the fulfilment of their requests. Additionally, technical partners had direct access to the wills of the end users which offered them the ability to make scientific progress and support innovative solution based on actual needs of potential end user of the system.

9.3.2 Economic Impact

The economic impact of the project is not easy to measure, since it can be derived only indirectly from the outcomes and the benefits of the project and without properly set and test the beAWARE system for a well-accepted period of time. Nevertheless, various participants noted that beAWARE can be a very useful and supporting tool that can coordinate multiple resources such as national and volunteer first responders, instantly and accompanied with the existing legacy tools (VHF/UHF radios) resulting to a better coordination and management of these resources. As a result, the effective management of all resources can reduce the economic impact of managing an extreme event. For instance, by having from the area of the event, analysed data of each incident, PSAP can dispatch specific and dedicated units reducing the cost as also the time of each operation.

Leading from the above, the positive impact for the economy is that thanks to the beAWARE system, the time of the operation is reduced resulting in lower operational costs. To add more, by having full knowledge of the situation, and the available resources, more sophisticated and dedicated personnel can be dispatched to its incident, resulting in better management of the crisis and reducing the overall costs of the operations.

9.3.3 Safety Impact

Safety impact is very important, from decision-makers to citizens. From governmental representatives and PSAP operators that will take the proper decision to dispatch specific first responders or national authority to answer the call, to citizens that are in danger. Safety played a vital role in the development of beAWARE. For that reason, data acquisition and analysis reached a high level in order to address the call with safety and minimize the danger of the citizens, with special alerts, as also of the first responders with analysed data from the incident that are dispatched.

Finally, beAWARE system successfully answered the expected impact of citizens' protection and saving lives, targeting the needs of users, from different authorities such as national and local law enforcement, climate and weather services, volunteer organizations of civil protection mechanism as well as private operators. This combination of different end users gives a significant added value to the beAWARE because it answers different approaches and needs from all end users' perspectives from different countries of Europe.

9.3.4 Training Impact

beAWARE system can help PSAP operators, decision-makers, first responders and citizens to use it and in training events in order to maximize their efficiency and be more effective in order to tackle more success incidents in real life. beAWARE has already set a well-established procedure having a detailed script based on specific Use Cases and User Requirements for each pilot, external participants and an evaluation methodology contacted after each pilot. This procedure that beAWARE consortium followed can be used as a guideline for future training exercises from organizations or authorities that want to use beAWARE system in order to carry out exercises to meet their needs and increase their operational capacity.

9.3.4.1 Societal Impact beAWARE project has a different impact on each target group in the society. This is easily understood because there are different needs and requirements of each target group. From the target groups, the impact on PSAP operators, policymakers and government representatives is positively large. This can be proven from the evaluation results and from the fact that beAWARE enables them to be able to prepare and interact directly with citizens and first responders before, during and after an event enhancing the communication and flow of information between all parties involved.

On the part of first responders and citizens, the impact of beAWARE is also positive here. Those groups are directly connected with the PSAP through the application and exchange important data from the event. Moreover, they also inform at any time their position, their situation and any changes that may take place in the area. Citizens are given the opportunity to participate in the overall crisis management system, giving information about the phenomenon and the various risks they identify.

9.3.4.2 Governmental Authorities/PSAP Operators Governmental authorities that are responsible to operate PSAPs, such as police, fire brigade or generally the civil protection mechanism of a country, have certain requirements in order to operate a PSAP at their station. The most important thing for a PSAP is to be easy to use to handle incoming information smoothly and to allow outgoing commands to be sent as quickly and easily as possible, as well as understandable to the sender and to the recipient. In addition, it is important to consider that the PSAP may need as few operators as possible and therefore equipment to operate and be flexible in its management.

To add more, the representatives from governmental authorities stated that 'beAWARE is a very interesting tool and will help in managing situation, as a support system', 'it can be a useful tool for us', 'has many potentials', 'it has a good and constant flow of data', 'platform map has good interaction time'. Those expressed opinions of the players showing that generally beAWARE system can be a supportive tool next to their already established decisional system.

Thanks to the continuous feedback to the technical partners from the end users and the pilots' participants, the final beAWARE platform that was created meet their needs and requirements. The main PSAP screens, which are the task manager screen and the main event map screen, offers the opportunity to the operators to have a clear view of the entire area of the event, different map layers and information that are required and to have the ability to assign tasks to the field rescue teams and inform citizens for an imminent danger. As a result authorities can manage risks more productively and take appropriate decision affecting first responders and citizens in order to reduce their exposure to danger.

9.3.5 Policymakers

Policymakers and governmental representatives in general, who are involved in crisis management as well as in search and rescue operations, are mostly interested to receive reliable and as comprehensible as possible information on time in order to take action and make decisions as correct as possible. In addition, as it was extracted from the questionnaires and the discussions with the participants from the pilots, it was noted that governmental authorities need to be immediately informed of any change in the situation. This is very important, due to the fact that everything should be clearly presented on the platform, in order proper decisions and actions take place depending on the situation of the event. Finally, the processing of digital information coming to PSAP should be fast in order to properly evaluate the situation and if necessary make appropriate adjustments again.

Additionally, the project's partners in order to promote the project to policymakers and strengthen its impact made a lot of meetings with national authorities or entities to inform them about beAWARE project and its abilities.

Various meetings with representatives of local, regional and national authorities, even from countries outside the project's consortium, took place in order to promote the beAWARE system and its potential to support the Crisis Management mechanism in each respective area.

In all those meetings, beAWARE system was presented in each different stage to representatives from various entities and authorities. Politicians especially recognized the need for tools like beAWARE risk management and all have recognized and confirmed the great impact of beAWARE system to citizens and the community. To add more, it was stated that since this system can be easily customized and integrated on existing systems, it can upgrade and enhance existing operations and services. In addition to the above meetings, consortium members made significant efforts to disseminate beAWARE project with well-established media coverage, such as publications in local newspapers, interviews in radio broadcasts as well as in their social media accounts resulting to positive interaction, visibility and impact for the project.

beAWARE platform, through the pilot exercises and online presentations after each pilot, has succeeded in proving that fast processing and analysis of digital information is possible. Data such as meteorological info, photos, videos, audio messages, texts from Twitter are processed, analysed and presented on the platform without long delays. This gives the ability to the policymakers to take decisions, faster, more accurate and in the correct directions as also to receive crucial information from the field, even from citizens.

Last but not least, as it was demonstrated in the final pilot of the project beAWARE system can be potentially used to connect a local or national emergency with the EU civil protection mechanism, potentially with an easy and direct way of communication. However, having that in mind, it is too early to estimate the potential impact on EU policies from the use of such a system and if in the future such a feature will be adopted. Nevertheless, the system can be adapted with little efforts to new crisis scenarios (such as heavy storms, snowstorms, etc.) beyond those covered in beAWARE offering an adaptable solution to policymakers that can used in different crisis regardless the scenario.

9.3.6 First Responders

First responders from public authorities or voluntary organizations, which are mainly involved in SAR missions in the field, act as the intermediary between the PSAP and citizens in need. The primary concern of first responders is to get the right and necessary information from the PSAP as soon as possible in order to know the exact location of the area of the event, what's the event about and in general acquire as much information as possible. In addition, interactive communication with the PSAP and the ability to send information quickly and immediately is a very critical and crucial need for first responders. Additionally, one of the biggest problems in rescue operations is the difficulty of the continuous online communication because of the 'blind' spots from which rescuers cannot send or receive data with new information about the event, due to lack of mobile coverage.

beAWARE platform and mobile application that have been tested in all three pilots, have proven to be able to send and receive a large amount of data from the PSAP to the field rescue teams and vice versa. As a result, the system can support exchange of information among first responders efficiently.

To support that, some statements that first responders made during the evaluation of the pilots were:

- 'beAWARE app could be a great support to the radio but not a total replacement',
- 'detailed information and time saving',
- 'useful supporting tool with high quality of data',
- 'a good assisting tool'

Those positive feedbacks from the first responders signify that beAWARE can assist them before and during an operation in order to maximize their efficiency.

Additionally, in all three pilots, the participants noted down that the mobile application menu is easily understandable, all the actions that should be taken can be achieved without problems and the data upload procedure was very familiar for them. Moreover, it is easily handled and offers only the required information without any unnecessary or irrelevant information that may cause unwilling concerns and queries. Finally, the majority of the first responders reported that the beAWARE mobile application can easily work seamlessly with existing technologies such as VHF/UHF communication, whereas the beAWARE mobile application was overall a good one and the overall reactions and use of it was in a satisfactory level.

It can be easily understood that the impact of a system such as beAWARE, to first responders can be significantly high, especially taking into account the fact that the system can be used both as a stand-alone system and complimentary to legacy tools. It can support operation management in a productive way, which allows first responders to be more efficient of the field.

However, it is worth mentioning that national services such as fire brigade, civil protection and police authorities besides their positive opinion for the system state that in order to integrate such a system in their own mechanism, a political decision is required.

Therefore the impact on policymakers and authorities is crucial to the adaptation of the system by first responders.

9.3.7 General Public (Citizens)

In a state of emergency, the main concern of citizens is to be able to transfer the information as quickly and easily as possible to emergency services that they or other people are in danger. In addition, they want to transfer as much information as possible regarding their location, and if a photo, video or audio can be sent so as to make rescuers' work easier.

Through the continuous communication of technical partners with end users to create a user-friendly and easy understandable mobile application, as well as the continuous improvements and changes that took place after each pilot thanks to feedback from pilots participants, the final version of the application that was created to fulfil all the requirements of the participants based on the existing technology. Thanks to the beAWARE mobile application citizens can easily be informed of dangerous events and the affective area.

Additionally, the ability to attach files such as photos, videos or voicemails to send information to PSAP and rescue teams about and an emergency is easy and understandable based on the participants' opinion after the pilots. Last but not least is worth mentioning that participants with the role of the citizen stated that beAWARE mobile application makes them part of the participative process of emergency management, as they contribute with their information, such as a photo, video or a voice message of the area of the event, while today they are almost totally outside of this process.

To add more, societal impacts were also noted and documented in all three pilots by the participants, especially citizens. As previously mentioned, when citizens participate in all the crisis management systems, especially when they send the important data (photos, videos or voice-calls) to the PSAP, feel that they too are involved in solving the problem and are a very important link in the rescue chain. Additionally, citizens when they know that PSAP and first responders have a combined coordinated system that will be used to support them, feel safer and more secure because operations will be organized and methodologically prepared and executed. Moreover, social media analysis by beAWARE adds and promotes another way of communication between citizens and authorities. Knowing that relevant posts are spotted by authorities it is safe to assume that citizens might increase the use of social media during an emergency, in order to support the crisis management mechanism as well as to provide information to the authorities even at a very early stage of the event.

Moreover, adding another form of communication which offers modern tools such as photo/video or voice-messages exchange between citizens and PSAP gives the opportunity to crisis management operators to have useful analysed data from the area of the event that can be proven important for the operation.

9.4 Conclusion

The impact of a project is the project's 'legacy'. More specifically, how a project affects the environment that it exists and the people with which it interacts directly and indirectly. beAWARE being a project that aims to enhance the ability of authorities to manage events that are related to extreme weather conditions, hence to enhance the safety of citizens, provides a holistic approach to support decision-makers throughout the management of such an event.

beAWARE successfully met all impacts referred to the call through all the separate tools that were developed that provide a holistic approach in the crisis management of events related to extreme weather conditions. The system has demonstrated through the execution of three pilots that it can support more effective and faster responses with faster risk anticipation by using forecasting and analysis of various data with the use of cyber technologies in order to improve the coordination of responders on the field. Additionally, beAWARE has also demonstrated its capability to improve the capacity of first responders and the efficiency of their reactions, all in the support of the overall goal to protect citizens by meeting the needs and requirements of all potential end users and decision-makers.

However, the impact of beAWARE was not limited only to decision-makers but it expanded to the scientific community. With the publication of 43 papers and participation in more than 60 events in order to promote the project and its results, the dissemination of the outcomes was extensive and reached various audiences. Dissemination to the scientific community was also supported by various events, most notable the workshops that took place during ISCRAM 2018 and 2019, that allowed to further influence the scientific community with the developments made during the implementation of the project by putting beAWARE in the centre of discussions.

Furthermore, society and citizens were a major target of beAWARE. The system allows the involvement and faster notification of citizens during an emergency event that could potentially support the saving of more lives. Since the system proactively targets to meet the requirement of all responders that are involved in an emergency situation related to the management of extreme weather-related events, it increases the capacity and the effectiveness of first responders.

Additionally, beAWARE system supports and improves the coordination of first responders improving the efficiency of existing resources and therefore better management of an event. Taking also into account the fact that beAWARE does not aim to explicitly replace legacy tools rather than operate as an additional tool whenever and wherever necessary, the impact in managing events and thus the impact on society as well as first responders can be significantly positive.

Moreover, beAWARE support authorities and first responders to improve their capacity through task management and better use of resources, offering an indirect impact in the economic aspect to the emergency management. beAWARE supports quicker reaction time through its early warning system and its crisis classification module that supports a more holistic approach to emergency management as well as shorter reaction time and higher efficiency of reactions.

Finally, the impact on policymakers is worth mentioning as well. Through face to face meetings as well as the participation of policymakers in the project's pilots, set the ground for beAWARE to disseminate its results and present its operational impact to those that make the decisions leading to perhaps changes in policies towards managing events related to extreme weather conditions.

To support all the above mentioned activities and in order to strengthen its impact, beAWARE implemented an overall dissemination plan that addressed different audiences from national and local authorities, first responders, PSAP operators, citizens to the scientific and industrial community. This address was made in three periods. In the first period, the project awareness was created by addressing targeted audiences and beginning communicating and recruiting users that were important to support the project. In the second period, there was an active invitation of the user groups to participate in the project's events and continuous engagement and training of the users. In the third period and until the end of the project, there was an attempt to produce a first business interest to potential customers and generally promote the beAWARE solution.

Furthermore, the use of social media and the production of online and offline communication material allowed beAWARE to spread its message and to increase the awareness for the project and its objectives. Additionally, other dissemination products such as the project website, online newsletters, press releases in newspapers, audiovisual material, publications, workshops, networking and direct communication with potential users or clients and finally with relevant projects and initiatives took place in order to maximize the positive the overall impact of beAWARE.

Throughout its implementation, it can be argued that the overall impact of beAWARE can be characterized as positive to various parts of the society and specifically the defined target groups of the project. Taking that into account, this positive impact allowed and even pushed the consortium to formulate an exploitation plan after the project's duration in order to invest on that. Local authorities welcomed that approach which set the ground for the development of beAWARE system to continue beyond the project's duration. Moreover, technical partners recognized this opportunity and agreed to support the system the period after the project in order to enhance the efforts of further exploitation of the system.

Acknowledgement

This work is supported by the EC-funded research and innovation program H2020 beAWARE: "Enhancing decision support and management services in extreme weather climate events" under the grant agreement No.700475.

References

Antzoulatos, G., Karakostas, A., Vrochidis, S., and Kompatsiaris, I. (2021). The crisis classification component to strengthen the early warning, risk assessment and decision support in extreme climate events. *Dynamics of Disasters: Impact, Risk, Resilience, and Solutions* 39–66.

beAWARE (n.d.). beAWARE Enhancing decision support and management services in extreme weather climate events. https://beaware-project.eu/

Hertweck, P., Moßgraber, J., Kontopoulos, E. et al. (2018). The backbone of decision support systems: the sensor to decision chain. *International Journal of Information Systems for Crisis Response and Management (IJISCRAM)* 10 (4): 65–87.

Kontopoulos, E., Mitzias, P., Moßgraber, J. et al. (2018). Ontology-based representation of crisis management procedures for climate events. *ISCRAM*, May 2018.

Mordecai, Y., and Kantsepolsky, B. (2018). Intelligent utilization of dashboards in emergency management. *ISCRAM*.

Moumtzidou, A., Giannakeris, P., Andreadis, S. et al. (2018). A multimodal approach in estimating road passability through a flooded area using social media and satellite images. *MediaEval*, October 2018.

Shvets, A., Mille, S., and Wanner, L. (2018). Sentence packaging in text generation from semantic graphs as a community detection problem. *Proceedings of the 11th International Conference on Natural Language Generation*, November 2018, pp. 350–359.

10 Innovative Visual Analysis Solutions to Support Disaster Management

Emmanouil Michail, Panagiotis Giannakeris, Ilias Koulalis, Stefanos Vrochidis, and Ioannis Kompatsiaris
Centre for Research and Technology Hellas (CERTH), Information Technologies Institute (ITI), Thessaloniki, Greece

10.1 Introduction

Visual analysis components in the framework of beAWARE project are responsible for processing images and videos obtained from mobile, static and aerial cameras. The outcomes of the analysis process are intended to contribute to the detection of emergency events and the enhancement of contextual understanding, through the recognition of emergency indicative situations such as traffic bottlenecks, flooded areas, fires, elements at possible risk (e.g. people, vehicles) etc. For that purpose, various computer vision and machine learning techniques have been deployed for image classification, object detection, semantic segmentation, dynamic texture recognition and localization and motion analysis. The design and development of the components were informed by requirements set by end users of the system including first responders and authorities, who were actively involved throughout the project's development phases. Thus, in order for example, to provide authorities the ability to detect the presence of people and vehicles in impacted areas (e.g. cars and people in a flooded area) from video and images sent from mobile phones and social media accounts, object detection and tracking technologies have been developed and integrated. Furthermore, image classification techniques, based on deep neural networks, have been deployed in order to address first responder needs for early detection of emerging disasters, such as floods and fires, and for an overall assessment of an ongoing incident. Also, on top of accurate weather

Responding to Extreme Weather Events, First Edition. Edited by Daniel Sempere-Torres, Anastasios Karakostas, Claudio Rossi, and Philippe Quevauviller.

predictions and creation of risk maps, there is also a need for monitoring rivers and embankments in order to early detect water level elevation and possible overflow. To address this requirement, a visual component for river monitoring was developed, and able to perform visual analysis on videos captured by static surveillance cameras installed near a river. The component is designed to estimate water levels and generate alerts if the thresholds are exceeded. Moreover, to enable authorities to monitor current traffic situations so they can direct first responders or inform them of routes to avoid, a traffic analysis framework using visual material from static surveillance cameras has been developed. This framework has the capability to monitor traffic conditions and detect traffic anomalies such as car accidents.

Finally, the potential benefits of incorporating unmanned aerial vehicles (UAVs) were recognized and utilized in the system. UAVs possess unique capabilities, such as the ability to access remote locations and fly at low altitudes, host multiple sensors and capture high-resolution imagery from multiple perspectives. These features make UAVs ideal for performing small-, medium-, and large-scale monitoring tasks. To this end, a drone platform has been developed and integrated into the system, able to simultaneously perform various visual analysis tasks, providing assistance throughout all stages of disaster management.

10.2 Related Work

In this section, we examine relevant works that have employed visual analysis techniques in the domains of fire and flood detection in images and videos, object detection, traffic flow analysis, along with works that have employed UAV technologies in the field of disaster management.

Fire and flood detection is usually handled as an image classification problem. Deep convolutional neural networks (Krizhevsky et al. 2017) constitute the state-of-the-art in image classification. Further investigation on deeper models began with Simonyan and Zisserman (2014), which examined smaller stacked kernels, leading to the development of the Inception (Szegedy et al. 2015) and ResNet (He et al. 2016) architectures. Image classification has been applied in several studies in the field of fire and flood detection. In Bischke et al. (2017) for example, authors fuse visual and textual features extracted by deep neural networks in order to detect floods in images, and also extended existing network architectures for semantic segmentation to include both RGB and infrared (IR) channels. Avgerinakis et al. (2017a) used visual and textual analysis with deep convolutional neural networks and performed late fusion of their similarity scores to detect floods in social media images. Muhammad et al. (2018) proposed an early fire detection framework using fine-tuned CNNs for static surveillance cameras, capable of detecting fire in various indoor and outdoor environments. The authors also proposed an adaptive prioritization mechanism for cameras in the surveillance system and a dynamic channel selection algorithm. Finally, in Sharma et al. (2017) authors use two pretrained state-of-the-art Deep CNNs, VGG16 and Resnet50, enhanced with fine-tuning based on a fully connected layer, to perform fire detection in images.

Another visual analysis task with numerous applications is object detection, which has been employed in various fields such as autonomous vehicles, smart video surveillance, facial detection, ambient assisted living etc. Deep CNN architectures have been thoroughly examined for this task. Early works, such as Girshick (2015) used multi-scale bounding box proposal generation techniques like Selective Search, as a feeding mechanism of candidate boxes to deep classifiers. Later it became common practice to incorporate this function into single-shot object detectors, using end-to-end deep architectures (Liu et al. 2016; Redmon et al. 2016). Those models achieved a better trade-off between accuracy and speed. Finally, in Avgerinakis et al. (2017b) a novel hybrid shallow-deep representation was proposed for detecting vehicles and pedestrians from traffic surveillance cameras.

Smart city technologies for assistive transportation and safe driving have received significant attention during the last decade. Video surveillance and various other types of monitoring infrastructure provide a huge amount of exploitable data for extracting optimal traffic management rules, increasing safety in the streets and detecting or preventing accidents. Automated visual traffic data analysis is necessary for extracting valuable information in a reasonable amount of time with limited human involvement in these cumbersome and extremely time-consuming tasks. Many computer vision algorithms have been developed for the automated analysis of traffic video data. A big challenge, however, lies in the development of fast and computationally efficient methods to be used for real-world scenarios that require near real-time solutions.

One of the major tasks in traffic flow analysis is vehicle speed estimation. Methods performing speed estimation can be categorized into semi-automatic and fully automatic. In the first case, most of the calculations are performed automatically, but a user's manual input is required usually in the form of some known distance in the scene. In a method from this category (Nurhadiyatna et al. 2013), authors detected and tracked the vehicles using GMM background subtraction and Kalman filters, calibrated assuming a zero-pan pinhole camera model. In He and Yung (2007), the calibration is based on patterns of lane markings on the road and image rectification to cope with perspective projection. A simple method using optical flow to compute displacement of pixels and relaxation of the perspective projection effect in Lan et al. (2014) measured the speed inside a rectangle region of interest using known lane width as reference. There are fewer works on fully automatic camera calibration methods. In Dubská et al. (2014), vanishing point detection from vehicle movement using a diamond space accumulator is performed and scale inference is computed by matching statistically detected vehicles' dimensions to mean dimensions of real vehicles. An evolutionary algorithm for camera parameter extraction is used in Filipiak et al. (2016), assuming constant speed of vehicles, and the accuracy of the method is improved through license plate detection.

Another major field of research in traffic analysis is traffic anomaly detection, which aims at detecting unusual traffic patterns, such as traffic collisions. Relevant methods can be separated into two categories. The first one consists of methods using as input raw image data, such as pixel location or other low-level features. In (Cheng et al. 2015) for example, authors use hierarchical feature representations and a Gaussian Process Regression framework to build a low-level and a high-level codebook respectively. Anomalies are then detected, after the integration of local and global anomaly detectors. Probabilistic topic models are also proposed in a variety of works to capture spatiotemporal changes in traffic scenarios. Some characteristic examples are the new

Markov Clustering Topic Model in Hospedales et al. (2012), the Probabilistic Latent Sequential Motifs introduced in (Varadarajan et al. 2013) and the Dependent Dirichlet Process-Hidden Markov Model (DDP-HMM) framework proposed in (Kuettel et al. 2010). The main drawbacks of these methods are the computational cost and that they deal with modelling at pixel level, ignoring more complicated structures such as the objects themselves, thus missing important information. The second category involves methods based on trajectory extraction and analysis. Object, or even pixels are first localized and tracked to obtain their patterns, which are then clustered or modelled to represent the dominant underlying motions. A work in this category is (Jiang et al. 2011), where three different levels of semantics are considered after tracking all moving objects in the video. Rules of normal events are automatically extracted at each level and anomalies are defined as the events deviating from these rules. Finally, in (Yang et al. 2013) trajectory segmentation and multi-instance learning are used for the detection of local anomalies.

In the field of disaster management, there have been numerous studies and projects that utilize unmanned aerial vehicles (UAVs) to take advantage of their capabilities over traditional methods in various aspects of disaster management. A review of several UAV systems was conducted in (Giordan et al. 2017) with a focus on (i) data acquisition of pre- and post-environmental events, (ii) operational support during catastrophic events and (iii) damage monitoring in critical infrastructures. Another review of the potential use of drones in disaster management was presented in (Restas 2015) covering five types of disasters. The majority of works examine the use of UAVs in specific fields of disaster management and their systems focus on and improve specific UAV tasks.

One of the tasks of disaster management that UAVs have been particularly useful is Search and Rescue (S&R) operations, as they can reach inaccessible or hazardous places for humans and cover a large area in a short time. Furthermore, advances in Deep Learning for Computer Vision, have greatly enhanced UAV technology, granting new capabilities such as automatic object detection and tracking, image segmentation, image classification etc. Several works, for example, have addressed the issue of automatic victim detection through a variety of sensors and techniques. In (Rudol and Doherty 2008) for example, trained classifiers are used for human detection based on Haar-like features. In (Lee et al. 2016) authors propose a drone architecture for locating victims in GPS-denied unknown environments. The system is based on a combination of a LiDAR sensor and an inertial measurement unit (IMU) for global mapping (overall floor plan with 3D mapping) and an infrared (IR) depth camera for local mapping (victim detection). However, manoeuvring is performed manually by an operator and there is a need for specialized and expensive sensors. In (Rivera et al. 2016), a drone platform for human detection and geolocation is presented. Human detection is based on image processing performed on data from thermal and optical cameras, while geolocation is performed using triangulated-adjusted GPS data and integration of Google Maps. The study is focused on object detection with a single class (human), whereas it does not apply object tracking in order to track and distinguish targets in consecutive frames.

UAVs can also be particularly useful in assisting and monitoring evacuation missions, either by inspecting and determining possible evacuation routes or by automatically overseeing the operations. In (Maher and Inoue 2016) for instance, authors implemented a drone as an end device and data source able to automatically track pedestrians and scan the impacted area, after an earthquake occurrence, in order to assist in the generation of alternative evacuation routes. However, the tracking process is based on the colour of the target and works by training simple thresholds in the RGB colour space, whereas the targets must have specific colour properties.

In general, most of the existing works employing UAVs focus on a single task and develop solutions to address the needs of a specific problem of disaster management. Additionally, many works demand specialized and expensive equipment for performing more elaborate tasks. On the contrary, our work aims at developing a versatile overall UAV system, able to simultaneously perform various tasks, in near real-time, in order to assist all stages of disaster management, from early warning to emergency response and monitoring, in different disaster scenarios. Analysis component employs image and video processing based on machine learning techniques for multi-object detection, localization, tracking and counting for several types of objects simultaneously. Additionally, we have employed machine learning for disaster detection (smoke, fire and flood incidents) through drone footage by using a deep-learning image-classification model. Furthermore, we have developed a workflow for drone assistance on evacuation missions. Our system is able to effectively perform multi-object tracking even in low frame rates, by using post-processing morphological analysis in order to enhance object tracking in cases of rapid relative movement between the drone and the objects. Finally, contrary to other studies that use more expensive and specialized sensors (LiDAR, multispectral cameras, laser sensors), our UAV system performs all tasks by using a simple RGB camera.

10.3 Methodology

Visual analysis in the framework of the beAWARE project is performed by image and video analysis components with the aim of concept extraction from visual content. The following components have been developed and integrated to beAWARE platform:

a. Disaster detection, in order to assist in early detection of emerging incidents such as floods and fires, or to monitor and assess an ongoing incident. Image classification techniques have been developed in order to determine which images/videos contain an incident. Components have been tailored to work with both ground and UAV cameras.
b. Object Detection and Tracking, so as to find people and vehicles that exist in impacted locations. Similarly, separate components have been designed for both ground and UAV cameras.
c. River level monitoring component which performs visual analysis on videos from static surveillance cameras installed on a river, in order to estimate the water level and generate alerts, in case of threshold exceeding.
d. Drone analysis component, which is able to analyse drone footage in order to detect people and vehicles in danger, detect emergency incidents and facilitate evacuation missions.
e. Traffic analysis and monitoring which is able to estimate traffic conditions in impacted areas and detect anomalous traffic events.

10.3.1 Disaster Detection

A disaster detection component capable of identifying flood, fire and smoke from visual content has been developed to facilitate recognition of emergency events in images and videos. This component can serve as an early warning system during periods of heightened fire risk through drone patrols and also support disaster response by estimating the extent of the disaster, detecting affected areas and determining proximity to individuals and critical infrastructure.

To fulfil the requirements of the disaster detection task, an image classifier was developed by fine-tuning a pre-trained deep convolutional neural network (CNN) in order to classify each image into four possible classes: 'flood', 'fire', 'smoke' or 'other'. Specifically, by using as a basis the VGG-16 architecture trained on Places365 dataset, we fine-tuned the parameters of the model. In order to fine-tune the model, a public dataset was collected of over 20,000 flood, fire and smoke images, from online databases and images generated in the framework of beAWARE project. The full collection can be found in Table 10.1. Around 20% of the collected dataset was reserved as a testing set for evaluation and the rest was used for training. In general, flood images have been taken from flooded areas like city streets and overflowing rivers. Fire and smoke images in those datasets have been captured during forest wildfires, explosions, fires in urban environments and fires caused by riots. Fire images may contain flame and smoke textures. The presence of flames is sufficient to classify an image as a 'fire' instance, while the presence of smoke alone classifies the image as a 'smoke' instance. The 'other' category serves as the negative class and is given to instances that do not show any of the three emergency events. Note that different negative ('other') samples are provided

Table 10.1 Gathered datasets for flood, fire, smoke detection.

Dataset name	Flood images	Fire images	Smoke images	Other images
Bowfire[1]		118		107
Corsican fire database[2]		1135		
MediaEval 2017 multimedia satellite task[3]	1920			3360
European flood 2013 dataset[4]	3151			
Fire-detection-image-dataset[5]		109		537
Fire-smoke-dataset[6]		1000	1000	999
ForestFire[7]		223	45	420
FiSmo (part of)[8]		1885	369	3583
beAWARE data	380	25	18	262

[1] https://bitbucket.org/gbdi/bowfire-dataset/src/master/.
[2] http://cfdb.univ-corse.fr/index.php?newlang=english&menu=1.
[3] http://www.multimediaeval.org/mediaeval2017/multimediasatellite/.
[4] https://github.com/cvjena/eu-flood-dataset.
[5] https://github.com/cair/Fire-Detection-Image-Dataset.
[6] https://github.com/DeepQuestAI/Fire-Smoke-Dataset.
[7] https://dataturks.com/projects/qdyzl1013/Forest%20Fire.
[8] https://github.com/mtcazzolato/dsw2017.

for each category, as non-flood, non-fire, non-smoke, since the deep CNN features that are going to be extracted need to discriminate between the important visual clues that uniquely characterize flood, fire and smoke textures and non-flood, non-fire, non-smoke instances. For example, the non-fire class includes images of sunsets. This approach helps 'instruct' the model to avoid learning colour as an important visual trigger for fire detection and give weight to texture features instead.

A unified framework has been deployed to handle both images and videos, sharing a common function. More specifically, we have wired the CNN classifier to analyse images as well as sequential video frames and assign an emergency label for each of them. In case of video analysis, a majority vote scheme is conducted over all video frames in order to decide the emergency class of the full video.

10.3.2 Object Detection

In addition to the identification of disastrous events, the detection of impacted targets is also an important aspect of crisis management, especially in rescue operations. Thus, a separate component was developed, using deep-learning object-detection techniques, in order to detect and track humans and vehicles, estimate their number and provide relevant reports. Object Detection (ObD) works in two stages. At stage 1 the task of detection is carried out in order to provide candidate bounding boxes that may contain the individuals or the vehicles. At stage 2 the task of recognition is carried out in order to classify the boxes based on their visual content. For the detection stage, a deep CNN feature extractor is used to extract appearance features for a number of candidates bounding boxes and then, for the recognition stage, Region of Interest (RoI) pooling is performed in order to refine bounding box coordinates and finally classify their content. Based on (Huang et al. 2017), we use Faster R-CNN architecture (Ren et al. 2015), pre-trained on Microsoft COCO dataset (Lin et al. 2014). As feature extractor, we use Inception Resnet in atrous mode (Szegedy et al. 2017) for the extraction of deep features. From all the predicted classes only human and vehicle object classes are kept as valid predictions.

This process is more complicated when dealing with videos, especially in the case of multi-object detection. Several pedestrians or vehicles may appear in multiple sequential video frames and consequently, simply detecting all the instances at each video frame would be redundant and inaccurate as the same targets would appear in multiple frames. As a result, visual tracking techniques of the detected bounding boxes are also integrated in the module. The task of object tracking is cooperatively performed by two different modalities, the object detection algorithm, described above, and a tracker, which accepts the detected bounding boxes and attempts to predict the position of each bounding box in the next frames. Tracking is based on the Kernelized Correlation Filters (KCF) algorithm (Henriques et al. 2014). Initially, object detection is applied in order to detect new candidate objects. In the next step, the tracker is using the detected bounding boxes in order to predict their coordinates in subsequent frames. Object detection is periodically applied again every a certain number of frames in order to rectify object trajectories. This repetitive synergetic procedure leads to the incremental construction of object trajectories.

In order to estimate the number of unique detected targets as accurately as possible and to provide informative reports about the number of affected people and vehicles, there should be an object distinction process, in order to differentiate between different objects. Therefore, each object is assigned with a different id by applying a certain counting process. The first object that is detected is assigned an id equal to 1. During the tracking phase, objects retain their assigned id. During the detection phase, the newly detected bounding boxes are compared with the previous ones and, if there is a sufficient overlap between them, they retain the same ID. In a different case, the object is considered as a new object and a new id is assigned to it, increased by 1.

10.3.3 River Level Monitoring

This component performs visual analysis on videos from static surveillance cameras installed by the river, in order to monitor the water level and create alerts, in case some predefined thresholds are exceeded. The component is used both as an early warning system for floods and for assisting authorities assess the situation during an ongoing flood. The water level estimation algorithm is based on edge detection for the detection of the visible part of a marker (rod), which is installed on the river bank. After detecting the visible part of the marker, the algorithm estimates its length in pixel units and then corresponds it to real length in meters, by using calibration information (measurements of the marker from a reference image). Subsequently, the length of the visible part corresponds to water level. If the water level exceeds some predefined thresholds, three different types of alerts are generated respectively: 'Moderate', 'Severe', 'Extreme'. For the implementation of the edge detection algorithm, Canny operator is applied using OpenCV framework (OpenCV Canny Edge Detection 2019). The algorithm was developed and evaluated using footage from static surveillance cameras captured during past flooding events in the Bacchiglione river (Vicenza, Italy) in February 2016.

Figure 10.1 depicts the result of the application of the algorithm for the detection of the marker on footage captured at Bacchiglione river. The algorithm for the water level estimation is easily adaptable and can easily be configured for a new camera system, since the only requirements for its calibration are a single frame depicting the marker and the length of the marker in this frame.

10.3.4 Drone Analysis

In addition to the use of ground cameras, the benefits and advancements of UAVs have been leveraged through the deployment of similar disaster detection models for aerial missions. To this end, a drone platform has been integrated in the overall beAWARE platform, which serves both as a data source and an analysis component. It was developed on the framework of beAWARE project but it is also able to function as a standalone service. Drone platform is a service to easily configure, launch and monitor drone-related missions. Within beAWARE it acts as an additional content provider with a unique point of view which is able to provide content fast, in a safe manner and even in areas that are difficult to reach by humans and ground vehicles. The drone

Figure 10.1 A captured frame from Bacchiglione river (on the left) and an image patch of the captured frame containing the marker along with the result of the edge detection algorithm (on the right).

platform is developed to be able to act independently as a possible supporting generic plug-in for collaborating components.

The drone platform captures media from on board equipment and through the message bus passes it to its analysis component called drone analysis (DA). DA is responsible for the visual analysis of the drone footage. DA is registered as a message consumer at the message bus in order to obtain messages from the drone platform. In case it receives an analysis request, DA performs one of the predefined analysis tasks according to an input variable. In order to organize and speed-up analysis, DA functionalities have been grouped into four distinguished analysis tasks, which can be performed separately or in combination at each request. The first task is 'Object Detection' during which DA performs object detection and object tracking with the aim of detecting people and vehicles and tracking their trajectories. For this task, the same detection and tracking process is used as the one designed for object detection from ground cameras, with some adaptations in order to take into account the rapid movement of the camera or of the targets. The second task is 'Disaster Detection' during which DA performs image classification in order to classify each image or video frame into four possible classes: 'flood', 'fire', 'smoke' or 'other'. Likewise, the same CNN is used as the one designed for disaster detection from ground cameras adapted for the needs of this task. The third analysis task, 'Evacuation', has been developed to facilitate and monitor evacuation processes during crises. During this task DA performs object detection and tracking to detect if any person is left behind and create relevant alerts during ongoing missions and create overall reports at the end of the mission. The distinguished tasks can be performed separately or in combination, according to an input variable.

10.3.4.1 Drone Disaster Detection This functionality is able to detect flood, fire and smoke from drone footage. It can be used for early warning, for the early detection of forest fires through drone patrols, during periods of high risk for fire. It could be used also during the response phase, for the estimation of the extent of the disaster, the detection of burning areas and the proximity to people or critical infrastructures. This functionality uses the same detection and tracking process as the ones used for ground cameras. During drone patrol flights, when the 'Disaster Detection' task is selected, drone footage is forwarded to the 'Disaster Detection' module, either as a video sequence or as an image sequence. In the case of an image sequence, each image is classified separately. In the case of video transmission, each video segment is analysed by individually classifying each frame into one of the four classes and subsequently, the whole video segment is classified according to a majority voting scheme. A dedicated field in the output message of the analysis component indicates whether or not the video segment contains an incident.

Regarding the localization of the disaster, after collecting drone coordinates of all the frames that were classified as containing an incident, average coordinates are estimated. Subsequently, the frame with coordinates closest to the average coordinates is identified and the rest of the drone's location parameters (altitude, heading, gimbal pitch) are extracted from the metadata of this frame, in order to use them for drone relocation, if needed. The reason for using the average coordinates instead of e.g. the coordinates of the first or last frame containing a disaster, is to try to keep the disaster in the center of the drone's view as much as possible.

10.3.4.2 Drone Object Detection This functionality is used to detect and track humans and vehicles, estimate their number and provide relevant reports. It uses the same detector as the one used in ObD. In order to keep a low computational load and network payload and achieve near-real time analysis, the drone platform has the ability to adjust the frame rate of the transmitted material in order to keep low frame rates, such as 1 fps. However, in low frame rates, in case of rapid motion of the tracked objects or the drone camera, the tracker fails to correlate objects in consecutive frames, if the objects have moved so far that their bounding boxes do not overlap. This erroneously increases the number of tracked objects, which is not desirable in the case of multi-object tracking. Thus, an additional post-processing step was added on top of the KCF tracker in order to fine-tune results. All detected bounding boxes in a new frame are compared to the valid bounding boxes of the last frame, even if they do not overlap. The matching process includes three criteria: (i) the class of the detected objects; (ii) their similarity score, which is defined as their Intersection over Union (IoU) when their bounding boxes are shifted so as their top left corners coincide; and (iii) the proximity of the objects, which is estimated as the minimum distance between all pixels of their bounding boxes. In case two compared objects fulfil these criteria, they are considered as the same object, no matter if they do not overlap and the new object retains the same id. Other post-processing operations include the rejection of scattered positive object detections (objects of a certain class that are detected only in a single frame and not in adjacent frames), as probable false positives, and objects with size more than a certain portion of the image, which are probably false detections.

After detecting a target, the drone sends its location to the platform, while it can also 'lock' the target in order to continuously monitor, e.g. a person in danger until a rescue team can reach the place. Regarding the localization of detected targets, a good approximation would be the coordinates of the drone. However, in order to fine-tune results, camera orientation and altitude metadata are also utilized to obtain a good approximation of the actual position of the target. By using the projection of the detected objects in the image plane and the above metadata, a good approximation of the object location is estimated.

10.3.4.3 Evacuation Monitoring This feature is designed to facilitate and monitor evacuation processes during crises. Specifically, when a drone is sent on an evacuation mission, footage from the area under evacuation is analysed in order to detect any remaining people. Alerts are created every time a person is detected in an evacuated area during the mission. At the end of the mission, the drone sends a notification that the mission is over and a relevant report is created for the whole drone mission, indicating that the mission is over and whether people were detected or not. This feature uses the same deep-learning model as the one used in the 'Object Detection' task with the difference that only humans are reported. In order to produce the final report for the evacuation mission, DA keeps a log with all detected human instances and corresponding location metadata, while the drone is in the middle of an evacuation mission. At the end of the mission, the Drone Platform sends a notification that the mission is complete and an overall report is produced in the selected language of the user.

10.3.5 Traffic Analysis and Management

A separate component has been developed for traffic monitoring and detection of anomalous traffic events during crises in the framework of beAWARE. The process of detection and tracking of moving vehicles from surveillance cameras is based on the traffic flow analysis system, whereas traffic anomaly detection is based on detecting anomalous car trajectories.

10.3.5.1 Traffic Flow Analysis Passing vehicles have to be successfully detected in each frame and then subsequently tracked as they follow their full trajectory on screen. The task is simultaneously performed in a cooperative manner by two separate modalities: (i) a generic object detector, specifically trained to discover bounding boxes of vehicles at an adjustable detection rate; and (ii) a tracker which accepts new image patches as input queries and is assigned to discover the most probable position of each query in subsequent frames. This finally leads to the incremental construction of the trajectory of each individual vehicle that has been captured by the detector at some point in the frame sequence.

To accurately compute the velocities of a detected vehicle, known pixel coordinates in the image plane have to be back-projected in real-world coordinates in the road plane. Then, given a vehicle's travelled distance in meters and the time duration in seconds, velocities can be calculated. Since the frame rate of the recorded videos is known, it can be used to compute time intervals in seconds between consecutive frames. What is not directly available, however, is a way to translate displacement of pixels in the image plane to the real distance in meters a particular vehicle has travelled in a given amount of time. In order to calculate this precisely and effectively, an algorithm for automatic camera calibration is needed. Once camera parameters have been discovered, pixel coordinates in the image plane can be successfully back-projected to real-world coordinates.

In order to automatically obtain camera parameters of a traffic surveillance scene, the algorithm proposed by (Dubská et al., Automatic Camera Calibration for Traffic Understanding 2014) was deployed, which is based on detection of two vanishing points. The description of the algorithm that was used in order to extract the location of the vanishing points is described in (Giannakeris et al. 2018). Once the first two vanishing points have been found, the third vanishing point, the focal length and the road plane normal vector, which defines the road plane up to a scale are calculated. Subsequently, back-projection of a 3D image plane point onto the road plane is possible following the procedure described in (Schölkopf et al. 2001).

For the purpose of detecting vehicles in video frames a deep CNN object detector was used that extracts deep image representations from a CNN and predicts pixel coordinates of bounding boxes. Similarly to ObD, the aforementioned object detection and tracking architecture was adopted here. Figure 10.2 depicts bounding boxes and trajectories of vehicles successfully detected and tracked using this methodology.

In order to estimate the velocity of a tracked vehicle at a certain frame the tracker is fed with points inside the previous box instance of the previous frame and several displacement calculations are produced for each point. Then, back-projection of all the displacement pixel pairs is applied to the road plane according to the calibration parameters that have been found on that specific scene and the median displacement is selected as the true value. To calculate the velocity of the vehicle the true distance (D) is divided by the time duration of a frame (which equals fps^{-1} seconds), resulting in a velocity of D·fps meters per second.

Figure 10.2 Visualization of detected vehicles and their trajectories. *Source*: AI CITY CHALLENGE.

10.3.5.2 Anomaly Detection in Traffic Scenes

In order to detect any anomalies in traffic videos, such as traffic collisions, we focused on detecting anomalous car trajectories. To this end, the Support Vector Method For Novelty Detection of (Schölkopf et al. 2001) was adopted. Support vector machines (SVMs) are chosen, as they generally exhibit good performance relative to other machine learning methods at a low computational cost, while they are able to handle large data sets, which generally appear in real life situations. The proposed method is described in (Giannakeris et al. 2018).

10.3.5.3 Inference of Severity Levels

On top of the main visual components, a feature for the overall severity level estimation was developed in order to fuse their outputs and provide qualitative levels for the overall situation in an image or video and construct quantitative severity indicators. For the severity level of an individual image or video, the possible outcomes are defined as:

a. 'Unknown': When some targets have been detected by ground or aerial cameras, but there is no disaster detected by the disaster detection component, the Drone Analysis or the River Level Monitoring component.
b. 'Severe': A disaster is reported but no living targets or vehicles have been detected.
c. 'Extreme': If both a disaster and living targets or vehicles have been detected.

The severity level estimated by visual components is subsequently forwarded to the overall Crisis Classification component of beAWARE in order to be fused with the output of other components and provide a more comprehensive assessment of the overall situation.

10.4 System Evaluation

The described system was quantitatively evaluated through performance measurements by using existing and generated datasets, in order to estimate the accuracy of the analysis components. Apart from these results, the function and the effectiveness of the system was also validated through two large-scale field tests, with the wide participation of stakeholders. It should be noted that data produced during the field tests have also been used for the quantitative evaluation of the system.

In this section, the system is evaluated using existing and generated datasets regarding the performance of the individual visual analysis components. Evaluation of visual analysis is focused on the performance of the system in (i) disaster detection, by evaluating the ability of the system to detect disastrous events in images and videos; (ii) object detection, so as to evaluate the effectiveness of the system in finding people and vehicles that exist in impacted locations; (iii) water level estimation, by evaluating the accuracy of the algorithm in determining the river level in order to timely raise a flooding alarm and finally; (iv) traffic analysis, with respect to vehicle tracking and speed estimation and anomaly detection.

10.4.1 Disaster Detection

This component is responsible for the detection of disastrous incidents such as floods, fires or smoke from images or videos through image classification techniques. As already mentioned, we have trained the classifier with over 20,000 images taken from public datasets and beAWARE data. Around 80% of the collected dataset was used for training and the rest was reserved as a testing set for evaluation. The mean accuracy of the classifier on those data is 0.98%. Figure 10.3 shows the confusion matrix of the emergency classification task along with its normalized version. As we can see, there are very good accuracy rates in general. Most of the errors are false positive cases, i.e. images where no emergency is taking place but are classified as flood in 81 cases, fire in 51 cases and smoke in 13 cases. Some false negatives also appear. Interestingly, there is very little confusion between the emergency classes. There are just 31 fire instances that have been misclassified as smoke and only 11 smoke instances misclassified as fire. Moreover, flood is almost never confused with fire or smoke instances. This means that even though there is one unified classification model for all emergency cases, the visual analysis will rarely confuse emergency events.

By examining some false positive cases in Figures 10.4–10.6, we can observe that the flood false positives are images where water bodies are found but the water level is low. However, even though they are false positive cases, the thematic content of the images are not entirely foreign to the context of a flood event, which is a good indicator of the quality of the support that the system can provide to authorities even when a wrong prediction happens. Some hard examples are also shown for the fire use case and the smoke use case. The fire texture can sometimes be found in fireworks and smoke is sometimes found in areas with high humidity, like clouds or fog, which can resemble smoke textures.

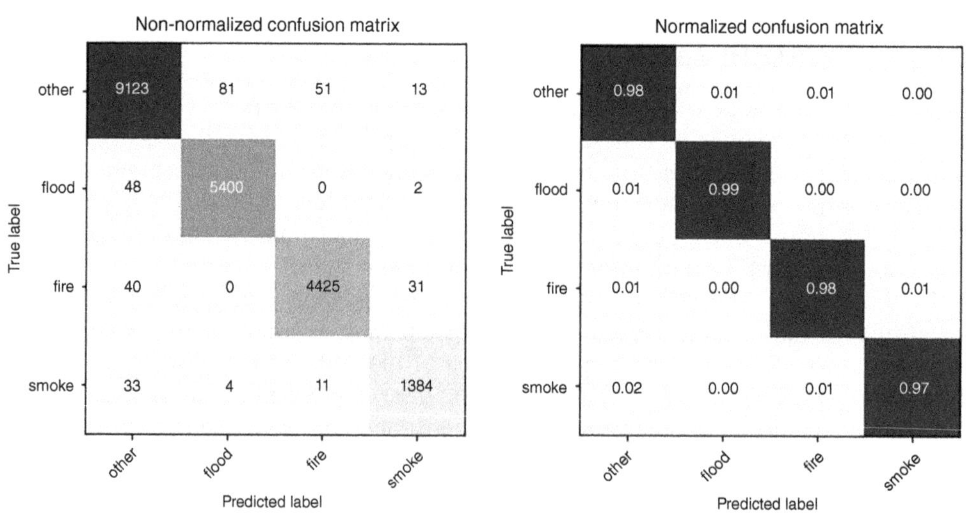

Figure 10.3 Non-normalized (left) and normalized (right) confusion matrix of the EmC.

Figure 10.4 False positive flood cases. *Source*: (b) Jdeeringdavis / Flickr / CC BY 2.0.

Figure 10.5 False positive fire cases.

10.4.2 Object Detection and Tracking

For Object Detection and Tracking (ObD), the *Precision – Recall* graph and the *mAP* metric will be presented to evaluate its performance. Evaluation is performed on images collected during the pilot tests of the project. Those images were manually annotated with ground truth bounding boxes of vehicles and pedestrians so as to make a comparison with the detected bounding boxes. Vehicles were categorized in five different classes (car, bicycle, motorcycle, bus, truck), mainly because the detector is also the core component of traffic analysis framework, where the information about the

Figure 10.6 False positive smoke cases. *Source*: peasap / Flickr / CC BY 2.0.

Table 10.2 Object detection performance during the 1st and 2nd pilot tests.

		Class						mAP (%)
		Car (%)	Bicycle (%)	Motorcycle (%)	Bus (%)	Truck (%)	Person (%)	
Average precision	1st pilot	60.73	10.15	48.23	50.70	39.74	59.21	44.79
	2nd pilot	73.34	—	66.67	—	—	47.74	62.58

Figure 10.7 Pedestrians and vehicles detected in flooded areas and near a fire scene in various scales, viewpoints and occlusion rates. *Source*: (a, b, d) Associated Press / Alamy Stock Photo.

type of detected vehicles plays an important role. Table 10.2 shows the detection performance for each class. Figure 10.7 provides some qualitative examples of ObD depicting detected vehicles and pedestrians in various scales, viewpoints and occlusion rates in order to showcase the detector's robustness.

10.4.3 River Level Monitoring

As already described, River Level Monitoring component performs visual analysis on videos from static surveillance cameras installed on a river, in order to monitor the water level and create alerts in case some predefined thresholds are exceeded. The water level estimation is performed by an edge detection algorithm, which detects the visible part of a marker (rod), as has been described in the Methodology Section 10.3.3. After

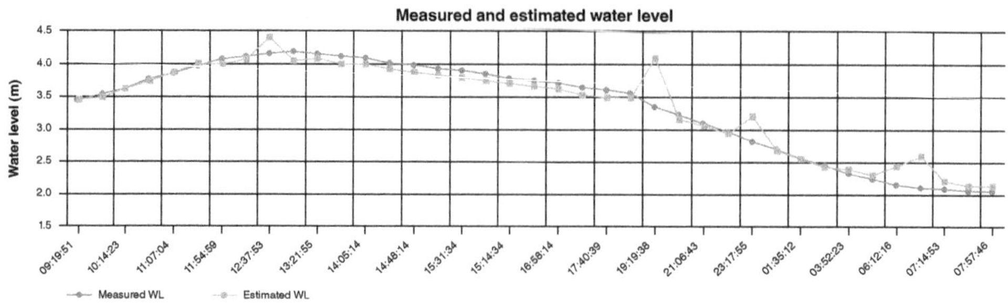

Figure 10.8 Comparison of water level values measured from the sensor (green circled line) and estimated by the water level estimation algorithm (red squared line).

detecting the marker, the algorithm estimates its length by using calibration information. Subsequently, the length of the visible part is translated into water level. If the water level exceeds some predefined thresholds, an alert is created with possible values: 'Moderate', 'Severe', 'Extreme'. In order to evaluate the water level estimation algorithm, there was a need for an annotated dataset captured during different points in time with different water levels. For this reason, the algorithm was evaluated using existing footage from a static surveillance camera captured during past flooding events in the Bacchiglione river (Vicenza, Italy) in February 2016, for which there are water level measurements available. The dataset consists of 39 videos, spanning from 29 February 2016 09 : 19 a.m. to 1 March 2016 07 : 57 a.m. Annotations are water level measurements from a sensor installed on a bridge. It should be noted that the video resolution of the training dataset is relatively low (640 × 340). Higher resolutions can be used during real-time application in order to improve accuracy. For the evaluation of the accuracy, the percent error between the sensor measurements and the corresponding estimated water level was used as a performance indicator. Results are presented in Figure 10.8.

From the results it was observed that during the day, the detection of the marker was very accurate, resulting in a low percent error. In particular, the average percent error from 09 : 19 : 51 to 18 : 02 : 27 was 2.62%, with only one outlier. Figure 10.9 shows two characteristic examples of an accurate and an inaccurate detection of the marker. The first was recorded during the day, at 09 : 47 : 36 a.m. However, during the night, the selected parametrization of the detector, i.e. the canny edge detection threshold parameters, completely failed to detect the marker, resulting in a blank image, due to totally different lighting conditions. To solve this issue, the algorithm was modified in order to use a different parametrization (that seemed to perform well during the night) in case no pixel is detected from the initial parameterization. This modification leads to a high detection accuracy even during the night. It can be concluded that, while the detection is, in general, highly accurate, some minor inaccuracies can be noted during dusk (e.g. 19 : 19 : 38) and dawn (e.g. 06 : 12 : 16, 06 :53 : 46), due to particularities in the lightning conditions. Contrary to videos captured at night, the problem in these specific cases is that even though the detection is not highly accurate, however some pixels are still being detected from the initial parameterization, so the algorithm does not change parametrization. In Figure 10.9 we can also see an example of an inaccurate detection

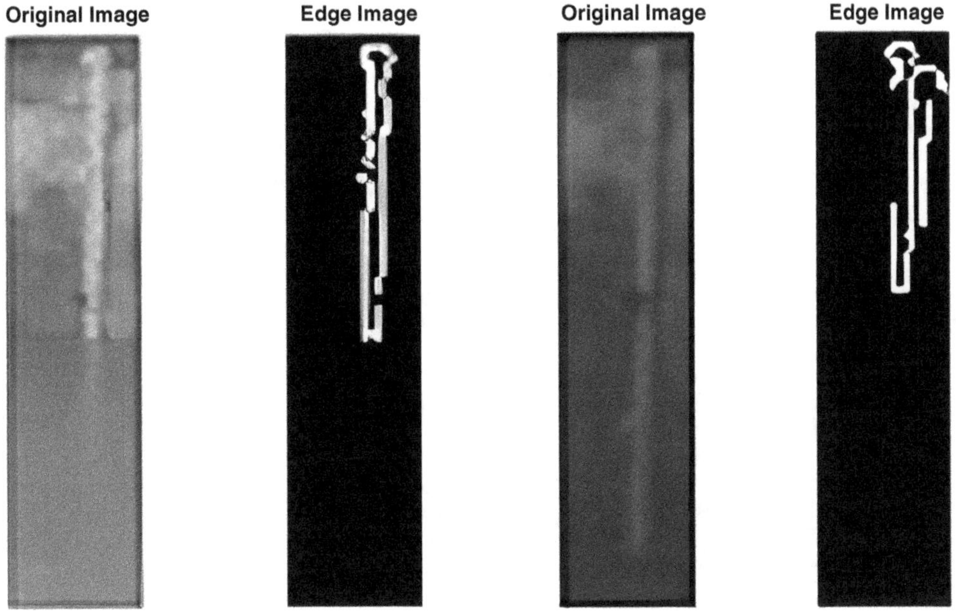

Figure 10.9 Left images: example of an accurate estimation of water level during the day (09 : 47 : 36 a.m.). Right images: example of an inaccurate estimation during dawn (06 : 12 : 16 a.m.). Both cases depict the cropped image of the rod and the corresponding edge detection result.

of the marker during the dawn. In order to take into consideration these types of inaccuracies, the component could become more robust in the future by investigating different edge detection parametrizations, which could lead to more global results or, if this is not possible, to take into account the timestamp metadata of the video in order to adjust the detector accordingly.

10.4.4 Drone Analysis

Drone components were also evaluated using existing and generated datasets regarding the performance of the individual analysis components. Evaluation of visual analysis is focused on the performance of (i) disaster detection by evaluating the ability of the system for early detection of disastrous events from aerial cameras; (ii) object detection, so as to evaluate the effectiveness of the system in finding people and vehicles that exist in impacted locations; and (iii) performance in evacuation missions.

In order to evaluate the ability of Drone Analysis to detect flood, fire and smoke, the classifier that was developed for ground cameras and tailored to the needs of Drone Analysis, was evaluated on footage from actual drone flights that took place during the pilot tests of the system. For the evaluation 48 video files from a drone patrol flight containing smoke were used. 231 frames were annotated as 'smoke' frames and 489 were annotated as 'other', which means that they did not contain any hazard. Figure 10.10 depicts the normalized Confusion Matrix of the classifier. Results show that 75% of the

Normalized confusion matrix

	Smoke	Flood	Fire	Other
Smoke	0.75	0.01	0.09	0.15
Other	0.1	0.1	0.0	1.0

Figure 10.10 Normalized confusion matrix of image classification on a drone video sequence containing smoke.

Figure 10.11 Example of a 'smoke' image mis-classified as 'other'.

frames containing smoke were correctly identified, whereas there was a 15% of FNs and 10% of miss-classifications, with the highest portion of these to be classified as 'fire' (as expected, since usually most fire images contain smoke). The problem with this sequence is that, since it was a simulated fire for the needs of the field test, the scale of the fire and consequently the spread of the smoke was limited, thus in many frames the proportion of fire in the image was low, which makes it easy to be misclassified as 'other', as can be seen in Figure 10.11. This could justify the relatively low performance compared to the performance on the dataset collected from existing databases that was presented Section 10.3.1. However, despite these restrictions, still 84% of the frames of this sequence containing smoke were classified as either 'fire' or 'smoke', which, after the application of the post-processing rectification step, increased to 89%. Moreover, by taking into account that the generation of a fire alarm depends on the aggregated result of all these separate detections in each frame, this implies that sometime during the drone mission an alarm is generated for possible fire.

For the evaluation of the object detection functionality of the component, the detector was evaluated on footage from drone flights captured during the pilots' tests of the

Table 10.3 Average Precision results of the object detection classifier.

Car (%)	Bus (%)	Truck (%)	Person (%)	mAP (%)
63.4	51.20	40.01	53.31	51.98

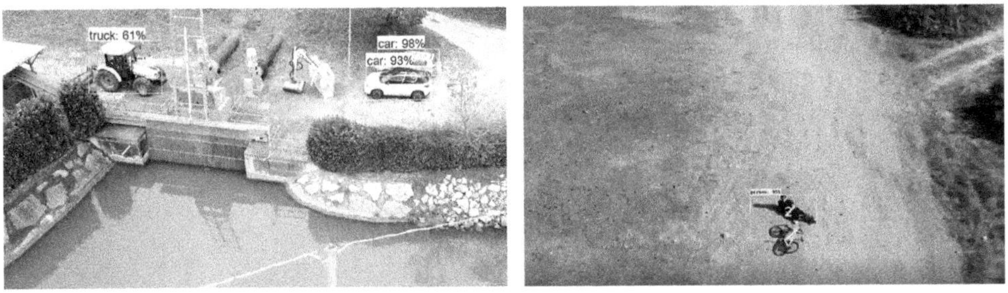

Figure 10.12 Examples of object detection results on drone footage.

project. Out of 20 video sequences, 250 frames, containing more than 600 objects, were extracted. Selected frames were manually annotated for the generation of ground-truth bounding boxes. Subsequently, the Average Precision was calculated for each class. Quantitative results are presented in Table 10.3. Additionally, Figure 10.12 shows indicative examples of correctly detected objects. It should be noted that, even though miss-detections in single frames do exist, results in frame sequences are significantly improved, since they act cumulatively. For example, in a search mission for the detection of a missing person, even if the person is missed in a few frames, the successful detection in subsequent frames guarantees the generation of an alarm.

Finally, the ability of the drone platform to assist evacuation missions was validated during one of the pilot tests of the project, where, according to the scenario, a fire was heading to an educational institution, which had to be quickly and effectively evacuated. Once the evacuation was ordered, the drone was sent to scan the external area of the building in order to identify if there were any remaining people. During a first scan, the drone analysis component identified some people in the area, who were visualized at the drone dashboard. Then, a team was sent back to the school to assist the remaining people. The drone confirmed that the area was empty. The school evacuation was then determined to be complete. At the end of the mission, the drone platform sent a message to the analysis component marking the end of the mission. Subsequently, the following overall report was generated: 'Evacuation mission completed. No people have been detected'.

10.4.5 Traffic Analysis and Management

We evaluated our traffic analysis methods in the NVIDIA AI CITY challenge (The NVIDIA AI City Challenge 2017). Teams that entered the competition had the opportunity to experiment and evaluate their algorithms on three separate traffic analysis challenge tracks: (i) traffic flow analysis that focused on speed estimation of vehicles, (ii) anomaly detection for the detection of anomalous events such as car crashes or stalled vehicles and (iii) multi-camera vehicle detection and re-identification. In the next subsections, we present the experiments that took place and our evaluation scores for the first two tracks of the challenge.

10.4.5.1 Vehicle Tracking and Speed Estimation Evaluation The dataset provided for the speed estimation track consists of 27 videos, each containing 1800 high-definition frames. The recorded videos depict highway traffic at various locations and viewpoints. Most of the cameras were static, with the exception of some locations where small tremors were observed, likely due to windy conditions. The overall score $S1$ for this track takes into account the quality of the detections and the accuracy of speed measurement and is defined as follows:

$$S = \mathrm{DR}(1 - \mathrm{NRMSE})$$

where, DR is the detection rate and NRMSE is the normalized root mean square error (RMSE) of speed estimation across all submissions, obtained via min-max normalization. A vehicle is considered to be detected if it is localized in at least 30% of frames it appeared in. Our vehicle detector and tracker managed to achieve a high detection rate of 89%. The RMSE score of our speed estimator was 27.30, higher than other teams participating in the challenge.

10.4.5.2 Anomaly Detection Evaluation The NVIDIA dataset for the anomaly detection track comprises 100 videos, each with a duration of 15 minutes and a resolution of 800 × 410. The dataset poses a substantial challenge due to the presence of diverse real-world traffic scenarios, substantial camera motion, varying weather conditions, changes in illumination, occlusions, and low-resolution shots. To address the issue of video sequences featuring a range of camera motions, such as zoom-in/out or view changes, the videos were divided into sub-clips based on the stability of the view. Subsequently, distinct models were developed for each sub-clip. The evaluation of the performance of the anomaly detection is based on the F1-score, which measures the effectiveness of the detection, and the RMSE, which measures the detection time error. The final score is computed as:

$$S2 = F1(1 - \mathrm{NRMSE})$$

where, for the estimation of $F1$, as true-positive (TP) detection is considered a predicted anomaly with an overlap of five minutes with the true anomaly that has the highest confidence score, as false-positive (FP) is considered a predicted anomaly that does not correspond to any existing anomaly and false-negative (FN) is an existing anomaly that was not predicted. RMSE is calculated between the ground-truth anomaly time and the predicted time for all TP predictions. NRMSE is the normalized RMSE obtained from min-max normalization across all submissions. Our algorithm reached an $F1$-score of 0.33 and the detection time RMSE was 227.

10.5 Conclusions

Visual analysis components that were developed in beAWARE project incorporate advanced machine learning techniques to extract and deliver actionable insights to the project's decision support system. Evaluation of these components was performed using existing visual databases and data generated within the project. Specifically, the Disaster detection component was evaluated on a large-scale database sourced from various sources. The results show a high accuracy rate, with a few false positive cases that were analysed further. Most of the false positives were attributed to inter-class texture similarity in the samples. The Object Detection component was evaluated both quantitatively and qualitatively to assess its robustness to variations in scale, illumination and occlusion. The River Level Monitoring component demonstrated high accuracy in estimating water levels, with a few exceptions during certain hours of the day due to lighting conditions. The Drone Analysis component can analyse image and video sequences for object detection, tracking and classification, enabling the detection of people, vehicles and emerging disasters.

To address more demanding scenarios and produce more robust analysis results, we plan to further improve our methods and techniques by incorporating additional visual features. This will provide first responders and authorities with robust and reliable computer vision-based tools for disaster management. In regards to fire and flood detection, our future steps include further and more customized training of the deep CNN architecture that is deployed, with more data related to fire and flood scenarios, so as to better tailor our algorithms to respond well in those types of challenges. Additionally, an approach for object detection based on occluded models would greatly aid that purpose since most of the targets we are trying to detect are occluded from flood or fire pixels. To improve our traffic analysis and management, we plan to investigate advanced techniques for automatic camera calibration to enhance our speed estimation results. In addition, we will also create and evaluate a sophisticated descriptor that will encode motion information patterns from the vehicles found in the scene based on optical flow. A new model will be then trained in order to estimate traffic levels of highways based on motion patterns. With these efforts, we anticipate delivering more robust and reliable results for disaster management operations.

References

AI City Challenge (2017). https://www.aicitychallenge.org (14 September 2023).

Avgerinakis, K., Giannakeris, P., Briassouli, A. et al. (2017a). Intelligent traffic city management from surveillance systems (CERTH-ITI). *NVIDIA AI City Challenge* 2017.

Avgerinakis, K., Moumtzidou, A., Andreadis, S. et al. (2017b). Visual and textual analysis of social media and satellite images for flood detection@ multimedia satellite task mediaeval 2017. *MediaEval* 1984. https://ceur-ws.org/Vol-1984/Mediaeval_2017_paper_31.pdf.

Bischke, B., Bhardwaj, P., Gautam, A. et al. (2017). Detection of flooding events in social multimedia and satellite imagery using deep neural networks. *MediaEval* 1984. https://ceur-ws.org/Vol-1984/Mediaeval_2017_paper_51.pdf.

Cheng, K.-W., Chen, Y.-T., and Fang, W.-H. (2015). Gaussian process regression-based video anomaly detection and localization with hierarchical feature representation. *IEEE Transactions on Image Processing* 24: 5288–5301.

Dubská, M., Herout, A., and Sochor, J. (2014). Automatic camera calibration for traffic understanding. *BMVC* 4: 8.

Filipiak, P., Golenko, B., and Dolega, C. (2016). NSGA-II based auto-calibration of automatic number plate recognition camera for vehicle speed measurement. *Applications of Evolutionary Computation: 19th European Conference, EvoApplications 2016,* Porto, Portugal (30 March–1 April 2016), pp. 803–818. *Proceedings, Part I 19.*

Giannakeris, P., Kaltsa, V., Avgerinakis, K. et al. (2018). Speed estimation and abnormality detection from surveillance cameras. *Proceedings of the IEEE Conference on Computer Vision and Pattern Recognition Workshops*, pp. 93–99. IEEE.

Giordan, D., Manconi, A., Remondino, F., and Nex, F. (2017). Use of unmanned aerial vehicles in monitoring application and management of natural hazards. In: *Use of Unmanned Aerial Vehicles in Monitoring Application and Management of Natural Hazards*, vol. 8, 1–4. Taylor & Francis.

Girshick, R. (2015). Fast R-CNN. *Proceedings of the IEEE International Conference on Computer Vision*, Santiago, Chile (7–13 December 2015), pp. 1440–1448. https://doi.org/10.1109/ICCV.2015.169.

He, X.C. and Yung, N.H. (2007). A novel algorithm for estimating vehicle speed from two consecutive images. *2007 IEEE Workshop on Applications of Computer Vision (WACV'07)*, Austin, TX, USA (21–22 February 2007), pp. 12–12. https://ieeexplore.ieee.org/document/4118741/

He, K., Zhang, X., Ren, S., and Sun, J. (2016). Deep residual learning for image recognition. *Proceedings of the IEEE Conference on Computer Vision and Pattern Recognition*, pp. 770–778.

Henriques, J.F., Caseiro, R., Martins, P., and Batista, J. (2014). High-speed tracking with kernelized correlation filters. *IEEE Transactions on Pattern Analysis and Machine Intelligence* 37: 583–596.

Hospedales, T., Gong, S., and Xiang, T. (2012). Video behaviour mining using a dynamic topic model. *International Journal of Computer Vision* 98: 303–323.

Huang, J., Rathod, V., Sun, C. et al. (2017). Speed/accuracy trade-offs for modern convolutional object detectors. *IEEE Transactions on Pattern Analysis and Machine Intelligence* 37 (3). https://doi.org/10.1109/TPAMI.2014.2345390.

Jiang, F., Yuan, J., Tsaftaris, S.A., and Katsaggelos, A.K. (2011). Anomalous video event detection using spatiotemporal context. *Computer Vision and Image Understanding* 115: 323–333.

Krizhevsky, A., Sutskever, I., and Hinton, G.E. (2017). Imagenet classification with deep convolutional neural networks. *Communications of the ACM* 60: 84–90.

Kuettel, D., Breitenstein, M.D., Van Gool, L., and Ferrari, V. (2010). What's going on? Discovering spatio-temporal dependencies in dynamic scenes. *2010 IEEE Computer Society Conference on Computer Vision and Pattern Recognition,* San Francisco, CA, USA (5 August 2010), pp. 1951–1958. https://doi.org/10.1109/CVPR.2010.5539869.

Lan, J., Li, J., Hu, G. et al. (2014). Vehicle speed measurement based on gray constraint optical flow algorithm. *Optik* 125: 289–295.

Lee, S., Har, D., and Kum, D. (2016). Drone-assisted disaster management: finding victims via infrared camera and lidar sensor fusion. *2016 3rd Asia-Pacific World Congress on Computer Science and Engineering (APWC on CSE)*, Nadi, Fiji (5–6 December 2016), pp. 84–89. https://doi.org/10.1109/APWC.

Lin, T.-Y., Maire, M., Belongie, S. et al. (2014). Microsoft COCO: common objects in context. *Computer Vision–ECCV 2014: 13th European Conference*, Zurich, Switzerland (6–12 September 2014), pp. 740–755. *Proceedings, Part V 13.*

Liu, W., Anguelov, D., Erhan, D. et al. (2016). Ssd: single shot multibox detector. *Computer Vision–ECCV 2016: 14th European Conference,* Amsterdam, The Netherlands (11–14 October 2016), pp. 21–37. *Proceedings, Part I 14.*

Maher, A. and Inoue, M. (2016). Generating evacuation routes by using drone system and image analysis to track pedestrian and scan the area after disaster occurrence. *Proceedings of the 10th SEATUC (South East Asian Technical University Consortium) Conference,* Tokyo, Japan, pp. 22–24.

Muhammad, K., Ahmad, J., and Baik, S.W. (2018). Early fire detection using convolutional neural networks during surveillance for effective disaster management. *Neurocomputing* 288: 30–42.

Nurhadiyatna, A., Hardjono, B., Wibisono, A. et al. (2013). Improved vehicle speed estimation using gaussian mixture model and hole filling algorithm. *2013 International Conference on Advanced Computer Science and Information Systems (ICACSIS)*, Sanur Bali, Indonesia (28–29 September 2013), pp. 451–456. https://doi.org/10.1109/ICACSIS.2013.6761617.

OpenCV Canny Edge Detection (2019). https://docs.opencv.org/3.4.6/da/d22/tutorial_py_canny.html (14 September 2023).

Redmon, J., Divvala, S., Girshick, R., and Farhadi, A. (2016). You only look once: unified, real-time object detection. *Proceedings of the IEEE Conference on Computer Vision and Pattern Recognition*, Las Vegas, NV, USA (27–30 June 2016), pp. 779–788. https://doi.org/10.1109/CVPR.2016.91.

Ren, S., He, K., Girshick, R., and Sun, J. (2015). Faster r-cnn: towards real-time object detection with region proposal networks. *Advances in Neural Information Processing Systems* 28: 91–99.

Restas, A. (2015). Drone applications for supporting disaster management. *World Journal of Engineering and Technology* 3: 316.

Rivera, A.J., Villalobos, A.D., Monje, J.C. et al. (2016). Post-disaster rescue facility: human detection and geolocation using aerial drones. *2016 IEEE Region 10 Conference (TENCON)*, Singapore (22–25 November 2016), pp. 384–386.

Rudol, P. and Doherty, P. (2008). Human body detection and geolocalization for UAV search and rescue missions using color and thermal imagery. *2008 IEEE Aerospace Conference*, Big Sky, MT, USA (1–8 March 2008), pp. 1–8. https://doi.org/10.1109/AERO.2008.4526559.

Schölkopf, B., Platt, J.C., Shawe-Taylor, J. et al. (2001). Estimating the support of a high-dimensional distribution. *Neural Computation* 13: 1443–1471. https://doi.org/10.1162/089976601750264965.

Sharma, J., Granmo, O.-C., Goodwin, M., and Fidje, J.T. (2017). Deep convolutional neural networks for fire detection in images. *Engineering Applications of Neural Networks: 18th International Conference, EANN 2017,* Athens, Greece (25–27 August 2017), pp. 183–193. *Proceedings*.

Simonyan, K. and Zisserman, A. (2014). Very deep convolutional networks for large-scale image recognition. *arXiv preprint arXiv:1409.1556* https://doi.org/10.48550/arXiv.1409.1556.

Szegedy, C., Liu, W., Jia, Y. et al. (2015). Going deeper with convolutions. *Proceedings of the IEEE Conference on Computer Vision and Pattern Recognition,* Boston, MA, USA (7–12 June 2015), pp. 1–9).

Szegedy, C., Ioffe, S., Vanhoucke, V., and Alemi, A. (2017). Inception-v4, inception-resnet and the impact of residual connections on learning. *Thirty-First AAAI Conference on Artificial Intelligence,* vol. 31 no. 1. https://doi.org/10.1609/aaai.v31i1.11231.

Varadarajan, J., Emonet, R., and Odobez, J.-M. (2013). A sequential topic model for mining recurrent activities from long term video logs. *International Journal of Computer Vision* 103: 100–126.

Yang, W., Gao, Y., and Cao, L. (2013). TRASMIL: a local anomaly detection framework based on trajectory segmentation and multi-instance learning. *Computer Vision and Image Understanding* 117: 1273–1286.

11

Social Media Monitoring for Disaster Management

Stelios Andreadis, Ilias Gialampoukidis, Stefanos Vrochidis, and Ioannis Kompatsiaris

Centre for Research and Technology Hellas (CERTH), Information Technologies Institute (ITI), Thessaloniki, Greece

11.1 Introduction

Natural disasters severely affect places across the world and can lead to great damage of infrastructure and loss of human lives. Since social media are widely used in our daily lives, they can be proven valuable also during a disaster crisis, where people share critical information. Thus, the beAWARE project, aiming at enhancing situational awareness with heterogeneous data, involves social media monitoring as one of the many sources of information. Social media posts from Twitter are retrieved in a real-time manner, while fake news and irrelevant content are filtered out with machine learning. Then, spatiotemporal clustering is applied to group tweets that refer to the same incident, so that an incident report can be produced for reasons of compactness. This group of tweets serves as a summary of the available Twitter posts to support first responders in decision-making. Moreover, the collected and analysed tweets are visualized in dedicated user interfaces with advanced functionalities that enable users to filter and view.

Big natural disasters occur often around the globe and their frequency is expected to increase in the following years due to climate change. In moments of such crises, communication between affected people intensifies, since they ask for help or exchange information about shelters, transportation, dangers, etc. Social media are widely used in daily life for communicating and disseminating information, thus they can become vital during a natural disaster event.

Responding to Extreme Weather Events, First Edition. Edited by Daniel Sempere-Torres, Anastasios Karakostas, Claudio Rossi, and Philippe Quevauviller.

As a result, social media data are considered important in natural disaster management and literature has proposed three main ways to use them: (i) preparing for a natural disaster, (ii) responding during the natural disaster and (iii) recovering from the natural disaster (Velev and Zlateva 2012). Four dimensions of social media data, i.e. space, time, content and network, have been identified as useful to mining valuable information that can significantly assist the improvement of situational awareness and disaster response (Wang and Ye 2018).

In beAWARE, social media is one of the many sources of information to be fused by the system for crisis management in the case of a natural disaster, mainly during the emergency response phase of the crisis. Social media monitoring includes two core components: Social media analysis (SMA), responsible to collect and further analyse tweets, and social media clustering (SMC), responsible to collect analysed tweets and group them by spatiotemporal criteria. Complementary to these components, relevant user interfaces that display and filter the social media posts have been developed.

The chapter is organized as follows. Section 11.2 presents the SMA component. First, an overview of the flow of execution is given and the integrated sub-components are presented (Section 11.2.1). Next, the section describes in detail how social media data are collected from Twitter (Section 11.2.2) and how they are analysed in order to filter out the irrelevant posts and improve the quality of incoming information (Section 11.2.3). Moreover, the structure of collected and analysed data is presented, as well as the necessary edits before storing them in a database, to support faster queries (Section 11.2.4). Section 11.3 continues with the SMC component. After an evaluation of a multitude of spatial clustering techniques (Section 11.3.1), the spatiotemporal clustering proposed in beAWARE is described (Section 11.3.2). Next, Section 11.4 demonstrates two user interfaces that have been implemented mainly for displaying the collected social media data. Finally, Section 11.5 concludes with a summary of the work done and discusses future work.

11.2 Social Media Analysis

11.2.1 Framework Overview

The SMA component aims at crowdsourcing information from connected citizens and authorities concerning natural disasters. During the project's lifetime, the platform of Twitter has been selected as a source, due to its high popularity (leading to rich information) and its available real-time streaming API. The topics of interest, derived from the project's use cases, are floods, fires and heatwave incidents in areas where English, Greek, Italian and Spanish are spoken. Tweets that concern the combinations of the above topics and languages are stored in separate collections in a database and, after they are analysed, they are ingested into the beAWARE platform, to serve as a complementary source of information that is valuable to disaster management. The complete flow of the SMA component is demonstrated in Figure 11.1.

The starting point is a collection of specified keywords (Section 11.2.2) per use case, stored in a MongoDB database named 'Feeds'. These keywords are fed to a Java HTTP client, which constantly crawls tweets from the Twitter API that contain any of these

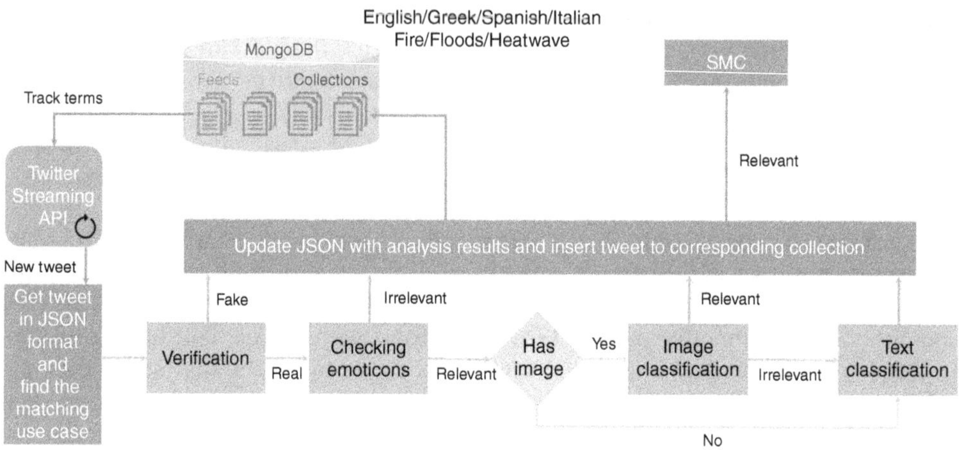

Figure 11.1 Complete flow of the SMA framework.

terms. When a new tweet is collected, it arrives in JSON format. After using its text content to detect the collection it is targeted to (e.g. a crawled tweet that includes the keyword 'flooding' will be stored in the English floods collection), the analysis phase commences, aiming to filter out the non-valuable posts.

First, a verification service (Section 11.2.3.1) tries to identify whether a tweet is real or fake, e.g. hoax news. If it is identified as fake, it is not examined further for its relevance to the topic. If it is considered as real, its content is checked for certain emoticons (Section 11.2.3.2). In the case that unrelated emoticons are found (e.g. a smiley or a heart), the analysis phase stops again. Otherwise, the relevance of the tweet to the examined use cases is estimated. If a picture accompanies the tweet, the image classification service is called to determine whether it is relevant (Section 11.2.3.4). If the result is negative or the tweet does not contain an image in the first place, then a text classification service is called to assess (Section 11.2.3.5). In whichever step the analysis phase concludes, the tweet's JSON is updated in order to include the outcomes of the analyses (Section 11.2.4) and it is stored in the respective collection of the MongoDB database. In case that the post is finally considered as relevant, it is pushed to the SMC component, where it is clustered with other posts that refer to the same incident and a Twitter report is created (more in Section 11.3.2).

11.2.2 Data Collection from Twitter

In order to gain access to Twitter's global stream of data, we have exploited the Streaming APIs,[1] a streaming client that receives tweets the moment that they are published. Compared to Twitter's REST APIs,[2] this option offers a real-time stream of tweets instead of constantly making requests and thus overriding any rate limiting, i.e. maximum number of requests. The only limitation when using the Streaming APIs is that each account is allowed to create only one standing connection.

There are various streaming endpoints that can be divided into the following categories: Public streams, User streams and Site streams. In our case, the 'POST statuses/filter' endpoint of the public streams is the most suitable, since it focuses on public data flowing through Twitter that matches one or more filter predicates. Specifically, the 'track' field can be used to define up to 400 search keywords, combined with an OR operator, so that the API will return tweets matching any of these keywords.

To consume Twitter's Streaming API, we chose to adopt the Hosebird Client (hbc),[3] an open-source and easy-to-use Java HTTP client. A required parameter is the user account's credentials, while an optional parameter is the track keywords. For each combination of pilot and language, we sustain a separate collection in a MongoDB database to store the crawled tweets. In addition, there is a 'Feeds' collection where we define a set of relevant keywords to serve as track terms during the crawling procedure.

After connection with the Streaming API is established, the client constantly receives newly created tweets in JSON format. We choose to maintain the structure provided by the API, since the JSON format fits well with a MongoDB database. Every time a new tweet is retrieved, we examine which one of the track terms exists in the tweet's text. In this way we can match the tweet to the corresponding language and pilot (e.g. 'inundación' matches to Spanish and floods) in order to insert it to the respective collection.

As mentioned above, the option to retrieve statuses that contain any keyword of a predefined list fits our goal the most, since a set of keywords is a usual practice to detect data that refer to specific use cases, and thus it was adopted in this framework. The initial set of keywords was composed of words suggested by the user group members of beAWARE, for the languages of interest within the project's use cases, namely Greek, Italian and Spanish, as well as English, for an additional control and demonstration language. However, after using the proposed keywords in practice, it was noticed that some of them were bringing a large number of irrelevant tweets, so eventually it was decided to ignore them. The complete list of keywords is shown in Table 11.1, separated by language and use-case scenario. It can be noticed that some of the listed keywords are aggregated words (e.g. allertameteo), since (i) it is a common format in Twitter due to hashtags or the character limit, and (ii) these words separately are expected to return more irrelevant than relevant tweets. In addition, the removed keywords are denoted by strikethrough.

SMA was actively running from March 2017 (M3 of beAWARE) until the conclusion of the project, i.e. December 2019 (M36). The ultimate size of the collections, i.e. the number of tweets that have been collected for each use case and language, can be found in Table 11.2.

11.2.3 Analysis of Social Media Data

11.2.3.1 Verification Targeting to solve the recent problem of hoax news circulating online and to improve the quality of incoming information into the beAWARE system, a verification methodology is the first part of the analysis phase. A mature solution has been included instead of implementing a completely new technique in the frame of beAWARE, since its development would be outside the scope of this project.

Table 11.1 Search keywords per language and pilot.

Language	Use case scenario		
	Floods	Fires	Heatwave
English	Flooding	Forest_fires	Heatwave
Greek	πλυμμήρες πλημμύρες GSCP_GR	πυρκαγιά πυρκαγια πυρκαγιες πυρκαγιές Pyrosvestiki GSCP_GR	καύσωνας Κελσίου θερμοκρασία_ρεκόρ GSCP_GR
Italian	Alluvione Alluvionevicenza Allagamento Bacchiglione Fiumepiena Allertameteo Sottopassoallagato Alluvione2017 Allertameteovicenza Esondazione Livellofiume ~~allerta~~ ~~pioggia~~ ~~maltempo~~ ~~sottacqua~~ ~~soccorso~~ ~~Vicenza~~ ~~protezionecivile~~ ~~regioneveneto~~ ~~piena~~ ~~veneto~~	Fiamme Vigilidelfuoco Piromane Pompieri ~~incendio~~ ~~forestali~~ ~~fumo~~ ~~allertaincendio~~	Ondatedicalore Allertacaldo Emergenzacaldo Altetemperature Troppocaldo ~~protezionecivile hipertermia~~
Spanish	Fuertesprecipitaciones Gotafría Inundación Inundaciones Desbordamiento Riada Riadas Lluviastorrenciales Tormentas Caudaldesbordado ~~corte de carretera~~	Incendio Llamasdefuego Bomberos Focodeincendio ~~fuego~~ ~~humo~~ ~~quema/quemas~~ ~~conato~~ ~~quemada/quemado~~ ~~extinción~~ ~~inhalación de humo~~ ~~intoxicación de humo~~ ~~evacuación/confinamiento~~	Oladecalor Altastemperaturas Nochetropical Golpedecalor Sequía

Table 11.2 Number of collected tweets per use case and language.

Language	Use-case scenario		
	Floods	Fires	Heatwaves
English	6,251,140	81,777	1,429,743
Greek	41,102	241,235	75,226
Italian	89,589	534,659	1,483
Spanish	2,785,541	11,267,727	876,289

The solution deploys a robust approach for detecting in real time whether a social media post in Twitter is fake or real by using a variety of content-based and contextual features. The categories and subcategories of the features are:

1. tweet-based:
   a. text-based (e.g. how many characters are uppercase)
   b. language-specific (e.g. slangs)
   c. twitter-specific (e.g. is retweeted)
   d. link-based (e.g. existence of an external URL)
2. user-based:
   a. user-specific (e.g. how many followers)
   b. link-based (e.g. existence of an external URL in a profile description)

Leveraging part of its own predictions for retraining, the approach follows a semi-supervised learning paradigm, to adapt the model to unseen content, and experiments on a public annotated corpus of multimedia tweets have demonstrated its effectiveness (Boididou et al. 2017, 2018).

11.2.3.2 Checking Emoticons The second analysis step is checking inside the tweet's text for emoticons (a pictorial representation of a facial expression using characters) and emojis (actual picture instead of typographic) that are not expected to be present in posts during a crisis event. The main idea is that important tweets that forewarn or describe the current situation of a natural disaster are not supposed to contain smileys, winks or hearts.

Based on the official list of emoticons in Wikipedia,[4] certain categories, which are popular in social media and frequently used, were selected (Table 11.3) and in the case that a tweet includes any of these, it is labelled as irrelevant. Please notice that negative emoticons, such as sad faces or expressions of horror and shock, are excluded from this check, since it is consider normal for them to be used during a crisis.

11.2.3.3 Relevance Classification Classification is the problem of identifying to which of a set of categories (i.e. classes) a new instance belongs to, based on a training set of data containing instances whose class membership is known. Most algorithms describe an individual instance whose category is to be predicted using a feature vector that is comprised of measurable properties (i.e. features) of the instance. In case the

Table 11.3 List of emoticons and emojis that are considered irrelevant.

Category	Emoticons	Emojis
Smiley, happy face	:-) :) :-] :] :-3 :3 :-> :> 8-) 8) :-} :} :o) :c) :^) =] =)	🙂😊😀😁
Laughing, big grin, laugh with glasses, wide-eyed surprise	:-D :D 8-D 8D x-D xD X-D XD =D =3 B^D	😃😄😆😍
Tears of happiness	:'-) :')	😂
Kiss	:-* :* :×	😘😗😚😙😍
Wink, smirk	;-) ;) *-) *) ;-] ;] ;^) :- ;D	😉😜😙
Tongue sticking out, cheeky/playful, blowing a raspberry	:-P :P X-P XP x-p xp :-p :p :-Þ :Þ :-þ :þ :-b :b d: =p >:P	😛😝😜
Heart	<3	♥

instance is an image, the feature values might correspond to the pixels of an image; if the instance is a piece of text, the feature values might be occurrence frequencies of different words.

In beAWARE, the relevance of collected social media posts is estimated using both textual and visual (when the tweet includes an image) information, aiming to filter out irrelevant social media posts. In the sequel, both frameworks for visual classification and textual classification are presented, together with the best results of their evaluation.

11.2.3.4 Image Classification Image classification involves the use of visual concept detection algorithms based on low-level features and classifiers for deciding whether an image shows evidence of the beAWARE use cases, namely flood, fire and heatwave.

In the employed framework, we trained a 22-layer GoogLeNet network (Szegedy et al. 2015) on 5055 ImageNet concepts (Pittaras et al. 2017), which are a subset of the 12,988 ImageNet concepts. The subset of the 5055 concepts was produced by considering the tree structure of the ImageNet and the following assumptions:

1. Concepts that were very similar were merged. For example, all different dog breeds (e.g. 'Shih-Tzu', 'Pekinese', 'Maltese dog') were removed and only the concept 'dog' was kept. The same philosophy was followed for other animals and plants as well.
2. Concepts that correspond to scientific terms were removed; for example, biological terms such as 'eukaryote', 'prokaryote', 'sporozoite' etc.
3. Concepts with very few positive images were removed.

The network was applied on the TRECVID SIN 2013 development dataset and we used as a feature (i.e. a global image representation) the output of the last pooling layer with dimension 1024.

Next, we used a dataset annotated in the frame of beAWARE for training and validating an SVM classifier per concept (i.e. flood, heatwave, fire). The end users of the project made a significant effort annotating manually thousands of tweets with the

Table 11.4 Number of tweets annotated within beAWARE.

Language	Use case scenario		
	Floods	Fires	Heatwaves
English	1053 560 T 493 F	—	—
Greek	—	3871 1973 T 1898 F	7169 2509 T 4660 F
Italian	37,889 6584 T 31,305 F	1904 635 T 1269 F	760 259 T 501 F
Spanish	1075 966 T 109 F	4073 1357 T 2716 F	—

Table 11.5 Results of the evaluation of the visual classifier.

Descriptor	SVM parameters		Precision	Recall	Accuracy	F-score
	t	g				
Dcnn-based	1	0.5	0.7139	0.2738	0.5461	0.3957

developed annotation tool (Section 11.4.1) and the outcome can be seen in Table 11.4. The SVM classifiers were tuned by setting different t and g values in order to achieve maximum performance. t parameter in SVM classifier defines the kernel type, while g stands for the gamma in the kernel function.

The evaluation experiments ran on a subset of the annotated beAWARE dataset and in particular on 1594 Italian tweets about floods that contained an image. Best results were achieved with the combination of DCNN-based features and the SVM classifier with $t = 1$ and $g = 0.5$ and the evaluation metrics can be found in Table 11.5.

It is evident that the recall achieved is rather low, which also leads to a low F-score. This could be explained by the fact that the annotators marked the tweets as relevant or irrelevant based on both their provided texts and images, thus it is quite possible that while a tweet actually refers to a flooding/fire/heatwave incident, its attached image does not depict something related to these cases, and vice versa. The low performance is expected to be compensated by the text classification.

11.2.3.5 Text Classification Social media text classification involves the use of text classifiers that consider textual features for deciding whether a tweet shows evidence of flood, fire or heatwave. In the employed framework, we follow the steps below:

1. We pre-process the collected tweet messages by removing punctuation, all non-characters and stop words from the collected text and by performing word stemming.

2. As far as text representation is concerned, we select Term Frequency (TF), Term Frequency – Inverse Document Frequency (TFIDF), or word2vec. Various experiments were realized for different feature length and n-gram values for the first two representation methods, and different corpus and vector dimensions for the third method.
3. We serve each text feature vector as input to a classifier (i.e. SVM, Naïve Bayes or Random Forests), which is tuned in order to achieve maximum performance.

The comparison experiments ran again on a subset of the annotated beAWARE dataset of Table 11.4; this time 10,063 Italian tweets about flood (duplicates were removed). The best runs from all combinations of text representation and classifiers can be found in Table 11.6, along with the specific parameters. We can deduce that the best performing method is the TFIDF method with Random Forest (F-score = 0.657). On the other hand, it is quite noticeable how low the performance of word2vec is. A possible explanation for this lies to the quality of the corpus used for the word2vec representation, or the size was not enough to produce a good vector space.

For further details on the visual and textual classification and the experiments, the reader is referred to our respective works (Avgerinakis et al. 2017; Moumtzidou et al. 2018).

11.2.4 Data Representation

As mentioned before, the tweets that are received from the Twitter Streaming API are encoded in JSON format, which is based on key-value pairs, with named attributes and associated values. We prefer to keep this provided structure while storing the tweets in our database, because JSON is indicated for MongoDB installations.

Plenty of information is offered by Twitter for a single post,[5] but there are some fundamental fields that play a significant role in the social media analysis procedure. To begin with, *id_str* serves as a unique String identifier that can be used in different stages of the beAWARE framework in order to refer to a certain tweet. *text* is the main content of the Twitter post and it is used in text classification (Section 11.2.3) to estimate the tweet's relevance, while *entities.media.media_url* (periods should be interpreted as 'has subfield') refers to the link of an attached image and it is used in image classification (Section 11.2.3), again for relevance estimation. *created_at* is the date when the post was published and *geo.coordinates* (if available) indicates the location where the tweet originated. Sometimes a tweet might exceed the maximum character limit and then a field named *extended_tweet* is included. In that case, the subfields *extended_tweet.full_text* and *extended_tweet.entities.media.media_url* are used in text and image classification, respectively.

Apart from the aforementioned fundamental fields, there are also some fields that are not included originally in the JSON format of a consumed tweet but are later added by our framework. In particular, the field *verification,* with subfields the Boolean *predicted* and the Double *confidence,* contains the outcome of the verification technique (Section 11.2.3) and the Boolean field *emoticon_relevancy*, the outcome of checking emoticons (Section 11.2.3). Furthermore, the Boolean field *estimated_relevance* refers

Table 11.6 Best parameters from TF, TFIDF, word2vec text classification methods.

Method	Text input	Text representation parameters	Classifier	Classifier parameters	Precision	Recall	F-score
TF	Text with stop words removed + stemming applied	n-gram = 1 min_ df = 0.001 features length = 1313	Random forest	Features num for best split = auto number of trees = 1000	0.81081	0.54364	0.65087
TFIDF	Text with stop words removed	n-gram = 2 min_ df = 0.001 features length = 2762	Random forest	Features num for best split = auto number of trees = 200	0.79968	0.55806	0.65737
Word2vec	Text with stop words removed	Vector dimension = 100 words window = 3 training algorithm = 1	SVM	Penalty parameter = 5.0 Kernel type = rbf	0.95855	0.06842	0.12772

to the result of the multimodal classification, while the Boolean field *relevant* to the human annotation (Section 11.2.3). The field *concepts* includes the extracted concepts during text classification and *entities.media.dcnn_feature* is the image feature vector during image classification; these fields assist the analysis by making the information flow more efficient. In order to decrease the response time of common queries to the database, *is_retweeted_status* defines whether the Object field *retweeted_status* exists or not. Finally, aiming to support geospatial queries on the MongoDB, the '2dsphere' index had to be created. For this, all geolocations needed to comply with the GeoJSON format,[6] which is not the case for locations of tweets that are provided in form of a bounding box by the Twitter API. These locations are provided as an array of four coordinates, which are the four corners of the box. However, in order to comply with the GeoJSON format, they need to be arrays of five points, where the first and the last points are the same (to close the box). As a result, the first point of each bounding box is copied and pasted as a fifth.

11.3 Social Media Clustering

The constant crawling of social media posts in a real-time manner results to a very large amount of individual tweets that can be confusing to the end user. The need to receive information that is more compact can be satisfied by grouping the various posts according to their location, expecting that tweets sent from a specific region refer to the same incident. To this end, various spatial clustering methodologies have been compared to select the most suitable in the context of beAWARE and, based on the results, a straightforward spatiotemporal technique was proposed and deployed in the beAWARE system.

11.3.1 Evaluation of Spatial Clustering Techniques

The target of the following experiments was to determine which clustering technique is the most suitable for the spatial grouping of tweets in the frame of disaster incidents. In this quantitative evaluation, the dataset consisted of 88 synthetic Spanish tweets about fires, which were associated with locations in the regions of El Saler and the natural park of Albufera in Valencia. The tweets have been assigned to clusters with human annotation (Figure 11.2), while four of them were considered as outliers and thus were not matched with any cluster (labelled with the white cross in the figure).

Having the dataset and the ground truth, a comparison was realized between 16 clustering indices: CH (Calinski and Harabasz 1974), Duda (Duda et al. 1973), Pseudo t2 (Duda et al. 1973), C-index (Hubert and Levin 1976), Ptbiserial (Davies and Bouldin 1979), DB (Davies and Bouldin 1979), Hartigan (Kaufman and Rousseeuw 1990), Ratkowsky (Ratkowsky and Lance 1978), Ball (Ball and Hall 1965), McClain (McClain and Rao 1975), KL (Krzanowski and Lai 1988), Silhouette (Kaufman and Rousseeuw 1990), Dunn (1974), SDindex (Halkidi et al. 2000), SDbw (Halkidi et al. 2000) and DBSCAN (Ester et al. 1996). The evaluation metric was the Normalized Mutual Information (NMI) score[7] and the results can be seen in Table 11.7. The first column is the clustering index evaluated, the second is how this index is called

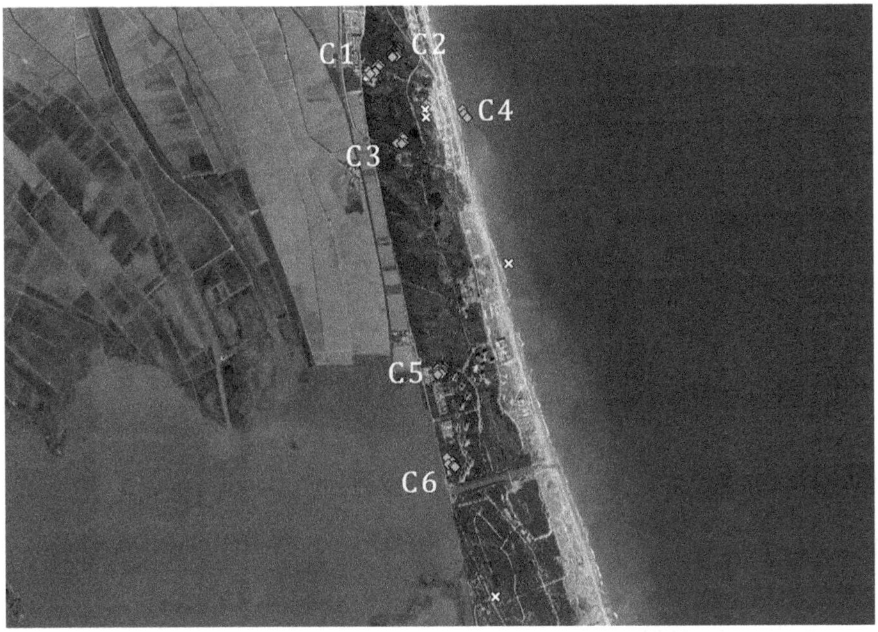

Figure 11.2 Clustered tweets and outliers in Valencia.

Table 11.7 Estimated number of clusters and NMI score of each clustering index.

Index	Index in R	Number of clusters	NMI score
CH	ch	30	0.6647121
Duda	duda	2	0.5180138
Pseudo t2	pseudot2	2	0.5180138
C-index	cindex	3e+01	0.6647121
Ptbiserial	ptbiserial	3	0.5355914
DB	db	8	0.8860461
Hartigan	hartigan	9	0.9853166
Ratkowsky	ratkowsky	2	0.5180138
Ball	ball	3e+00	0.5355914
McClain	mcclain	9	0.9853166
KL	kl	9	0.9853166
Silhouette	silhouette	8	0.8860461
Dunn	dunn	9	0.9853166
Sdindex	sdindex	7	0.8650745
SDbw	sdbw	30	0.6647121
DBSCAN minPts = 2 eps = 0.001	dbscan	7	0.9901629
DBSCAN minPts = 3 eps = 0.001	dbscan	6	1.0
DBSCAN minPts = 4 eps = 0.001	dbscan	5	0.9825724

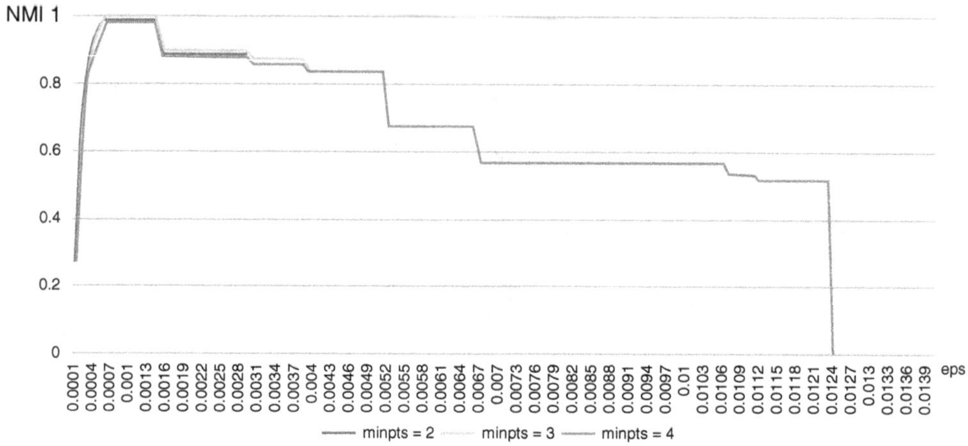

Figure 11.3 NMI score for different values of eps, for minPts 2, 3 and 4.

in R language (in which the experiment ran), the third is the estimated number of clusters and the last column is the NMI score.

The Hartigan, McClain, KL and Dunn indexes achieved a very high score of 0.98 and above, but only a version of DBSCAN managed to predict the correct number of clusters and achieve an NMI score of 1.0. This is due to its ability to detect outliers and avoid assigning them to clusters.

Especially for the case of DBSCAN, several variations were investigated by tuning the *eps* (the local radius for expanding clusters) and the *minPts* (the minimum number of points required to form a dense region) parameters. Eps ranged between 0.0001 and 0.0140 in a step of 0.0001, while the integer values of minPts were 2, 3 and 4. The fluctuation of the NMI score is displayed in the line chart of Figure 11.3, where the best scores occurred for eps between 0.0006 and 0.0014, for all three values of minPts. However, scores for minPts = 3 were slightly better than the others.

11.3.2 The Proposed Spatiotemporal Clustering

For the implementation of clustering tweets in the beAWARE system, the DBSCAN algorithm was selected, because it achieved the best performance in the evaluation described above. The tweets are crawled in a real-time manner, so time is another important factor that needs to be considered. The final deployment is a spatiotemporal clustering and the respective workflow inside the beAWARE system is shown in Figure 11.4.

The SMC component consumes the messages that the SMA component is producing. When x tweets are collected (named bag of tweets) or a timer counts more than y seconds since the last received tweet, SMC performs spatial clustering, empties the bag and resets the timer. The idea behind the first condition is to have a sufficient number of tweets to group and the idea behind the second condition is to

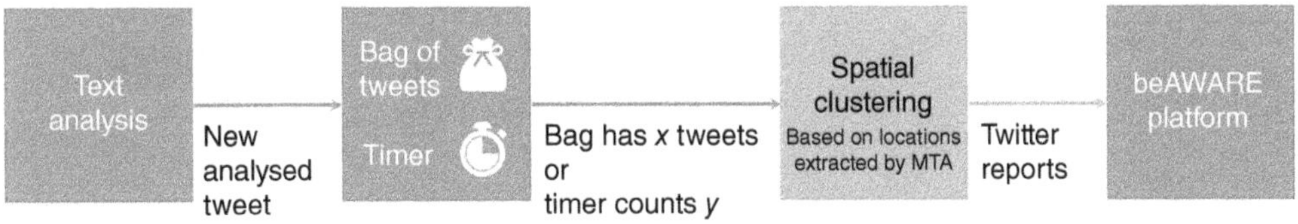

Figure 11.4 The flow of SMC in the beAWARE system.

be responsive, in case that it takes a long time between incoming tweets. The variables x and y can change; by default clustering is initiated when 10 tweets are collected or 60 seconds (1 minute) passes. When clustering is completed, the clusters are considered as Twitter Reports, where each Twitter report contains the list of tweets (i.e. their text, image and date) it comprises. Finally, these reports are sent to the main beAWARE platform, which is used by the first responders, to assist them in the decision-making process.

11.4 Visualizations

During beAWARE's lifetime, different needs of the project have prompted the implementation of different user interfaces, namely the *annotation tool* and the *demonstration tool*, which are thoroughly described in the next subsections.

11.4.1 Annotation Tool

The beAWARE annotation tool serves a twofold purpose: (i) it can be used to view and search the complete collection of tweets that are crawled by SMA, and (ii) it can be used to label tweets as relevant or irrelevant, thus creating an annotated dataset for classification training (Section 11.2.3).

The homepage of the web tool (Figure 11.5) offers the end users the ability to select the type of Twitter posts that they would like to be displayed, based on two criteria: language, i.e. English, Greek, Italian and Spanish; and pilot case, i.e. fire, flood and heatwave. Each combination of language and pilot case defines a different collection of crawled tweets. In a minimal, but efficient way the users are able to set their preferences; first, by a drop-down list on the upper left corner of the website that provides all the available languages and then by clicking one of the three pilot buttons in the middle of the page. This will navigate to the presentation of the respective tweets.

Figure 11.6 depicts a set of tweets that are relative to floods and are written in Italian. As it can be easily seen, the page consists of two main components: a pagination header that also includes other useful utilities and a panel of textboxes in vertical direction where each box displays a Twitter post.

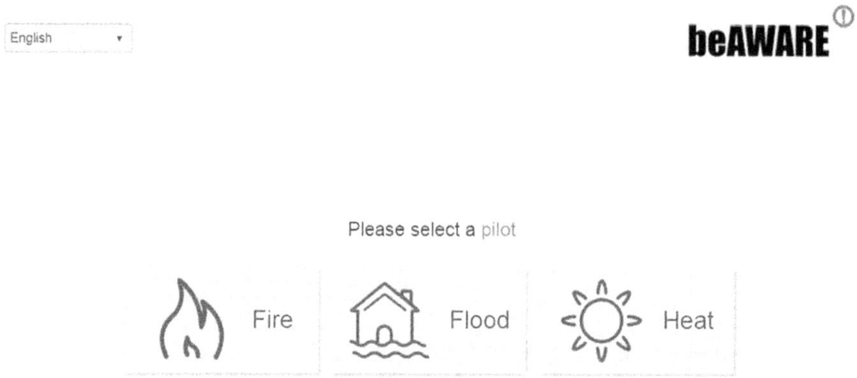

Figure 11.5 beAWARE annotation tool – homepage.

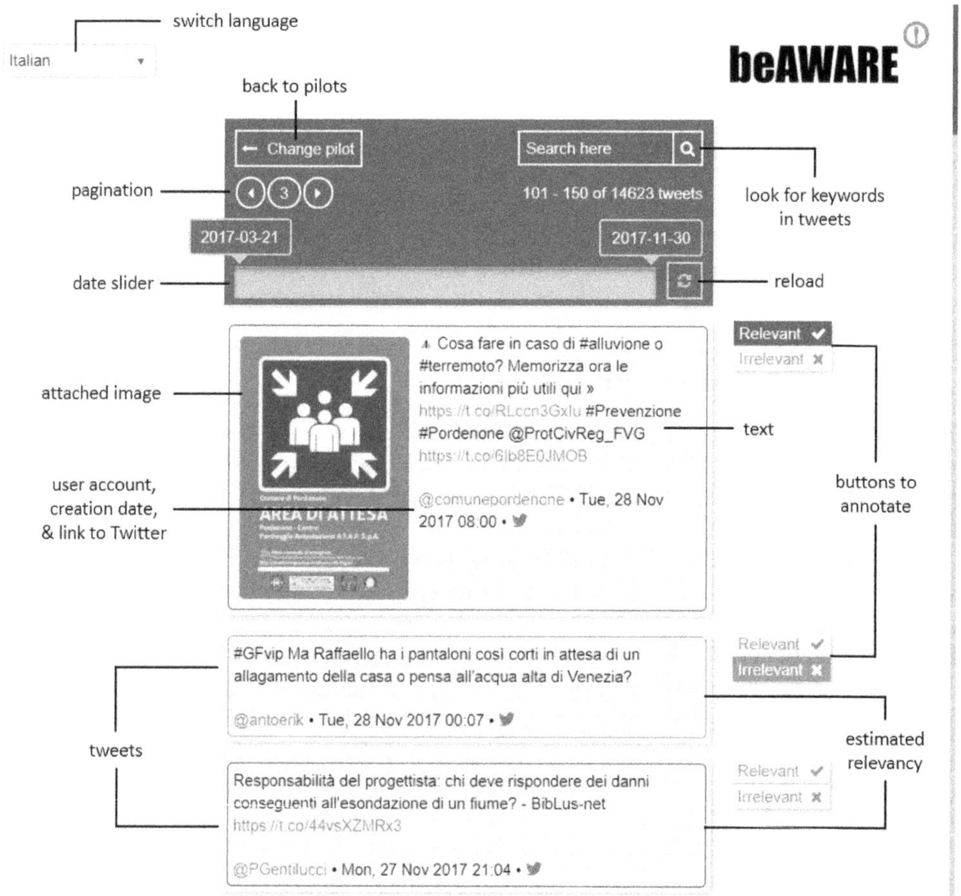

Figure 11.6 beAWARE annotation tool – tweets presentation and buttons to annotate.

In order to avoid presenting thousands of tweets in a single page, the user interface gives a pagination option that divides the posts into pages of fifty. Using the left and right arrows inside the header, end users are able to navigate back and forth through pages, while the page numbering and the total number of tweets are always shown. In addition, there are two filtering options: posts can be filtered either by their date of creation or by the existence of certain keywords inside their text. For the former, a date slider serves to define the time period during which posts were created and then pressing the reload button, which lies next to the slider, is necessary to return the new results. For the latter, users can type an unlimited number of words into a text input field on the upper right side of the header and by clicking on the magnifier icon or pressing the Enter key, results will be updated with tweets that contain at least one of the given words in their text. In order to switch use cases, the 'Change pilot' button navigates back to the homepage, while switching languages is possible through the drop-down list that was previously described and is always visible in the annotation tool. In addition to this header, a more compact version of it, which includes only the pagination functionality, was added under the panel of tweets, according to feedback from real end users.

Regarding the crawled Twitter posts, they are represented by a list of boxes and they are sorted from most recent to oldest. For each tweet, a variety of information is available. In detail, the main text of the post, along with images or active links if existing, the username of the author, which links to the user account's Twitter page, the time and date when the tweet was published, and a link to the original post in Twitter. Moreover, the coloured border around each tweet indicates its relevance to the pilot case as estimated by the multimodal classification framework (Section 11.2.3) during the crawling phase. Green border refers to relevant texts and red violet border to irrelevant. Finally, next to every tweet box, there are two buttons; by clicking the 'Relevant' or the 'Irrelevant' button, users are able to annotate all posts and the annotation is instantly saved in the database where the crawled tweets are stored.

In summary, the online beAWARE annotation tool provides two basic functionalities: (i) a presentation of all crawled Twitter posts per language and pilot, together with filter options (e.g. getting tweets that were published in a specific date or that contain certain words), and (ii) a simple way to mark if a tweet is relevant or not, creating in this way a training set. In the frame of the project, the tool was used by Spanish, Greek, and Italian users, and more than 5000, 11,000 and 40,000 tweets, respectively, have been successfully annotated.

11.4.2 Demonstration Tool

The second interface is the beAWARE demonstration tool, which was used during the project's pilots and shows the tweets at the same instant they are collected by SMA. As it can be seen in Figure 11.7, the left side contains a set of simulated tweets that have been proposed by beAWARE user partners. These tweets are actually posted in Twitter and successfully crawled. To initiate the process of the social media component, the 'Insert to DB' button has to be clicked. After the workflow (described in Section 11.2.1) is completed, the tweets that are estimated as relevant and are sent to SMC will be displayed on the right side. Clicking the 'Empty the DB' button will clear the consumed tweets, making it possible to repeat the demonstration.

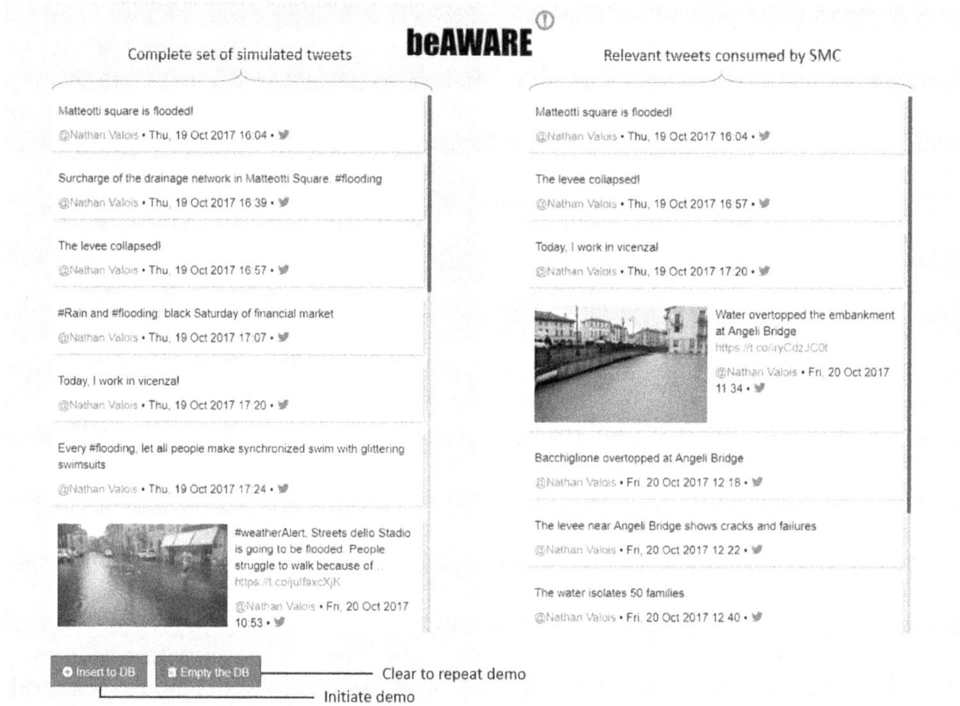

Figure 11.7 beAWARE demonstration tool.

11.5 Conclusion

This chapter presented the work that has been done in the frame of the beAWARE project with regards to social media monitoring, which aims at providing crowdsourced content as an alternative source of information for raising awareness during a natural disaster crisis.

The main two components of this work are SMA and SMC. The first concerns the means to retrieve data from social media in real time, filter out fake or irrelevant posts by using machine learning and store the analysed information in a proper and reusable format. The latter involves the spatiotemporal clustering of tweets that are close in time and space and are expected to refer to the same incident. After a comparison between different clustering techniques, DBSCAN has been selected as the most appropriate. Finally, the chapter continued with the presentation of the annotation tool and the demonstration tool, two user interfaces that covered the needs of the project in terms of visualization.

Future work will consider an event detection methodology that could trigger the SMC component, only when there is an incident detected. In addition, an issue that has been noticed is that the majority of tweets do not include geographical information. Thus, a Named Entity Recognition technique needs to be applied in order to identify words that refer to locations and associate them with coordinates (e.g. through

OpenStreetMap). Lastly, the developed user interfaces will be extended to show the tweets also on a map as pins, offering in this way a better visualization of the areas affected by a natural disaster.

Notes

1. https://dev.twitter.com/streaming/overview.
2. https://dev.twitter.com/rest/public.
3. https://github.com/twitter/hbc.
4. https://en.wikipedia.org/wiki/List_of_emoticons.
5. https://developer.twitter.com/en/docs/twitter-api/v1/data-dictionary/overview.
6. https://geojson.org.
7. https://en.wikipedia.org/wiki/Mutual_information.

References

Avgerinakis, K., Moumtzidou, A., Andreadis, S. et al. (2017). Visual and textual analysis of social media and satellite images for flood detection@ multimedia satellite task mediaeval 2017. *MediaEval,* Dublin, Ireland (13–15 September 2017). http://slim-sig.irisa.fr/me17/Mediaeval_2017_paper_31.pdf.

Ball, G.H. and Hall, D.J. (1965). *ISODATA, a Novel Method of Data Analysis and Pattern Classification.* Stanford research inst Menlo Park CA. https://apps.dtic.mil/sti/pdfs/AD0699616.pdf.

Boididou, C., Papadopoulos, S., Apostolidis, L., and Kompatsiaris, Y. (2017). Learning to detect misleading content on Twitter. *Proceedings of the 2017 ACM on International Conference on Multimedia Retrieval*, Bucharest, Romania (6–9 June 2017), pp. 278–286. https://doi.org/10.1145/3078971.3078979.

Boididou, C., Papadopoulos, S., Zampoglou, M. et al. (2018). Detection and visualization of misleading content on Twitter. *International Journal of Multimedia Information Retrieval 7* (1): 71–86. https://doi.org/10.1007/s13735-017-0143-x.

Caliński, T. and Harabasz, J. (1974). A dendrite method for cluster analysis. *Communications in Statistics-theory and Methods 3* (1): 1–27. https://doi.org/10.1080/03610917408548446.

Davies, D.L. and Bouldin, D.W. (1979). A cluster separation measure. *IEEE Transactions on Pattern Analysis and Machine Intelligence 2*: 224–227. https://doi.org/10.1109/tpami.1979.4766909.

Duda, R.O., Hart, P.E., and Stork, D.G. (1973). *Pattern Classification and Scene Analysis*, vol. *3*, 731–739. Wiley. http://www.svms.org/classification/DuHS95.pdf.

Dunn, J.C. (1974). Well-separated clusters and optimal fuzzy partitions. *Journal of Cybernetics 4* (1): 95–104. https://doi.org/10.1080/01969727408546059.

Ester, M., Kriegel, H.P., Sander, J., and Xu, X. (1996). A density-based algorithm for discovering clusters in large spatial databases with noise. *kdd 96* (34): 226–231. https://www.aaai.org/Papers/KDD/1996/KDD96-037.pdf.

Halkidi, M., Vazirgiannis, M., and Batistakis, Y. (2000). Quality scheme assessment in the clustering process. *European conference on principles of data mining and knowledge discovery*, Lyon, France (13–16 September 2000), pp. 265–276. https://doi.org/10.1007/3-540-45372-5_26.

Hubert, L.J. and Levin, J.R. (1976). A general statistical framework for assessing categorical clustering in free recall. *Psychological Bulletin 83* (6): 1072. https://doi.org/10.1037/0033-2909.83.6.1072.

Kaufman, L. and Rousseeuw, P.J. (1990). *Finding Groups in Data. An Introduction to Cluster Analysis*. Wiley. *Wiley Series in Probability and Mathematical Statistics. Applied Probability and Statistics*. https://doi.org/10.5860/choice.28-4558.

Krzanowski, W.J. and Lai, Y.T. (1988). A criterion for determining the number of groups in a data set using sum-of-squares clustering. *Biometrics* 23–34. https://doi.org/10.2307/2531893.

McClain, J.O. and Rao, V.R. (1975). Clustisz: a program to test for the quality of clustering of a set of objects. *Journal of Marketing Research* 12: 456–460. https://www.jstor.org/stable/3151097.

Moumtzidou, A., Andreadis, S., Gialampoukidis, I. et al. (2018). Flood relevance estimation from visual and textual content in social media streams. *Companion Proceedings of the The Web Conference 2018*, Lyon, France (23–27 April 2018), pp. 1621–1627. https://doi.org/10.1145/3184558.3191620

Pittaras, N., Markatopoulou, F., Mezaris, V., and Patras, I. (2017). Comparison of fine-tuning and extension strategies for deep convolutional neural networks. *International conference on multimedia modeling*, Reykjavik, Iceland (4–6 January 2017), pp. 102–114. Springer International Publishing. https://doi.org/10.1007/978-3-319-51811-4_9.

Ratkowsky, D.A. and Lance, G.N. (1978). Criterion for determining the number of groups in a classification. https://publications.csiro.au/rpr/pub?list=BRO&pid=procite:f205b61b-c1c6-44c1-a2ee-99a4cf6883d3 (accessed 24 June 2023).

Szegedy, C., Liu, W., Jia, Y. et al. (2015). Going deeper with convolutions. *Proceedings of the IEEE conference on computer vision and pattern recognition*, Boston, MA (7–12 June 2015), pp. 1–9. IEEE. https://doi.org/10.1109/cvpr.2015.7298594.

Velev, D. and Zlateva, P. (2012). Use of social media in natural disaster management. *International Proceedings of Economic Development and Research 39*: 41–45. https://doi.org/10.4018/978-1-5225-6195-8.ch011.

Wang, Z. and Ye, X. (2018). Social media analytics for natural disaster management. *International Journal of Geographical Information Science 32* (1): 49–72. https://doi.org/10.1080/13658816.2017.1367003.

12
Human-Centred Public Warnings

Claudio Rossi and Antonella Frisiello
LINKS Foundation, Turin, Italy

12.1 Introduction

The Public Warning is an institutional duty, as it must come from official sources and provide authoritative, accurate and timely information to the population potentially affected by risks and hazards at any level. As a core component of the disaster risk management (DRM) cycle, the public warning system (PWS) includes and implements the sets of practices, tools and procedures that enable experts and authorities to transfer and share information on hazards and safe behaviours with the intended audience. An effective PWS must guarantee to reach the interested population: individuals, communities, productive activities, public services and all the subjects potentially affected by an adverse event. At the same time, the PWS enables all public authorities involved in the emergency management cycle to cooperate successfully in the field. The alert's core content is timely and clear information about the incoming hazards and the countermeasures to adopt. Besides multiple communication capabilities required to spread the alerts widely, the PWS shall interface with hazard monitoring, forecasting and risk assessment systems, which can effectively work if both technical and organizational aspects are harmonized.

The current PWS, even though they might profit from the most advanced technologies, still need to be expanded by traditional unidirectional approaches, which do not

Responding to Extreme Weather Events, First Edition. Edited by Daniel Sempere-Torres, Anastasios Karakostas, Claudio Rossi, and Philippe Quevauviller.

foresee a feedback loop with recipients. For instance, data about alerts' correct reception or application of the provided instructions are still not supported by current PWS or related processes. For this purpose, the UNISDR promotes the implementation of a **people-centred PWS.** According to this perspective, the public is not an undefined group of people that passively receive information, warnings and instructions. Still, it represents a heterogeneous community with various characteristics and vulnerabilities that must be addressed to guarantee that the alert contents will be received, understood and effectively followed. People centrality is also at the core of a slightly different approach, the **Community-Based DRM**, which widens the perspective on the whole process by promoting the involvement of potentially affected communities in DRM at the local level (UNDRR 2021). This approach aims to drive organizations to improve local disaster preparedness and risk reduction strategy. In this view, an articulated knowledge of citizenship is the base for sustaining and fostering community engagement. At the same time, local authorities role is critical to successfully implement the Community-Based DRM approach.

The progressive relevance of the citizen role in DRM processes is also envisioned by the **Situational Theory of Public** (Grunig 2005). In particular, the theory suggests different engagement and communication strategies based on the diverse population segments. According to this view, people's behaviour is context-dependent, and three main variables shape it: (i) problem recognition, (ii) constraint recognition and (iii) the level of involvement, which represents how much people are connected with the situation (e.g. hazard or crisis). According to this theory, a high-risk perception combined with low constraint recognition results in complex behavioural patterns, such as more active information seeking and less effective information processing. The level of involvement in the crisis pushes people to search for information continuously. In such situations, people are more willing to communicate and cooperate with response organizations.

Crisis and disasters reveal and make evident the need to review the traditional role assigned to citizens, especially given the undergoing social changes. In such an evolving scenario, technology can play an essential role because the public is no longer a passive recipient of messages generated by authorities or domain experts. The people actively enter the information flow, producing data, developing content and distributing it. Citizens are active individuals who **directly and indirectly interact** with information, playing a (un/conscious) role in risk prevention and emergency response. Acknowledging the public as a **responding community** is a step forward that drives the DRM to implement more effective strategies to manage increasingly complex hazard scenarios. Inclusive and participatory approaches can also reframe the traditional PWS linear 'warning chain', where the population is the end point, opening the opportunity to re-design and integrate innovative processes fitting the broad DRM ecosystem (Figure 12.1).

An **integrated DRM** approach can tackle this complexity, implementing a PWS that includes people in the process, even though they are not part of any formal organizations, who can benefit from the voluntary contributions from active citizens, the scientific community and organizations of digital volunteers. New socio-technical systems can facilitate the paradigm shift toward distributed and people-centred processes improving the four pillars of the DRM (Figure 12.2).

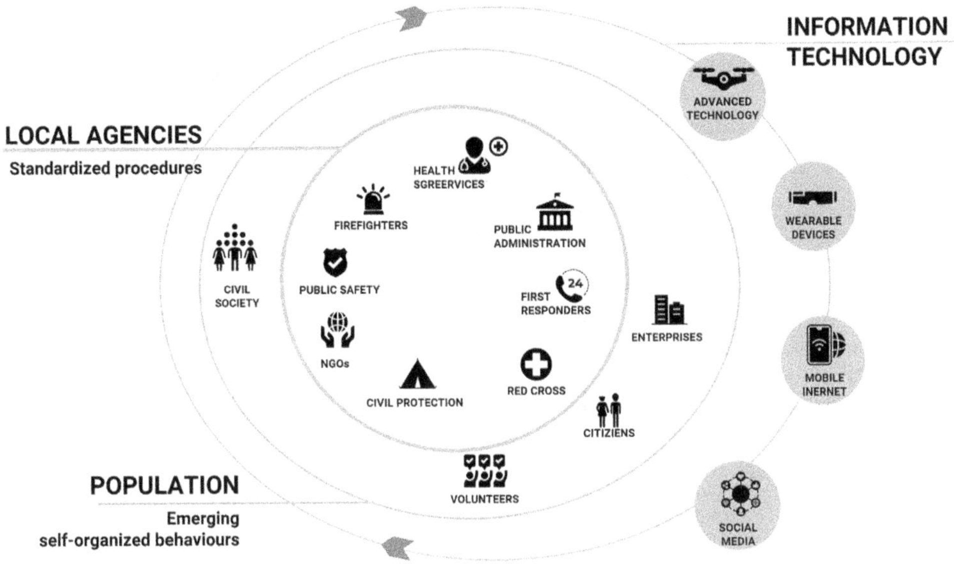

Figure 12.1 DRM ecosystem. *Source*: Nguyen et al. (2019).

RISK KNOWLEDGE knowledge of the relevant hazards, and of the vulnerabilities of people and society to these hazards	MONITORING AND PUBLIC WARNING technical capacity to monitor hazard precursors, to forecast the hazard evolution and to issue understandable alerts for heterogeneous targets
DISSEMINATION AND COMMUNICATION early and continuous learning on specific risks preparedness and channels to listen and replies with citizens	RESPONSE CAPABILITY knowledge, plans and capacities for timely and appropriate action by authorities and those at risk

Figure 12.2 The four elements of systematic people-centred DRM. *Source*: Adapted from Basher (2006).

12.2 Risk Communication

The integrated DRM stresses the relevance of improving information exchange and dialogue with citizens. This function pertains to **Risk Communication** processes, which aim to inform the population about an ongoing hazard or an incoming adverse event, informing on possible consequences (impacts) of hazards (CDC – US Centers for Disease Control and Prevention 2012). The PWS specifically entails Risk Communication, providing emergency alerts that include hazard-related information, instructions and behavioural guidance. It is relevant to highlight that Risk

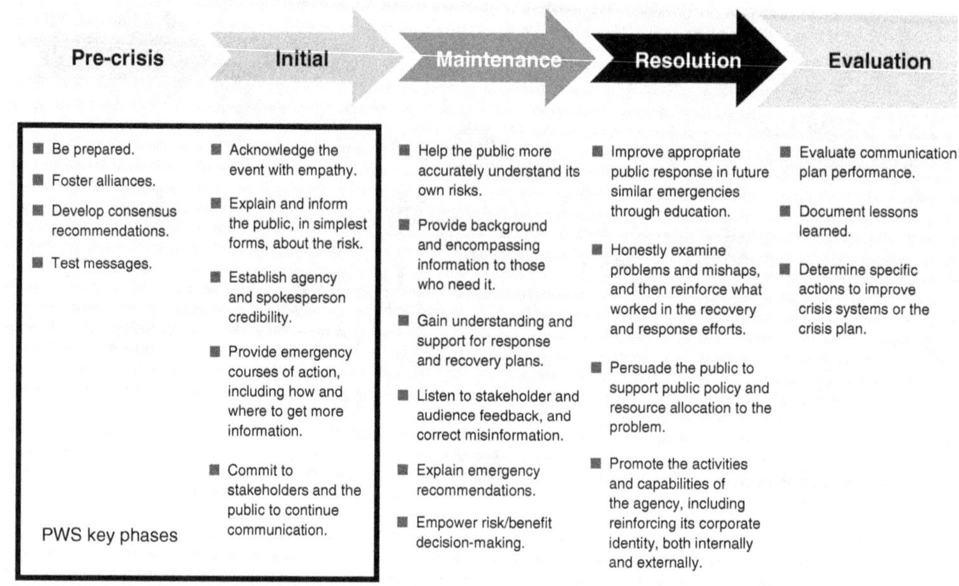

Figure 12.3 Crisis and emergency risk communication (CERC) cycle. *Source*: Adapted from US CDC (2012).

Communication works on a broad timespan and is a dynamic process that follows the emergency management life cycle (Figure 12.3). As the scenario evolves, the communication updates and adapts to it, trying to anticipate as much as possible the required information.

12.2.1 Risk Communication Key Aspects

Information sharing and training on hazards, prevention measures and safe behaviours are essential in the risk communication life cycle. A solid and updated knowledge of risks and prevention actions increases the effectiveness of public alerts and people's safety during crises. In addition, an articulated risk communication strategy includes initiatives and resources to educate people and keep them informed about risks, safe behaviours and preventive measures for disaster risk reduction. These actions can help to keep a fruitful dialogue between the population and the local authorities. They are also engagement opportunities for diverse population segments (young, workers, elderly, companies), and they can benefit from exploiting multiple offline and digital communication channels.

Literature investigates some gaps that limit the effectiveness of PWSs. Among them, the capability to **adequately reach the target people, with particular attention to frail segments of the population, the accessibility of the provided information,** and **the coherence of the messages conveyed by different organizations** represent areas of improvement (Kasperson 2014; Sufri et al. 2020). To achieve effective communication, it is critical to consider all stakeholders (e.g. official institutions, responders, citizens, etc.) because they have very different needs and skills. To tackle these differences, the PWS shall be part of an alerting strategy considering the high complexity of all interplaying factors. The alert must be designed as a communication act, creating trust, raising awareness, providing easy-to-understand information and motivating the public to act safely. Sometimes, alerts include technical data and forecasts designed to address crisis and operation manager needs. Therefore, they should be adapted to better match the population's needs.

Previous experiences and studies offer clear indications to improve alert generation and delivery (Bean 2015). In the following, the essential aspects are summarized.

- *Hazard*: the alert must describe the physical phenomenon and its risks. This information set is crucial to harmonize subjective differences and facilitate a correct risk perception. In front of unclear or ambiguous messages, people tend to minimize the risk and become more prone to unsafe behaviours.
- *Location*: geographical boundaries and environmental details about the area affected by the hazard shall be explicitly described. In addition, the nearest safe areas should be indicated. Care should be put when the estimated affected area is displayed on a map because map-based visualization is beneficial if it provides only the information required to understand the situation, thus limiting the complexity for the intended readers (Dransch 2010). It is known that reading a map requires specific skills that not all population segments equally have (van der Ham et al. 2020).
- *Temporal validity*: the temporal validity of the alert must be specified with clear reference to the time zone (ISO 8601, 2004). In particular, two dates are usually included: the so-called start date or effective date, specifying when the alert enters in force; and the end date or expiration date, stating when the alert expires. Whenever possible, the estimated temporal validity of the event to which the alert is referred should be specified.
- *Guidance*: effective alerts shall include clear instructions to stay safe. Literature shows that the more instructions the alert provides, the more likely people comply with them (Párraga Niebla et al. 2016). This is especially important for risks people are less prepared to cope with. Complete instructions should include the onset date, meaning the exact moment they shall be followed. This information is crucial to reinforce the risk's urgency and activate the response when needed.
- *Source*: the organization emitting the alert is essential to make it trustable for receivers. Due to the multiple sources of information, we are exposed to, the credibility of the alert source is crucial to set up an effective communication process. Official sources are considered trustworthy, even though they have to share the audience's attention with other sources, e.g. citizens-generated content on social media platforms.
- *Anticipation and timeliness*: as expected, the sooner the message is received, the better the citizens will organize. In the case of flash floods, a positive correlation between the warning lead time (anticipation) and the damage reduction has been observed, i.e. from a minimum of 20% at 1 hour lead time up to 60% at 12 hours

lead time (Rogers and Tsirkunov 2010). Still, at the same time, the warning reliability decreases with increasing lead time, i.e. from a maximum of 0.65 at one hour lead time to about 0.1 at six hours lead time where the reliability is measured between 0 and 1 (Rogers and Tsirkunov 2010). Therefore, a correct balance between anticipation and reliability shall be found to reach optimal response.

Repetition is another factor that can affect the warning outcome. For instance, it has been observed that the same message received many times on multiple channels is more effective since it is perceived as more reliable and urgent. For example, increasing alert frequency raises the likelihood of evacuation. According to the literature, it is highly recommended to communicate early, often and clearly throughout the DRM cycle and not only during the response phase (Lundgren and McMakin 2018).

- *Style*: regardless of the communication channel, the alert contents shall be brief, clear and avoid redundancy. This guideline addresses the need to comply with human attention and information processing mechanisms (Mayhorn et al. 2006). The emotional response to the incoming or ongoing risk can decrease the receivers' understanding. Textual shortness helps to convey understandable messages, inspire credibility and transmit a proper sense of severity or urgency. These requirements suggest that the design of an alert shall fully consider the target characteristics, including their context of life and cognitive capabilities, as well as the platforms or services they use to receive it.

Both human and contextual factors strongly affect compliance with the alerts and instructions to be followed, especially in emergencies. The Human-Centred approach applied to the design of PWS can incorporate an articulated knowledge of the target population regarding socio-demographic differences, needs, skills and habits that affect the capability to correctly interpret and follow alert instructions.

Humans need time to process and assimilate information, mainly if it contains critical messages. The main steps of the alert communication processing have been described as follows (Bean 2015):

1. receiving the alert (hearing/reading);
2. understanding the alert message;
3. believing the reliability;
4. personalizing the alert;
5. confirming the alert is true;
6. and taking protective action.

Several studies investigate the contextual factors that can interfere with or improve this process. Relevant **differences in how an alert is received and interpreted depend on the socio-demographic characteristics of the audience**. Standard segmentation variables are age, socio-economic status, gender, education level, impairments and frailty factors. In addition to these, habits, family and professional roles should be kept into consideration because they can influence the message interpretation and the actual response. Moreover, **the experience of prior disasters affects the response** because people might have learnt about protective actions and warning systems, but they could also be hindered by emotional reactions (Mayhorn and McLaughlin 2014).

From the **receiver perspective**, an alert message activates cognitive and emotional mechanisms, most of which are unconscious. This assumption allows us to reframe the **alerts as emotional contents**: they shall convey a message about a substantial risk but not raise panic or adverse emotional reactions. Additionally, alerts must motivate people to behave correctly to stay safe. Messages announcing a possible incoming hazard or suggesting how to act during an ongoing crisis may trigger basic emotions, such as fear and anger, that could be directed towards the alert or even to the emitter, decreasing the probability of activating safe behaviours. The emotional reactions that alerts may trigger include hyper-activation, panic, apathy and resignation. **Trustworthiness** is essential to help people modulate and mitigate negative emotions: in emergencies, people select and follow only reliable sources. The lack of consistency or disagreement among messages delivered by official sources is perceived as a lack of coordination and can generate confusion, **biases and misinterpretations.** To maximize the expected impact, values such as empathy, commitment, competence, transparency and equity shall drive the public warning creation and delivery. To this end, a growing number of toolkits for crisis communications design are available online to help authorities and organizations to plan and design a human-centred public warning strategy and long-term risk communication (COE – Council of Europe 2015; UNDP 2016; UNDRR 2021).

12.2.2 United Nation Guidelines

Since 1999, the UN has been working to promote the implementation of the International Strategy for Disaster Reduction (ISDR) and raise global commitment to a more aware and effective risk assessment and management. Political, scientific and economic institutions have created synergy and facilitated the implementation of the proposed measures at the national and local levels. In 2015, the UNISDR released the **Sendai Framework**, a worldwide agenda for Disaster Risk Reduction for 2015–2030. The document summarizes decades of experience in DRR, redefining it from technical and humanitarian perspectives (2015). The Sendai grounds on three main pillars:

- Focus to shift the current paradigm: **from disasters to risks**.
- The implementation of a **multi-risk perspective**, including both small-scale and large-scale disasters, frequent and sudden events, natural hazards and anthropogenic disasters related to environmental, technological and biological causes.
- The **people-centred DRM approach**.

The implementation guidelines prioritize reaching **well-designed and impact-oriented Early Warning Systems** to address new risks and rapidly changing vulnerabilities. Nowadays, PWS implemented in many countries leverages scientific results and reliable ICT, often including low-cost technology, thanks to the applications of Standard Operating Procedures combined with impact-based forecasting and risk-based warnings (WMO – World Meteorological Organization 2019).

12.3 Technical Standards and Recommendations

The importance of adhering to and correctly implementing standards is crucial for the assurances they offer in terms of the effectiveness of the foreseen procedures and data while allowing a meaningful comparison with other actors (e.g. public bodies at the local, regional and national level) adopting the same standard. Such comparison can be useful to assess the goodness of the techniques and the adopted methodologies, as well as to accelerate the innovation process towards better approaches that could enhance the resiliency against the faced problems.

12.3.1 Standards and Requirements for Public Warning Systems Implementation

For the PWS management, this document identifies two primary standards: ISO 22322 (2015) of the International Organization for Standardization (ISO) and ETSI TS 102182 (2010) of the European Telecommunications Standard Institute (ETSI).

ISO 22322:2015 – 'Social security – Emergency Management – Guidelines for public warning' provides guidelines for developing, managing and implementing public warning before, during and after incidents. It applies to any organization responsible for public warning, from local to international. It identifies two main phases in the public warning system: hazard monitoring and warning dissemination. The monitoring phase is led by the risk assessment and should involve all the interested public authorities, which shall cooperate to enhance public awareness. The dissemination phase includes all the activities that aim to identify the correct communication process (i.e. which is the most effective message, which communication channel to exploit, which groups of people to alert).

ETSI TS 102 182 (2010) – 'Emergency Communications (EMTEL); Requirements for communications from authorities/organizations to individuals, groups or the general public during emergencies' gives an overview of the technical requirements to manage the communications from authorities to citizens in all types of the emergency management cycle. It defines the effectiveness of an emergency notification[1] system in terms of objectives it can achieve (e.g. detailed message that focuses only on strategic information, full accessibility of the information by the people) and features it provides (i.e. requirements to satisfy). The main features regard:

- *The delivery capability of the notifications*: people coverage should be at least 50% of the citizens in the relevant area within three minutes and 97% of the citizens in the surrounding area within five minutes. Of course, in case of rapid emergencies, such as earthquakes or tsunamis, the notification should be delivered to as many interested citizens as possible in seconds (e.g. 10 seconds for earthquakes). For this reason, an emergency notification system should have high availability and should offer geographic redundancy of the components and recovery plans for broken parts.

- *The geographical coverage capability of the notifications*: citizens need to have information specific to their location and the emergency location. An effective notification should cover an area of:
  - 1 km inside community boundaries;
  - 5 km outside community boundaries;
  - 30 km in rural areas;
  - 60 km over the sea of the desert.
- *The delivering methods of the notifications*: multiple methods shall be supported, such as telephone voice, SMS, broadcast radio and TV, sirens, email; and instant messaging service to mobile phones, mobile apps and web pages.
- *Message accessibility*: messages should be delivered even to people with special needs, such as hearing and vision impaired, and even in different languages (e.g. to visitors from other countries).
- *Security and data protection*: the emergency notification system should provide user authentication and protect the sensitive user data it collects.
- *The recognizability of the emergency message*: an emergency notification system should be intrusive. This concept means that the reception of an emergency message should be enabled by default. Moreover, the message should be immediately recognizable and put in the foreground as soon as possible.

ETSI TS 102 182 clearly specifies that none of the available technologies met all the requirements to fulfil the high ambitions in terms of wide-ranging distribution of highly effective messages and their delivery in the shortest possible time. Instead, combining different methods would make it likely to achieve the specified goals. To accomplish message recognizability, adopting a standard communication protocol that can be unambiguously parsed by the organization responsible for message dissemination is necessary. In its standards, ETSI proposes the Common Alerting Protocol (CAP) as a notification message-delivering protocol, also adopted by the International Telecommunications Union X.1303bis (2014).

12.3.2 The Common Alerting Protocol

The **CAP standard** (CAP, Common Alerting Protocol Version 1.2) is a digital message format for PWS notification promulgated by OASIS, the Advancing open standards for the information society (O.A.S.I.S. 2010). It follows the serialization rules of the XML standard, as specified in (Bray 2008). The CAP is not related to any specific application or telecommunication method, and it can be easily used for any hazard.

The CAP standard is based on the following principles:

- *Interoperability*: the CAP message provides a means for the interoperable exchange of alerts and notifications among different information systems;
- *Completeness*: the CAP message provides all the elements for generating effective PWS messages;

- *Simple implementation*: the CAP standard does not place burdens of complexity on technical implementers;
- *Simple XML and portable structure*: the CAP message implementation is an XML document, but its format is sufficiently abstract to be adaptable to other coding schemes (e.g. JSON);
- *Multi-use format*: the message supports multiple types (e.g. alert, update, cancellation, acknowledgements, error message) and can be exploited in several applications (actual, exercise, test, system message);
- *Familiarity*: the data and the code values must be meaningful also to non-expert recipients;
- *Interdisciplinary and international utility*: the design allows a wide range of applications in public safety and emergency management and is applicable worldwide.

A CAP message is composed of segments with different meanings. Its structure can be expressed with the Document Object Model (DOM), which gives a tree representation of the message structure, as described in W3C (2002).

The core segment is the <alert>, which can contain one or more <info> segments, each of which includes one or more <area> and/or <resource> segments. Each segment is composed of several elements that characterize the message, some of which are mandatory.

The CAP template is simple but precise and allows its adoption in different scenarios. A single message can be manually forwarded to linked warning systems, which convert it in the form suitable for their technologies (e.g. text caption on TV, audio on radio and telephone, triggering sirens). Alternatively, automatic procedures can be implemented to enable the warning systems components to generate alerts (e.g. a wireless sensor co-located with a siren creates and disseminates a CAP message that includes the location and the sensed data after the siren activation). Examples of CAP alert messages are available in the standard (CAP, Common Alerting Protocol Version 1.2).

12.3.3 Recommended System Architecture

A standard-based approach can simplify and improve communication among stakeholders who accomplish different functions in DRR. For instance, neighbouring countries may face extreme hydro-meteorological events, including windstorms, heavy rains, floods or coastal surges over a large geographical area in a very short timescale. Situations like these need frequent and efficacious **cross-country communication.** To reply to this need, a PWS must offer secured and accessible **Application Program Interfaces (API)** to share the alerts with other countries (e.g. border countries) and higher-level entities (e.g. EU, UN bodies). In general, **centralized systems might better integrate alerts coming from different sources. A unique and secure gateway deployed on the Internet** could easily enable interoperability, allowing multiple subsystems to be connected for reading and spreading alert messages.

The usage pattern of a PWS can be subject to strong bursts when an emergency is expected or ongoing and suffer from long periods of underutilization when no hazard

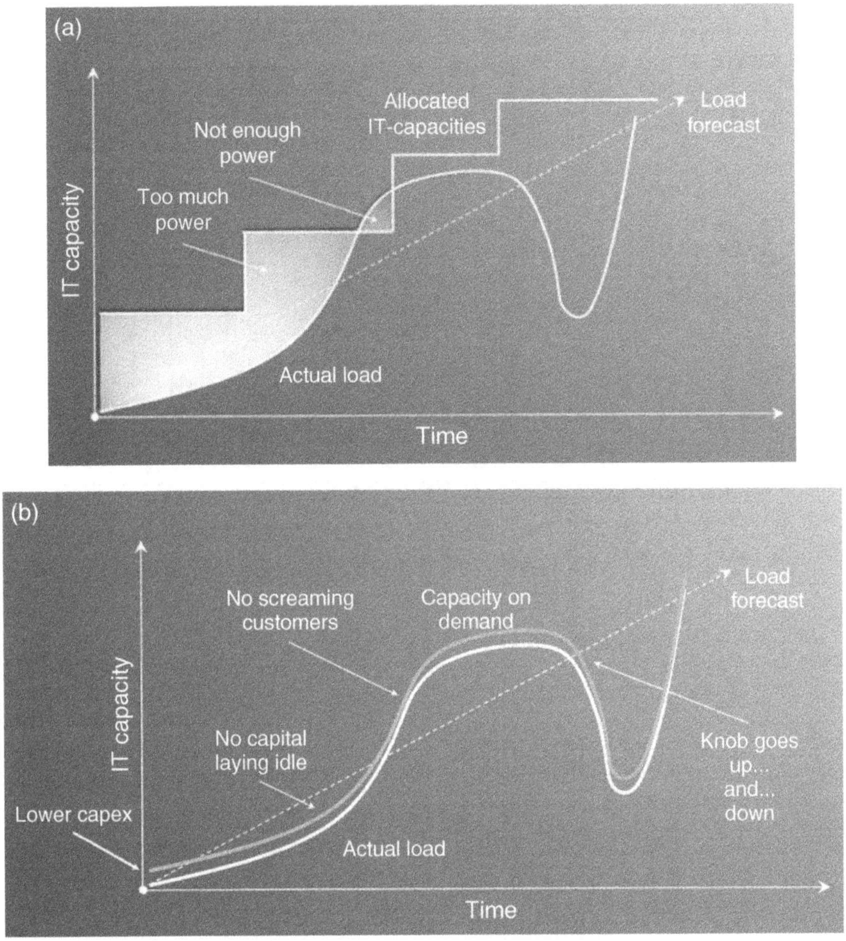

Figure 12.4 Comparison between (a) the CAPEX model and (b) the OPEX model.

is expected. This aspect suggests the usage of modern cloud-based services that, according to the definition given by NIST (2011), must feature *broad network access and rapid elasticity*. The latter is a priority to keep OPEX aligned with the actual usage patterns, as shown in Figure 12.4.

The PWS system architecture shall be included in technical and functional specification documents and be kept updated. Figure 12.5 proposes a suitable macro architecture consisting of three main components:

- *ALERTING AUTHORITIES INTERFACE (AAI)*: this component manages the interface with a country's public authorities. According to the internal regulation related to emergency management, the authorities can belong to the national, regional and local levels. All the authorities shall have valid and certified credentials to access the PWS. An authorization system shall be implemented to enable

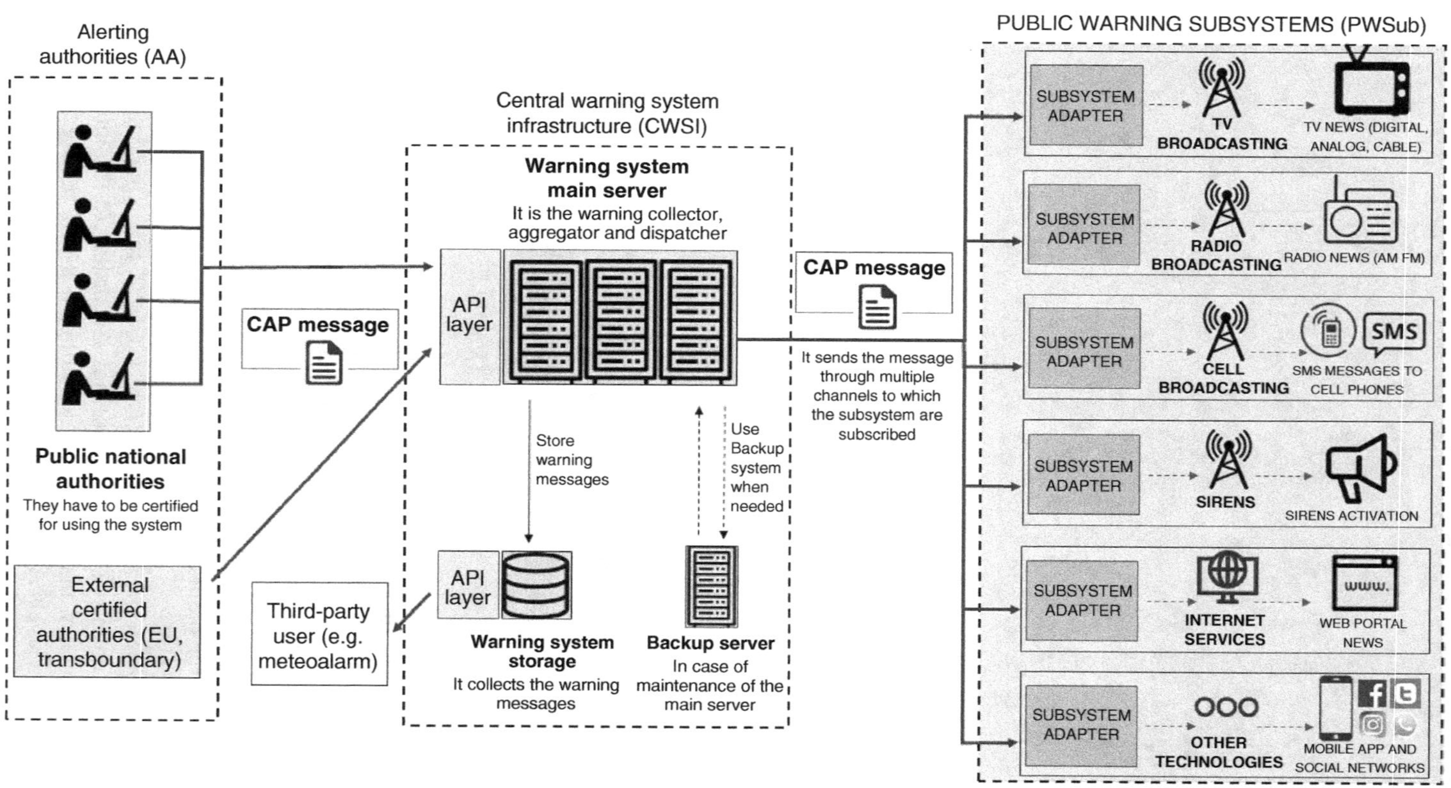

Figure 12.5 Suggested Public Warning System architecture.

permissions configuration according to the policies and governance chain in force. Each authority included in the AAI component must also be able to send a message using a standard template that, as identified in Section 12.3, could comply with the CAP standard. The AAI component should be extendable, allowing new certified authorities to be included in the system. This feature also allows the inclusion of external organizations (from abroad the country). This is the case of cross-boundary authorities, whose public warning messages can be useful in case of cross-border crises.

- *CENTRAL WARNING SYSTEM INFRASTRUCTURE (CWSI)*: it is the core of the architecture, and it comprises several sub-components that guarantee the communication between the authorities that create the warning message and the dissemination channels. The CWSI contains the **Warning System Main Server,** which is the collector and dispatcher of messages. To collect the messages, the Main Server exposes a set of application programming interface (API), through which the authorities of the AAI component can send the message. The set of APIs shall be reachable from the Internet and should be used only by authorized organizations. For this reason, the CWSI must also include a user authentication and authorization system. When an alert comes to the Main Server, it stores the message in the internal **Warning System Storage** (i.e. a single database or a set of linked databases) to allow future inspections (e.g. internal statistics, etc.). If allowed, the Warning System Storage content could also be made available to third-party users or other PWS subsystems through the concept of federation or by using a single sign-on system. The CWSI must also include a Backup Server, which acts as the Main Server when this last is under maintenance.

After a message is stored, it is **delivered to multiple dissemination subsystems**, each controlling one or multiple homogeneous communication channels. This communication can be achieved through a communication bus (e.g. RabbitMQ, Apache Kafka) that implements the publish/subscribe paradigm, where each receiver can subscribe to topics of interest that the publisher uses. Each subscription can be subject to specific rules on the message content that enables message delivery. For example, if the CAP standard is used, such conditions can be based on the hazard (the CAP *<category>* field), the severity (the CAP *<severity>* field) or the geolocation (the CAP *<polygon>* field). Using this rule-based approach, every dissemination subsystem can be activated only if certain conditions are met (e.g. maximum severity), thus allowing the dissemination segment to be used according to local regulations and needs.

- *PUBLIC WARNING SUBSYSTEMS (PWSub)*: it contains all the subsystems aiming to disseminate the CAP warning message (Figure 12.5). Each subsystem should have an **Adapter**: a software component that listens (subscribe) to the dedicated channel (topic) to receive warning messages. When the Adapter detects a message on its channel, it reads the communication, parses the message to extract only the fields that are needed to compose the final content (text, audio, video), it creates the final content following pre-defined style and templates and spreads it to the attached communication channel. The PWSub must be customizable and extendable, allowing additional subsystems, which could be required as new technologies are adopted, to be easily integrated. There can be more than one subsystem using the same type of dissemination channel: TV and radio subsystems can be attached to multiple channels and operate at different levels (national, regional or local). For the social media subsystem, it is recommended to cover multiple platforms (e.g. Twitter, Facebook, Instagram, etc.) and be linked with institutional accounts corresponding to all relevant agencies (e.g. environmental monitoring, Civil Protection, firefighters, Red Cross, etc.) to automatically post the adapted content on all appropriate pages.

As an example of a PWSub, Figure 12.6 outlines a solution for a national siren system to alert the population in case of extreme risk, requiring immediate reactions. The system includes an adapter, which parses and filters the CAP message, a communication channel and the receiver technology, precisely the siren. The Adapter can receive the CAP message both from the CWSI, as previously described, and directly from the authority in case the CWSI infrastructure is unavailable or is under maintenance, and the backup server is unavailable.

Once the Adapter receives the message, it is filtered and translated to be reproduced by the siren. The siren should be **multi-tone** and should also be able to **play voice-recorded messages.** The multi-tone feature can be based either on the message's hazard or severity; in this case, the Adapter should check the CAP fields <category> or <severity>, respectively. The voice-recorded feature can instead reproduce a standard message depending on the hazard or severity of the warning communication. In this case, the Adapter should check the CAP fields *<category>* or *<severity>* and in addition, one of the fields *<uri>* (i.e. the link to the resource voice-recorded message to reproduce) and *<derefUri>* (i.e. the base-64 encoded content of the resource file). All the unnecessary fields must be discarded by the Adapter, to minimize the message's size sent to the network infrastructure.

The network infrastructure should be constituted by a primary secure (i.e. with encrypted messages) channel, which can be based either on 3G/4G cellular technology or on SATCOM technology and by a secondary backup channel to be used only if the primary channel is not available. The backup channel can be based either on the TETRA (ETSI 2018) standard, which guarantees message encryption or on the UHF channel, which is commonly used for television broadcasting and, for this reason,

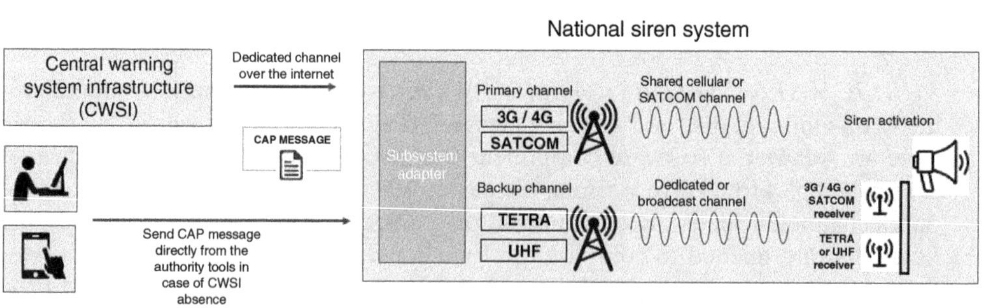

Figure 12.6 Suggested public warning subsystem for sirens.

needs to be secured through an encryption key exchange process on an auxiliary channel. Finally, after the message goes through the channel, it is reproduced by the siren.

To summarize, a national PWS should be technically designed to support the following scopes:

- **To be used by all agencies involved in DRR** across all administrative units (e.g. commune, region/department/prefecture, national). The PWS must be able to be configured and map users according to the organization structure of such agencies, associate them with the geographical area under their responsibility and implement in a configurable rule-based manner the alerting procedures (e.g. escalation to a higher level).
- **To be used for all kinds of alerts** and hazards and across all DRR process phases.
- **To receive standard alerts** from monitoring systems (e.g. sensors) as well as agencies (e.g. operators of environmental agencies, police, etc.) **and broker them** to public warning subsystems**. The CAP standard is the suggested system to structure the alert content.** The brokerage should be done according to the publish-and-subscribe pattern to allow the subsystems to subscribe to the topic of interest (e.g. if a siren system is used only for nuclear alerting, it will not subscribe to topics related to natural hazards) and according to the area they cover.
- **To publicly publish every alert generated** at a given URL containing complete information. This URL, eventually shortened, should always be disseminated despite the content adaptation required to propagate the alert on the selected distribution channels. Note that there are exceptional cases in which this may not be feasible, e.g. in the presence of tonal sirens. In such cases, the population must know a priori how to find the complete alert information.
- To feature a **reliable deployment** that can guarantee availability, reliability and fault tolerance according to clearly defined service levels. A Cloud-based deployment (private, public or hybrid) is suggested to comply with these requirements;
- **To allow Public Warning subsystems to interface with the national PWS** to make it capable of adapting the content to the media and target recipient.
- To support **standard data exchange with other (external) systems** through secured API and open the message broker to PWS of neighbouring countries and supranational entities (e.g. EU). In this way, the alerting in cross-border emergencies will become manageable.

12.3.4 Use of Technical Standards

As previously described in the document, we **recommend using the CAP standard to structure the alert content** and the ETSI standard for the requirements that the technologies involved in a PWS should satisfy. However, these two standards show some gaps and are not always up to date, considering the availability of new technologies.

Due to its flexibility and customizability features, the CAP protocol considers many important fields as optional. Still, for effective warning message generation, some of them should always be included in a message. Below we report a shortlist of considerations regarding the CAP fields that are not mandatory for the protocol but that, in our opinion, must always be present:

- The ***<polygon>*** field, which is a list of (WGS 84) coordinates pairs, is considered optional, but it is very important for the generation of effective warning messages. To generate the best communication possible, it is necessary to geolocate the interested groups of people according to the severity of the disaster effects in their surroundings. To warn people on the move, who may reach the estimated affected area within the alert temporal span, it is strongly suggested to extend the alert area and **create a bigger communication area**, e.g. taking the bounding box that inscribes the alert polygon and extending it according to the maximum ground travel speed and the temporal validity of the alert. Alternatively, and/or in parallel, the same alert can be repeated over time with a periodicity that should be computed according to the alert severity and certainty. The **alert should be automatically disseminated to all subsystems/agencies with coverage/responsibility on areas having a non-null interception with the computed communication area**.
- In a multicultural society, the ***<language>*** field, which indicates the language of the warning message, is very important. A CAP message can include, in fact, more *<info>* segments with the same information (*<severity>*, *<urgency>*, *<certainty>*, *<category>* etc.). Still, with the textual or media fields (*<instruction>*, *<event>* description, *<derefUri>* for the media attachment) in different languages, in this way, the communication end-point, such as television, radio or even sirens, can reach the larger audience as possible, including even ethnic minorities. It is essential also that the values of the *<language>* field belong to a defined standard, one of which can be, for example, the RFC3066 of the Internet Engineering Task Force (IEFT 2001);
- The ***<uri>*** and ***<derefUri>*** fields indicate, respectively, the full absolute URI of a resource (typically an URL over the Internet) and the base-64-encoded data content of the resource (in this case the media content is directly attached to the CAP message). Both are optional for the CAP protocol, but at least one of them should always be included in the message. If the *<uri>* is included, then the message has a lower size, but the end-point needs to be connected over the Internet to download and show the media content. On the other hand, if the *<derefUri>* is specified, no internet connection is requested to the end-point for downloading the resource, but the message size will be higher;
- The ***<audience>*** field, which is optional in the CAP standard, is also crucial for message effectiveness because it can be helpful to specify the recipient's category. It can be used as a discriminator by the subsystems of the PUWSub for disseminating messages to a specific target of people (either professionals or citizens, people with special needs, etc.);
- The ***<effective>*** and ***<expires>*** fields indicate the effective time of the information of the alert message and the expiry time of the information of the alert message, respectively. Either of them is not mandatory, but it is crucial to include this information in all the CAP messages. In addition, the ***<onset>*** field, which indicates the expected time of the beginning of the event of the alert message, can be essential to provide additional information about the context of the alert communication.

All these dates and time field values should follow a standard (e.g. ISO 8601, 2004). Finally, we notice the absence of a field indicating the end of the hazardous event, and we think this can represent an important lack of the CAP protocol.

- The **<parameter>** field can be handy for adding information that the other CAP fields cannot map, for example, for information that has been set out within the organization. However, it is recommendable to avoid the abuse of such a field and try to map as much information as possible with the corresponding CAP fields.

It is worth noting that the ETSI standard does not include any reference to the technologies that nowadays are commonly used by all the people in their daily life, such as **mobile applications**, and **social networks** (e.g. Facebook, Twitter, Instagram etc.) and **chatbots** (e.g. as services built upon Telegram or WhatsApp). For this reason, following the ETSI requirements, we have tried to map and check how the three technologies mentioned above fulfil the requirements. The result is shown in Table 12.1, reporting that many of the requirements for an effective PWS are met by mobile applications, social networks and chatbots.

We must highlight the absence of recognized standards providing guidelines on how to present the warning message to the user, especially for social media and mobile application technologies. Furthermore, these standard documents are not regularly updated, so they often do not reflect the arising needs of society and the opportunities provided by the most recent and emerging technologies.

Table 12.1 Requirements vs technologies – mobile application, social media and chatbot based on ETSI TS 102 182 requirements. Legend: ✓ compliant; ⊘ not compliant. In certain regions of work access to some applications or social media can be blocked.

Emergency notification systems shall	Mobile applications	Social media	Chatbot
Be able to reach citizens in their dwelling	✓	✓	✓
Be able to reach citizens at their place of work	⊘	⊘	⊘
Be able to reach citizens in public venues	✓	✓	✓
Be able to reach citizens on foot	✓	✓	✓
Be able to reach citizens in a vehicle	⊘	⊘	⊘
Be able to reach a citizen visiting another European country	✓	✓	✓
Provide sufficient instructions regarding actions to be taken	✓	✓	✓
Provide identification of the message/notification originator	⊘	⊘	⊘
Deliver messages within a planned specified time	⊘	⊘	⊘
Allow simultaneous delivery to targeted, large audiences or geographies	✓	✓	✓
Offer sufficient details of the emergency	⊘	⊘	⊘
Be able to retry delivery when the initial message delivery fails	⊘	⊘	⊘
Support delivery of notification messages to those with special needs and unique devices, like terminals of hearing and speech-impaired persons	⊘	⊘	⊘
Have the capability to deliver messages in multiple languages	✓	✓	✓
Be capable of addressing congestion management across the various networks used	⊘	⊘	⊘

12.3.5 Media Adaptation and Usability of Alerts

The plethora of channels and media that a PWS include implies careful management to provide reliable, clear and effective messages considering how human factors and the context can affect the communication flow. In addition, **each channel poses specific capabilities and limitations, which shall be correctly addressed** to guarantee a viable and effective content design. In addition, the delivery of alerts heavily depends on contextual factors. For instance, even though an SMS can reach mobile phones, i.e. most adult people with an active mobile phone, there might be phones in 'not disturb' mode (or other conditions) that reduce the target effectively reached. A multi-channel communication plan could be the countermeasure to this simple but realistic scenario. Table 12.2 recaps the main PW channel, describing their key strengths (pros) and weaknesses (cons).

While applying multiple-channel PWS, we recommend to:

- **Include the use of social media** in parallel with traditional media, such as TV and radio, creating specific pages to disseminate the alerts. Given the multitude of agencies and territorial segregations, we suggest that **every agency** (e.g. water, environmental monitoring, responder) **should have its own social media page**, which should be known by the centralized PWS and the page scope in terms of hazard and territorial coverage. The alerts should be inserted by the agencies, using

Table 12.2 Public warning channel overview.

Technology	DRR Phase	Pro	Cons	Notes on CAP
Sirens	• Prevention and preparedness • Detection and Response	+ Volume regulation + Multi-tone + Customizable sounds (and voice messages) + Remotely controlled	– Do not provide instructions – Coded audio messages. Recipients must know the codes to understand the meaning – Indoor coverage may be limited	Only a few CAP fields can be used, mainly the category
Public displays	• Prevention and preparedness • Detection and Response	+ Site-specific + Remotely controlled	– Text only – Brevity – Local target audience	Only brief texts can be visualized
Broadcast comm. (radio, TV)	• Prevention and preparedness • Detection and Response	+ Wide audience + Plain language + Images allowed	– Target needs to be tuned on the channel when the alert is sent	Full content possible
Cellular com. (SMS)	• Prevention and preparedness • Detection and Response	+ Geolocated distribution + Wide audience	- Text only - Brevity	Only brief texts can be visualized

Table 12.2 (Continued)

Technology	DRR Phase	Pro	Cons	Notes on CAP
Mobile App	• All	+ Geolocated distribution + Data collection (Crowdsourcing) + Target specific + Bidirectional. Allows feedback from the recipient	– Internet connection dependent – Digital divide prone – Needs to be downloaded by the user	Full content possible
Website	• All	+ Accessibility + Rich text and media allowed + Multi-hazards + Multilevel information + Bidirectional. Allows feedback from the recipient	– Internet connection dependent – Digital divide prone	Full content possible
Social media	• All	+ Wide audience + Rich text and media allowed + Bidirectional	– Internet connection dependent – Digital divide prone – Heavy coordination and control required – Need the registration to the page/channel – Emergency-related messages are not prioritized to other ones	Only selected textual content can be visualized
Chatbot	• All	+ Natural interface, natural language + In-app already in use + Multimedia content allowed + Data collection (Crowdsourcing) + Target specific + Bidirectional. Allows feedback from the recipient	– Internet connection dependent – Digital divide prone – Need the registration to the channel	Full content possible

the client of the centralized PWS system according to the selected standard. The alerts will be brokered and automatically delivered coherently to all media, agencies and social media pages after a proper adaptation.

- **Include the use of dedicated apps or chatbots** to enable a bidirectional communication flow and target specific capabilities, allowing recipients to receive feedback and customize the content considering the human and contextual factors. For example, colours shall be adapted to be accessible to colour-blind people or poor lighting conditions.
- **Adapt the alert content to the media used.** Starting from a complete common standard that contains all the required information, develop automatic adapters to compose ad-hoc messages and styles according to the media on which the alert is propagated. For example, SMS cannot deliver multimedia content, only text is allowed, and long messages may be difficult to read if split into multiple SMSs. Some social media (e.g. Twitter) have limitations on content (limited characters and multimedia content per post).
- Build **templates for styles in a coherent way** across the different media. Alerts should always be easily readable and understandable.
- Develop **multi-agency alerting plans and templates** for every possible hazard according to the severity level, so the plan can be quickly applied when an emergency strikes. A single alert database containing pre-defined alerts should be designed and shared in advance among all entities involved in emergency management across different administrative areas and agencies.
- Alerts are effective only if understood. Plan **prevention campaign** to inform people about the risks they are exposed to, the situations they may face and explain the standard alert codes/colours and, in case of special media such as sirens, which are the sounds used for the different hazards and define a clear behaviour, e.g. if you hear the siren, immediately go home and turn on your TV or radio to get the instructions. Design easy-to-understand leaflets customized per target audience. A good example of prevention is the 'IO NON RISCHIO' initiative (2022).

Next, we provide some reference examples of alerts issued on selected channels, including sirens, SMS, web and social media.

SIRENS: as an outdoor warning system, sirens are used in most European countries. Mainly used in the military context, they are also used to alert the population on high-level hazards, such as critical storms. The message that sirens convey must be known *a priori* to be correctly decoded. The sequence of activation or the sounds – if many tones are supported – may have different meanings. In general, sirens are associated with a severe incoming threat. People might perceive and interpret the sounds differently (e.g. evacuate the area or stay at home). The use of sirens needs to be complemented by clear instructions that allow people to recognize the audio message and reply correctly to the warning.

Therefore, a dedicated section on local authorities' websites or apps would aim to:

- Inform people about the existing siren systems and their location;
- Explain their use and the meaning of each tone;

- Provide instructions about what to do in case of sirens activation, covering the different contexts in which people can be (e.g. driving, at night, with children, for enterprises);
- Allow to hear the alerting sounds with a sound preview;
- Provide contacts of the authority.

SMS/CELL BROADCASTING: the ETSI 102 900 (2012) provides exhaustive guidelines to help PWS to pre-set a portfolio of messages to handle different levels of alert and types of communication. For example, the NL Alert system is a valuable example of effectively arranging SMS communication by applying the CAP standard. Note that this example does not show any temporal information. This is a shortcoming because the temporal validity of the alert is key information (Figure 12.7)

In general, it is recommended to keep the message short, including the basic information highlighted in Section 12.2.1 and using only selected CAP fields to compose the message. The incoming hazards, the interested area, the sender, the response type, the severity, the certainty and the urgency are fundamental.

Instruction for citizens cannot miss and must fit with the identified severity level. This latter is not enough to suggest and promote the safest actions to be taken, which must be explicitly indicated. Finally, the URL (or URI) where to get the full message should be provided. These recommendations are also valid for social media communication.

Web Platform: nowadays, several online platforms allow receiving map-based alerts on multiple hazards worldwide. They automatically aggregate data from multiple sources and present the warning communication to the public. An excellent example of such a system is Google Crisis Response (2023). When showing an alert, a good visualization should focus on the key information, such as the affected area of the hazardous event and all the mandatory CAP data in a well-formatted way. The communication should be reported on the map through its geographic coordinates and the click action on this localized warning on the map could trigger the opening of a pop-up (right side of the figure) with detailed information from the alert. Figure 12.8 shows how a CAP-compliant message disseminated by the UK Flood Environmental Agency is displayed in a web application by people and authorities alike.

Social Media: social media adoption is very large, and it is still growing. Such technology is widening the range of possible interactions among players and enabling powerful

"A <event> alert has been issued for <areaDesc> by <senderName>. Persons in this area are encouraged to <responseType>, and <instruction>. This event is rated as <severity>, and is <certainty>. Responsive action should be taken <urgency>. For more information about this event, visit <URI> or call <contact>."

Figure 12.7 An SMS alert structure derived from CAP. Note that temporal information is not included in this example.

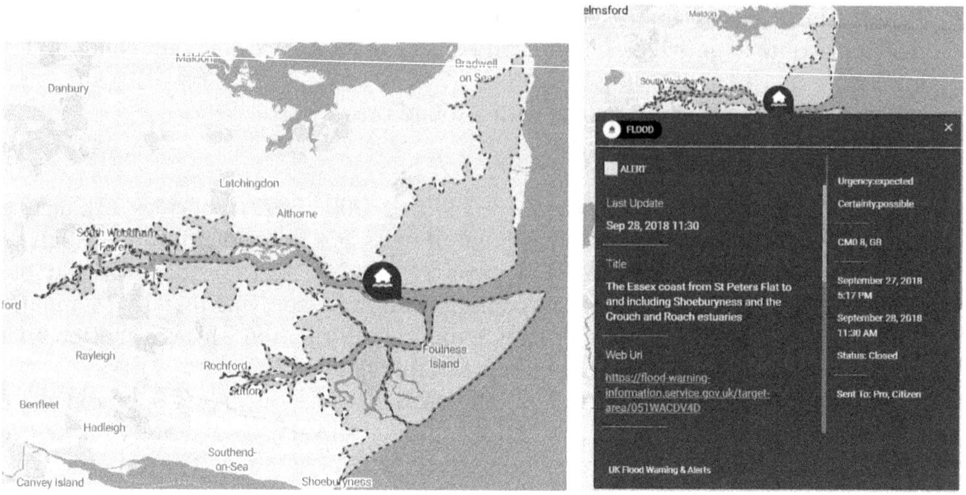

Figure 12.8 CAP warning message showed on a web platform. This example refers to a flood alert issued by the UK Environmental Agency and integrated into the I-REACT system. *Source*: I-REACT project.

and effective Public Warning practices. During a crisis, people immediately start using mobile networks to communicate both among them and with authorities. This happens in a wide range of situations and information, as shown in Table 12.3. Social media play a prominent role in an undefined number of life and professional sectors. In terms of communication, they are relevant channels that cannot be neglected, even in the DRM. Social media represent a digital environment **where user-generated content and official sources of information co-exist**. Authorities, volunteers and citizens can find there a space to cooperate and interact to promote prevention and safe behaviours, monitor expected or ongoing events and stay in contact during the response. Moreover, they are excellent channels to **offer a dynamic online presence on behalf of emergency organizations** and authorities at any level. According to EENA (2015), 'warning systems shall consider social media in their concepts as additional means to reach out to specific citizen groups', due to the possibility to set up a dynamic dialogue with connected people (Table 12.3).

Effective social media communication for DRR requires transferring and adapting the communication and coordination protocols to the digital sphere. Over the institutional accounts, which can work as authoritative sources from where to **send out alerts on any hazards**, additional dedicated services/accounts can be very effective in creating communities, informing and training people on prevention and preparedness in more 'quiet times'. A good example comes from the **I-REACT Project** (2017) (Improving Resilience to Emergencies through Advanced Cyber Technologies), funded by the H2020-EU.3.7. Program on *Secure Societies – Protecting freedom and security of Europe and its citizens*. The project created a social media page to actively manage a growing community on natural hazards events and disaster risk reduction (Against Disasters on Twitter 2016). Moreover, the exploitation of social media demands that organizations and institutions define and apply a suitable communication style, providing clearness

Table 12.3 Functions of social media in disasters.

Disaster social media use	Disaster phase
Provide and receive disaster preparedness information	Pre-event
Provide and receive disaster warnings	Pre-event
Signal and detect disasters	Pre-event → Event
Send and receive requests for help or assistance	Event
Inform others about one's own condition and location and learn about a disaster-affected individual's condition and location	Event
Document and learn what is happening in the disaster	Event → Post-event
Deliver and consume news coverage of the disaster	Event → Post-event
Provide and receive disaster response information; identify and list ways to assist in the disaster response	Event → Post-event
Raise and develop awareness of an event; donate and receive donations; identify and list ways to help or volunteer	Event → Post-event
Provide and receive disaster mental/behavioural health support	Event → Post-event
Express emotions, concerns, well wishes; memorialize victims	Event → Post-event
Provide and receive information about (and discuss) disaster response, recovery and rebuilding; tell and hear stories about the disaster	Event → Post-event
Discuss socio-political and scientific causes and implications of and responsibility for events	Post-event
(Re)connect community members	Post-event
Implement traditional crisis communication activities	Pre-event → Post-event

Source: Houston et al. (2014).

and accountability simultaneously. A social media strategy integrated with the PWS is then essential to communicate with the public, the press, authorities and other organizations. Each social network works as a specific environment and has its mechanism. A style guide is necessary to set content creation and communication management rules, considering technical constraints and the different public to reach. Due to their interactive nature, a **social media response protocol** is also highly recommended. It shall be managed by dedicated professionals and/or by an automatic AI algorithm to address the responses (feedback) that are received on the channel. However, the risk of an automated approach is failing to create engaging and relevant communication concerning both the single messages and the organizations' profiles. Figure 12.9 allows to compare two different crisis communication approaches on Twitter: the example on the left provides more articulated and empathic messages, including videos, links to external sources and hashtags (#). In contrast, the example on the right shows a text-only approach providing automated messages appearing briefer but less rich and captivating.

Considering the role that digital technology plays in crisis communication, even though the number of people connected is growing every day, few or small segments of the population that have no access to digital channels still exist, as well as people that are not familiar with digital content and interaction. It has been shown that ICT effectively reduces fatalities only in the presence of higher levels of educational attainment and **digital literacy**. Latter is a paramount factor in misinformation, increasing in social

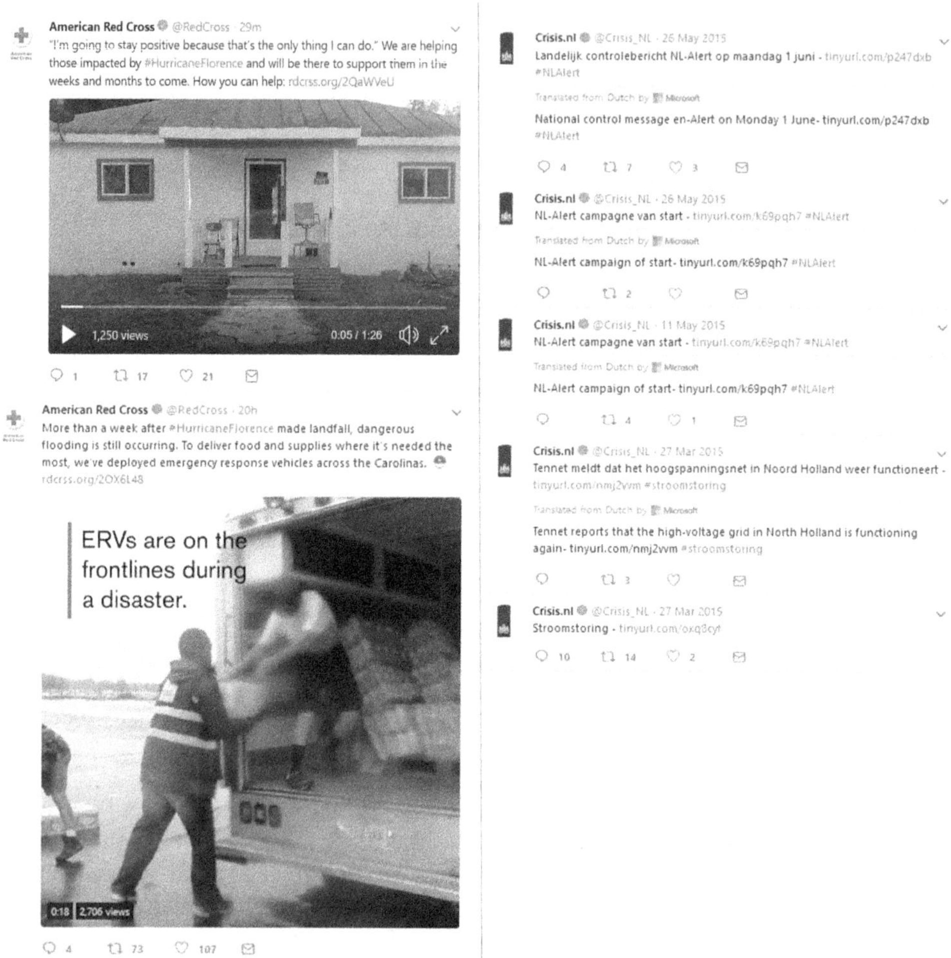

Figure 12.9 DRR messages via Twitter in comparison.

media due to their distributed nature and the impossibility of performing a content check for all content. The best practices and the state of the art of PWS recommend combining **different channels** in parallel, possibly managed by a centralized system to **harmonize the messages** on the base of the audience and the media. This allows to tune communication with the evolving status of risks or hazards and maximizes the possibility of reaching different audiences in different locations and situations.

Having teams of skilled personnel to manage risk communication processes, from the engagement and alphabetization strategy to the design and management of warning messages and supportive communications products for several channels, is essential. It allows to establish and manage bilateral communication with citizens, implementing a full DRR people-centric approach.

12.4 Future Outlooks in Public Warning and Risk Communication

As highlighted, the people-centric approach can evolve and be integrated into well-designed and dynamic crisis communication strategies, adopting data-driven methods such as continuous community mapping to create effective alerts and promote a safety culture. The **people-centric** approach can benefit and profit from traditional systems as well as the newest information and communication technologies.

12.4.1 Crowdsourcing Approaches

Thanks to the growing adoption of digital technologies, collaborative paradigms are becoming closer to being integrated into DRR processes. For instance, a **crowdsourcing approach** can be used to collect and share geospatial information from people witnessing a common context via internet-enabled mobile devices (Goodchild 2007). This approach leverages people's will and capabilities to provide real-time and geo-localized information during a crisis, contributing to the dissemination of warnings and help requests (Schulze et al. 2015). This cooperative process is compelling in establishing a bidirectional communication channel with the public. People tend to produce and share information, enabling data collection that can support the situational awareness of all stakeholders (e.g. authorities, responders, etc.) and better disseminate official information and alerts. Thanks to mobile technologies, citizens can act as digital volunteers contributing to a **participatory form of surveillance** (Albrechtslund 2008) **based on the** cooperative collection, validation and sharing of **peer-generated data**. Data collected from the field represent a tangible resource able to affect emergency management meaningfully. Such dynamic data collection can support more detailed action plans at a control room level and improve the in-field agents' situational awareness, promoting optimal resource allocation (Whittaker et al. 2015). Moreover, crowdsourcing can raise individuals' **cognitive alertness**, stimulating them to learn and improve their proactive attitude towards collective safety (population, environment, common goods) (Tallacchini et al. 2014). But this integration poses some constraints to be considered:

- The **quantity of the data collected** is crucial: emergency-related processes need data to be reliable. Moreover, these data need to be pre-processed before working as input for the control rooms and to allow operators to search, filter and explore them in many ways. The information provided does not have to overflow and negatively affect decisions and operations (Meier 2013). From the operators' perspective, new sources and data types can be an added value: the physical space monitored by traditional surveillance tools becomes a **data-space**. Technological innovation technology and new communication channels provide authorities with additional resources, but they also change work procedures, introducing complexity. Automatic data filtering procedures based on machine learning techniques could support the human agents to cope with the quantity, velocity and variety of data being collected and help decision-makers to focus on actionable data while excluding uninformative or misleading content. Data validation is mandatory, and the possible management approaches range from automatic systems to human verification. Again, the crowdsourcing approach can help support the peer-validation by integrating techniques such as social tagging and reputation systems (Pennacchiotti and Popescu 2013).

- **Accuracy and reliability of data** are crucial to avoid introducing biases into the action planning and execution (Park and Johnston 2015), while crowdsources can suffer from disinformation, false judgements and social biases (Guo et al. 2014). Moreover, their expertise can influence the collected data's accuracy and further use (De Longueville et al. 2010). In addition, the self-generated communication exchange via social media is extremely fast, informal, and subjective.

One of the most known crowdsourcing platforms is **Ushahidi** (2017). Initially created for monitoring the Kenya election, it was used for gathering reports by voluntary groups during natural disasters. Originally, it was based on SMS by which citizens reported any information or resource on the territory, enabling the possibility to provide feedback on requests and instructions. Raised to global attention during the Haiti Earthquake in 2010, Ushahidi effectively shows how such platforms of crisis mapping work for DRR organizations and the public simultaneously. Other interesting examples to cite are **Crisis Mappers** (Crisis Mapping 2017), leveraging mobile platforms, computational linguistics, geospatial technologies and visual analytics to power rapid crisis response; Mobile4D (2012), a crowdsourced disaster reporting and alerting service then published as a Community Climate & Weather Journal in the **I See Change**, the community climate and weather journal (2018). These examples show how crowdsourcing can integrate traditional monitoring systems (e.g. video cameras, sensors, satellites and other sources), augmenting the informative potential of **public reporting**. Crowdsourcing enhances the data collection and monitoring of natural events and fosters the population's awareness of hazards, risks and safer behaviours. The main challenge is validating and integrating such public data into existing DRR ecosystems.

EU-funded projects proposing public reporting in DRR have been developing. An example is WeSenseIT (2016), presented under the FP7 EU research program, which aims 'Developing citizen observatories of water and flooding to facilitate citizen engagement in planning, decision making and governance'.

Using a dedicated mobile application, people can report pictures about the water behaviour in their communities and the information can be viewed on a free access internet website. This way, people can be informed and aware of what is happening in a specific location. These crowdsourced data are also included in HydroNet, a professional control room dashboard allowing decision-makers to combine the data submitted by the people through the mobile app with other data sources (e.g. forecasting and nowcasting models, measurements etc.).

A similar approach applied to multi-hazard scenarios is proposed by the I-REACT project. The framework, among the many other features, allows the people to create reports with pictures and associated measures to depict the ongoing situation, either dangerous or not (Figure 12.10).

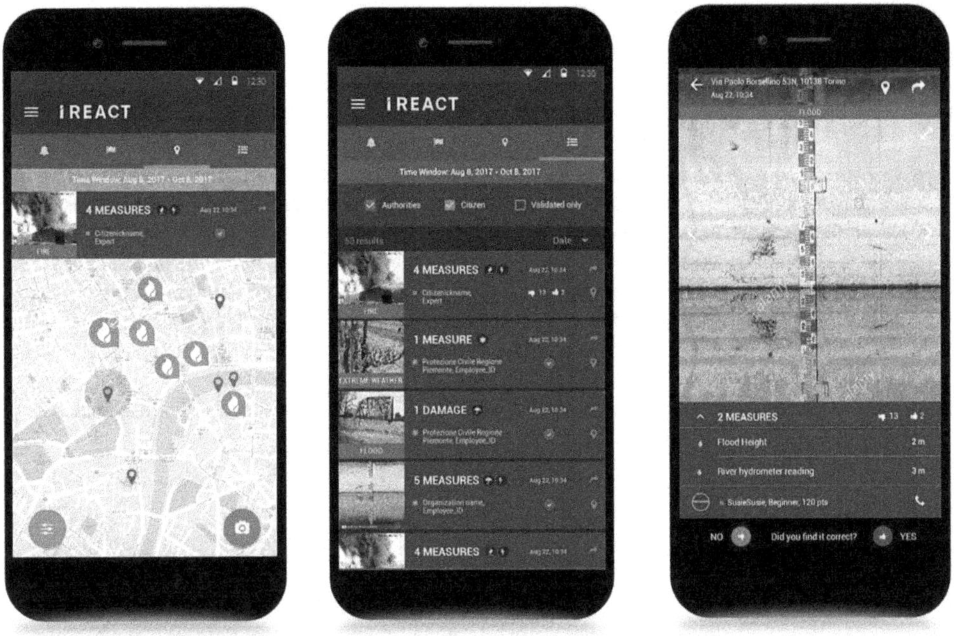

Figure 12.10 I-REACT reporting system. *Source*: I-REACT project.

I-REACT provided emergency management professionals in the control room with a web-based application to visualize data from in-field reporters (volunteers and professionals), featuring a double validation process that allows peers and professionals to validate or rejects the reports, preventing the proliferation of fake data.

12.4.2 Organizational Best Practices

Besides the technical capacity, **non-technical requirements and capabilities** are worth considering when implementing innovative solutions. The PWS is highly dynamic: data and information, inbound and outbound, are subject to frequent changes and updates. In addition to this, different human agents need specific information and representations. **Organizational measures** that implement the official procedures might enhance and improve the public warning and response at a different scale, from national to the local levels and vice versa. The full potential of the PWS can be better exploited if ruled not only by emergency protocols but supported by a well-designed risk communication strategy that entails the design and the implementation of a decision tree that specifies what to communicate, when, how and to whom. The following table summarizes some organizational best practices for implementing a people-centred PWS process (Table 12.4).

Table 12.4 Recommended best practices for people-centred PWS design and maintenance.

<table>
<tr><th>Step</th><th>Actions</th></tr>
<tr><td>1</td><td>CREATE A DEDICATED TEAM FOR PUBLIC WARNING COMMUNICATION
It is crucial to dedicate human resources to warning communication design and management. The suggestion is to create, train and support a dedicated group of people in charge of designing and assessing the strategy and the operations of handling inbound and outbound information and alerts. The group's mission shall be to effectively convey and correctly disseminate information to the proper receivers through the most appropriate channels while verifying the impact of the communication strategy. It also aims to harmonize the messages from other official sources and work with local organizations to improve the dissemination and compliance with alerts through citizen engagement.

Such Public Warning Communication Team should include some specific profiles, such as:
• A crisis communication manager to lead and manage the information to be disseminated according to the communication strategy and to guarantee intra- and inter-organizational coordination and cooperation.
• User experience and graphic specialists to adapt the messages and their style to different targets and channels.
• Social media specialists to manage social networks exploiting several benefits. First, use them as an additional communication channel for citizens, organizations and authorities. Second, to collect data and extract meaningful information that could complement the understanding of the ongoing scenario or perform retrospective analysis.
• ICT engineers to implement the communication strategy and monitor the ICT infrastructure.
• A legal consultant to ensure liability and compliance with regulations.</td></tr>
<tr><td>2</td><td>CONTINUOUS COMMUNITY MAPPING
The official maps provided by the competent authorities that monitor the expected risks should be complemented with specific maps integrating additional data about the local communities. This integration will help the decision-making process in each DRR phase, which must be based on the actual social structure, including the accurate knowledge of human resources and frail targets on the territory. The community mapping shall monitor:
• Formal groups, e.g. associations and other emergency-related organizations active on the field.
• Local population meaningfully segmented, including impaired and frail people, e.g. people with disabilities, such as speech and hearing impaired, non-local language speaking, elderly and children.
• Trained staff in social and health services (hospitals, rest homes, schools of every level, etc.), who could be helpful to support during emergencies.
• Productive bodies, e.g. enterprises.
• Informal groups, e.g. support groups, leisure groups etc.
• Transitory people, e.g. visitors, usually don't know the areas and the specific risks.
• Commuting people, namely people that are daily commuters due to work or other reasons.
• Local media.
• Other groups.

For each of the above groups, the reference persons and their preferred communication channels (e.g. phone, email, social media) shall be collected, kept updated and used in the alerting strategy.</td></tr>
</table>

Table 12.4 (Continued)

Step	Actions
3	**DESIGN and KEEP UPDATED THE COMMUNICATION ASSETS** A good Risk Communication strategy has to exploit the multi-channel capabilities of the PWS infrastructure according to the technical standards and the audience characteristics. The definition of such a strategy has to be coordinated by the Public Warning Communication Team and involve different stakeholders, agencies, local communities and group leaders, who can help in the definition of: • **Multi-hazard scenario descriptions under different target audience** perspectives and the mapping with the systems and channels to be used in the identified cases. • **The main communication procedures** to be followed. After a consolidation of the above, communication experts and designers will then: • **Design the informative resources** needed to cover and serve all DRR phases, the different channels supported by the PWS, and the population segments, while being compliant with the standards. • **Iterative test of the resources** created with selected representatives of the targets and integrate the feedback collected to optimize the initial design. • Prepare a **dissemination plan** according to all scenarios and channels selected. • Prepare and implement **impact assessment tools** to refine the communication strategy based on a reality check.
4	**IMPLEMENT THE PEOPLE-CENTRED PWS** The implementation of new practices and systems must be assessed over time. Some recommended initiatives are: • To plan periodic training of the operators through practical exercises on PWS, user evaluation techniques and tools. • To organize education initiatives to raise risk awareness and train the population on prevention and response measures. These latter shall contribute to establishing effective cooperation with official authorities in case of incoming or ongoing crises. • To keep the communication channels alive and curated, publish good quality content. Besides social media, local TV and radio, it is still helpful to include offline channels (e.g. paper materials, screens) to reach less digitalized people. • To involve the public in two-way communication, e.g. by creating a mobile application that allows citizens to provide selected information and feedback. • Assess the timeliness, effectiveness and impact of the PWS and the risk communication strategy introduced.

Note

1. In the context of the standard, notification has the equivalent meaning of alert.

References

Against Disasters on Twitter (2016). https://mobile.twitter.com/againstdisaster (accessed June 2023).

Albrechtslund, A. (2008). Online social networking as participatory surveillance. *First Monday 13* (3). http://firstmonday.org/article/view/2142/1949https://doi.org/10.5210/fm.v13i3.2142.

Basher, R. (2006). Global early warning systems for natural hazards: systematic and people-centred. *Philosophical Transactions of the Royal Society 364*:2167–2182. https://doi.org/10.1098/rsta.2006.1819.

Bean, H.S. (2015). The study of mobile public warning messages: a research review and agenda. *Review of Communication 15* (1): https://doi.org/10.1080/15358593.2015.1014402.

Bray, T.P.-M. (2008). *Extensible Markup Language (XML) 1.0. W3C REC-XML-20081126*, 5e. W3C. https://www.w3.org/TR/xml (accessed June 2023).

CDC – US Centers for Disease Control and Prevention (2012). crisis emergency risk communication. https://emergency.cdc.gov/cerc/ (accessed June 2023).

COE – Council of Europe (2015). Major hazards and people with disabilities. A Toolkit for good practices.https://www.coe.int/en/web/europarisks/major-hazards-and-people-with-disabilities.-toolkit-for-good-practices (accessed June 2023).

Crisis Mapping (2017). http://crisismapping.ning.com/ (accessed June 2023).

De Longueville, B., Luraschi, G., Smits, P. et al. (2010). Citizens as sensors for natural hazards: a VGI integration workflow. *Geomatica 64* (1): 41–59.

Dransch, D.R. (2010). The contribution of maps to the challenges of risk communication to the public. *International Journal of Digital Earth* 3:300. https://doi.org/10.1080/17538941003774668.

EENA – European Emergency Number Association (2015). *Public Warning*. EENA Operations Document.

ETSI (2010). TS 102 182. Emergency Communications (EMTEL); Requirements for communications from authorities/organisations to individuals, groups or the general public during emergencies.

ETSI (2012). TS 102 900, European Telecommunications Standards Institute. Emergency Communications (EMTEL); European Public Warning System (EU-ALERT) using the Cell Broadcast Service. v1.2.1.

ETSI (2018). TETRA. https://www.etsi.org/technologies/tetra: https://www.etsi.org/technologies-clusters/technologies/tetra (accessed October 2018).

Goodchild, M. (2007). Citizens as voluntary sensors: spatial data infrastructure in the world of web 2.0. *International Journal of Spatial Data Infrastructures Research 2* (2): 24–32.

Google Crisis-Response (2023). https://crisisresponse.google/ (accessed June 2023).

Grunig, J.E. (2005). Situational theory of publics. In: *Encyclopaedia of Public Relations*, vol. 2 (ed. R.L. Heath), 778–780. University of Houston https://doi.org/10.4135/9781412952545.

Guo, B., Yu, Z., Zhou, X., and Zhang, D. (2014). From participatory sensing to mobile crowd sensing. *IEEE International Conference on Pervasive Computing and Communications Workshops (PERCOM Workshops)*., Budapest, Hungary (24–28 March 2014), pp. 593–598. IEEE. https://doi.org/10.1109/PerComW.2014.6815273.

van der Ham, I., Claessen, M., Evers, A., and van der Kuil, M. (2020). Large-scale assessment of human navigation ability across the lifespan. *Scientific Reports* 10 (1): 3299. https://doi.org/10.1038/s41598-020-60302-0. PMID: 32094394.

Houston, J., Hawthorne, J., Perreault, M., Park, E., Goldstein Hode M, Halliwell, M.Griffith, S. (2014). Social media and disasters: a functional framework for social media use in disaster planning, response and research. *Disasters*(39), 1–22. doi:https://doi.org/10.1111/disa.12092

IEFT (2001). Tags for the identification of languages. https://www.rfc-editor.org/rfc/rfc3066 (accessed October 2018).

Io non rischio (2022). https://iononrischio.protezionecivile.it/it/ (accessed November 2022).

I-React-project (2017). https://cordis.europa.eu/project/id/700256/it (accessed November 2021).

I-See-change (2018). https://www.iseechange.org/ (accessed November 2021).

ISO 22322 (2015). Societal security – emergency management – guidelines for public warning.

ISO 8601 (2004). Data elements and interchange formats – information interchange – representation of dates and times.

ITU – International Telecommunication Union Recommendation (2014). X1303bis. Secure applications and services – emergency communications, common alerting protocol (CAP 1.2). https://www.itu.int/rec/T-REC-X.1303bis-201403-I/en (accessed October 2018).

Kasperson, R. (2014). Four questions for risk communication. *Journal of Risk Research* 10: 1233–1239. https://doi.org/10.1080/13669877.2014.900207.

Lundgren, R. and McMakin, A. (2018). *Risk Communication: A Handbook for Communicating Environmental, Safety, and Health Risks*. Oxford: Wiley.

Mayhorn, C.B. and McLaughlin, A.C. (2014). Warning the world of extreme events: a global perspective on risk communication for natural and technological disaster. *Safety Science* 61: 43–50. https://doi.org/10.1016/j.ssci.2012.04.014.

Mayhorn, C.B., Yim, M., and Orrock, J. (2006). Warnings and hazard communications for natural and technological disasters. In: *Handbook of Warnings* (ed. I.M.S. Wogalter), 763–769. Lawrence Erlbaum Associates Publishers.

Meier, P. (2013). Human computation for disaster response. In: *Handbook of Human Computation* (ed. P. Michelucci), 95–104. New York: Springer.

Mobile4D (2012). Crowdsourced humanitarian aid systems. http://www.capacitylab.org/ (accessed October 2019).

Nguyen, Q., Frisiello, A., and Rossi, C. (2019). The design of a mobile application for crowdsourcing in disaster risk reduction. In: *Proceedings of the 16th International Conference on Information Systems for Crisis Response and Management (ISCRAM)* (ed. Z. Franco, J.J. González, and J.H. Canós). Spain: València https://doi.org/10.1145/3152896.3152898.

NIST – US National Institute of Standards and Technology (2011). *The NIST Definition of Cloud*, 800–145. Special Publication. https://nvlpubs.nist.gov/nistpubs/legacy/sp/nistspecialpublication800-145.pdf.

O.A.S.I.S., S. (2010). Standard O.A.S.I.S. Common alerting protocol. Version 1.2. http://docs.oasis-open.org/emergency/cap/v1.2/CAP-v1.2.pdf (accessed October 2019).

Park, C. and Johnston, E. (2015). Crowdsourced, voluntary collective action in disasters. Crowdsourced, voluntary collective action in disasters. *16th Annual International Digital Government Research Conference*, Phoenix Arizona (27–30 May 2015), pp. 329–330. https://doi.org/10.1145/2757401.2757458.

Párraga Niebla, C., Mulero Chaves, J., and De Cola, T. (2016). Design aspects in multi-channel public warning systems. Wireless public safety networks 2. A systematic approach. In: *Wireless Public Safety Networks 2. A Systematic Approach* (ed. D. Câmara and N. Nikaein), 227–261. Elsevier.

Pennacchiotti, M. and Popescu, A. (2013). A machine learning approach to Twitter user classification. *Fifth International AAAI Conference on Weblogs and Social Media*, Barcelona, Catalonia, Spain (17–21, 2011), pp. 281–288.

Rogers, D. and Tsirkunov, V. (2010). Costs and benefits of early warning systems. Global assessment report on disaster risk reduction. GAR 2011. UN-ISDR and the World Bank. https://www.preventionweb.net/english/hyogo/gar/2011/en/bgdocs/Rogers_&_Tsirkunov_2011.pdf (accessed June 2023).

Schulze, K., Lorenz, D., Wenzel, B., and Voss, M. (2015). Disaster myths and their relevance for warning systems. *Proceedings of the 12th International Conference on Information Systems for Crisis Response and Management (ISCRAM),* Kristiansand, Norway (24–27 May 2015).

Sufri, S., Dwirahmadi, F., and Rutherford, S. (2020). A systematic review of community engagement (CE) in disaster early warning systems (EWSs). *Progress in Disaster Science* 5: https://doi.org/10.1016/j.pdisas.2019.100058.

Tallacchini, M., Boucher, P., and Nascimento, S. (2014). *Emerging ICT for Citizens' Veillance. Theoretical and Practical Insights*. JRC90334. European Union https://doi.org/10.2788/11828.

UNDP (2016). Climate information & early warning systems communications toolkit. https://www.undp.org/publications/climate-information-and-early-warning-systems-communications-toolkit (accessed June 2023).

UNDRR (2015). Sendai framework for disaster risk reduction 2015–2030. *UN World Conference on Disaster Risk Reduction*, Sendai, Japan (14–18 March 2015). Geneva: United Nations Office for Disaster Risk Reduction. http://www.wcdrr.org/uploads/Sendai_Framework_for_Disaster.

UNDRR (2021). *Disaster Risk Reduction to Achieve the Sustainable Development Goals. A Toolkit for Parliamentarians*. Geneva: United Nations Office for Disaster Risk Reduction Inter-Parliamentary Union. https://www.undrr.org/publication/disaster-risk-reduction-achieve-sustainable-development-toolkit-parliamentarians.

Ushuahidi (2017). https://www.ushahidi.com (accessed June 2023).

W3C – World Wide Web Consortium (2002). The W3C document object model (DOM). https://dom.spec.whatwg.org/ (accessed June 2023).

We Sense It (2016). We sense it. https://cordis.europa.eu/project/id/308429 (accessed June 2023).

Whittaker, J., McLennan, B., and Handmer, J. (2015). A review of informal volunteerism in emergencies and disasters: definition, opportunities and challenges. *International Journal of Disaster Risk Reduction* 13: 358–368.

WMO – World Meteorological Organization (2019). Multi-hazard early warning system (MHEWS), a tool for effective ocean prediction and services. https://public.wmo.int/en/resources/bulletin/Products_and_services/MHEWS_for_Ocean_Prediction (accessed June 2023).

13

A DRM Solution for Professionals and Citizens

Claudio Rossi, Antonella Frisiello, Gianluca Marucco, and Marco Pini
LINKS Foundation, Turin, Italy

13.1 A Novel Mobile Application for DRR

Nowadays, ICT and mobile technologies are multiplying the possibilities to interact among different actors, enabling richer information exchange and novel forms of collaboration. If properly addressed, the digital engagement of citizens can contribute to overcoming the lack of information before, during and after a disaster (Foresti et al. 2015; Witten et al. 2016). Recent environmental disasters have shown that besides the national and local governments and the field organizations, communities and individuals actively participate in the response, taking initiatives, regardless of their distance from the interested area. In any emergency, most people actively seek and share information to establish and stay in contact. Mobile technologies allow for fulfilling this need. For instance, sharing real-time and geolocalized information related to affected areas can enhance communication, increase the effectiveness of warnings and advice and facilitate requests for help and assistance (Schulze et al. 2015). In every DRR phase (DRMKC 2017), different ICT systems and services are linked and work as an infrastructure for specific strategies and procedures. Among them, tools and services dedicated to the population will increase: collaborative platforms propose solutions supporting cooperative microtasks as part of distributed processes, in which citizens play as active local agents at the service of the DRR. Crowdsourcing – as defined in

Responding to Extreme Weather Events, First Edition. Edited by Daniel Sempere-Torres, Anastasios Karakostas, Claudio Rossi, and Philippe Quevauviller.

Chapter 12 – is one of the most common socio-technical paradigms where technology supports individuals and communities to perform volunteer tasks (i.e. provide information, collect economic resources, rate services etc.) for a common goal (Goodchild 2007). Thanks to mobile devices and networks and services, and applications used as sensors, people can gather and exchange geo-located information (text, videos, images) about events and act as digital volunteers in environmental hazard monitoring (Starbird and Palen 2011). Crowdsourcing is promising for the DRR due to the capability to bring local knowledge from the territory and citizens to the decision-makers and back to the field agents. All the actors involved can benefit from a richer situational awareness in all phases of the disaster cycle.

Nevertheless, these potential benefits imply some constraints: crowdsourcing needs large and distributed cooperating crowds to be effective and accurate. The more significant the voluntary contribution, the more reliable and useful the data collection is. In addition, the effort to make crowdsourced information validated and suitable for the existing procedures must be considered: the volunteer contribution must be integrated with a rule-based process, including solid and redundant validation mechanisms (Guo et al. 2014). To this purpose, automatic data filtering techniques based on machine learning are advisable to manage the high amount and variety of data collected and guarantee proper levels of accuracy, preventing the risks related to an unregulated stream of information (Narvaez 2012). The automated validation techniques can be further integrated by manual validation methods through which peers and/or professionals can refine and share criteria and information. Finally, to exploit the crowdsourcing capabilities, good design solutions for data visualization and interaction are fundamental to provide actionable hints enhancing the professionals' decision-making (Chen et al. 2021). These issues must be considered when designing and introducing crowdsourcing solutions in safety-critical processes like DRR.

13.2 The I-REACT Co-Design Approach

Innovation can be designed and developed following different approaches. A technology-driven approach suggests starting from the available features to move towards technological advancement that would overcome state-of-the-art or meet market needs. A complementary perspective is design-oriented, leveraging experts' knowledge and skills to find useful, usable and acceptable solutions.

In this vein, the Human Centred Design perspective focuses on the target people to assume their needs, goals, real context and activities as a reference (Abras et al. 2004). In this view, considering innovation technology to be integrated with real-life contexts like tools is limiting. Tools radically change the way we behave and work. According to this perspective, technological innovation cannot be designed and implemented without taking into account the individuals who are, in contrast, considered part of the system to be defined. One method used to involve end-users and relevant stakeholders in the design process directly is co-design, a participatory method that aims to create a collaborative path from the user needs elicitation and problem-setting to the definition of desired solutions. The output takes birth from and is negotiated among many heterogeneous stakeholders (citizens, professionals, representatives of communities, institutions, experts etc.). The co-design concretizes the claim 'Building With, Not For' people the solution is intended for (Stepanek 2015). It happens by stimulating and collecting local knowledge from multiple points of view and providing people with expressive techniques to shape and refine their sustainable and effective solutions. More specifically, the co-design considers and empowers the involved people to think and work as designers in a collaborative process (Forlano and Mathew 2017). The expected output of co-design activities is a knowledge base including the description of the target users and their real context, activities, existing systems and procedures, gaps to be filled and expected solutions. These assets can be collected in different formats like stories, data, cultural probes and low-fidelity prototypes (pen and paper, collage, or simulated mock-up). Whatever the amount and format of the user-generated data, the empirical base is the starting point and will be the reference of the system design and development process. The elaboration of data will result in functional requirements, system architecture, information architecture and interaction specification; the evolving specifications and releases of the system dynamically will bring to the final solution.

A design-oriented approach implies the elaboration of ideas, requirements and recommendations into actionable ideas, enabling an iterative test phase that is fundamental to early detect possible improvements and problems, so increasing the possibility of releasing efficient, effective and satisfactory solutions. These dimensions are at the core of an optimal User Experience (UX), the whole of the user's perceptions resulting from interacting with a system/service. The UX in the DRR domain plays a role in facilitating the acceptance and adoption of new solutions, if able to support actual processes and tasks with usable solutions. It includes the expectations, the use and memorability (Franzreb and Franzreb 2016).

13.2.1 The Co-Design Process in the I-REACT Project

The participatory design approach was adopted throughout the I-REACT project from the beginning to define the early concept and collect the requirements of the systems. The I-REACT approach was defined and implemented as a process consisting of three main phases, where – from the early conceptual design to the validation of the system – participatory activities interested both the project consortium and external stakeholders(Figure 13.1).

User research is a fundamental design activity that aims at identifying and better knowing the system's stakeholders and end users in terms of needs, expectations and skills. Qualitative and quantitative research techniques profitably serve this aim. The I-REACT user research consisted of two surveys designed for and distributed to different stakeholders. The first survey, addressing the project stakeholders, intended to map the basic operational needs related to the hazards focused on by the project, the current risk management practices and tools, the interest in crowdsourcing, social media and advanced cyber technologies as integrative tools in the DRR processes. The second survey aimed to investigate the population's environmental awareness and attitude

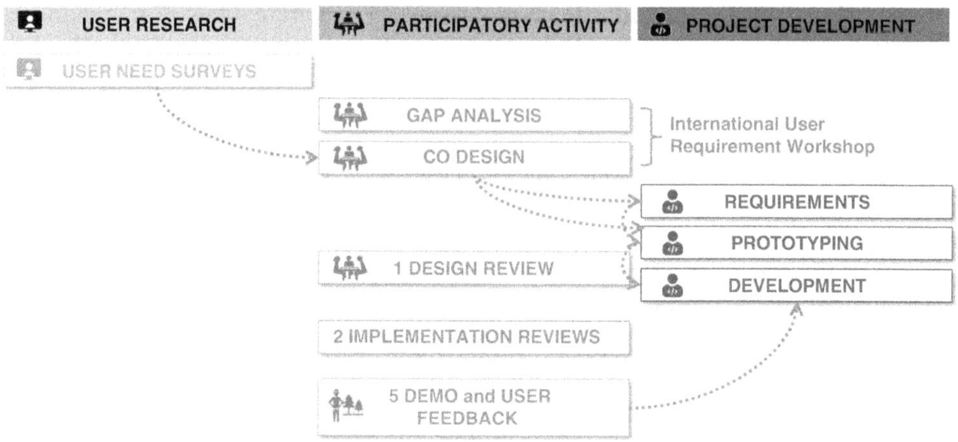

Figure 13.1 I-REACT co-design strategy, from user research to conceptual design and validation of a DRR solution. *Source*: I-REACT project.

towards digital collaborative practices and crowdsourcing tools. The surveys provided the stakeholders' overview and an early set of needs useful to plan the following steps.

Participatory activities are at the core of the co-design approach. In I-REACT an ad-hoc methodology in four steps was defined. For each step, specific stimuli (including documentation, infographics, . . .) and design tools were defined to allow participants to work together as designers on common goals and share their best practices, needs and proposals. These materials are precious for designers since they work at the same time as systematic and structured data collection tools, enabling further and deeper analysis and supporting the active engagement of people in the discussions and collaborative phases.

The I-REACT co-design workshop phases are presented in the following.

- *Phase 1*: **Introduction and initial needs.** The goal of the early step is to welcome participants and introduce the activities. In the specific case of I-REACT, two questions were posed as a file-rouge of the workshop: (i) how to enhance the emergency process with real-time information gathered from the field? (ii) how to engage citizens as active players able to provide relevant, reliable and actionable data to be used by crisis managers and operators? After that, this phase is very important to provide time for participants to introduce themselves to the plenary audience. The introduction has to focus on their activities, needs and expectations related to the DRR (Figure 13.2, box 1). A dedicated template to be provided before the workshop helps participants to keep their time and focus.
- *Phase 2*: **Data Scouting.** The goal of this phase is to make an inventory of the data, parameters, information and communications that are useful in the current DRR cycle. Participants divided into small groups are asked to freely list all the data used in DRR and sort them per type of hazard (Spencer 2009). A second sorting iteration is required to associate the generated items with the specific DRR phases and to specify the involved users (in this case: citizens and/or professional users). In this

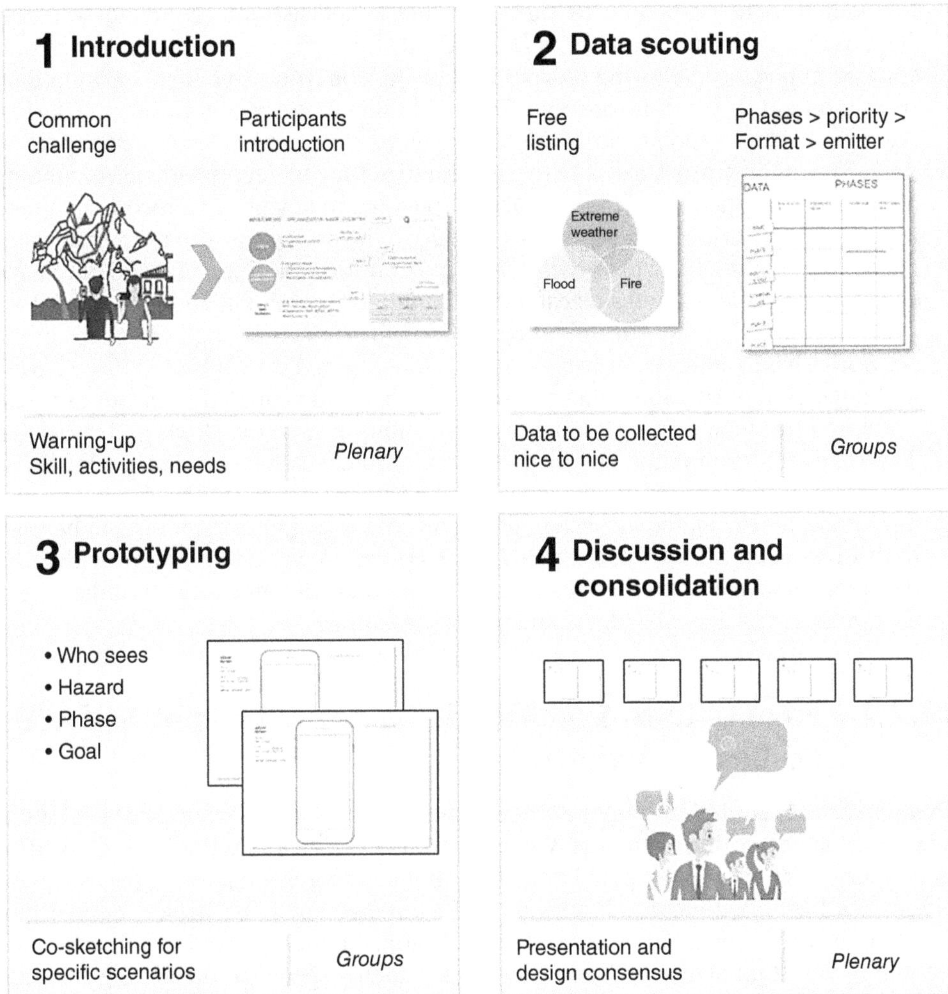

Figure 13.2 I-REACT Co-design workshop overview.

phase, it is not needed nor required to specify data properties such as source, format, acquisition media, accuracy etc. Properties that will be deepened afterwards. To support this phase, ad hoc canvases such as Venn diagrams and tables (Figure 13.2, box 2) are strongly recommended to facilitate the generation and sorting of data.

- *Phase 3*: **Prototyping**. After having focused and prioritized the data serving the DRR process, the participants are called to envisage and concretize a conceptual model representing how they expect to use (receive, produce, exchange) those data. Prototyping is a core step in a design process: it allows to put concepts and ideas into practice and visualize them – even roughly – to stimulate further refinement, possible evolutions, as well as limitations and constraints (Warfel 2009). To enable participants to visually express their ideas in a low-fidelity prototype, providing ad hoc scenarios helps people to turn their high expertise into concrete solutions (Figure 13.2, box 3) (Carroll 1999). The I-REACT design goal was to ideate a crowdsourced application allowing citizens and professionals to exchange accurate data from the field to the control room. To enable people to create a user interface, a paper wireframing toolkit with mobile stencils was designed to facilitate the participants to work in groups, co-sketching ideas and getting to negotiated ideas, entailing all different perspectives. It is very useful to assign to any group a recorder to take notes about the criteria and decisions that drive the project, as well as the criticisms explaining solutions and proposals. This type of information shall be added to each sketch to be validated and suggest implementation recommendations.

- *Phase 4*: **Discussion and consolidation.** The visual artefacts created by groups in the previous phase (collages, drawings, notes) are showcased to stimulate the sharing and discussion during the plenary session. Each group is invited to present its proposal, and the audience to add more specifications or raise questions and criticisms. This is an important phase in which designers and developers can learn a lot about the features and behaviours expected by the future system. At this point, the plenary validates the viable ideas, functions and solutions. Different tools can be used in this phase: polls tool, form. In I-REACT co-design workshops, we provided the participants with coloured adhesive dots to put onto the mock-ups to indicate the user interface preferred widgets and priority solutions.

13.2.2 From Data to Specifications: The Results of I-REACT Co-Design Activities

This presented co-design methodology was put into practice in the I-REACT International User Requirements Workshop (IURW), held in Paris, at UNESCO Headquarters, where a large panel of 52 stakeholders representing emergency organizations (end users), domain advisors and ICT experts, including project consortium members, has been involved in a two-day workshop, that has allowed to collect a huge amount of data, that elaborated, provided requirements, specification and high-fidelity mock-up of the whole system. Twofold the goal proposed to the engaged stakeholders: co-design a mobile solution (i) to empower operators with real-time information gathered from the field and (ii) to engage citizens as additional informative resources of the DRR process.

The User Research: Each I-REACT consortium member was invited to formulate closed questions and reply options to the project stakeholders. In total, 166 questions on several topics were expressed to investigate the user and technical requirements. Clusterized in topics, the resulting survey focused on informative needs related to the most impactful hazards (flood, fire, extreme weather conditions), risk management practices and tools to be enhanced, crowdsourcing opportunities, the use of social media and advanced cyber-technologies in the different phases of DRR. Launched online, the survey collected replies from 11 different organisations in Europe, in particular: crisis managers (46%), policymakers (18%), technical experts (18%) and in-field agents (9%). Results provided an early knowledge base to set up the following design activities. A positive attitude toward the crowdsourcing approach as an additional capability to receive data from the field was recorded. In particular, the responders identified Preparedness as the phase that most would benefit from the crowdsourcing, motivated by the high potential of implementing a distributed human-based monitoring network exploiting the ubiquity of connected mobile devices to fetch geo-localized multimedia content. Nowadays, citizens are already involved in DRR, mainly as volunteers who, together with professional responders, are the more critical profiles in an emergency scenario. The volunteer is the role that would best facilitate the crowdsourcing introduction in the DRR processes. Thanks to their training, volunteers are considered a reliable and skilled source of information from the field. They would be engaged to detect hazard-specific parameters (such as water level, water speed, and water volume) and weather measures (such as air temperature, wind speed, rain, and hail). On the other side, the sample appeared much more undecided concerning the importance of involving 'common citizens' in this reporting.

The Participatory Activity: To move from the project challenge and the stakeholders' needs, toward a concept of solution, an international design workshop was held at the UNESCO Headquarters in Paris. The described methodology was implemented in two full days of workshops. Emergency responders, domain experts, and end-user representatives were selected to interplay with the I-REACT system designers and developers. In total, 52 participants contributed: the 23 stakeholders, including 11 emergency organisations and 12 advisors, plus 29 members from the project consortium, including eight facilitators.

- *Phase 1*: **Introduction and initial needs**. According to the methodology previously described, the participants received an overview of the I-REACT project before starting with the first phase to set up a common ground of interest. After that, every stakeholder gave a presentation (prepared on the provided template), focusing on their activities, role, key competencies, stakeholders/end-users, and the main expectations from the I-REACT project. The Data scouting step was carried out by five previously defined five groups, including both project members and stakeholders. The grouping also considered expertise and geographic proximity to minimize cultural and language-based barriers and possible similarities in procedures or regulations. One recorder per group was appointed to take notes about the key discussed points, including divergences and agreements. Photos of the work in progress allowed us to rebuild the path of each group.

The analysis of the material collected in this phase allowed to focus on the advancements expected by the research – and in particular by the project, covering a broad spectrum, including space assets, climate change, standardisation, prevention, risk, mitigation, security, training, environment and information. Maps and Copernicus EMS are frequently among the most used tools. Real-time reporting is confirmed to be the most relevant requirement for end-users, followed by the availability of dynamic risk maps, which is crucial for the capability to track the risk evolution in the early warning phase and better manage the operations in the response and post-event phases. Greater anticipation of early warnings and social media content synthesis were finally ranked

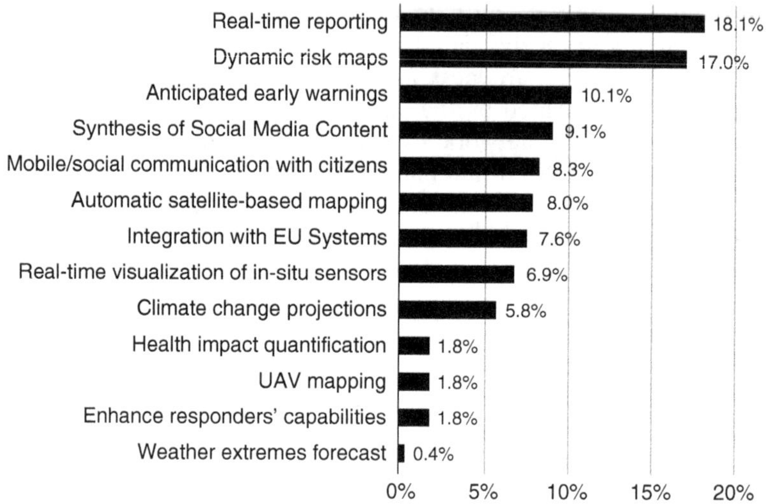

Figure 13.3 Initial stakeholders' needs and wants.

as interesting. Advisors highlighted the importance of integrating information from different sources, such as satellites and sensors. A mobile app seemed the most effective way to communicate with citizens and highlighted the importance of integrating local solutions with European systems. The aggregated stakeholders' needs are ranked in Figure 13.3.

- *Phase 2*: **Data Scouting.** In this phase, it is relevant to reserve an introductory activity to set a common ground. Moreover, the provided tools facilitated high productivity in generating lists of data, then collaboratively selected and described in relation to the scenario elements (hazards, DRR phase and actors' roles). Afterwards, the collected information was furtherly analysed and organized into two main clusters:
- WHO, concerning the information related to the field reporters. This analysis unexpectedly confirmed that while decision-makers do not require the personal identity of the in-field agent in the control room, the user role (e.g. citizen, first responder, organization) and their health status are the core elements to estimate the trustworthiness of the received information.
- WHAT, regarding the information related to the specific scenario. Here are included: data useful to precisely identify hazards (e.g. extreme weather, fire, flood), damages (e.g. a collapsed building), available resources (e.g. nearest safe place), and hazard-related measures (e.g. direction of wind and speed, water level). These latter emerge as the most requested data to support the DRR operations.

Almost half (45%) were marked as mandatory or essential for all data named and processed in this phase. Participants declared that citizens could provide one-third of this base (Figure 13.4). This result is promising to start designing an active role of 'common' citizens in providing critical data to emergency management. On the other side,

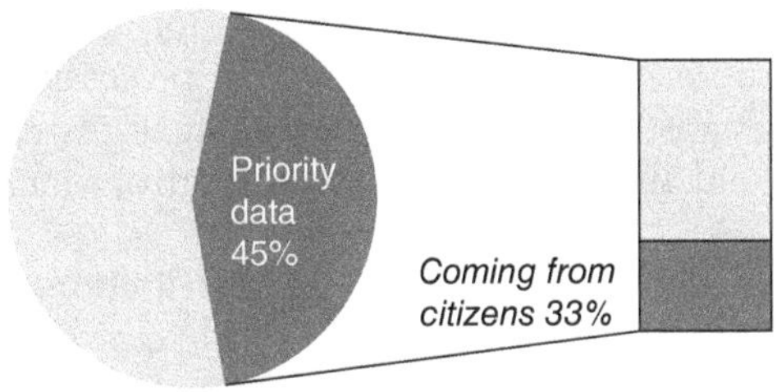

Figure 13.4 Acceptable crowdsourced data (by priority).

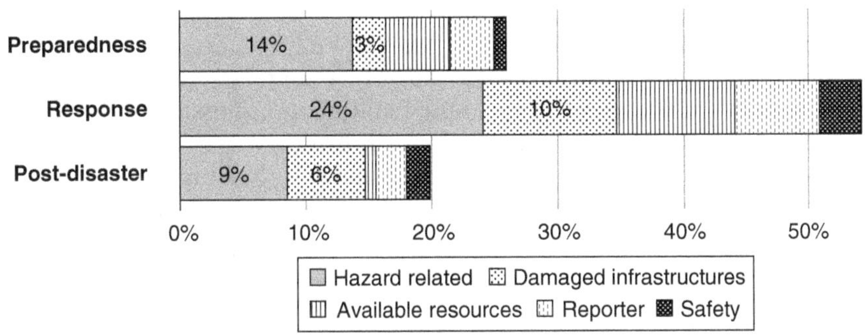

Figure 13.5 Informative needs per DRR phase.

as expected, first responders are considered the most trustworthy source, especially when collecting specific hazard-specific measures that might require training.

According to the stakeholders' panel, the Response is the one that can profit more from crowdsourced data, followed by the Preparedness and finally by the post-disaster phase, confirming the outcomes of the initial survey. Generally, the most requested data type is hazard-related, while other data types vary according to the emergency management phases (Figure 13.5).

- *Phase 3*: **User-Generated Prototypes**. Furtherly divided into 10 smaller groups (including at least one end-user, one expert, and one consortium member inside each group). Each one received a specific scenario, specifying the hazard, end-user role (professional or citizen) and phase of the DRR cycle. The micro-scenarios worked as a design brief for the groups, asked to get to a paper mock-up, concretizing the information from the previous activities into a mobile solution. The realized mock-ups were considered as a particular type of content. We isolated and separately analysed all the UI patterns generated, as suggested by Constantine and Lockwood, (1999). The visual analysis highlighted commonalities and differences in outlined functionalities, inserted data, layout and structure of the screen, and labelling. The first distinction relevant in terms of design is between:
- An informative level is expected to be provided by a centralized system and displayed via an app, on the mobile devices of the in-field agents. Such information is

generated by the authorities managing the disaster and includes warnings, maps (hazard, risk, delineation, forecasts), and tasks for professional users;

- A generative level, enabled by the crowdsourcing features, refers to the information gathered from the field and to be routed to the control room.

The visual analysis of the sketches shows the distribution of informative features expected by the system (green areas in Figure 13.6) and crowdsourcing features (dotted line in Figure 13.6). Generally, we can observe that the informative level is always placed in the upper part of the screen, while the crowdsourcing functionalities are put in the lower one. Apart from one singular sketch, a large section of the screen, from ½ up to ⅗, is expected to be dedicated to content generation (reports from the field).

- *Phase 4*: **Discussion and consolidation.** After two days of intensive design activities, the final discussion held in plenary confirmed the great stakeholders' interest in adopting new technologies to enable the real-time transmission of data (including mission-critical information from and towards the field) for risk reduction purposes. A mobile app is considered a highly reliable and easy-to-adopt solution because of its universal use by the potential target and in consideration of the capabilities to create geo-localized multimedia content. At the same time, the stakeholders warn about the hindrance of adding a new tool to manage, which might affect the final efficacy of usage both at individual and at the organisational level. Introducing new technologies requires training efforts addressing the staff and the public. This is considered mandatory to successfully implement collaborative paradigms like crowdsourcing. On the one hand, mobile applications allow the collection of a large amount of valuable information by citizens and first responders. On the other hand, the receivers are forced to implement some validation mechanisms, which can require resources in terms of time and staff. This point highlights the need for automatic validation techniques based on intelligent algorithms, able to reduce the amount of data to be manually validated.

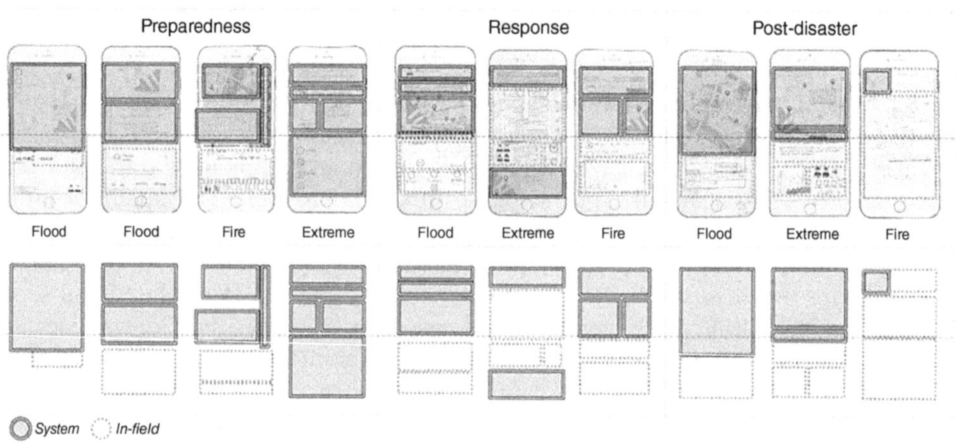

Figure 13.6 Visual analysis of the collaborative sketches.

The co-design workshop was the early stage of a participatory design approach that drove the further mobile solution implementation, described in the following section.

13.3 The Development and Implementation of the I-REACT Mobile Solution

In this section, we describe the design of the I-REACT app, describing it as a walk-through of the main functionalities and UI details.

Access: I-REACT is a mobile app accessible to registered users. After the user logs in, they land on the home screen that displays the app's main sections, made accessible through a grid of big tiles (Figure 13.7). The most important commands concerning the Reports are placed in the bottom part of the screen to make them usable even when handling the phone with one hand. The calls-to-action drive the user to two different reporting modes: on-demand (reports asked by the control room) and spontaneous (respective flows are described in the following figure) (Figure 13.7).

Map-based visualization: the app provides location-based content that can be read on the map and in a list view. The two visualizations are complementary since the map provides a geo-referenced overview while the list provides more details about every event. A further level of detail is also provided: by selecting a specific item of the list or pin on the map, an overlay window with more information is displayed. By default, the content and the map are set on the user's area of interest, set in their settings. In addition, information can be sorted for a given temporal interval, selected via a

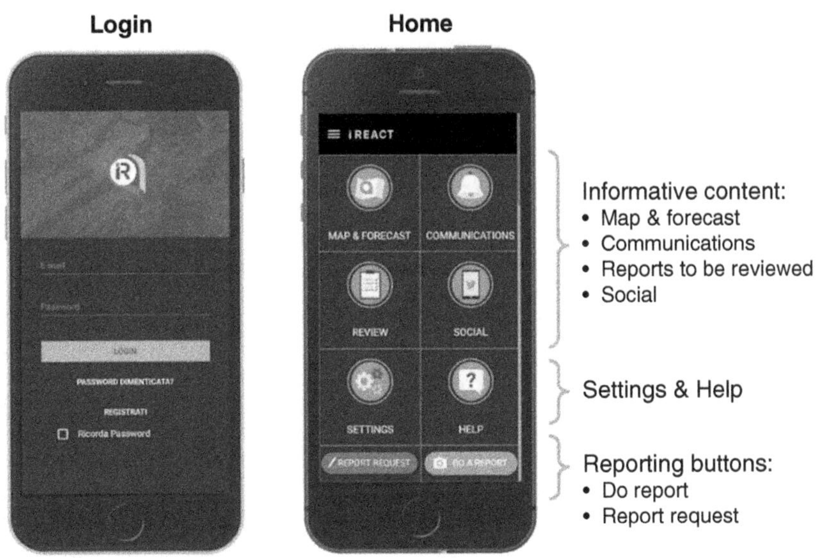

Figure 13.7 Log in (left) and Home (right) of the app.

dedicated widget. Every content managed via the app (both incoming and outgoing information) is geo-localized and includes either a timestamp or a validity period. On the map, specific and relevant commands are displayed (besides the typical map controls), to allow the users to start the report-creation procedure from their current position (Figure 13.8). The data visualisation is designed to prevent information overload: visual solutions such as coloured bubbles refer to clusters of geolocalized events. Moreover, the map focuses by default on data available for the user's Area of Interest (dotted line) within the selected time window (Figure 13.8).

Layers. A set of layers enriches map-based data visualisation, including meteorological forecasts. When a layer is selected, the data are plotted on the map, showing the dynamic time evolution (Figure 13.9). Moreover, the metadata layer is provided. A dedicated legend facilitates understanding chromatic coding.

gi: the app enables the exchange of specific types of bidirectional messages: (i) those from the authority to the field, such as Warnings (Alerts), weather bulletins from regional/national authorities and Report Requests. In particular, Warnings implement the Common Alerting Protocol (CAP) (O.A.S.I.S. 2010), which is a standard format enabling the interoperability and the integration of alerts coming from different sources (Figure 13.10 – left). Concerning the report requests, they are sent to the in-field agents according to their location and the Area of Interest specified by the sender (Figure 13.10 – right).

Social Media: a specific section is dedicated to displaying hazard-related Tweets. The social media content is personalized based on the user selection (Figure 13.11) by an algorithm that extracts and classifies hazard-related tweets (Rossi et al. 2018).

Figure 13.8 Map & Forecast view: location-based contents visualized on the map together with map controls.

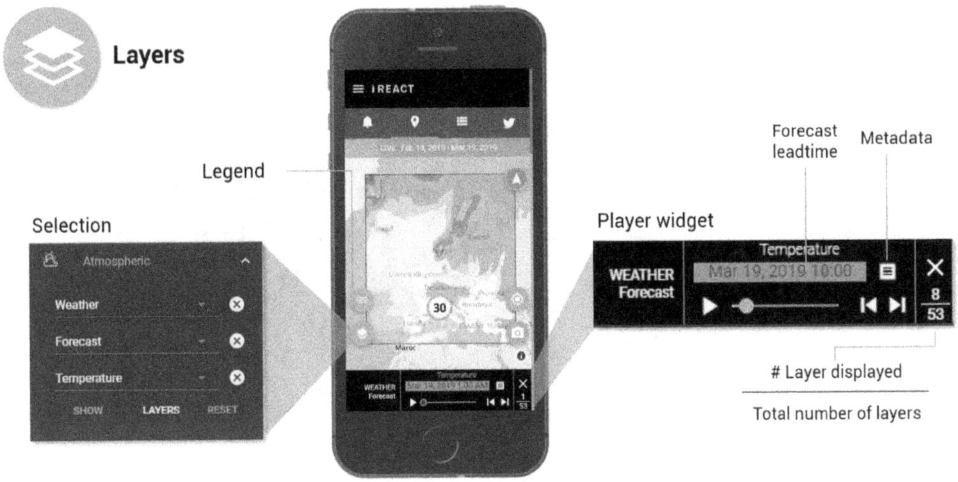

Figure 13.9 Map view of a forecast layer (temperature). Layer selection (left), map view (centre), player widget (right).

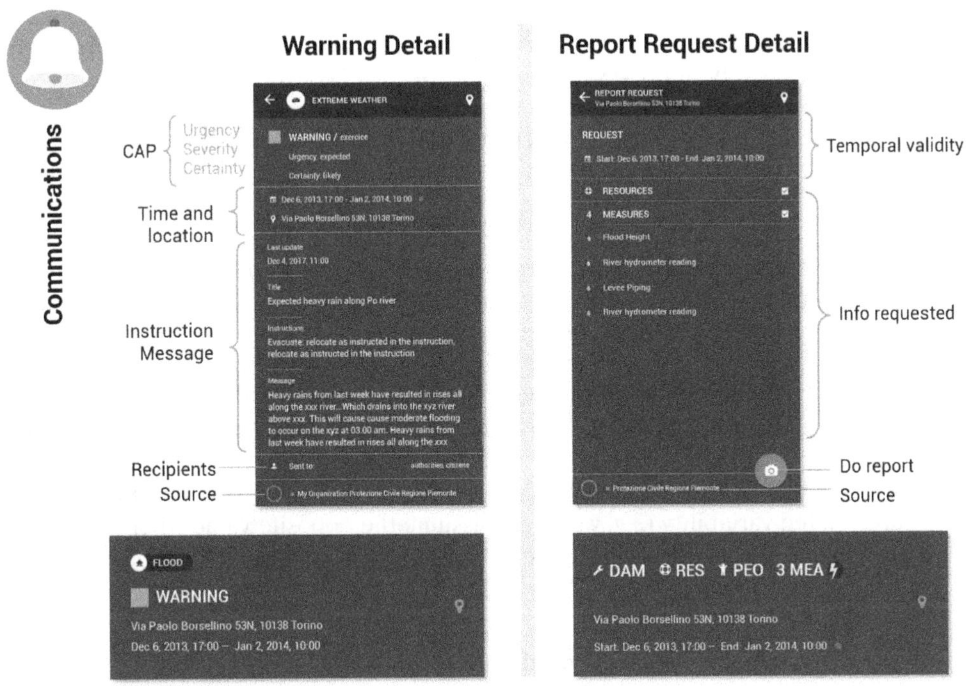

Figure 13.10 'Warning' (left) and 'Report Request' (right): card (bottom) and Detail (top) elements.

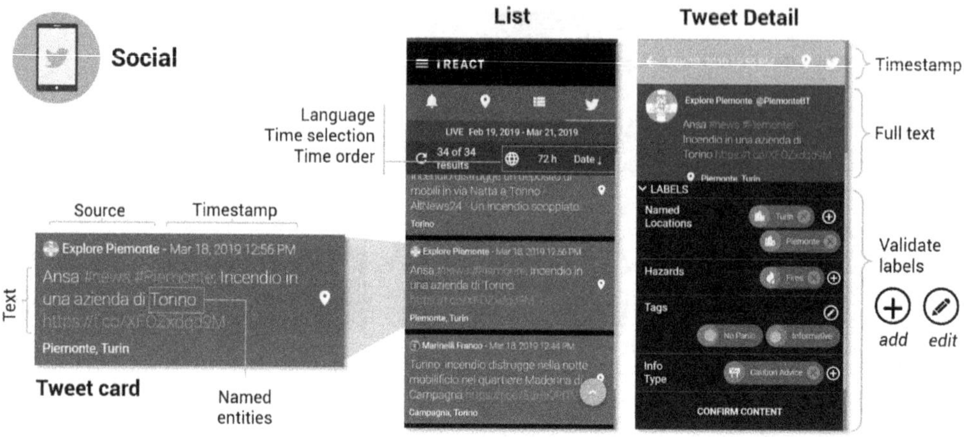

Figure 13.11 Social media information: tweet card (left), List (centre) and Detail (right).

Quick filters allow selecting language, time range, and time sorting. Each Tweet is also classified by automatic labels that may be supervised (confirmed or amended). Categories that may be associated are locations, hazards, information type (Affected Individuals, Caution and Advice, Donations, Emotional Support, Infrastructures, Volunteering), Informativeness (yes /no values), and Panic (yes/no values). Over the citizens' validation, professionals can also comment on posts. Their feedback is valid as validation, visible to all users. Over supporting user engagement, the collaborative validation iteratively improves the accuracy of the classification algorithm, making the system even more accurate.

Report creation and validation: reports enable data collection from the field. The report creation is a guided procedure in five steps: (i) Picture capturing (photos were indicated by responders organizations as the core format to receive to understand the situation from the field); (ii) Geo-localization; (iii) Hazard description; (iv) Information category; (v) Information details (Figure 13.12). The structured sequence replies to the need for pre-categorized and systematic data collection and minimizing typing errors. Each report is geo-localized on the device position (GPS) or manually selected from the map. This option allows the reporter to eventually keep a safety distance from hazardous points while creating the Report. A Report can include quantitative and qualitative measures, provided as options, depending on the hazard type. Reports consist of standard scales (e.g. Beaufort scale for wind velocity) and indications on damages, resources or involved people. Due to great hazard-related variability (e.g. water can assume the opposite value whether there is a flood or a fire), and missing a unique standard, ad hoc scales have been created after benchmarking scales applied by different emergency organizations.

The report validation works at two different levels: by accessing the Report detail, citizens can peer-review other citizen reports by upvoting or downvoting them, while professionals can validate or reject citizens' reports (Figure 13.13).

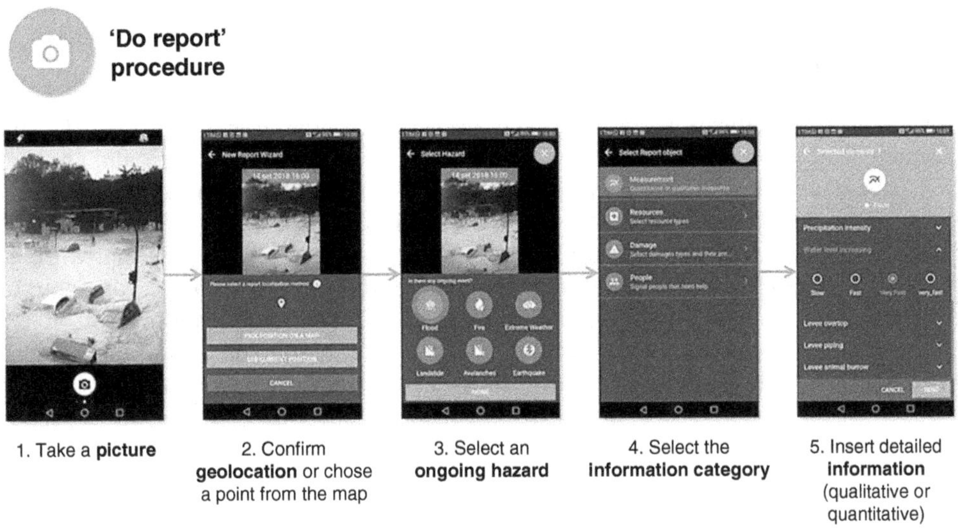

Figure 13.12 Report creation procedure made of five different steps.

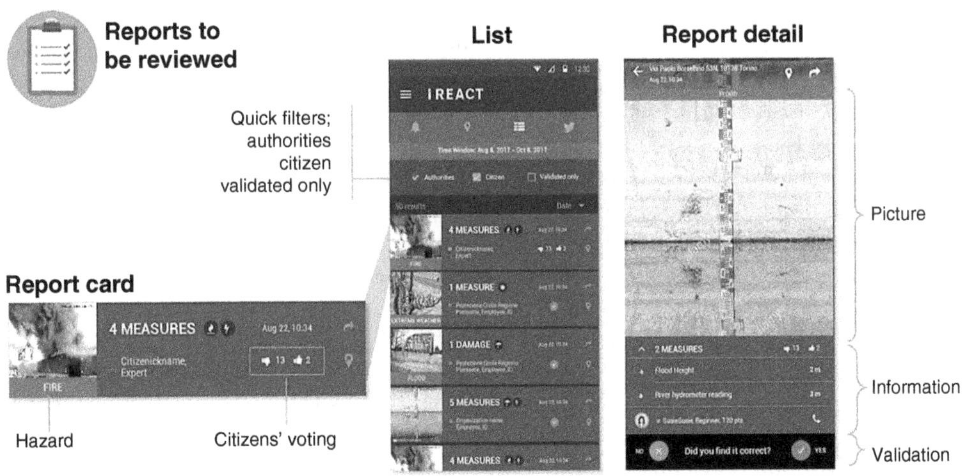

Figure 13.13 Reports to be reviewed: report card (left), list (centre), and report detail (right).

The functionalities and usability of the app were implemented and iteratively tested with real users. During two years, four training and simulations sessions in different European Countries (Sava Basin in Croatia, Cataluña in Spain and Helsinki in Finland) and one large scale exercise run by diverse European Civil Protections gathered in Italy were the stages of in-field user testing. Results allowed to progressively develop the app, which has been finally engineered and demonstrated with civil protections and other stakeholders such as public authorities and environmental agencies.

13.4 Gamified Crowdsourcing for Disaster Risk Management

The history of gamification is quite long but starting from Jane McGonigal's TED Talk 'Gaming Can Make a Better World' (2010), it has found growing interest and application in several fields. In literature, gamification is defined as the application of game thinking in non-game contexts to motivate desired behaviours (Deterding 2012). Leveraging the innate attitude to play, gamification repropose elements that work as native cues to education or entertainment and potentially in any domain. Compared to traditional games, which mainly aim to create enjoyment, gamification incorporates elements from game design, scientific grounds and empirical approaches (Mora et al. 2015). Gamification is not gaming. Gamification works like an additional layer of a system or service, able to enhance the engagement and experience of the users but not essential to the correct working of the system itself. To engage users in a meaningful and lasting experience, gamification tries to trigger human motivations and values. It is shown that environmental risks and emergencies activate different attitudes of people, even those not directly involved in the scenario: in these situations, people spontaneously tend to act, and ICT facilitate it (Olteanu et al. 2015). These attitudes can be mapped on the so-called RAMP model, one of the most known motivational models used in gamification (Marczewski 2013). The acronym refers to four intrinsic drivers that move players: (i) the relatedness, or the need to be connected with other people; aspects such as identity, socialization, membership, and sharing reply to it; (ii) the autonomy, or the need of expressing creativity, and experiment while being in control; (iii) the mastery, related to the need to know, learn and become keener on some topic or ability; (iv) the purpose concerns the meaningfulness of performed actions in terms of personal values. The RAMP is a simple model that was adopted in the early stages of game design and helped to shape the core gamification ingredients, such as challenges, incentives, rewards, events, aesthetics and all the fundamental design bricks of the gamification strategy. The so-called MDA is one of the first approaches to gamification design, and organizes this building block into three categories: (i) Mechanics, which include all the playful components enabling the (ii) Dynamics, the set of rules that characterize the gameplay. Finally, (iii) Aesthetics refers to the visual and narrative elements that trigger the experience's emotional responses (Hunicke et al. 2004). These components allow for shaping the gamified experience according to the goals, values, desired behaviours and skills to promote.

The I-REACT gamification strategy was designed by applying these fundamentals and then implemented on the mobile app for crowdsourcing (Figure 13.14). Three the goals of the strategy:

- Increase citizens' awareness and train their environmental monitoring skills
- Build and maintain the citizens' engagement, to act as reporters (crowdsources) from the field in cooperative environmental monitoring tasks; this goal intends to strengthen the crowdsourcing for the DRR
- Implement a peer-to-peer validation of crowdsourced information to provide reliable information, valuable for civil protection.

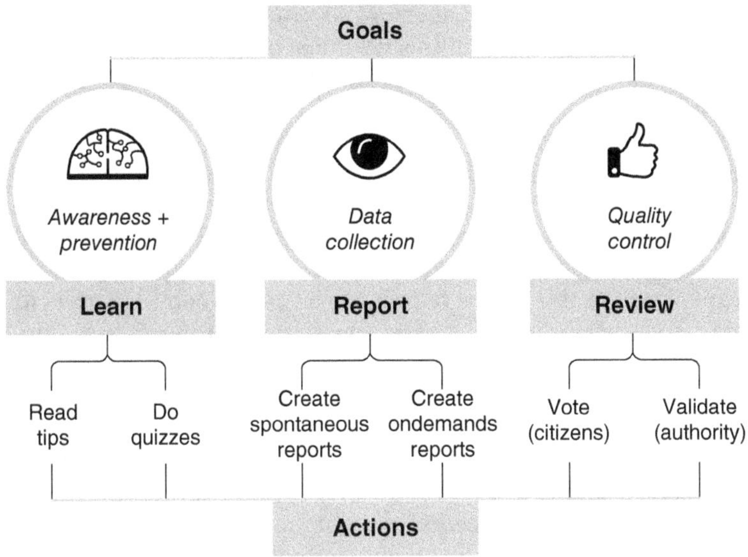

Figure 13.14 I-REACT gamification strategy: competencies, levels and required actions.

The goals mentioned above directly map into the core functions of the I-REACT mobile app and content, designed following the MDA approach, following some operational steps:

1. To outline the context and the target users in a systemic view;
2. to identify motivators and trace the narrative journey;
3. to build a system of Competences required to effectively contribute to the crowdsourcing objective, considering the core motivations;
4. to define for each competence a list of traceable Activities (quests) to be performed by players;
5. to define the Score system and the rules to assign points for performed activities;
6. to define the Reward system, articulated in Achievements and Awards, which are progressive and periodical rewards assigned according to performance indicators;
7. to define the Level System to update the user's status according to their points and Achievements;
8. to refine the Aesthetics (metaphors and look and feel of the UI).

The analysis of the stakeholders allowed to describe the target audience of active players and to define the engagement loop. The challenge for designers was to move citizens from the role of 'Observers', individuals who are part of the real ecosystem but not still engaged, to 'Players', active users involved as crowdsources. Designers also considered the so-called 'Spectators', the decision-makers or operation managers, receivers or beneficiaries of the collective monitoring working in control rooms (e.g. fire services, public authorities, Civil Protection, technical services, meteorological agencies). Their engagement was fundamental to designing the report creation flow and validation mechanisms via the app.

The gamified path to becoming a reliable I-REACT reporter implies some steps identified with core Competencies progressively achievable by performing some predefined Actions (quests).

The competencies focused on by the I-REACT gamification are:

1. Background in hazards and risks; this knowledge is offered in two primary modalities: by reading tips designed ad hoc and by answering quizzes. From the crowdsourcing perspective, this base is relevant to introduce and train users to know how to monitor some environmental parameters and – in the future – produce good quality data.
2. Reporting, or the capability of the user to create good reports and provide data from the field. Also, in this case, two modalities are provided to act as a reporter: users can generate spontaneous reports or take charge and reply to specific geolocalized requests from the control room.
3. Reviewing, core competence needed to assist the system in increasing the quality of reports collected, which has to be reliable for peers and emergency managers.

Different Mechanics allows running the I-REACT gamification (Figure 13.15): a Score system was defined, specifying the points to assign to the player's actions performed through the system (Zichermann and Cunningham 2011). In particular, the system assigns the points coherently with the importance of goals and activities: in this specific case, the Reporters' activities being the most valuable and complex, they allow to gain the highest amount of points; moreover, among the reports, the on-demand ones worth more than spontaneous ones. Over earning points, users can lose points

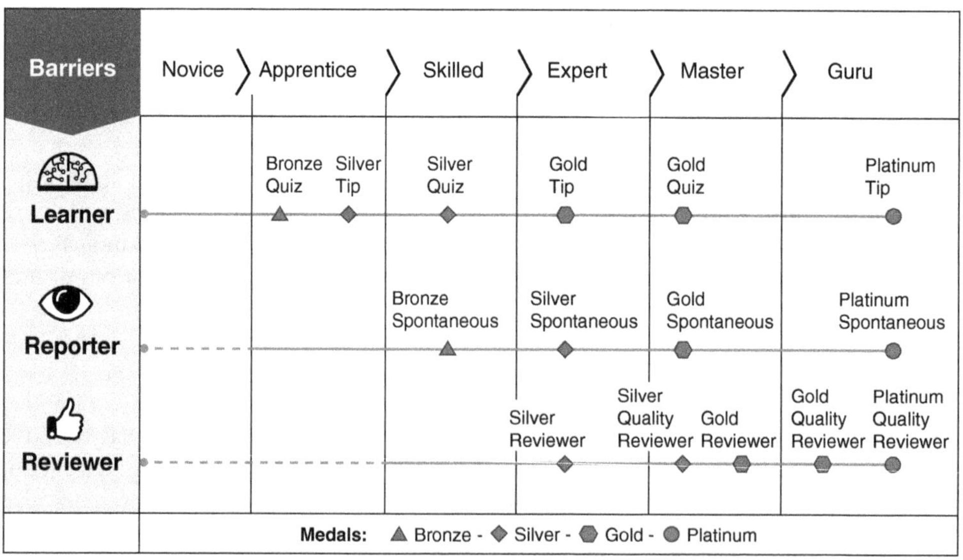

Figure 13.15 I-REACT mechanics: statuses, competencies, Achievements and barriers.

depending on their validation rate: a report validated by an authority provide an additional 50% of points previously gained. Conversely, a report rejected because not appropriate or out of scope or inaccurate (e.g. the image is blurred and/or badly cropped concerning the relevant content) causes a loss of 50% of points previously awarded. These rules are intended to encourage inaccurate reporters to improve the quality of the content and strongly discourage the generation of not appropriate. The points coming from the authority validations want to promote good reviewers to partners of the responders' organizations. Besides the level system, different rewards complete the journey. Two types of rewards were designed: the Achievements are recognized when a specific Competence is reached (according to thresholds on the related actions); they work as medals that enrich the player profile, making it more visible and better ranked. Visual elements and labels explain the achievement reached (e.g. Bronze Learner, Golden Reviewer, etc.). Achievements are also relevant in the gamification strategy to define barriers and controls to the players' progression, which shall proceed according to a balanced combination of activities.

Other rewards are the periodical badges earned by completing a set of hazard-specific activities. They recognize the user's specialization (e.g. 'Fire Reporter').

The Level System completes the gamification strategy: it depicts in detail the user's progression in their path to becoming an active and relevant part of the DRR process. The level system in I-Reach includes six steps, identified by Statuses (Novice, Apprentice, Skilled, Expert, Master, Guru). If the level system stimulates competition with themselves, the Leaderboard tries to connect the user to the virtual community of the I-REACT reporters. Leaderboards are often recognized as a negative element, overfocusing on the competition among people. This depends on how it is designed. The so-called absolute leaderboards focusing on the top-10 list would stress newbies or medium contributors in a huge community of users, making them feel frustrated. Conversely, relative leaderboards reduce the attention to top players and work more as a window on the community of players and their specificities (badges, achievements, skills etc.). Considering the specific domain of DRM, geographic leaderboards are strongly related to the hazards and risks of the different countries/areas.

Introducing gamification in a safety-critical domain might seem inappropriate, but over the apparent contradiction between emergency and gamification, the I-REACT mobile app aimed at exploring a different approach to engage people in DRR processes, stimulating an active attitude and self-protection behaviours. By designing a role of a contributor in environmental monitoring, people can develop useful protective skills for themselves, other citizens and the environment itself, hopefully contributing to reducing the human impacts and risks.

13.5 The I-REACT Wearable Solution for First Responders

In the case of professionals, it is worth enhancing the localization features of the app to improve the coordination and management of the first responders by the operations manager. This is achieved by pairing an external device (wearable) with the I-REACT App. This device uses multiple positioning systems to support outdoor and indoor scenarios. For outdoor ones, precise positioning and position validation are guaranteed through EGNOS. For indoor ones, the wearable is equipped with a ultra-wideband (UWB) chipset and a protocol that allows positioning using external anchors (i.e. outdoor) that can use EGNSS positioning. The I-REACT wearable is part of the I-REACT system being connected to the I-REACT centralized component: the I-REACTOR. Its relevant functionalities will be described in the following.

13.5.1 Ad-hoc Positioning Wearable Device for Enhanced Localization

The safety and effectiveness of rescue operations are hampered by not knowing the accurate location of personnel. I-REACT wearable aims to fill this gap by providing real-time indoor/outdoor operators' localization to the control room and, in the end, to support faster and better tactical decisions. From the first responders' perspective, it aims to improve in-field coordination and the operators' safety during emergency management. The wearable device can be associated with mobile phones and tablets via the Bluetooth Low Energy interface and connected to the I-REACT mobile application and I-REACTOR. These systems, in combination with the cloud-based Global Navigation Satellite System (GNSS) augmentation, can provide reliable information about the location of endangered personnel almost anywhere and anytime. Before dealing with the outdoor positioning based on the GNSS technology, we tackle the indoor positioning problem. Indoor positioning commonly relies on a combination of technologies: GNSS, Bluetooth, Wi-Fi sensors, and UWB. UWB is the best choice when reliability and accuracy are paramount because it brings range measurements with higher resolution than any other radio-based system and inherent communication links. Modern off-the-shelf UWB modules provide a ranging accuracy of 0.5 m (non-line-of-sight) down to 0.1 m (line-of-sight). The known UWB shortcoming is 300 m outdoors (in Europe) range limitations since it operates at low power to enable co-existence with other radios.

In order to achieve absolute positioning with UWB it is still necessary to rely on GNSS: this is done using the devices in the field with GNSS visibility as anchors for the indoor device. The indoor device's absolute position can be computed by measuring its distances from the anchors whose absolute positions are known from GNSS. However, the accuracy of low-cost GNSS receivers is below the UWB accuracy. Therefore, it is augmented by EGNOS. A cloud-based module part of the I-REACTOR performs the augmentation. The wearable device is implemented to provide GNSS raw data, which are used to compute the position by applying specific corrections. This means that the final position of the first responder is beyond the standard accuracy provided by the GNSS receiver itself. More in detail, wearable devices forward raw GNSS data (pseudo-ranges) over Bluetooth to the I-REACT mobile application, which acts as a gateway to the cloud-based augmentation algorithm. The data flow is depicted in Figure 13.16.

However, in case of severe infrastructure damage which affects cellular networks, a stand-alone device will still be able to independently compute positions even if it does not benefit from the augmentation module.

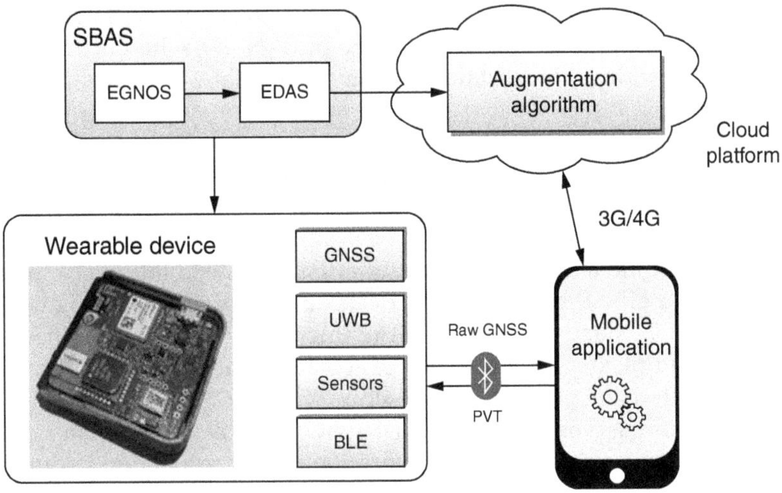

Figure 13.16 Wearable device data flow.

Advanced positioning is the core functionality, but it is enriched with multiple layers of sensors and positioning mechanisms. These allow location tracking and environmental and gas sensing thanks to combining advanced GNSS, UWB radio, Bluetooth Low Energy (BLE) and oxygen, temperature, humidity and pressure sensors. Also, thanks to motion sensors and custom-developed recognition algorithms it allows the detection of falls, steady and other specific activities. Two versions of the wearable device were realized: with and without an oxygen level sensor. To facilitate its use, the device was designed to be worn as a belt, a watch or – thanks to a clip – to be attached to a jacket.

To summarize, the I-REACT wearable device includes:

- advanced indoor/outdoor positioning;
- positioning GNSS pseudo-ranges to be sent to I-REACT mobile application – GPS/GLONASS/Galileo/SBAS;
- UWB-based micro-location positioning;
- localization algorithms;
- environmental monitoring;
- oxygen level;
- temperature, humidity, and pressure;
- accelerometers, rate-gyroscopes, magnetometers;
- activity monitoring (fall-detection, activity detection);

13.5.2 Operational Scenario

Wearable devices use multi-technology positioning. The main operational scenario assumes that devices are carried by first responders. Devices are paired with smartphones and dedicated applications over BLE link. The mobile application communicates with the backend server, which is used for emergency management.

A single I-REACT localization system consists of a maximum of **8 wearable devices**. Every system has a unique PAN_ID. The range of PAN_ID values is from 0 to 65535. Also, every device within a localization system has a unique ID -a number in a range from 0 to 7. Physical ranges of two or more localization systems shouldn't overlap; otherwise, systems will not correctly work.

The I-REACT wearable uses UWB radio to measure the relative distance to the other devices using the Time Difference of Arrival Technique (TDOA). The **devices exchange BLE messages every second.** When available, the device uses the1PPS signal from the GNSS module to synchronize the internal timer with the rest of the network. For this purpose, a GNSS receiver with advanced timing capabilities is used (uBlox NEO-M8T, where T- stands for timing). The internal timer is a reference for time-division-based multiplexing, in which devices communicate by transmitting messages only within dedicated and predefined time slots.

From the positioning point of view, the I-REACT wearable (shown in Figure 13.17) is part of the real-time system for the localization of first responders everywhere and anytime with an accuracy of 'a meter' level, with the following usage scenario:

1. The team of first responders is involved in firefighting and rescue operation
2. Each team member carries I-REACT wearable device

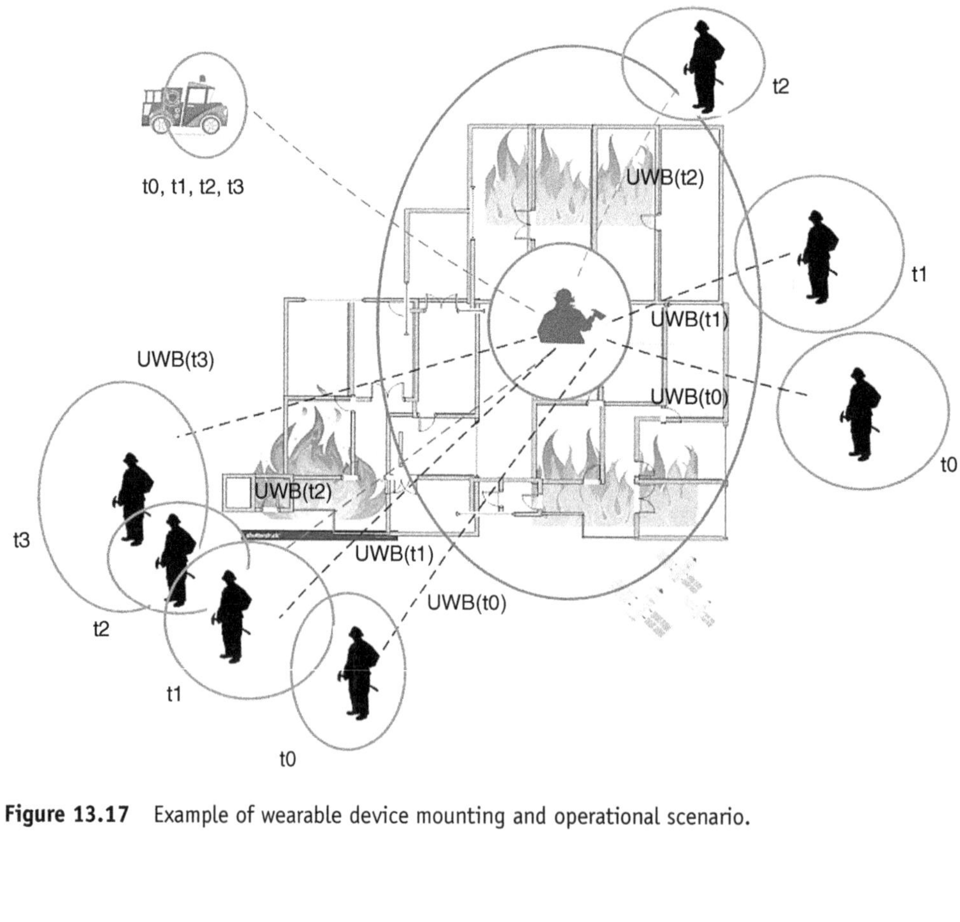

Figure 13.17 Example of wearable device mounting and operational scenario.

3. A team member enters the building (indoor environment, no satellite visibility).
4. Incident commanders lose their GNSS position.
5. I-REACT Wearable automatically senses the loss of GNSS position and potential fall event with its severity, loss of activity – no movement of a team member, which may indicate injury.
6. Wearable initiates the SEARCH procedure in cooperation with the surrounding devices (team members, utility vehicles, UAVs . . .) that are positioned outdoor and have knowledge of the position due to better satellite visibility. These outdoor devices are mobile 'radio anchors' for the positioning algorithm.
7. The SEARCH procedure relies on UWB radio, which enables fast and accurate ranging through walls and large buildings. I-REACT backend system performs a localization algorithm (statistical or deterministic, e.g. Least Squares optimization or Newton method) and tracks the indoor location of the lost member.

At one of the scenarios, assuming that the lost team member is quasi-static, the surrounding team members can repeat the radio ranging procedure from different outdoor positions and at different time instants (t_0, t_1, t_2, t_3, as seen in the figure below), and thus emulate a much higher number of radio 'anchors'. In case the lost member is inside the tall building, a UAV with a 'wearable' device can fly around it at various altitudes and thus emulate many team members.

13.5.3 Device Operating Modes

The I-REACT wearable devices operate in the following modes:

- IDLE,
- INITIATOR,
- RESPONDER.

The IDLE mode starts right after the power-up. This is required to synchronize the timer with the GNSS 1PPS signal. The device remains in this mode until GNSS is valid.

The device operates in **INITIATOR** mode while GNSS data is invalid. While in this mode, the device restarts the ranging process concerning the other reachable devices with valid GNSS data (devices in RESPONDER mode). This is done every second. In case of successful ranging, the results are sent to the smartphone as part of the BLE message. Every device in INITIATOR mode starts the ranging process using a dedicated slot. This allows 'simultaneous' operation of up to seven *indoor* devices (all devices without GNSS position fix are considered as 'lost' or 'indoor' devices). These 'lost' devices will thus be able to measure distances to the other devices that belong to the same I-REACT group and that have obtained valid GNSS data, assuming they are in the range of UWB radio. The range of UWB radio outdoor is roughly 200–250 and indoor around 25 m or more, depending on the specific environment. These are preliminary numbers depending on several external factors.

The device operates in **RESPONDER** mode while GNSS data is valid. In this mode, the device restarts the range process to other available devices with invalid GNSS data (that are in INITIATOR mode) at 125 ms. If ranging is successful, the results are sent in BLE message together with valid GNSS raw data (pseudo-ranges).

The ranging process is active only if:

1. at least one device has no GNSS position fix;
2. at least one device has GNSS position fix and
3. both devices can measure mutual distance using UWB signal.

In the text below, INITIATOR is the device in INITIATOR mode, and RESPONDER is the device in RESPONDER mode. The device in INITIATOR mode is a 'lost device' (device without known position).

The diagram in Figure 13.18 shows the automatic mode selection mechanism.

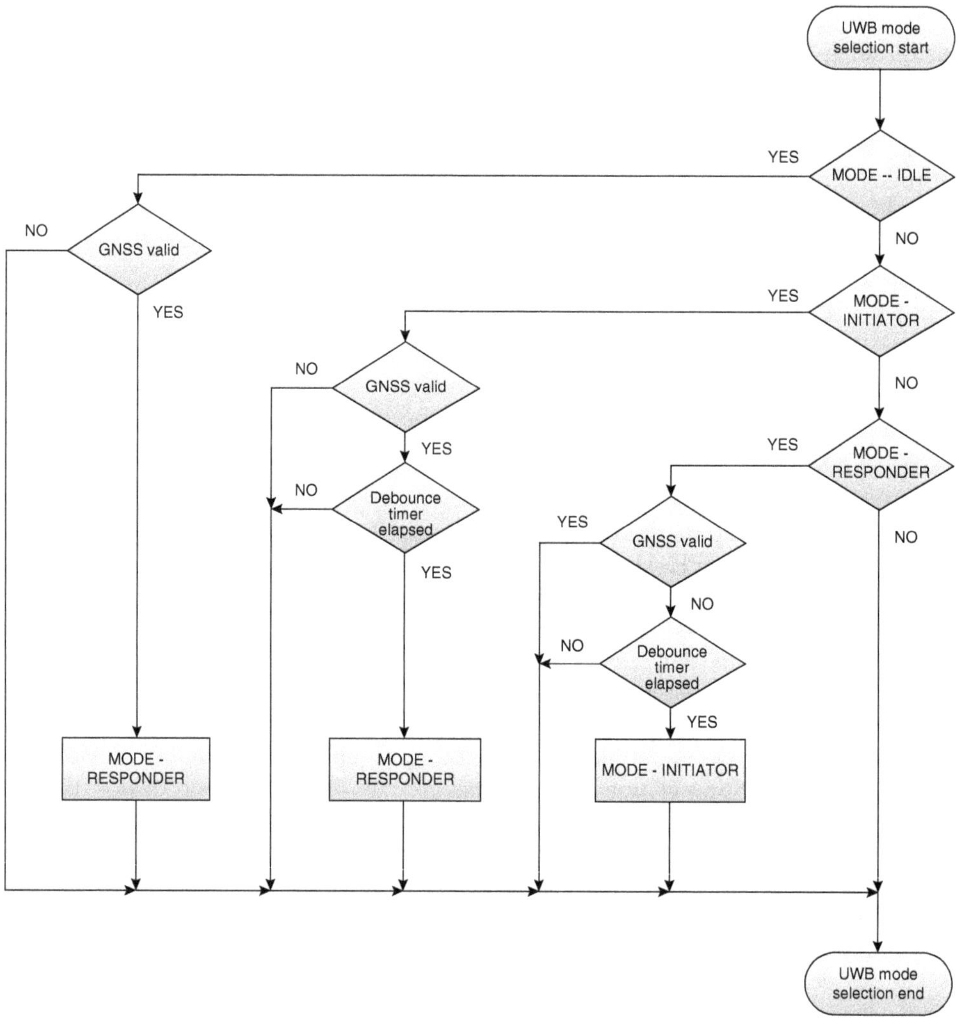

Figure 13.18 Device modes – internal logic.

13.5.4 Communication Flow

In order to measure the distance between two devices, there is the need to exchange three types of messages: POLL, RESPONSE and FINAL. Figure 13.19 shows the communication flow. The INITIATOR device sends POLL and FINAL messages and receives RESPONSE messages. The RESPONDER sends RESPONSE messages and receives POLL and FINAL messages.

One range process starts when the INITIATOR sends a POLL message while all RESPONDERS listen to receive a POLL message. Every RESPONDER sends a RESPONSE message at a predefined time if a POLL message is received with success. At the end of a range process, the INITIATOR sends a FINAL message, and when the RESPONDERS receive the FINAL message, they individually can calculate the relative distance to the INITIATOR.

There are eight range processes in eight predefined time slots in the timeframe of one second. The first range process is reserved for the device with ID 0 when it is the INITIATOR. The second range process is reserved for the device with ID 1 when it is INITIATOR and so on. The ranging process is active only if the device with a predefined ID is in INITIATOR mode. If all devices are in RESPONDER mode, there are no active ranging processes. With this mechanism, the distance computation refresh rate is 1 Hz.

13.5.5 Wearable Device Implementation and Prototyping Cycles

The I-REACT wearable was developed in three electronic design cycles. After each cycle, the I-REACT team carried out a deep functional and performance assessment. The results of this assessment were used for further design and performance improvement (Figure 13.20).

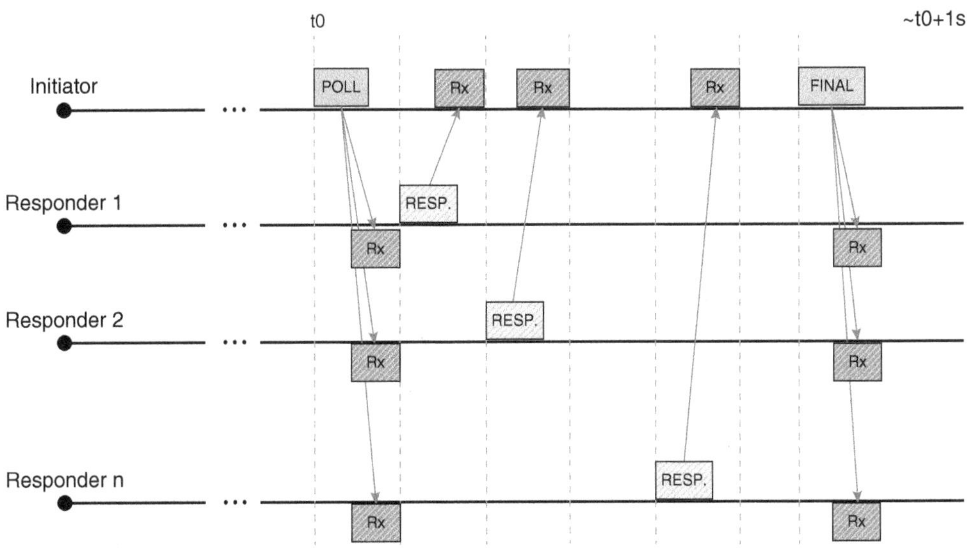

Figure 13.19 Communication flow.

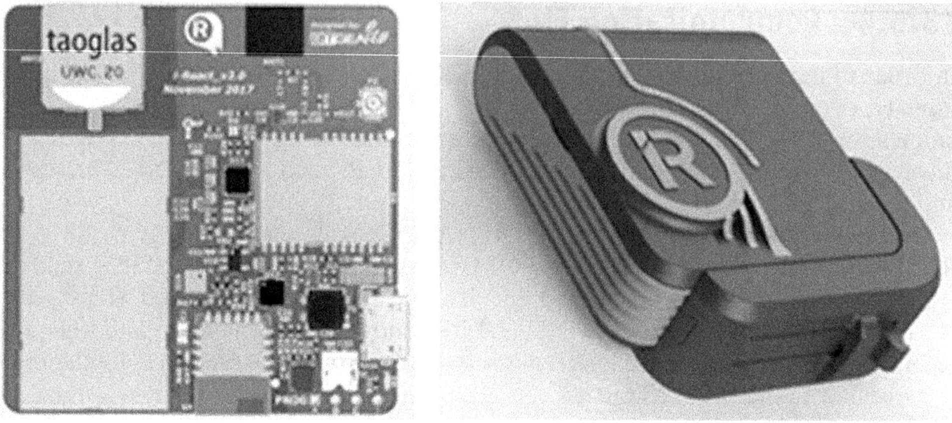

Figure 13.20 Proof of concept prototype – version 2 (PCB, enclosure, and electronics with oxygen level sensor).

The second revision of the wearable device has been a large-scale revision in which the UWB module has been replaced with a transceiver module and a new wideband antenna. The RF signal chain on 4 GHz has been custom designed with the inclusion of new power amplifiers. The enclosure was redesigned to support diverse options for the holder with a built-in cap for a micro-USB connector. This increased the protection of electronics from environmental effects. Both versions (1 and 2) foresaw the device in two forms – with and without the oxygen level sensor (in the absence of the sensor, the enclosure is waterproof and dustproof). The oxygen level sensor is in the form of an add-on printed circuit board (PCB). Other improvements included:

- Positioning:
  - GNSS – signal level improved by 14dB
  - optional UFL connector for the use of external GNSS antenna
  - Supported new GNSS RTK modules for future improvements toward high-precision outdoor positioning
  - Support of LAES – unique indoor positioning with custom UWB RF front-end and a new antenna
- Power/Battery:
  Battery capacity increased by 100%
- Other:
  - New motion sensors
  - New air quality sensors supported (VOC)
  - Single board solution with two assembling options – lower production costs
  - Better protection from dust and water drops
  - Better RF/EMC shielding
  - New mounting options (smaller, on jacket)

Version 3 – Final device with improved enclosure and common PCB design.

The third, final revision of the wearable device presented a single common PCB design for both 'oxygen' and 'non-oxygen' sensor devices. This solution is mechanically

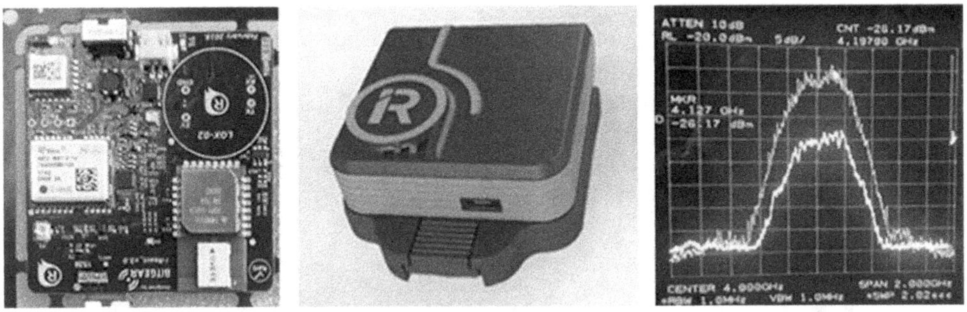

Figure 13.21 Final device version – assembled electronic device, the final version of the enclosure, and comparison of UWB signal spectrum on LAES and regular device version (lab measurement).

more robust, thus, more suitable for wearable devices and dynamic environments. Due to the potentially complex certification procedures of a custom UWB radio from version 2, the UWB module has been used again, but this time with the capability to provide additional transmission power. Two battery versions are offered – the smaller one for the device with the oxygen sensor and a larger one for the standard device version. In this way, the autonomy of the wearable device increased beyond initial requirements.

The final revision of the enclosure offers more robust and simpler mounting options. The first two revisions were built using a standard low-cost 3D printing process with ABS plastic. The third revision was made using a **selective laser sintering** process and polyamide 12 with glass fibres (material type PA 3200 GF). It has the following mechanical characteristics: density: $1.25\,\mathrm{g\,cm^{-3}}$, elasticity 3000/2500 MPa, Rockwell number $98\,\mathrm{N/mm^2}$, melting temperature: 180 °C. This significantly increased the robustness of the prototyping series but also offered a quick and affordable solution for low-volume production – from 100 up to 1000 units (Figure 13.21).

13.5.6 Wearable Device Performance Validation

The wearable device was evaluated in a planar field test in which it was capturing raw GNSS/SBAS carrier phase data and UWB at 5 Hz. UWB was used in non-LAES mode (low power). The outdoor UWB range achieved was 100–120 m with the usual obstruction by surrounding pedestrians. The indoor range was limited to 12–15 m (heavy indoor environment with reinforced concrete and aluminium walls), with a standard deviation below 0.1 m. ETSI LAES Directive permits the use of UWB with 20 dB higher transmission power for tracking firefighters in Europe. Consequently, with this modification, UWB radio should increase the indoor range significantly.

The field testing was done in a football stadium of size 68 × 99 m (Figure 13.22). The TARGET device in an 'unknown position' was placed on a 1.1 m high tripod, right on the centre of stadium at (34 m; 49.5 m). There were two kinds of ANCHOR devices – one with the internal GNSS antenna ('A' in Figure 13.22) and another device with the external survey grade GNSS antenna ('B', REF-GNSS in Figure 13.22). The position of the TARGET device was calculated using the GNSS positions of the ANCHORS

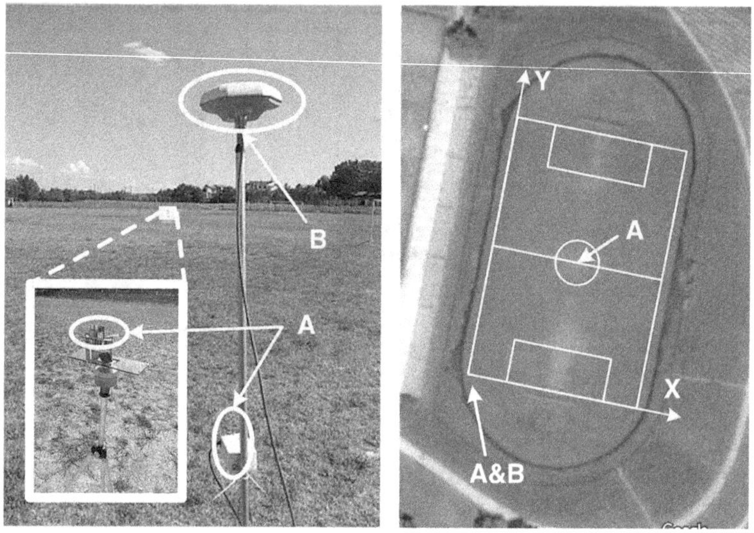

Figure 13.22 Test setup and configuration.

(considering adequate coordinate transformations), the UWB measurements, and the Gauss–Newton method to solve the least-squares problem. All measurements were first transferred to the local tangential East-North-Up frame, then to the local Cartesian X–Y frame fixed to one of the corners.

Two tests were carried out: the 'corner test' with the ANCHOR devices placed right at the four stadium corners, the 'edge test', where the ANCHORS were carefully moved along the edges. The results of the 'corner test' (Figure 13.23) show the key impact of GNSS on the result. The prototype devices were used without any antenna optimizations and any fine impedance matching. The error includes a systematic bias due to the coordinate transformations. In the 'edge test', the seven random GNSS-REF positions of the ANCHOR devices along the edge were combined with the UWB to estimate the TARGET location. The mean position error was 0.34 m, which is acceptable.

13.6 Improved Positioning of First Responders Using EGNSS Technologies

From the previous paragraphs, it is clear how positioning plays a key role in the I-REACT framework. Both the exploitation of crowdsourcing and the support to in-field first responders rely in some way on the positioning information. In fact, it is important to associate the correct position to each piece of information to improve the knowledge of the emergency scenario. More important is the accurate and reliable knowledge of the position of the first responders, which can be assisted and directed by the crisis management centre most effectively.

Current GNSS receivers can provide position in most environmental conditions, even with various levels of accuracy. Accuracy does not depend only on the

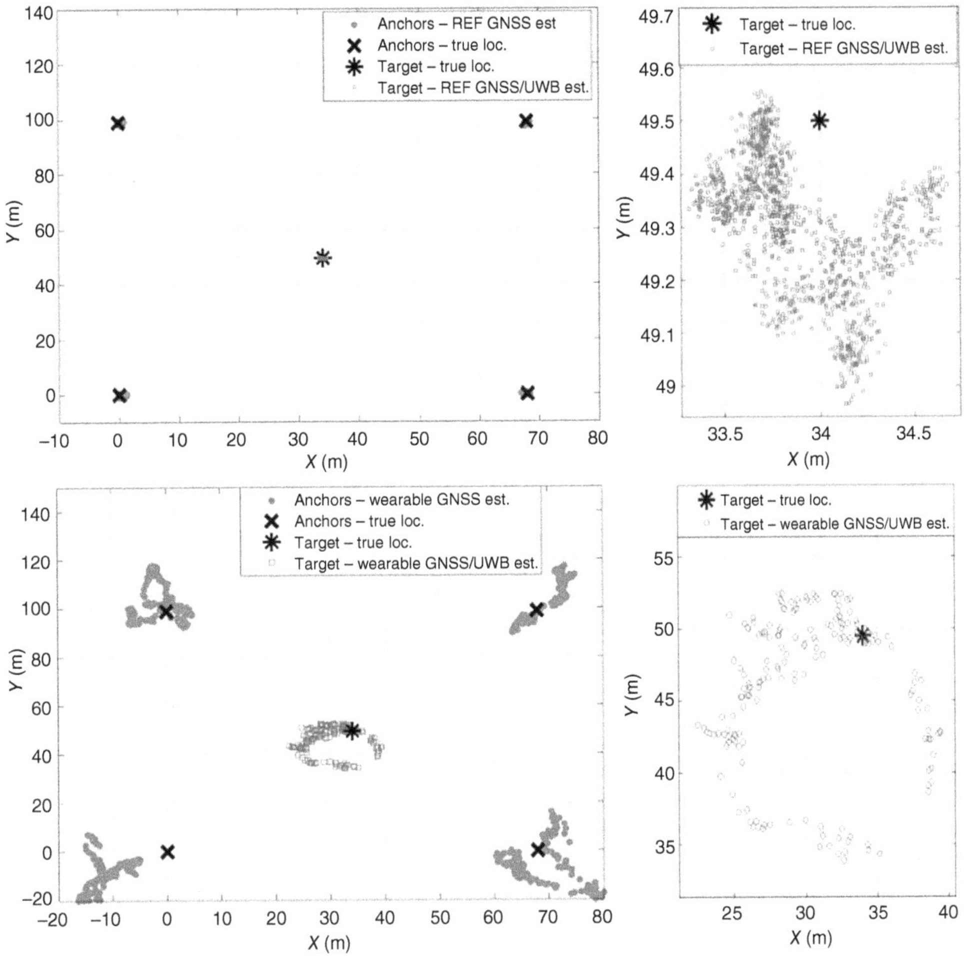

Figure 13.23 'Corner test': REF-GNSS (up), right-zoomed, wearable GNSS (down), right-zoomed.

environment but also on the GNSS receiver characteristics and their capabilities to exploit available services. However, accuracy is not the only key element because reliability is truly relevant when positioning information is exploited to operate emergency services and locate first responders. The work carried out in the frame of I-REACT focused on three distinct aspects:

- the improvement of positioning performance exploiting the European Geostationary Overlay Service in the frame of a service-oriented cloud-based architecture;
- the improvement of the positioning data quality, in terms of reliability, which has a key role when Safety of Life (SoL) is involved;
- the development of an ad-hoc positioning device for indoor localization based on UWB technology.

13.6.1 A Service-Oriented Cloud-Based Architecture for Mobile Geolocated Emergency Services (EGNOS in the Cloud)

Nowadays, the collection and dissemination of Earth Observation information play a key role in emergency management systems. The European Copernicus Programme for Earth Observation, Copernicus EMS3 provides geographical mapping services of natural disasters. The geospatial information based on satellite observations is complemented with available information from open data sources or direct observations (in situ). The Copernicus Emergency Management Service represents a powerful tool for land monitoring, and it can provide several products as a reference, delineation, and grading maps, generated prior to, during, and after the event, respectively. However, the product generation can still require some days because of administrative and technical operations. This delay represents a problem in case of events characterized by a rapid evolution, such as flash floods or wildfires.

The completion of existing information with the collection of geolocalized in-field reports that can be sent by both citizens and professionals through their mobile devices can be valuable for certain types of hazards, like floods and fires. Geolocated reports together with Copernicus maps provide useful information to compute near real-time extent maps with improved update frequency and accuracy.

To further increase reports' accuracy and reliability I-REACT built on top of existing European solutions, using cloud computing architectures and the mobile sensing paradigm to enable users to send in-field reports. This is done by implementing a system for positioning data validation and augmentation based on a custom cloud-based service that uses the Data Access Service of the European Geostationary Navigation Overlay Service (EGNOS) (EDAS). EDAS delivers EGNOS information through a specific Application Program Interface (API).

EGNOS was born as a system for enabling the automated approach of aircraft to the landing path without the use of any ground infrastructure. Beyond its original domain, it is suitable for safety-critical applications in general, having the main objective of improving GNSS based positioning from both the accuracy and reliability point of view. EGNOS provides corrections and integrity information for the GNSS signal over the European continental area. It is composed by two segments: the space one and the ground one. The space segment comprises some geostationary (GEO) satellites. They are usually three, of which two are always operational. They broadcast corrections and integrity information for GNSS satellites in the L1 frequency. The ground segment comprises a network of several stations and facilities. They collect measurements from the GNSS satellites and process them to compute a set of corrections transmitted to EGNOS satellites. They, in turn, broadcast them to users through the so-called Signal-In-Space. These corrections apply, for each GNSS satellite in view, to clock errors, ephemeris errors, and ionospheric errors, while integrity information refers to the goodness of the corrections themselves. The EGNOS system can also warn users in case of single GNSS satellite faults or anomalies in the GNSS system, which are detected and communicated in a very short timeframe (six seconds).

Here, it is worth better defining the integrity concept which is connected with system reliability. An informal but very effective definition was provided by Dr Brad Parkinson (2021): 'I know I'm getting this accuracy; the system is not lying to me . . .' However, the formal definition of position integrity is: 'Positioning integrity (or simply 'integrity') can be defined as a general performance feature referring to the level of trust a user can have in the value of a given position or velocity as provided by a location system'. (CEN/CENELEC, Space 2014)

Integrity will be further discussed in Section 13.6.3, while hereafter, the I-REACT-specific implementation for the exploitation of EGNOS is explained.

In conditions other than aviation ones, the signal shadowing caused by elements such as buildings, mountains or tree canopies can prevent the continuous EGNOS signal reception from the GNSS user devices. For that reason, EDAS provides an alternative way to deliver EGNOS data by implementing a service reachable by any ground-based network, including the cellular network. The high-level architecture of EGNOS/EDAS is shown in Figure 13.24, which illustrates the data provisioning scheme of the system.

I-REACT exploits GNSS and augmentation services through the so-called Augmentation Module (AM): a software component running externally to the I-REACTOR backend, which provides a precise and reliable positioning service exploiting EDAS and implementing the augmentation as foreseen by the standards (Convention on International Civil Aviation) and (Kaplan and Hegarty 2006), plus the computation of the augmentation as described in Section 13.6.6.3.

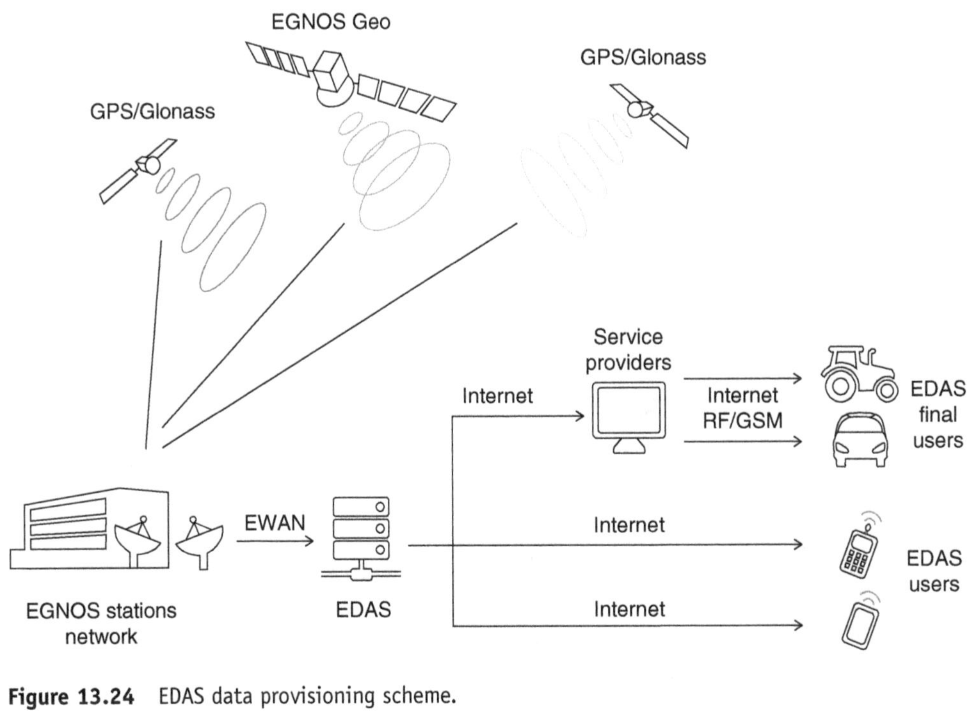

Figure 13.24 EDAS data provisioning scheme.

Through the I-REACTOR backend, the AM receives raw GNSS measurements from the users and provides the users' Position Velocity and Time (PVT), corrected with EDAS data. It also provides a confidence interval based on the 'local integrity' concept.

The AM can be seen as composed of two main parts the first manages the interface with EDAS, including EDAS data storage, and the second contains the core algorithms for the position computation and augmentation and manages the interface with the I-REACTOR backend.

The positioning augmentation process starts with a user making a request to the RESTful web services exposed by the AM. GNSS receivers' parameters (Time Of Week (TOW) and raw data) are made available through the web services (JSON format is used). After validating the request, the AM retrieves the required data (originating from EDAS) from the cloud storage and runs the augmentation algorithms. It returns the augmented position, including integrity information. The overall process of acquiring the data and providing the service is sketched in Figure 13.25.

13.6.2 EDAS Service Selector, Decoder and Storage

This software component runs on the Azure cloud as a web job, which can connect to the EDAS Client SW server and continuously receive the EDAS stream data from the EDAS Client SW Service Layer 2 (SL2) in RTCM format through an internet socket connection.

The service selector connects to the EDAS Client SW (part of EDAS) to request the SL2. Consequently, the EDAS data decoder algorithm parses the content to get the EDAS correction data to be stored.

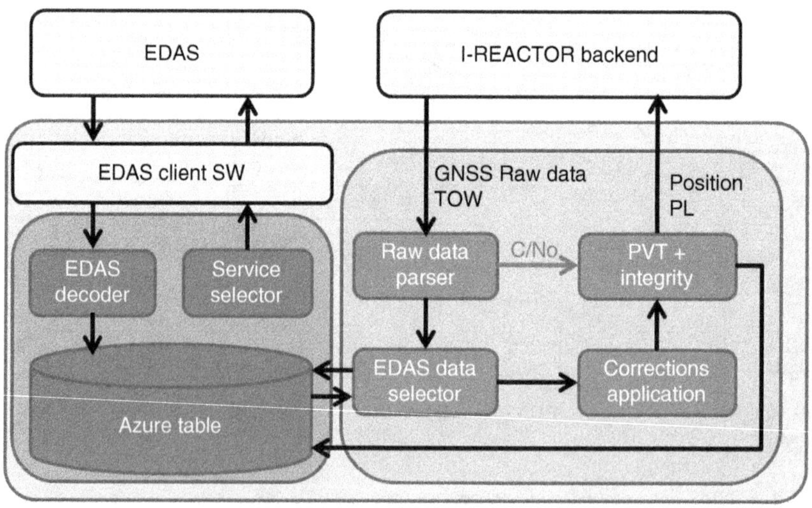

Figure 13.25 Augmentation module scheme.

Data storage is implemented with Azure Table, a service that stores structured NoSQL data in the cloud, providing a key/attribute store with a schemeless design. This design guarantees a good degree of flexibility, allowing fast adaptation to the algorithmic design evolutions.

13.6.3 Augmented PVT and Integrity Computation

The augmentation algorithm runs in the Azure cloud as a continuous web job. It is used to compute s' positions using the least square algorithm or Kalman filter, and it is able to augment position using EDAS correction data. It is also used to determine the confidence intervals around the computed positions.

The algorithm requires two inputs, namely a stream of GNSS raw data (i.e. pseudorange, Doppler measurements and signal carrier to noise ratio) and EDAS correction data. In real-time processing, the raw GNSS data come in JSON format from the I-REACTOR backend, and, in the same format, the results are sent back to the I-REACTOR backend. Depending on the received GNSS data (including time), matching EDAS data are retrieved from the Azure Table.

The algorithm for determining PVT using EDAS augmentation works step by step as follows:

- GNSS Raw data reception in JSON format and decoding;
- Retrieval of the EDAS correction messages from the Azure Table storage based on timestamps;
- Exclusion of pseudoranges from satellites declared by EGNOS as Do Not Use and Not Monitored;
- Determination of satellite clock corrections and the satellites' positions from the GPS ephemeris data provided by EDAS;
- Computation of EGNOS corrections using the three sets of corrections, which are broadcast to users:
  - Fast corrections – used to compensate for short-term disturbances in GPS signals attributable to satellite clocks;
  - Long-term corrections – used to compensate for the long-term drift in satellite clocks and the errors in the broadcast satellite orbits;
  - Ionospheric corrections – broadcast as vertical delays for a grid of points from which it is possible to determine a slant correction to be applied on each range measurement to compensate for the delay experienced by the signal as it passes through the ionosphere;
  - application of the corrections on the pseudoranges and satellite position for each GPS satellite;
- PVT computation using corrected pseudoranges and other data selected from the Azure Table (ephemeris, clock parameters). The computation can be done using Kalman filter or the least square method;
- Computation of satellites elevations: these data are used together with the Carrier to Noise ratio (C/N_0) estimated by the GNSS receiver and delivered along with the raw data, are used to estimate the signal reliability used to compute PLs. This step is

fundamental because it provides the input for implementing the local integrity algorithm (see the next step);

- Computation of the local integrity. This step represents the added value with respect to the standard implementation (i.e. aviation integrity). The output is confidence levels, here named HPL and VPL, along with the aviation nomenclature. In case the user motion is relevant, instead of the circle defined by the HPL, an ellipse with its axes oriented along and perpendicularly to the user path can be defined. These axes are the cross-track and along-track PLs and are computed considering user motion direction and satellites geometry and reliability;
- Delivery of augmented position and integrity information in JSON format to the I-REACTOR backend.

13.6.4 Implementation of the Architecture of the Cloud Software Module

To provide the architectural description, the system can be seen formed by the following software elements:

- EDAS Client SW Server – is a server that runs the Client Software (which is an external component provided by the EDAS service provider). It enables the connection to the EDAS server and ensures a controlled flow of the EDAS data through the network socket connection.
- 'EDAS decoder' is an algorithm running as a web job in the cloud. It can connect to EDAS client-server by providing the server IP address and port number to get the EDAS stream data (after the service selection operated through the service selector) and then decode it based on the message type and store the decoded data in the Azure table. It runs continuously as a web job in the Azure cloud.
- 'PVT' is an algorithm running as a web job in the cloud, used to compute the position and integrity information. It takes GNSS raw data from the user device ('Raw data parser') and EDAS correction data from the Azure tables ('EDAS data Selector') to obtain after the 'Correction application' an augmented position ('PVT + integrity'). Figure 13.25 shows the information about the Carrier to Noise ratio estimated by the receiver, which is one of the two pieces of information to be considered to include the local environment effects in the integrity computation. The second element is the satellite elevation, which is computed from the satellite ephemeris and a rough estimate of the user position. The augmented position is also stored in Azure table. This web job also runs continuously. However, it waits for the specific message to be written in the Azure queue storage and, based on the message, it selects the appropriate PVT estimation algorithm: Least square or Kalman filter methods.
- Web Service – it can send the GPS raw data in JSON format through REST service in real time and is able to receive the computed position in JSON format. This system also provides dissemination of parsed EGNOS/EDAS information through REST service based on timestamps.
- Azure Storage – the system uses Azure table and blob for storing and Azure queue to create communication between web job and the web service or web app.

The Augmentation Module (AM) has been implemented as a standalone service to perform its tasks independently from the capabilities (i.e. raw data provisioning or the exploitation of augmentation services as EGNOS and EDAS) of the single receivers. The AM leverages the diverse types of data made available by the receivers and takes advantage of the continuous availability of augmentation data provided by EDAS.

Figure 13.25 shows the three-tier architecture implemented, together with the main technologies used. The AM is deployed using the Platform as a Service (PaaS) approach. Since the GNSS raw data can be easily mapped into a key-value structure, a NOSQL database is used to exploit the greater speed of this kind of solution.

The .NET Entity framework as Object Relation Mapper (ORM) is adopted and the web service that serves the augmentation algorithm is implemented using the REST architecture, a JSON data format and the .NET WEB API 2.0 framework. The .NET Identity framework, which supports OAuth2.0, is used for authentication and login from mobile devices. log4net and Azure Application Insights are used for the log and the audit provider, respectively.

The EGNOS/EDAS augmentation algorithm was implemented in the business layer, following previous developments described in Favenza et al. (2014), while the local integrity described in the following was added as a new feature and embedded in the augmentation algorithm.

This version of the augmentation algorithm adds to the improvement of the position accuracy and the estimation of its reliability (confidence interval delivered as horizontal protection levels (HPL)). This is done considering also the local conditions.

13.6.5 Performance Evaluation of the Implementation

The system must be scalable to serve an increasing number of service requests: the present paragraph describes the tests carried out to verify the system's performance as implemented for I-REACT.

This section's focus is assessing the performance of the implemented cloud component, i.e. the AM, which is deployed using Microsoft Azure.

To measure the baseline performance, single client tests were run using the preliminary I-REACT mobile application, while the Visual Studio Online stress tool was used to assess the performance of the cloud services at scale.

For single client tests, all relevant and measurable delay contributions were considered, while stress tests were addressed to study the performance in terms of response time and requests per second in function of the user load and of the cloud deployment, which is characterized by the instance3 number and type. The specification of the instance types used is given in Table 13.1.

Table 13.1 Instance types specifications.

Type	Cores	RAM
P1/Small	1	1.75 GB
P2/Medium	2	3.5 GB
P3/Large	4	7 GB

For the stress tests, one parameter was varied, while the other was kept fixed. In this way, the influence of the varying parameter on the performance metrics could be better understood. For all the stress tests, a request timeout of 10 seconds was set, and each user repetitively called the tested service for 30 minutes, averaging the metrics with sampling intervals of 15 seconds.

Mobile users were assumed to send geolocalized reports, including raw data. To select the user load range, the Twitter datasets available at CrisisLex.org (n.d.) were analysed; they include the tweets obtained from a predefined set of keywords during natural disasters (floods, earthquakes, wildfires, etc.) and technological disasters (train crashes, bombing, explosions). Emergency scenarios are more significant because they are characterized by a peak of traffic, which occurs right after the event. Specifically, the maximum rate in terms of tweets per second during the 26 emergency events included in Olteanu et al. (2015) were considered, finding a maximum rate of 11 tweets/s, which was recorded during the Boston bombing in 2013. Assuming a service time of 2 seconds, 22 parallel users would be required to achieve such a rate, on average. Since emergency services using a crowdsourced approach are still in the initial stages, the parallel users were increased by one order of magnitude to 200 in the performance evaluation at scale for the AM. The cloud deployment is done at the Microsoft Azure west Europe data centre, which was also used for the agents (clients) generated through the stress tool. In this way, the contribution of the networking delay was minimized, being highly variable and not the focus of this evaluation. The assessment of performance variations depending on the data centres and cloud providers is outside the scope of this analysis.

The implementation of the AM was first validated in an urban environment with a single client. The AM was deployed in Microsoft Azure, using the web-app PaaS service with a single P1 instance. The first test was carried out by driving for 3.5 km at low speed along a heterogeneous path, passing along tall buildings, below underpasses, and crossing wide roads. The positioning rate was 1/3 Hz, and, at the same time, the position augmentation was performed by calling the AM web service.

Table 13.2 reports the average and the standard deviation of each delay contribution as well as the total delay suffered. The average total delay was always below one second, thus achieving the standard target performance.

The performance of the augmentation service at scale was evaluated, varying one at a time the number of concurring users, the instance number and the type. The evaluation started by assessing the impact of the user load on the deployment of 20 instances (at that time, the maximum allowed with a standard Azure subscription) of type P1.

As shown in the boxplot reported in Figure 13.26 (left), the averaged response time grows with the number of parallel users, scoring 0.341, 0.486, 0.519, 0.7692 seconds with 50, 100, 150, 200 users, respectively. The boxplot reported in Figure 13.26 (right) shows

Table 13.2 Delay contributions.

	Delay contributions (μ, σ) (ms)			
System	Network (3G)	DB extraction	AM algorithm	TOTAL
Cloud	(360,35)	(110,5)	(511,20)	(981,50)

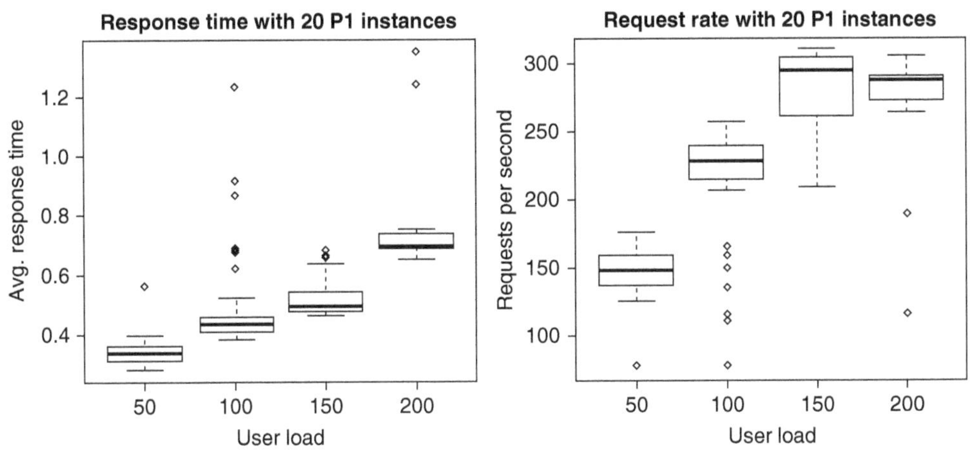

Figure 13.26 Response time with 20 P1 instances.

the number of requests per second increasing with users up to 282.4 req/s with 150 users and to fall to 171.1 req/s with 200 users. This behaviour indicates that a saturation condition is reached with this setup, i.e. a further increase of the user load would not result in an increased response rate.

Then, the user load was kept fixed at 200 while changing the number of instances from 1 to 20. As shown in the boxplot reported in Figure 13.27 (left), all requests experience a timeout with a single instance, while the mean response time decreases to 2.72, 1.217 and 0.77 seconds, with 5, 10, and 20 instances, respectively. This trend is also reflected in the boxplot of Figure 13.27 (right), which shows the request rate. Indeed, as the number of instances increases from 1 to 20, the average request rate increases from 19.88 req/s up to 270 req/s.

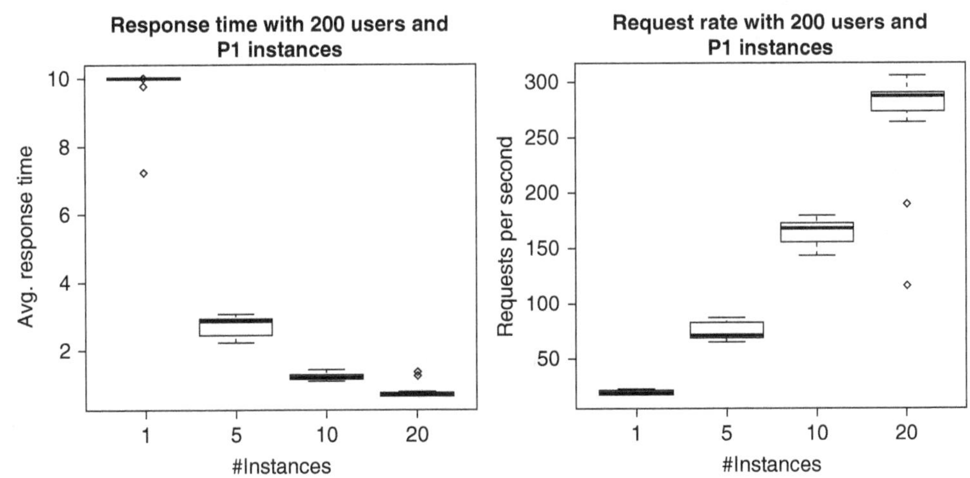

Figure 13.27 Response time with 200 users and P1 instances.

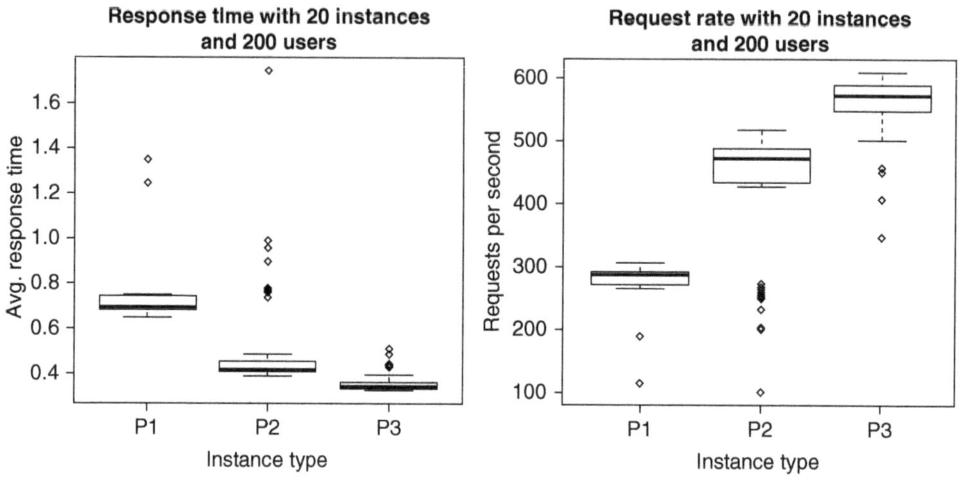

Figure 13.28 Response time with 20 instances and 200 users.

Finally, both the user load and the number of instances were kept fixed to 200 and 20, respectively, varying the instance type from P1 to P3. As expected, as the instance cores and RAM grow, the AM can respond faster, and in this case, the request rate increases with the instance size, as shown in the boxplots of Figure 13.28. The average response time is 0.77, 0.52 and 0.36 seconds; and the request rate is 270.2, 423.4 and 555 req/s, with types P1, P2 and P3, respectively.

13.6.5.1 AM Positioning Performance This section, more than providing further proof of the EGNOS effectivity, wants to show an example of the performances achievable thanks to EGNOS in a mountain environment.

To provide a performance evaluation from a quantitative point of view, a reference track was used to provide the ground truth. Exploiting a track as a reference foresees its surveying, which implies using specific measurement setup and instruments. In the particular case of this example, the test was performed on the road, climbing up to an alpine pass using a car. The car test relied on two different and independent dual–frequency and dual–constellation receivers connected to the same antenna through a signal splitter. The raw data collected by these two receivers were post-processed with the data of a virtual reference station generated by the Piemonte GNSS geodetic network. The second professional receiver was used to validate the results of the Post Processed Kinematic (PPK) obtained with the first. Both receivers gave the same positioning results: except few segments, the error of the reference track has been estimated to be less than 10cm by the commercial post-processing tool used.

An automotive-grade receiver was also used and enabled to process EGNOS SIS, while the 'user receiver' was logging raw data and, at the same time, autonomously computing the position without the use of EGNOS.

This test was performed along the road of the Colle dell'Agnello pass across the border between Italy and France. It lasted 22 minutes. The road environment and the

Table 13.3 GNSS receivers' performance comparison.

Error (m) Euclidean distance on the horizontal plane	Mean	Median	95%-ile
Professional - stand alone, dual freq.	1.2	1.0	2.9
Automotive - set to use EGNOS	1.4	1.3	2.0
UR - stand alone	2.6	2.2	5.2
UR - AM	1.8	1.6	3.6

absence of obstacles except the mountainside allowed very accurate measurements for the generation of a reference track using the post-processing method.

Table 13.3 compares the results of:

- The professional receiver set to use dual-frequency measurements for the correction of the ionospheric errors and no use of EGNOS corrections;
- The automotive receiver using EGNOS and set with an automotive-tailored Kalman filter. The reception and application of EGNOS corrections from SIS were declared to be active for the whole track, in fact the road ramps up on the south-looking side of the mountain (i.e. no mountain shadowing of the EGNOS satellites);
- The mass market receiver included in the CD performing a stand-alone solution;
- The AM output, based on the raw data measured by the receiver embedded in the 'user device.'

It is important to note that the 'user device' receiver used its own patch antenna, while other receivers relied on a geodetic grade antenna. The AM processing result is comparable to the result of the automotive receiver using a geodetic antenna. In this specific case, no Kalman filter was adopted, leaving space for improvements.

13.6.6 Positioning Integrity Computation for Consumer-Grade GNSS Receivers

The present section describes the developments performed within I-REACT for the local integrity computation. The idea at the basis of the development is the joint use of EGNOS information (through EDAS) with local information extracted from the processing of the received signal. This allows for the computation of a confidence interval of the computed position. Following the geomatics fundamentals (i.e. data and measurements associated with a geo-localized point of the environment), the confidence interval introduces an estimate of the level of reliability of geolocalized measurements.

13.6.6.1 Integrity Computation, as Described by the RTCA Standard Although integrity is a complex framework, its goal is to associate a confidence interval to any position estimate produced by the location system, provided that this confidence interval can be computed in the hypothesis that the operational conditions are properly monitored, modelled or estimated. The confidence interval and the probability inherently associated with it are typically mapped to the concepts of 'protection level' and 'integrity risk'.

The integrity of the navigation solutions is the basis for SoL applications, starting from aviation. In fact, algorithms and systems have been first developed to enable safe aircraft operations using GNSS. On the receiver side, the first method to provide integrity information belong to the Receiver Autonomous Integrity Monitoring (RAIM) algorithm family (Borre 2009). They enable proper GNSS receivers to detect signal problems, which can impair positioning performance. On the other hand, at a system level, the first receiver independent system developed was the Wide Area Augmentation System (WAAS), followed by EGNOS.

EGNOS provides integrity information in two diverse ways:

1. The GEO satellite broadcasts coarse integrity data only;
2. The GEO satellite broadcasts coarse integrity data plus wide-area corrections and related integrity information.

In the first case, the coarse integrity data include *use/do-not-use* information on all satellites in view of the applicable region. The receiver includes in the PVT computation only the satellites that are flagged as usable.

In the second case, for each satellite of the constellation, the SBAS provides information to correct satellite clock parameters, the satellite position and ionosphere effects. At the same time, the use of a model is foreseen to compensate for the effect of the troposphere. At the same time, supplemental information is provided to compute the reliability of such corrections and models. Obviously, SBAS provides neither information nor corrections to mitigate errors due to the propagation of the signal in the environment surrounding the receiving antenna.

Such supplemental information refers to the estimation of the residual errors after application of the corrections, which are expressed in form of variances:

- σ^2_{UDRE} is the variance of a normal distribution associated with the user differential range error (UDRE) for a satellite after the application of fast corrections and long-term corrections, excluding atmospheric effects. It is provided for all the satellites used for positioning.
- σ^2_{GIVE} is the variance of a normal distribution associated with the residual Grid Ionospheric Vertical Error (GIVE), and it is associated with a specific point (belonging to a defined grid). It is provided for all the grid points for the specific service area (for EGNOS it corresponds to Europe) used for positioning (through a specific interpolation method).

For the computation of the confidence interval on top of the estimated position (i.e. that in the aviation domain is called Protection Level), four parameters for each satellite *i* included in the position solution are needed. They are:

- $\sigma^2_{i,\mathrm{flt}}$, which takes into account σ^2_{UDRE} and its degradation in time.
- $\sigma^2_{i,\mathrm{UIRE}}$, which takes into account σ^2_{GIVE} interpolated for the user position.

- $\sigma^2_{i,tropo,}$ which takes into account the residual error following the application of the tropospheric correction provided by a specific model.
- $\sigma^2_{i,air}$ which takes into account the operational environment (*air* stands for aviation) and the GNSS receiver characteristics.

The data for the computation of the first two terms are broadcast by EGNOS, while the following two are defined in the DO-229D-Minimum Operational Performance Standards for Global Positioning System/Wide Area Augmentation System Airborne Equipment D (2006).

As defined in this document and explained in several books on GNSS (e.g. Kaplan and Hegarty 2006), for the satellite *i*, $\sigma^2_i = \sigma^2_{i,flt} + \sigma^2_{i,UIRE} + \sigma^2_{i,air} + \sigma^2_{i,tropo}$ provides the pseudorange measurement residual error variance after the application of EGNOS corrections.

The combination of these values with the satellite geometry and an aviation-specific requirement called *integrity risk* leads to the so-called Protection Levels (PLs). These are defined in the horizontal plane (HPL) and along the vertical (VPL) and provide an upper bound to the error affecting the onboard navigation equipment. The PLs, compared to the maximum allowed errors for each phase of flight, tell the user if the navigation system is trustable or not. If the PL is bigger than the allowed error (the so-called Alarm Limit), an alarm shall be raised within the so-called Time To Alarm (usually six seconds).

The two PLs define a cylinder centred in the user position computed by the GNSS receiver, while the true position is assured (with the probability requested by the specific application) to be contained in the cylinder volume, as shown in Figure 13.29.

What is reported above stands for aviation applications. Aeronautical GNSS signal propagation models are not usable as they are in other applications because:

- they assume open-sky satellite visibility;
- they assume wings and tail multipath or ground diffused multipath.

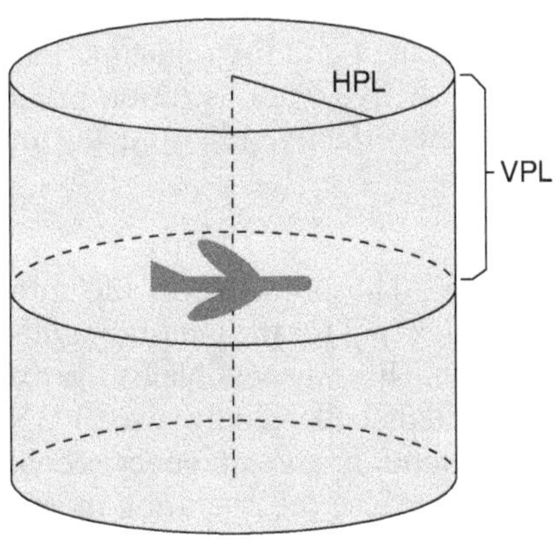

Figure 13.29 Protection levels.

In the other terrestrial applications, it is still possible to exploit the information made available by EGNOS, because it is related to the satellite clocks and orbits and signal disturbances due to ionospheric and tropospheric propagation. This is independent of the specific environment.

However, the local environment has important effects on signal propagation and, in turn, on the final accuracy and reliability of the computed position. Usually, one of the most detrimental sources of error is the multipath, which refers to the reception of the same navigation signals after one or more reflections from nearby objects. In the case of an aircraft, the effects produced by multipath are considered by models specifically developed for aviation. If other applications are considered, aviation-specific models are unfit for the last part of the signal path, dealing with complex environments like urban ones or where vegetation is present. This means that proper strategies able to evaluate the degradation due to the local environment shall be implemented. This is an open frontier of research, especially for safety-related terrestrial applications that estimate the user's position with GNSS.

The same concept is exploited in I-REACT. Although the application cannot be considered safety-critical, the knowledge of a confidence interval on top of the estimated position is considered valuable for the whole service.

13.6.6.2 Local Integrity: Latest Developments The work performed in the frame of I-REACT has been based on the local integrity concept introduced in Margaria and Falletti (2016). In its turn, this study was triggered by previous works (Cosmen-Schortmann et al. 2008; Pullen et al. 2011) about the limits of applicability of the aviation-born integrity to other transportation fields and the need for a deep reconsideration of the approach to effectively exploit it in non-aviation operations. This novel cooperative integrity monitoring concept was conceived to be suitable for automotive applications in urban scenarios. The idea is to take advantage of a collaborative Vehicular Ad hoc NETwork (VANET) architecture to perform a spatial/temporal characterization of degradations of Global Navigation Satellite System (GNSS) signals. Such characterization enables the computation of the so-called Local Protection Levels, considering local impairments to the received signals. Starting from theoretical concepts, this paper describes the experimental validation by means of an intense measurement campaign and the real-time implementation of the algorithm on a vehicular prototype. A live demonstration in a real scenario was carried out successfully, highlighting the effectiveness and performance of the proposed approach.

13.6.6.3 Proposed Algorithm The objective of the reference algorithm was the determination of a reliability index (σ^2_i) for the measurements obtained by any receiver for any specific time and location. This was possible by the exploitation of a very populated database composed of the data collected by several GNSS receivers mounted on board the vehicles travelling around. In an emergency scenario, this approach to local integrity may not be applicable because the operation theatres or the areas of interest can be everywhere and an extensive preventive mapping (and creation of a corresponding database) of the quality of the signals is unfeasible. The approach followed for the development of this generalized version of the local integrity algorithm foresaw four main phases:

- Datasets are not associated with a specific time and location; but they are just associated with distinct groups of environmental conditions. The assumption to be verified is that the effects on the signals can be associated with the environment where the user is operating;
- To classify the kinds of environment based on the elevation of the satellites in which signals are received;
- To define a signal degradation model for each environment to compute effective protection levels;
- To verify the goodness of the proposed approach by checking if errors are bounded by the confidence intervals.

The objective of the new algorithm is a probabilistic evaluation of the measurement's reliability, which is based on parameters as C/N_0, azimuth, and elevation that are commonly available. The intended output is the variance of the error component due to the local environment surrounding the antenna (i.e. a σ^2_{local}) to be used jointly with the other variances (i.e. σ^2) from the EGNOS estimates concerning GNSS systems and atmospheric errors.

The algorithm's development started from analysing some datasets taken in different environments (open sky, country, hill mostly covered with vegetation and urban) with a consumer-grade GNSS receiver (a u-blox LEO 8U). Among the measurements made available by such a receiver, the analysis was addressed to the so-called **pseudorange residuals**.

In principle, it is possible to compute such values once the receiver has computed a position. It is worth recalling that the pseudoranges are the raw measurement performed by the receiver on each incoming GNSS signal. These are ranging measurements, but they are called pseudoranges because of a common term affecting all of them by the same amount; this bias is due to the misalignment of the receiver clock with respect to the GPS time (tied to the Universal Time Coordinated). Any GNSS receiver can compute a position using the known pseudoranges from at least four satellites and solve a non-linear system of four equations to determine its coordinates and its time bias (unknowns). In case more measurements are introduced, the achieved redundancy can be exploited to improve the position estimation; this is obtained by minimizing the root mean square of the differences between the pseudoranges (once the time bias has been removed) and the distance between the estimated position and the related satellites. Such differences are the pseudorange residuals related to the errors affecting the measurements. Ideally, all the residuals should be equal to zero. However, the different error sources introduce some biases to the measurements, which affect the position computation. Therefore, it is important to highlight that the true pseudorange error does not equal the residual: in fact, the first is the difference between the real satellite-receiver distance and the measurement, while the second, being estimated by the receiver, derives from the computed position that is affected by the pseudorange errors. In any case, the pseudorange residuals provide a piece of useful information about the quality of the measurements: when their values are low, it is likely to have a good position estimate, whereas, in general, high values are a sign of errors in the measurements and consequently of worse position quality.

The idea at the basis of the algorithm uses the information about residuals to estimate the error variance of the user position. For each epoch, the set of residual variances is then combined according to the satellites geometry to provide an estimate of the position quality and, consequently the protection levels along with the different directions (along-track and cross-track).

Instead of resorting to generic analytical models for a User Equivalent Range Error (UERE) (providing $\sigma_{\mathrm{UERE},\,i}$ values for each i-th satellite as it happens in the case of SBAS), without considering the real effects of noise and multipath in specific conditions, it is possible to define an 'effective UERE' parameter ($\sigma_{\mathrm{UERE,\,eff}}$) as an *ensemble average* of several 'instantaneous' estimates of the covariance of the residuals. As detailed in Margaria and Falletti (2014), the vector **w** containing the pseudorange residuals can be put in relation with the (non-observable) pseudorange errors ε through a projection matrix **S**:

$$w = S\varepsilon$$

It can be demonstrated that it is possible to obtain an estimate of the effective UERE through

$$\hat{\sigma}^2_{\mathrm{UERE},eff} = \sqrt{\frac{\sum diag\left\{\bar{\Sigma}_w^T \bar{\Sigma}_w\right\}}{N_{\mathrm{sat}} - 4}} \tag{13.1}$$

where N_{sat} is the number of satellites used to compute the position and Σw is an estimator of the residual's covariance matrix defined as

$$\bar{\Sigma}_w = \frac{1}{M}\sum_{n=1}^{M}\left(\mathbf{w}_n \mathbf{w}_n^T\right) \tag{13.2}$$

where M is the number of independent observations and $\mathbf{w}_n$ is the n-th observations vector.

Starting from (1), confidence intervals can be computed considering the satellite geometry and the motion direction of the user. They are the Along Track Protection Level (ATPL) that bound the error in the direction of movement, the Cross Track Protection Level (CTPL) and the Vertical Protection Level. These values can be used to define an ellipse centred at the computed user position and containing the true user position with a certain probability. It is oriented according to the direction of motion.

At this point, it is worth highlighting two facts that drove the algorithm's practical implementation. The first is that the information about residuals is seldom provided by GNSS receivers (at least consumer-grade ones). From here, the necessity to exploit other information about satellites and signals more easily available. For this reason, one of the main activities performed to set up the algorithm was addressed to characterize different environments using as inputs the elevation of the satellites above the horizon and the signal strength, namely the C/N_0 and making residual ensemble averages for a set of elevation and C/N_0 values intervals. For example, all residual values collected for all the satellites in a specific environment and with an elevation of 10° are averaged. Note that satellite elevation resolution usually provided by GNSS receivers is 1°. During the current operation, the algorithm receives satellites elevation and C/N_0 values as input and associates with each satellite the averaged residual value for the specific environment; given this set of residuals, it is possible to compute the PLs by blending the residual variances coming from the evaluation based on the elevation and on the C/N_0.

Figure 13.30 shows how two different data collections yield averaged values of residuals for different satellite elevations (a) and C/N_0 values (b).

The second fact that drove the algorithm's practical implementation is less manifest, being related to the possibility of system disruptions that originates from problems at the system level (global): these events, even though extremely rare, can have a significant impact on position integrity. Indeed, SBAS is conceived to provide timely warning in case of these events involving one or more satellites: this is achieved through the constant real-time monitoring of the GNSS constellations. The development of the algorithm was addressed to the estimation of signal degradations at the local level. These degradations are independent of the system-level problems that, if present, are filtered out. In fact, the characterization of signal reception in the different environmental conditions was based on the averaging of the residual variances over extended data collections. For these reasons, local and global effects are simultaneously considered as it is done in the classical integrity, using EGNOS information (through EDAS), but replacing the variables which consider the local effects (σ^2_{air}) with the new ones (σ^2_{local}).

13.6.6.4 Validation of the Algorithm

To validate the algorithm, four trials were performed along with four paths representative of the different environments: urban area, hill with vegetation, country and open sky. They lasted from 30 minutes to 1 hour each. Three GNSS receivers were used for the validation:

- u-blox LEO 8T (configured to use GPS and Galileo);
- u-blox LEO 8U (configured to use GPS and Galileo);
- NV08C-CSM (configured to use GPS and GLONASS).

Further activity will be performed with the prototype wearable device in its final version.

The validation has been performed using a reference path (ground truth) computed with a Novatel SPAN-CPT receiver, which integrates GNSS measurements with an inertial sensor and provides sub-decimetric accuracy, and using Google Earth orthophotographs to have visual information.

The difference between the computed position with the ground truth was computed and compared with the confidence interval values. The orthophotographs show that each point belonging to the path covered during the test is contained within the ellipse defined by the computed position and the PLs as shown in Figure 13.31 in the case of the vegetated hill and Figure 13.32 in the case of the urban track.

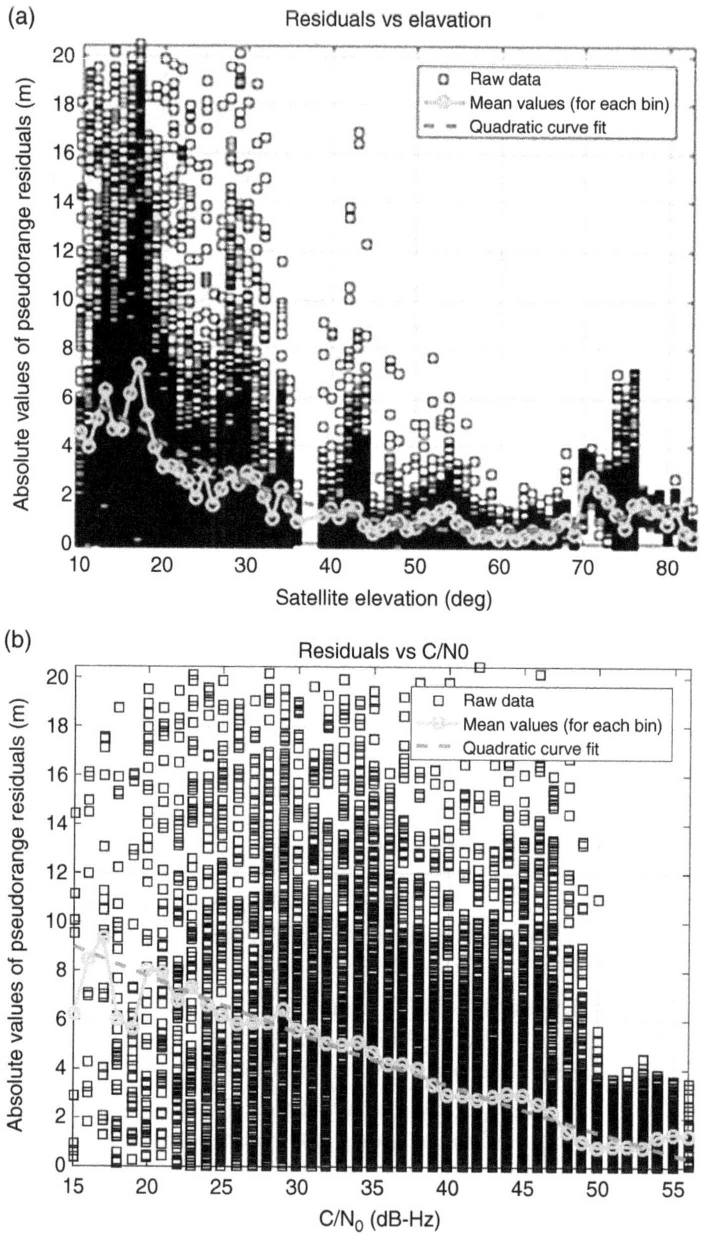

Figure 13.30 Results of data collections for the algorithm tuning. (a) and (b) show residuals' single values, their average and a quadratic fit to the average values in the case of a vegetated road on the hills, respectively, towards satellite elevation and C/N_0. Augmentation algorithm flow chart.

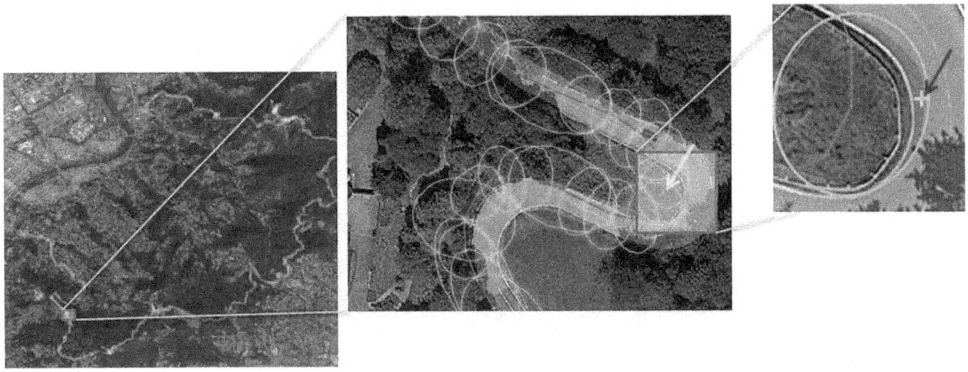

Figure 13.31 A 25 km long track zoomed in on the most critical point (i.e. worst off-track). The green arrow points to the computed position, while the red arrow points to the real position. The real position falls correctly inside the ellipse.

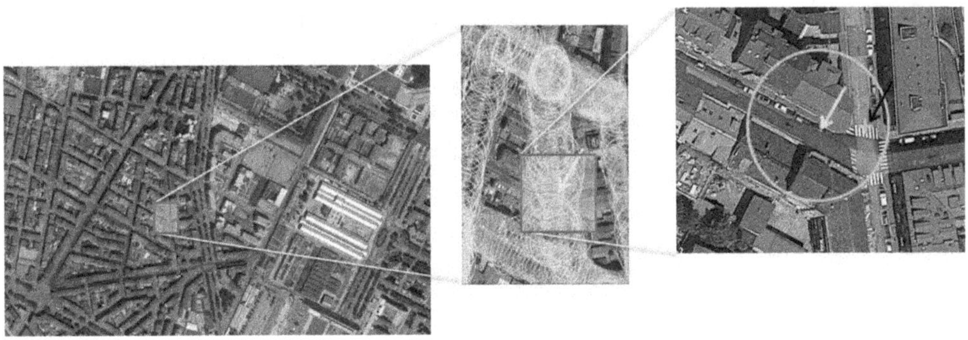

Figure 13.32 A 10 km long urban track zoomed in on the most critical point. The green arrow points to the computed position, while the red arrow points to the real position. The real position falls inside the ellipse. In this case, the data were collected using a bike passing more than once along the same streets: this explains the high ellipses density.

Table 13.4 reports an example of the results used for the algorithm validation.

The processing of the data acquired during the validation test did not show any critical points. From the analysis of data, it is possible to see that in the case of the test along the vegetated track (wood), the error was bounded by the confidence level of only 41 cm (Figure 13.31). In contrast, in the other cases, the margin is more consistent, meaning there is still some possibility of reducing the values of confidence.

13.6.6.5 The Improvement Provided by Galileo The first version of the algorithm was based only on GPS. The addition of Galileo provides a sensible reduction of the confidence level size due to the increased number of satellites and better signal quality. Figures 13.33 and 13.34 display the computed protection levels without and with the use of Galileo, respectively.

Table 13.4 Data with ublox 8T.

	Open sky	Country	Wood (hill)	Urban
Duration (@ 1 Hz) (s)	1800	2451	2367	2842
Average error (m)	1.2	1.8	2.5	4.3
Maximum error (m)	5.1	7.5	8.2	14.9
Average HPL (m)	5.56	6.5	9.01	20.26
Maximum HPL (m)	21.5	27.81	35.32	54.75
Minimum HPL (m)	4.64	5.02	5.21	10.67
SV in view (average)	13	12	10	9.6
Minimum error-PL difference (m)	1.8	1	0.41	3.6

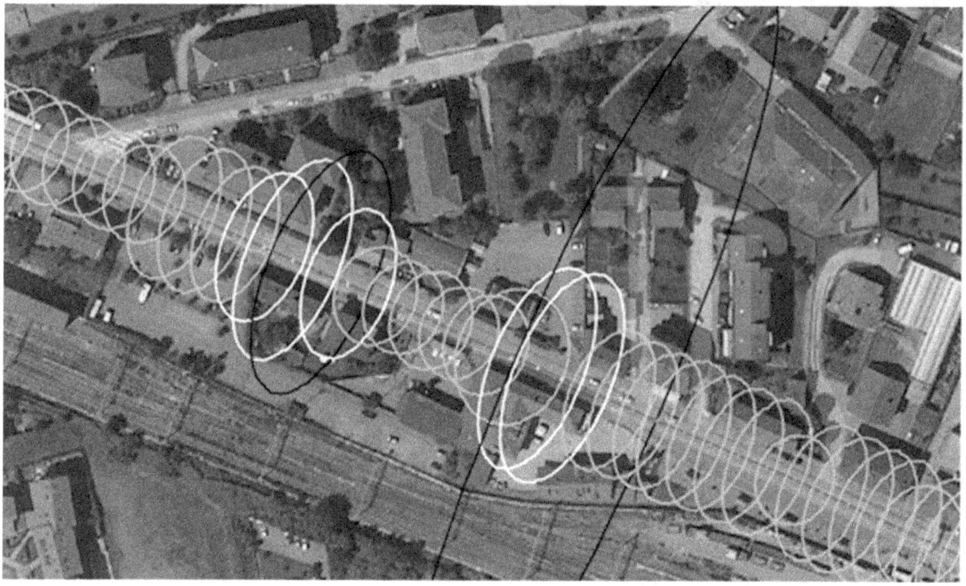

Figure 13.33 Protection levels computed without the use of Galileo.

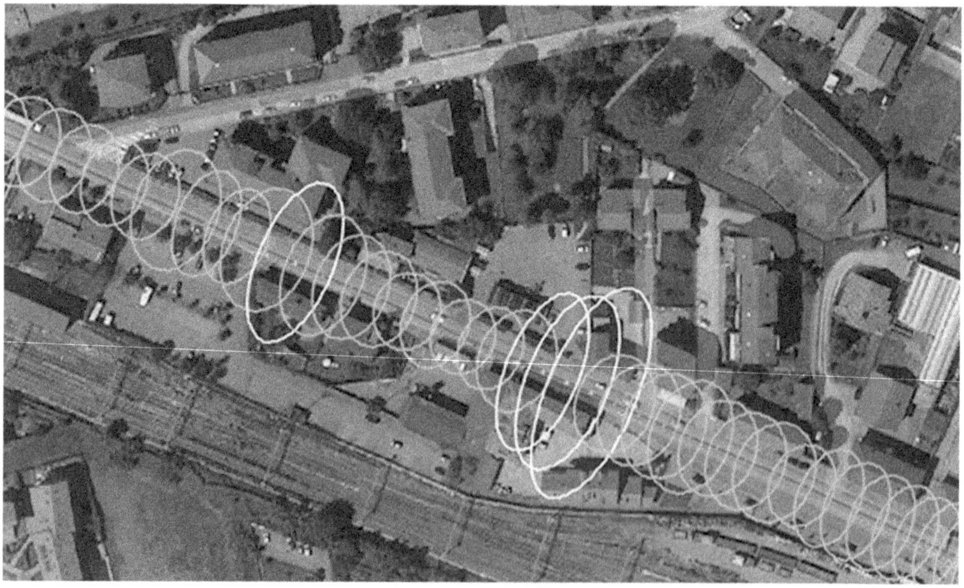

Figure 13.34 Protection levels computed with the use of Galileo.

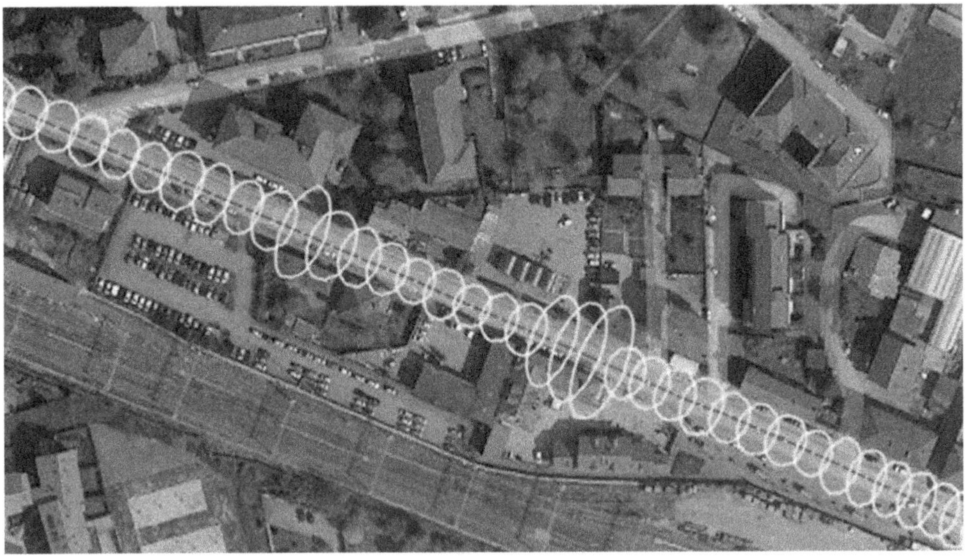

Figure 13.35 Protection levels computed after further optimization.

A further enhancement, achieved through the optimization of the parameters, allowed the sensible reduction of the Confidence Levels, which still bound the positioning errors, as displayed in Figure 13.35.

References

Abras, C., Maloney-Krichmar, D., and Preece, J. (2004). User-centered design. In: *Encyclopedia of Human-Computer Interaction* (ed. W. Bainbridge), 445–456. Thousand Oaks, CA: Sage https://doi.org/10.7551/mitpress/6918.003.0015.

Borre, K. (2009). GPS easy suite II – RAIM. *Inside GNSS* 4 (4): 48–51. https://www.insidegnss.com/auto/sepoct09-borre.pdf.

Carroll, J. (1999). Five reasons for scenario-based design. *Proceedings of the 32nd Hawaii International Conference on System Sciences*. 10.1016/S0953-5438(00)00023-0.

CEN/CENELEC, Space (2014. Use of GNSS-based positioning for road intelligent transport systems (ITS) – part 1: definitions and system engineering procedures for the establishment and assessment of performances. *DRAFT prEN 16803-1*.

Chen, F., Haritashya, U., Prakash, M., and Wang, L. (2021). Editorial: big earth data for disaster risk reduction. *Frontiers in Earth Science* 9: 727547. https://doi.org/10.3389/feart.2021.727547.

Constantine, L. and Lockwood, L. (1999). *Software for Use. A Practical Guide to the Models and Methods of Usage-Centered Design*. Boston: Addison Wesley.

Cosmen-Schortmann, J., Azaola-Saenz, M., Martinez-Olague, M., and Toledo-Lopez, M. (2008). Integrity in urban and road environments and its use in liability critical applications. *Location and Navigation Symposium*, Monterey, CA (5–8 May 2008), pp. 972–983. IEEE. https://doi.org/10.1109/PLANS.2008.4570071.

CrisisLex (n.d.). CrisisLex: crisis-related social media data and tools. http://www.crisislex.org (accessed 10 October 2021).

Deterding, S. (2012). Gamification: designing for motivation. *Interactions* 19 (4): 14–17. doi:10.1145/2212877.2212883.

DO-229D (2006). Minimum operational performance standards for global positioning system/wide area augmentation system airborne equipment D.

DRMKC (2017). DRM taxonomy. Retrieved from DRMKC – Disaster Risk Management Knowledge Centre. https://drmkc.jrc.ec.europa.eu/knowledge/drm-taxonomy#disaster-risk-management-taxonomy (accessed October 2022).

Favenza, A., Rossi, C., Pasin, M., and Dominici, F. (2014). A cloud-based approach to gnss augmentation for navigation services. *7th IEEE/ACM International Conference on Utility and Cloud Computing (UCC)*, London (8–11 December 2014). IEEE. https://doi.org/10.1109/UCC.2014.65

Foresti, G., Farinosi, M., and Venier, M. (2015). Situational awareness in smart environments: socio-mobile and sensor data fusion for emergency response to disaster. *Journal of Ambient Intelligence and Humanized Computing 6*: 239–257. https://doi.org/10.1007/s12652-014-0227-x.

Forlano, L. and Mathew, A. (2017). Codesigning urban technology for citizen engagement: from citizen-centered to collaborative cities. Retrieved from Institute of Design. https://www.id.iit.edu/wp-content/uploads/2015/03/Forlano-Codesigning-Urban-Technology-Final-3-27-1 (accessed October 2021).

Franzreb, D. and Franzreb, P. (2016). Designing with human centered usability standards. Retrieved from UX Booth. http://www.uxbooth.com/articles/designing-usability-standards (accessed October 2021).

Goodchild, M. (2007). Citizens as voluntary sensors: spatial data infrastructure. In the world of web 2.0. *International Journal of Spatial Data Infrastructures Research 2*: 24–32.

Guo, B., Yu, Z., Zhou, X., and Zhang, D. (2014). From participatory sensing to mobile crowd sensing. *IEEE International Conference on Pervasive Computing and Communications Workshops (PERCOM Workshops)*, Budapest, Hungary (24–28 March 2014). IEEE. https://doi.org/10.1109/PerComW.2014.6815273.

Kaplan, E. and Hegarty, C. (2006). *Understanding GPS: Principles and Applications*, 2e. Artech House.

Hunicke, R., Leblanc, M., and Zubek, R. (2004). Mda: a formal approach to game design and game research. *Proceedings of the Challenges in Games AI Workshop Nineteenth National Conference of Artificial Intelligence*, San Jose, California (25–29 July 2004), pp. 1–5.

Jewell, D. (2021). *Protect, toughen, augment: words to the wise from GPS founder*. http://gpsworld.com/protect-toughen-augment-words-to-the-wisefrom-gps-founder (accessed November 2021).

Marczewski, A. (2013). Understanding intrinsic motivation with ramp. Retrieved from Gamification.co.uk. http://www.gamification.co/2013/05/01/understanding-intrinsic-motivation-with-ramp/ (accessed November 2021).

Margaria, D. and Falletti, E. (2014). A novel local integrity concept for GNSS receivers in urban vehicular contexts. *Position, Location and Navigation Symposium*, Monterey (5–8 May 2014), pp. 413–425. IEEE. https://doi.org/10.1109/PLANS.2014.6851399.

Margaria, D. and Falletti, E. (2016). The local integrity approach for urban contexts: definition and vehicular experimental assessment. *Sensors 16* (2): 154. https://doi.org/10.3390/s16020154.

McGonigal, J. (2010). Gaming can make a better world. TED Talk. https://www.ted.com/talks/jane_mcgonigal_gaming_can_make_a_better_world?language=en (accessed November 2021).

Mora, A., Riera, D., Gonzalez, C., and Arnedo-Moreno, J. (2015). A literature review of gamification design frameworks. *International Conference on Games and Virtual Worlds for Serious Applications (VS-Games)*, Skovde, Sweden (16–18 September 2015), pp. 1–8. IEEE. https://doi.org/10.1109/VS-GAMES.2015.7295760.

Narvaez, R. (2012). *Crowdsourcing for Disaster Preparedness: Realities and Opportunities*. Geneva: Graduate Institute of International and Development Studies.

O.A.S.I.S, S. (2010). Standard O.A.S.I.S. Common alerting protocol. Version 1.2. http://docs.oasis-open.org/emergency/cap/v1.2/CAP-v1.2.pdf (accessed October 2019).

Olteanu, A., Vieweg, S., and Castillo, C. (2015). What to expect when the unexpected happens: social media ommunications across crises. *Conference on Computer Supported Cooperative Work; Social Computing – CSCW '15*, Vancouver, Canada (14–18 March 2015), pp. 994–1009. ACM. https://doi.org/10.1145/2675133.2675242.

Pullen, S., Walter, T., and Enge, P. (2011). SBAS and GBAS integrity for non-aviation users: moving away from "specific risk". *Proceedings of the 2011 International Technical Meeting of The Institute of Navigation*, San Diego CA (24–26 January 2011), pp. 533–545.

Rossi, C., Acerbo, F., Ylinen, K. et al. (2018). Early detection and information extraction for weather-induced floods using social media streams. *International Journal of Disaster Risk Reduction 30* (Part A): 145–157. https://doi.org/10.1016/j.ijdrr.2018.03.002.

Schulze, K., Lorenz, D., Wenzel, B., and Voss, M. (2015). Disaster myths and their relevance for warning systems. *Proceedings of the ISCRAM 2015 Conference*, Kristiansand (24–27 May 2015).

Spencer, D. (2009). *Card Sorting. Designing Usable Categories*. New York: Roselfeld Media.

Starbird, K. and Palen, L. (2011). "Voluntweeters": self-organizing by digital volunteers in times of crisis. *CHI '11: CHI Conference on Human Factors in Computing Systems*, Vancouver, BC, Canada (7–12 May 2011), pp. 1071–1080. https://doi.org/10.1145/1978942.1979102.

Stepanek, M. (2015). Building with, not for. Stanford Social Innovation Review. https://ssir.org/articles/entry/building_with_not_for (accessed November 2021).

Warfel, T. (2009). *Prototyping: A Practitioner's Guide*. New York: Roselfeld Media.

Witten, I., Frank, E., Hall, M., and Pal, C. (2016). *Data Mining: Practical Machine Learning Tools and Techniques*. Burlington, MA: Morgan Kaufmann.

Zichermann, G. and Cunningham, C. (2011). *Gamification by Design: Implementing Game Mechanics in Web and Mobile Apps*. Sebastopol, CA: O'Reilly Media.

14

Transforming Data Coming from Social Media Streams into Disaster-Related Information

Claudio Rossi[1], Edoardo Arnaudo[2], Dario Salza[1], Giacomo Blanco[1], and Lorenzo Bongiovanni[1]

[1] *LINKS Foundation, Turin, Italy*

[2] *Dipartimento di Automatica e Informatica (DAUIN), Politecnico di Torino, Turin, Italy*

14.1 Introduction

Using social media data during emergencies is challenging due to several factors, including data availability, reliability, completeness, and heterogeneity. However, social media providing open APIs, such as Twitter, can be effectively used as a complementary data source to automatically retrieve and classify relevant emergency-related information, detect events, and map their impacts.

In recent years, social media has become one of the most relevant sources of news and information worldwide. Thanks to their capillary diffusion and the inherent ease of use and content sharing, they represent an incrementally valuable source of information, available in real-time. This capability is essential during crisis events, such as wildfires or floods, thanks to the ubiquity of mobile devices, which allow for the reception and transmission of multimodal content to anyone or anywhere on the planet.

Moreover, social media can spread information extremely fast, sometimes even faster than official news sources. It is not uncommon for citizens to be the first ones to report unforeseen emergencies, such as in the case of the Manhattan and New Jersey bombings in 2016 (Saleem et al. 2014). This is also true for huge crisis events, whose magnitude makes it impossible to cover every aspect quickly. For instance, the latter was the case with Twitter during the Harvey and Houston hurricanes in 2017. As shown in Figure 14.1,

Responding to Extreme Weather Events, First Edition. Edited by Daniel Sempere-Torres, Anastasios Karakostas, Claudio Rossi, and Philippe Quevauviller.

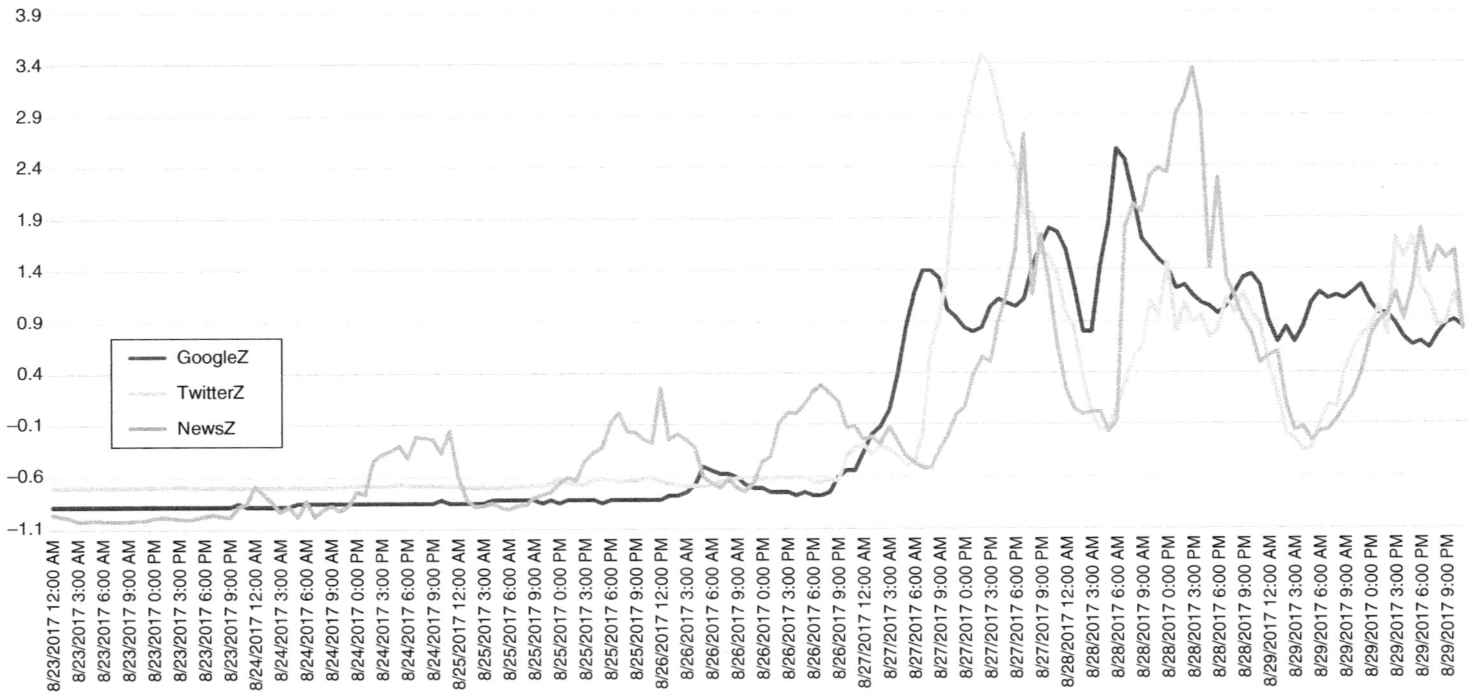

Figure 14.1 Google searches compared to tweets and news coverage regarding Hurricane Harvey and Houston. *Source*: Forbes (2019).

news coverage was already active days before, sharing advice and warnings concerning the upcoming emergency. However, status updates of the actual hurricane started to appear on Twitter hours before the official news coverage caught up.

While on the one side, this enables the diffusion of breaking news from people witnessing and experiencing the events firsthand, on the other side, the incredibly reactive scenario of social media provides users with an unmanageable continuous stream of fragmented, noisy, and sometimes outright wrong information that still requires careful filtering, selection, and aggregation in a coherent way (Piscitelli et al. 2021).

Therefore, exploiting data coming from social media during emergencies can be challenging due to several factors that can be summarized in data availability, reliability, completeness, and heterogeneity.

While the first point may sound contradictory, given the diffusion of social media thanks to the Internet and mobile devices, its usability for analytical purposes is often restricted or limited in terms of content and volume. Many platforms, such as Reddit and Twitter, expose a comprehensive set of developer API services that can be accessed freely. Although the quantity of available information is still highly relevant, some precautions must be taken to maximize the recall of retrieved content.

Reliability is the most crucial issue concerning social media content, given the amount of data shared limitlessly throughout the world. The source of information is often attributable to a standard user instead of an affirmed news channel; thus, its reliability might be highly questionable. Also, the problem of fake news has become so relevant that it is often challenging to discern between fact-checked content and fake information. To this day, many social platforms have introduced, with varying degrees of success, some control mechanisms, starting from simpler ones such as verified users on Twitter, which can be exploited to skim unwanted content from the stream further.

Completeness represents a limitation inherent to the social media format. While a news report is typically characterized by a single comprehensive article summarizing the whole event in a coherent document, social media information appears as a multitude of fragmented, incomplete, and noisy data scattered around the platforms. Sometimes, the causes can be traced back to human and technical factors, e.g. a single person reporting a sudden emergency is often unaware of the incident and its context. Moreover, social media platforms may impose some limitations regarding the amount of content of a single post. Twitter is a striking example of this: in November 2017, the length limit of a single tweet was extended to 140–280 characters. Despite such an update, more than the allowed text length is needed to summarize an emergency effectively.

Last, heterogeneity can be interpreted as both a strength and a weakness of social media platforms. Indeed, the presence of multilingual and multi-modal content being sent from various sources around the globe represents a drawback in terms of accessibility and usability for both users and analysis purposes. However, considering the analytical potential for event detection and impact assessment, dealing with many languages and modalities can benefit the final aggregation and estimation accuracy.

Given the plethora of data available on the internet, selecting which social media is the best fit for real-time analysis of incoming news is not a trivial task answer. While networks like Facebook most certainly present a more significant data volume, their content generally aims for personal or entertainment purposes. Thus, they are not particularly suited for real-time analysis aimed at emergency management. Moreover, we are mainly interested in platforms primarily distributing news and information.

For these reasons, we investigated two leading social networks, i.e. Reddit and Twitter, which are the most popular microblogging and news-sharing platforms to this date. For evaluation purposes, we compare them to other information service providers such as Google News using three main criteria of judgement: platform popularity, amount of daily content and format of the provided information, and the accessibility and ease of use of the provided API, if any.

Starting from the network popularity, despite the comparable number of users (around 330 million monthly active accounts). Twitter maintains a considerable upper hand compared with similar websites. In Figure 14.2, we report the research trends linked to the three candidates. It is clear from this graph that information services like Google News only display a small amount of the data volume that gets posted daily on social platforms. This is expected, as news websites typically present relevant and filtered information. However, that is not well suited for an automatic ingestion and analysis approach, where quantity is necessary to reduce uncertainty.

Considering the amount of daily content, similar figures can be obtained. Reports on daily usage on these three websites show a comparable daily amount of content on both Twitter and Reddit. On average, five hundred million tweets are posted every day, according to LiveStats.[1] Compared to Twitter, Reddit presents a smaller yet significant volume of data, with more than one million posts per day and around 10 million comments each 24 hours, according to the statistics platform SubredditStats.[2] It is also worth mentioning that the smaller numbers are also accompanied and justified by a different format. While Twitter allows for a maximum of 280 characters per tweet, Reddit does not impose hard limits on the length of the textual content, and posts can reach up to 40000 characters, therefore containing much more information when

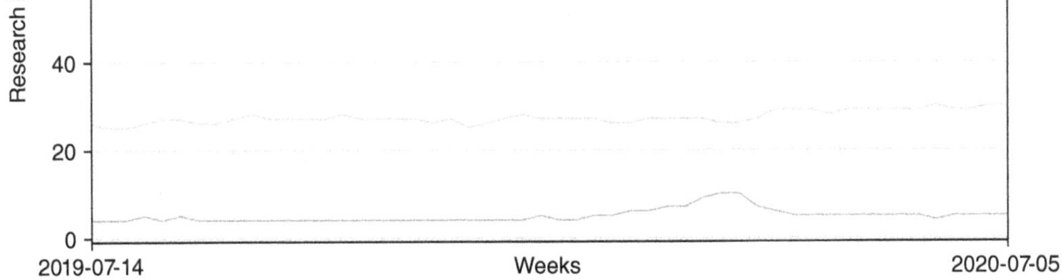

Figure 14.2 Weekly interest trends for the most popular social news platforms and services.

compared to the microblog format. Despite the difference, both platforms ingest a massive amount of content every minute, which is hardly comparable to news aggregator services like Google News.

Lastly, we analysed the development functionalities offered by these platforms, their ease of use and their relevance in the real-time detection of hazardous events. Among these services, Reddit APIs[3] provide the most comprehensive set of endpoints and operations, ranging from user details to thread updates. However, the platform exhibits two significant drawbacks for our use cases. First, the limited support for live notifications does not allow for fast and continuous updates of upcoming events. Second, the website's inherently compartmentalized structure, divided into branches named *subreddits*, only allows for a narrow overview of the situation. The same is true for news aggregator APIs such as Google News. The limitations are even more substantial due to the complete lack of live notifications and limited request counts for the free tier accounts.

The Twitter API service is the most popular among the considered websites and the most suitable for our use case. With a wide range of possible tools and solutions for different use cases, Twitter Streaming APIs are extremely valuable for real-time content analysis, continuously posted on the platform. Moreover, Twitter allows free access to a comprehensive, albeit limited, stream of posts with at most 1% of the current total volume, further allowing to filter the content with a maximum of four hundred keywords, languages, and geographical areas. This system provides a comprehensive view of what is happening almost everywhere, allowing one to easily track possible hazardous events by specifying carefully selected keywords. Given the reasons mentioned above, we adopted Twitter as a real-time information source. However, further investigations on other platforms are encouraged since they could provide supplementary data and integrate the knowledge extracted from simpler tweets.

The following sections will focus on using different processing techniques to use social media data from Twitter in emergency scenarios effectively. Specifically, we first investigate methods and solutions to pre-process and filter away useless content, even before the algorithmic steps. Then, we present different machine learning approaches applied through the years with increased complexity, highlighting the advantages and drawbacks of each one. The discussion first addresses the single documents, including text and image processing. Then the focus will shift to how to aggregate such information and the extracted annotations into a coherent summary of events and impact assessments.

A good portion of the drawbacks listed in the previous section can be contained and minimized directly in the data ingestion phase. In fact, the two most important aspects of social media content that differ from common news sources are the enormous amount of available data coupled with the extremely noisy information continuously shared among users and different platforms.

The first step towards an effective aggregation for event detection and impact assessment is thus represented by the initial filtering to exclude all the unwanted content, even before the actual categorization. Summarizing the available solutions to this day in the context of real-time text processing, we can subdivide the approaches in three main categories: rule-based methods, lightweight machine learning methods with manual feature engineering, and more recent deep learning-based approaches.

These categories are not necessarily disjoint: for instance, rule-based systems are often interpreted as machine learning tools, as well as heavy feature engineering can also be applied for newer deep learning solutions. Moreover, they can also be applied in the same pipeline, leveraging on the strength of each approach to maximize the goodness of the final outputs.

Considering platforms such as Twitter, filtering and categorization do not only involve text: in fact, images accompanying tweets are not always relevant and, most importantly, related to the textual content. Furthermore, the latter may only be telling one side of the story. Given the limited tweet length, it is pretty common to find ambiguous or unclear documents that can be misinterpreted as uninformative content by the automated pipeline or vice versa. In this case, the presence of images or videos can sometimes provide further insight on the overall situation, integrating or complementing the information extracted from the text. However, media entities also suffer from noisy and unwanted content and thus require a similar pipeline to maintain only the suitable and relevant portion of available data.

The solutions devised for both modalities are described in detail in the following sections, highlighting the pros and cons of every solution together with the challenges of dealing with a heavy throughput of noisy heterogeneous data.

14.2 Natural Language Processing Methods for Emergency-Related Text Processing

For what concerns the textual part of the tweets, we want to be able to analyse its content and solve essentially two tasks: a first task of deciding whether a tweet is about an actual emergency and a second task of classifying all crisis-related tweets into a predefined taxonomy of emergency categories. At a high level, both these tasks are naturally formulated as NLP (Natural Language Processing) sequence classification problems, where each tweet can be considered as a document on its own and its text can be interpreted as a sequence of words or sub-word tokens, to be mapped into two sets of labels. The first task concerns whether the tweet is about an actual emergency or talks about something else, referred to as *informativeness* from now on, and the second describes the kind of emergency, namely its *information type*.

Given the worldwide diffusion of Twitter, most of its textual content is posted in English. However, considering emergency scenarios, it is extremely common to retrieve content posted in the local language or even in multiple languages when the event becomes internationally relevant (i.e. pandemics). For these reasons, we also require our model to be able to work in a multilingual setting and scale well with the number of tweets while consuming a reasonable number of resources. Another important aspect is, in fact, the real-time scenario: it is crucial to detect sudden or growing emergencies in as little time as possible. This imposes a limit on the kind of processing that can be carried out on both text and media content, maintaining a good compromise between complexity, accuracy of the processing pipeline, and resource consumption. These requirements determine the architecture we choose and the series of experiments we perform.

14.2.1 Document Representation

The first step required for any automated text processing consists of selecting a representation format for the documents. This is a crucial step that heavily influences the outcome of any natural language processing task and the set of applicable methodologies.

In literature, traditional methods involve the extraction of specific features from the text, which can be carried out both manually and with automated approaches. By far, the most used technique is named by Bag of Words (BoW). Using this approach, the text of each document is represented by a vector, where each index is associated with a word in the dictionary of available tokens and each position represents the word count for that specific token. Despite the relatively simple solution, the BoW representation remains effective and extremely versatile, with a wide range of applicability (Boulis and Ostendorf 2005; Soumya George and Joseph 2014). Moreover, many other works exploit and further expand the representations using augmentations through carefully handcrafted features, such as Term Frequency – Inverse Document Frequency (TF-IDF) approaches, which substitute the simple word count with a more comprehensive value that considers the frequency of each token both at the document level and corpus level, thus mitigating the *stop words* problem. Other approaches involve considering more tokens at a time, namely bigrams or trigrams, or considering character-level information (Peng and Schuurmans 2003; Zhang et al. 2015).

Despite its simplicity and ease of use, BoW and its derivatives suffer from several issues: first, the resulting vectors usually suffer from the *curse of dimensionality* (Bellman 1966). Because of the large number of words typically present in a corpus, it is extremely common that the vocabulary size grows in the order of tens or hundreds of thousands of words. Such dimensionality, together with the resulting sparsity, becomes a burden both for geometrical and computational reasons. Second, BoW encoding only considers local or global term frequencies, ignoring other crucial aspects such as the structure of the sentence or the semantic meaning of words, resulting in similar tokens possibly having vectors orthogonal to each other.

Subsequent studies addressed some of these problems for text representation with the computation of *word embeddings*, dense vectors that intrinsically encode the meaning of the available tokens such that words with similar meanings appear geometrically closer in the vector space. These embeddings are in turn obtained through self-supervised learning, such as word2vec (Mikolov et al. 2013a) where a neural network tries to determine a single word from its context in the Continuous Bag of Words (CBOW) version, or vice versa predict the context from the current token in the Skip-Gram variant (Mikolov et al. 2013b). Another extremely popular and effective variant is GloVe (Pennington et al. 2014), where term representations are obtained as a ratio of co-occurrences probabilities between words.

Once learned, word embeddings are an extremely efficient alternative to BoW approaches for two main reasons: first, the obtained vectors are dense and contained in terms of the number of dimensions (typically hundreds of features instead of tens of thousands). Second, because of their offline training, they do not require additional computation. Thus, the conversion from a single word to its vector representation is equivalent to a simple lookup in the dictionary of available tokens.

One of the main drawbacks of these learned word vectors is their strong dependency on the corpus on which they were trained: it is most likely that even using the same starting vocabularies, vectors in different languages will be mapped in different latent spaces. To avoid model-specific approaches, which introduce a large overhead in terms of resource requirements and maintainability, many studies have been proposed over the years to extend word embeddings to multiple languages. These range from simpler transformations in latent space with singular value decomposition (SVD) (Henry and Hofrichter 1992) or using canonical correlation analysis (CCA) (Hardoon et al. 2004), or more complex systems based on supervised or unsupervised learning, such as MUSE (Lample et al. 2018).

Although it can be a benefit from an efficiency standpoint, the static nature of word embeddings constitutes, at the same time, their major weak point. It is not uncommon to find homonyms, namely words with the same spelling but having different meanings: this is outside the scope of static word vectors since they do not consider the inherent dynamism and irregularities of the language. Recently, many approaches have been proposed to tackle this problem with context-aware embeddings, thanks to the emergence of the Transformer architecture (Vaswani et al. 2017). These new techniques based on dynamic token representations through deep learning encoders have become the de facto standard for most of natural language processing (NLP) tasks. Among the plethora of available solutions, it is worth mentioning BERT (Devlin et al. 2019), GPT (Radford et al. 2018), RoBERTa (Liu et al. 2020) or eLMo (Peters et al. 2018). Particularly relevant for multilingual applications are the set of encoders trained as Cross-Language Models (XLM) on multilingual corpora, which provide both dynamic and language-agnostic vectors, effectively encoding the meaning of a given word regardless of the original language. For this project's scope, the most suitable language model available to this date is XLM-T (Barbieri et al. 2022), a multilingual network based on a cross-language RoBERTa model, directly fine-tuned on Twitter data.

Despite the undeniable breakthrough and leap in terms of prediction quality in most NLP tasks, the operationalization of transformer-based techniques remains a challenge, especially for streaming tasks, for two main reasons: first, the number of parameters is very high, even when compared to other deep architectures. Currently, even considering smaller distilled models, the smallest multilingual encoder contains 104 million parameters (Sanh et al. 2019). Second, inference speed is crucial for multilingual real-time tasks. Unfortunately, given their complex structure, most performant solutions still can't compete with simpler but faster techniques such as convolutional neural networks (CNNs).

Given the described techniques, in this work, we first adopt a relatively simpler yet robust baseline based on multilingual embeddings, representing the best compromise between model accuracies and resources consumed. Then, we compare it with an analogous evaluation of multilingual text processing using transformers, highlighting performances and differences from the former solution.

14.2.2 Document Classification

Once a document representation approach has been established, the resulting encoded information can be exploited for various tasks, including text categorization and sentiment analysis. Similarly to text representation approaches, document classification methods have progressively ranged from simpler probability-based techniques to more complex and resource-intensive deep learning architectures. Considering both sparse and dense word vectors, among classical methods, a robust solution remains the well-known Naïve Bayes approach, based on Bayes' theorem (Kim et al. 2006) which, given its simplicity and ease of use, it is typically paired with BoW representations. Other effective techniques commonly used in literature include K-nearest neighbours (Bishop 2007) or support vector machines (SVM) (Tong and Koller 2001). These methods are usually coupled with a BoW model and provided that an extensive fine-tuning and engineering phase is carried out, decent results can usually be obtained with minimal resource consumption. For this reason, they can usually serve as a strong baseline, which could be sufficient in some cases. However, better results can be obtained using deep neural networks.

Deep learning models have by now achieved state-of-the-art results in every NLP application, mainly due to their inherent ability to provide limited but dense information encoding from the document features. Apart from simpler multi-layer perceptron models, two main deep network architectures are typically applied in literature for text categorization tasks or sentiment analysis, namely recurrent neural networks (RNNs) (Mikolov et al. 2010) and CNNs (Lai et al. 2015), including all the possible variants. For instance, in Nguyen et al. (2017), a long-short-term memory (LSTM) network is employed to preserve the context information needed for the classification, as the inherent linear flow of documents represents the perfect input format for a sequential architecture.

Similarly, other works (Shi et al. 2019) show how a CNN can be used to provide an efficient and effective way to compress a whole sentence into a fixed-length vector, well suited for classification. A common rule of thumb applied for language tasks is the introduction of different kernel sizes to extract different information from the context around words, which can lead to good representations of the sentence or the whole document. However, this performance boost comes at the price of a higher number of parameters in the model and, therefore, greater resource consumption.

14.2.3 Named Entity Recognition

Named Entity Recognition is a key component in many NLP tasks where specific content type needs to be extracted from the text. In the last decades, many different methods have been developed and can be divided into rule-based approaches (hand-crafted rules based on the specific task), unsupervised learning techniques, feature-based supervised learning, and deep-learning-based methods. Among simpler solutions, conditional random fields (CRFs) remain quite effective and commonly used for tasks such as biomedical entity recognition or question answering.

As for the classification tasks, deep neural networks are the most widely used solution. In terms of neural architectures, some of these approaches comprise again the use of a CNN (Tang et al. 2020) with a fixed window length to capture the context around the word. This way, the network can represent each term. However, a fully parallel system cannot be implemented, as recursive sequential passes must be made to reduce the feature vector into a single output tag, emulating an RNN. For this reason, most studies use word-based or character-based RNNs such as Gated Recurrent Units (GRUs) and

LSTMs that inherently represent each input word as a vector storing the previous information. The advantage of recurrent networks is that they automatically maintain the context necessary for predicting the tag. Unlike CNNs, the context does not depend exclusively on the size of the window, and information can be maintained on the whole context seen until now.

As briefly mentioned in the previous section, transformer-based solutions can dynamically encode words and their surrounding context, usually obtaining better performances sacrificing inference speed and increasing resource consumption. A simple classification head is attached to the large encoder to obtain the named entity tags. A typical output layer is a single linear layer followed by a Softmax activation, which can be further improved, in a similar fashion to recurrent networks, by adding a CRF (Settles 2004).

We include here a preliminary test of classification and Named entity Recognition based on transformers, which will be assessed for deployment. Our selected approach also uses a standard LSTM network, as described in the following sections.

14.3 Model Architecture

As mentioned before, we will leverage a selection of pre-trained Language Models to map the text of the input Tweet, in any of the supported languages, into a sequence of word embeddings, representing the word meaning with a dense vector. This sequence of vectors will then be the input of a classifier that will learn the task of predicting the informativeness and informative type of the Tweet itself. In this sense, the classifier is decoupled from text and language, depending only on the word embeddings. We will consider different models that span from easier and lighter architecture to more complex in such a way that we are able to compare the performance and scalability of each one.

Support Vector Machines: Support vector machines (SVMs) is a class of models of relatively simple architecture and computationally light. SVMs learn the class boundaries in the input vector space during training, according to some chosen metric function. At prediction time, SVMs assign each vectorized input to a class according to the learnt boundaries. Despite being computationally much less demanding than most deep learning techniques, SVMs have shown to be very successful; therefore, we believe this model will offer a solid baseline to compare with more advanced models.

Convolutional Networks: In Figure 14.3 we present a schema of the CNN classifier architecture. Our proposed system consists of standard one-dimensional convolutional networks using a variable window length.

We apply a series of different kernel size filters to the input matrix so that each output focuses on various features while scrolling on the input matrix. Specifically, we opt for 60 filters with kernel size 2, corresponding to bigrams, with 60 filters of size 3, observing the trigrams and lastly, 60 filters with a bigger window of size 4 to focus on the overall context. After the convolution layer, we apply max-pooling over time with size 60 so that only the most representative features are maintained, and the variable-length input is forced into a fixed output sequence for the final classification. Repeating

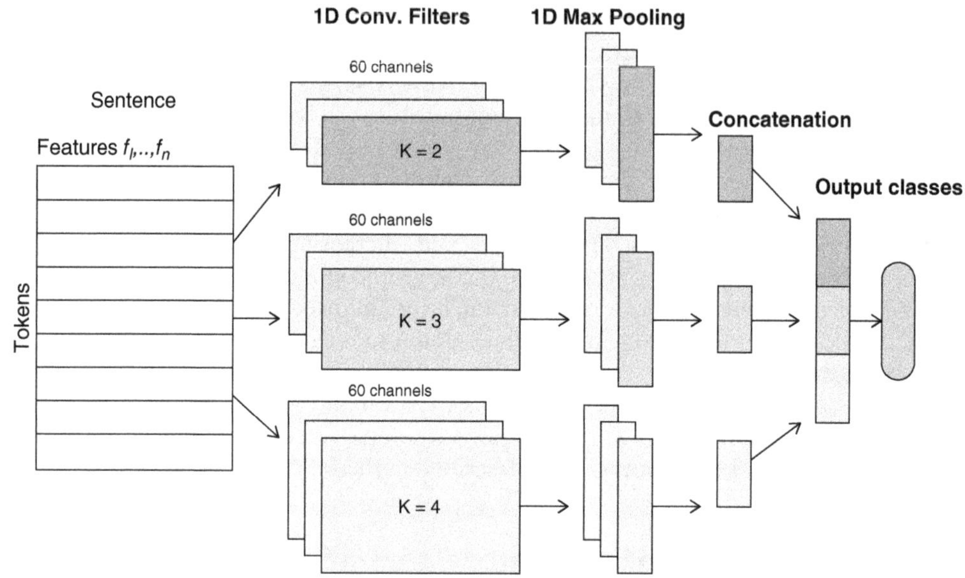

Figure 14.3 CNNfor text categorization from embeddings.

the same parallel operation in every case, all three outputs are concatenated into a single 180-dimensional vector. The latter is then used as input for the classification layer, which consists of a fully connected layer with N output neurons, where N indicates the number of classes.

14.4 Classification Results

We experiment and report results on three incrementally more robust setups: first, we build a baseline by considering a Bag of Word (BoW) approach, where the classifier is unaware of the order in which the words appear in Tweets, with an SVM classifier on top. As a second trial, we train a more advanced CNN-based classifier using multilingual FastText word embeddings. Last, we exploit the same CNN developed in the previous step but add XLM-T-generated contextual word embeddings to improve the multi-language performance.

14.4.1 Bag of Words with SVM

As a baseline, we use a Bag of Word model using handcrafted features based on the TF-IDF (Term Frequency – Inverse Document Frequency) technique. Tweets are represented by a vector with dimensions equal to the vocabulary size. We maintain a maximum vocabulary with the top 4000 terms found in the corpus, thus leading to high dimensionality and sparse representation. To obtain better representations, we eliminate the stop-words and considered only those tokens with a document frequency

Table 14.1 Results using a BoW with an SVM classifier.

Categories	Precision	Recall	F1-score
Caution and advice	0.71	0.46	0.56
Donation and volunteering	0.80	0.83	0.82
Infrastructures and utilities	0.64	0.44	0.52
Other information	0.69	0.86	0.76
People affected	0.76	0.59	0.66
People injured or dead	0.80	0.77	0.78
People missing or found	0.69	0.09	0.26
F1 macro-averaged = 0.59			
Informativeness	0.86	0.88	0.87

greater than 10. We apply an SVM classifier using the default parameters, namely a linear kernel with a tolerance set to $t = 10–3$. The results are shown in Table 14.1.

We can observe that although this baseline method is constructed from standard NLP methods in the pre-language model era, i.e. pre-word-embeddings and deep language models, the performance is quite good, especially for some classes, e.g. 'Donation and Volunteering' and 'People Injured or Dead'. We conclude that the method represents a strong baseline despite its simplicity.

14.4.2 CNN with Multilingual Word Embeddings

We then build upon the baseline by adding two primary improvements to the previous model:

1. We represent input documents as a bag of world of multilingual semantic word embeddings, namely FastText. Vectorization through word embeddings is supposed to capture the semantics of both words and documents better than the standard TF-IDF vectorization.
2. We replace the baseline SVM model with the more modern CNN-based classifier described in Section 15.1.2.

For the model training, we adopt an Adam optimizer with a learning rate set to 0.001, a dropout probability set to $p = 0.5$ and batch size equal to 64 and using a Cross-Entropy Loss with smoothed class weights inversely proportional to the label frequencies to maximize the recall. We iterate for 30 epochs, evaluating the performance on the test set. Training and test sets are multilingual with the same sets of languages. The results are summarized in Table 14.2.

We notice that, although the model and input features are more advanced, there is a little overall improvement as far as the macro-averaged F1 score goes. Nevertheless, we can observe that recall is more robust for most of the classes with this set up, which makes this model preferable over the baseline. This is expected, thanks to the improvement over the document representation approach and the underlying deep learning

Table 14.2 Results using a CNN classifier with multilingual FastText word embeddings.

Categories	Precision	Recall	F1-score
Caution and advice	0.71	0.30	0.42
Donation and volunteering	0.75	0.70	0.73
Infrastructures and utilities	0.66	0.47	0.55
Other information	0.63	0.85	0.73
People affected	0.84	0.68	0.75
Not relevant	0.63	0.29	0.40
F1 macro-averaged = 0.60			
Informative	0.84	0.81	0.82

model substituting SVMs. However, this approach's major drawback is the static nature of the embeddings. The following sections address this issue by assessing the effectiveness of contextual word embeddings.

14.4.3 CNN with XML-T Contextual Word Embeddings

Motivated by trying to improve the model even more, we will explore an additional set-up in which the input BoW of FastText embeddings is replaced by a sequence of contextual-word embeddings generated by the XLM-T multi-lingual deep language model. The contextual embeddings generated this way not only should better represent the semantics of the word itself but are also able to capture the context in which it is used. Other than that, we do not change the downstream CNN classifier, which is the same adopted in the previous section.

Although in principle, the XLM-T transformer network can be fine-tuned at training time together with the CNN, experiments have shown that this would only increase computational complexity without significantly improving performance. For this reason, we will freeze XLM-T's parameters and only use it to generate high-quality out-of-the-box contextual embeddings while focusing training only on CNN's parameters. This setup is similar to the previous ones, where we have changed the tokenization of the input text and the token embeddings while we have kept equal training of the CNN classifier. We train with Adam optimizer with a learning rate set to 0.001, a dropout probability set to $p = 0.3$ and a batch size equal to 64. As in the previous subsections, training and test sets are multilingual with the same sets of languages.

The results are summarized in Table 14.3, where performance improves for every class in terms of precision, recall, and F1 score. This is a direct consequence of the fact that transformers-generated contextualized word embeddings can better capture each word's semantics, resulting in better overall word embeddings. Obviously, downstream tasks, such as the CNN classifier trained here for hazard classification, equally benefit from having inputs that better represent the semantics of the text to be classified.

In conclusion, we explored three different architectures to approach the task of hazard type classification, ranging from simple statistical methods, simple and fast, to state-of-the-art deep language models that are more performing but slower and require bigger computational effort. We observe that, even with a simple architecture,

Table 14.3 Results using a CNN with multilingual XML-T contextual word embeddings.

Categories	Precision	Recall	F1-score
Caution and advice	0.73	0.55	0.63
Donation and volunteering	0.78	0.87	0.82
Infrastructures and utilities	0.65	0.59	0.62
Other information	0.75	0.80	0.77
People affected	0.84	0.78	0.81
F1 macro-averaged = 0.75			
Informative	0.88	0.88	0.88

it is possible to obtain good results on this task. However, employing more sophisticated NLP techniques brings some advantages. Specifically, we demonstrated that using contextual word embedding generated by XLM-T deep language model consistently yields the best results across every type of hazard.

14.5 Image Filtering and Classification for Contextual Awareness

During crisis events, multimedia content posted on social media represents a valuable source of information that can increase the contextual awareness of first responders involved in emergency management. The main drawback is the need to validate this content because of its varying nature. Most of the time, it is not related to emergencies or, even worse, not appropriate. The data validation process could be expensive when manually performed by a human operator and has significant limitations when dealing with large volumes of data, which are frequent during emergencies. For this reason, an AI-based approach is proposed, able to conduct this classification, albeit with some limitations, much faster than a human operator and which can therefore be used to process in real-time multimedia content posted on social media during emergencies.

The media classification is carried out in two steps: the first step aims to assess whether the image content is suitable for further processing or not. In detail, this preliminary step aims to prevent unsuitable content such as nudity and, in general, anything that requires to be treated as not informative. Images classified as *not suitable* in the first step are thus excluded from further processing in the subsequent model. Images that bypass this filter are instead forwarded to the multilabel model to classify the kind of hazard and place depicted in the image.

14.5.1 Filtering Unwanted Images

In order to filter not suitable content, an AI algorithm based on *MobileNet* CNN architecture (from now on called *NSFW* model) (Howard et al. 2017) has been trained on the *NudeNet Dataset*.[4] It contains almost 700 K labelled images obtained through web

Table 14.4 Label distribution in the NudeNet dataset.

Subset	NSFW	SFW	Total
Training	468511 (70%)	196281 (30%)	6,64,792
Validation	6248 (60%)	4058 (40%)	10,306

scanning and divided into the *nude*, *sexy* and *safe*, regrouped and respectively renamed into Not Safe for Work (*nsfw*) and Safe for Work (*sfw*). For this project, there is no distinction between *nude* or *sexy* images since both could not be considered informative content, so they have been aggregated in the first class, *nsfw*. The dataset is provided already split in a training and validation sets, whose label distribution is reported in Table 14.4.

During the training of the *NSFW* model, it was crucial to assess model performances on images collected during hazardous events and integrate this kind of images into the training process. For this purpose, the social media platform of the *I-REACT*[5] (Improving Resilience to Emergencies through Advanced Cyber Technologies) project has been used to retrieve almost 4K images that have been manually labelled as either *nsfw* or *sfw*. The obtained dataset contains images posted on social media during hazardous events between August 2020 and March 2021. In this dataset, *nsfw* images are below 1%. Since these images were attached to tweets describing some hazardous events, this measure could represent an approximate estimate of the real *nsfw* images prevalence in social media data. The downloaded images have been integrated with a sample from the NudeNet dataset and organized in *I-REACT* dataset. It consists of a balanced training set organized as follows: 5K *nsfw* images and 2.5K *sfw* images from the *NudeNet* dataset, 2.5K *sfw* images downloaded from the *I-REACT* platform. The remaining 1.5K images, again downloaded from the *I-REACT* platform, make up the *I-REACT* test set.

14.5.2 Methodology for NSFW Classification

In the realization of the *NSFW* categorization model, a deep convolutional neural network based on the *MobileNet* architecture has been adopted, using a version pre-trained on the *ImageNet* (Deng et al. 2009) dataset to speed up the training phase. Starting from this architecture, the last layer has been replaced with a binary output one. This approach allowed to exploit the high-level features representation learned by the model during the previous training and fine-tune this representation to better solve the problem under analysis.

The first set of experiments performed model fine-tuning on the training portion of the *NudeNet* dataset, optimizing the training hyperparameters to obtain the lowest false positive rate possible. In fact, in this application scenario, it is important to detect as many *nsfw* images as possible even at the cost of removing some suitable images, which is acceptable given the abundance of informative content constantly shared on social media. The second set of experiments aimed to further improve the first model by integrating images posted on social media during hazardous events into the training process so the model could get used to those images and better understand how they

	Validation set (10k) [60% NSFW, 40% SFW]				Test set (12 NSFW / 1492 SFW)			
	Val Acc	Val F1	Val AUC	Val Loss	Test Acc	Test F1	Test AUC	Test loss
Mobilenet pretrain	0.94	92.98	0.98	0.14	0.95	0.97	0.99	0.13
Mobilenet fine-tuning	0.93	0.91	0.98	0.17	0.98	0.99	0.98	0.07

Figure 14.4 Performances of the NSFW models on the validation and test set. For both cases, the accuracy, F1 score, and area under the curve (AUC) are shown.

represent informative content. This has been achieved by further fine-tuning the model obtained after the first set of experiments on the *I-REACT* training set.

The results of the first set of experiments led to a model achieving an *F1 score* of 0.92 on the validation portion of the *NudeNet* dataset, which counts more than 10 K samples. To assess model performances on real images posted on social media during hazardous events, the test set of the I-REACT dataset has been exploited. Results confirm the goodness of the obtained model since the scores on the test set are comparable with the ones obtained on the validation set, as reported in the first row in Figure 14.4. The results obtained after the second set of experiments are reported in the second row of Figure 14.4, reporting comparable results on the validation set and slightly better results on the test set in terms of *F1 score*.

14.5.3 Classifying Relevant Images

The second step of media classification is recognizing the kind of hazardous event and location depicted in the given image to provide detailed information about the crisis event and increase contextual awareness. To this purpose, a second AI algorithm, also based on *MobileNet* CNN architecture and from now on called the *Multilabel* model, has been trained. After a literature review, the *Incidents Dataset* seemed the most suitable dataset for this task. It consists of over 400 K scene-centric *class-positive* human-annotated images related to natural disasters. An additional set of almost 700 K *class-negative* images are part of the dataset and used to mitigate false-positive predictions. The released dataset has been downloaded referring to the original list of picture URLs; however, 200 K images were no longer available at the time of this writing. Therefore, the obtained dataset can be considered a subset of the original. It still covers the original 43 and 49 distinct types of incidents and places, respectively, whose distribution in class positive and negative class images is shown, respectively, in Figures 14.5 and 14.6. The images have been obtained through *Google Images*. The annotations

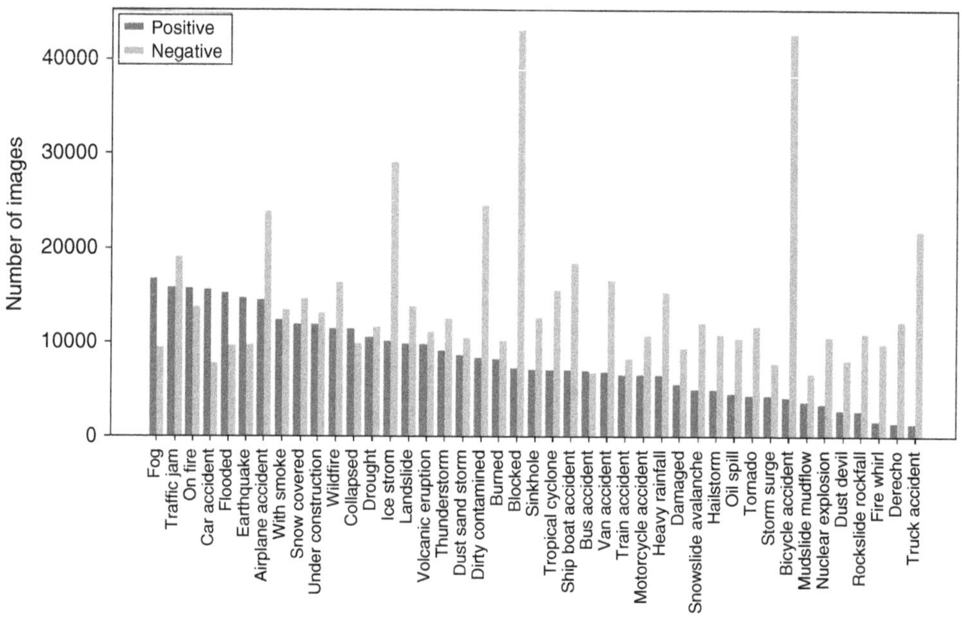

Figure 14.5 Distribution of incidents over class-positive and class-negative images in the incidents dataset.

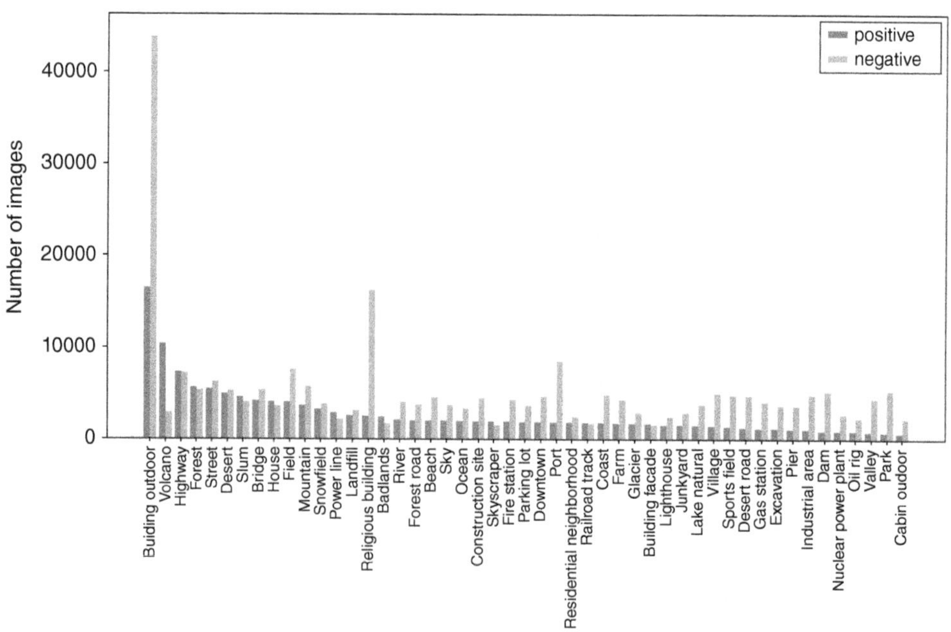

Figure 14.6 Distribution of places over class-positive and class-negative images in the incidents dataset.

result from an iterative process in which the images are labelled either as positive or negative with respect to the incident or place class associated with the query used to retrieve the image itself. For this reason, class-positive annotations are not exhaustive: the set of positive classes related to each image represents the classes that are indeed represented in the image, which, however, could still be associated with other classes not mentioned in the ground truth.

14.5.4 Methodology for Image Classification

The work presenting the dataset description also introduces a methodology to obtain a well-performing model on the dataset. The proposed approach has been applied with some modifications in the current use case. Given the two types of annotations associated with each image, incident and place, the authors propose using a single multi-task model. In this case, a single *Convolutional Neural Network* architecture, followed by two task-specific fully connected output layers, has been employed. This architecture offers efficiency as it can jointly recognize incidents and places, providing classification results for both tasks and forwarding images through convolutional layers only once. In the reference work, the authors obtained the shared convolutional architecture starting from *Resnet18* or *Resnet50* architectures and removing the last layer. Given the complexity of these architectures and the need for short inference time, we obtained our convolutional architecture starting from the MobileNet architecture. A schematic view of the used multi-task architecture is reported in Figure 14.7.

Once the model architecture has been defined, it has been trained following the methodology reported in the reference work. The authors proposed pre-training the basic convolutional architecture on the *Places365* dataset to improve the model's ability to extract visual features suitable for the place classification task. In our work, we wanted to assess the importance of model pretraining, so the entire methodology has

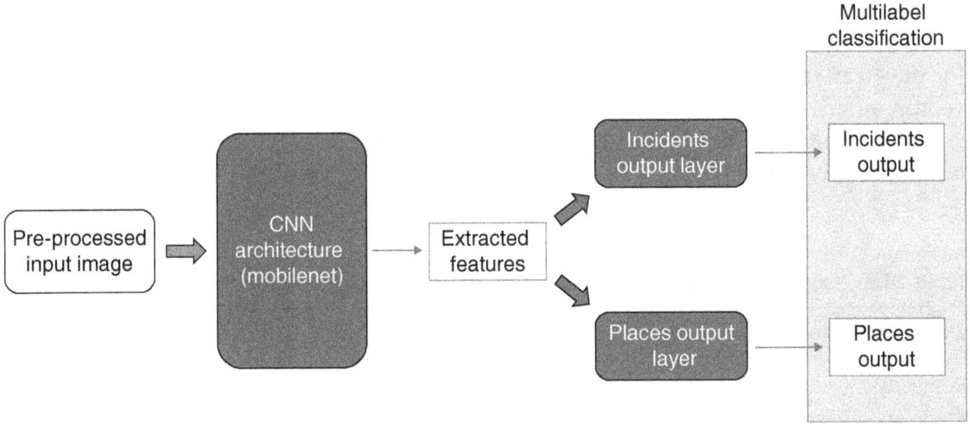

Figure 14.7 Detail of multilabel model. It exploits a single CNN architecture and two task-specific output layers.

been repeated, skipping this preliminary step and using the model pre-trained on *ImageNet* dataset instead. In the task-specific training methodology, the authors propose exploiting both class-positive and class-negative annotations during the training process to help the model find positive examples, which are usually a tiny portion out of a pool of many negative ones. A modified version of the *Binary Cross Entropy* (BCE) loss function is introduced, using class-positive and class-negative labels by means of a weight vector W associated with each training sample. For each training image with a class-positive label i, the values are set to $y_i = 1$ and $W = 1^N$ because it is possible to conclude all information is known. For a class-negative training image of class i, the values are instead reduced to $y_i = 0$ and $w_i = 1$. W is not set to 1^N in this case because there are no ground truth positive or negative labels for the rest of the classes (different incidents may or may not appear in the image). Hence, for any unknown class j, i.e., $j \neq i$, the weight is set to $w_j = 0$. The final loss is given by the sum of the *incidents* loss and the loss of the places, where the modified version of BCE gives both. We train on the same methodology, adopting a batch size of 64 images.

The obtained models have been evaluated on the *Incidents* (Weber et al. 2020) dataset validation set, using the mean average precision (mAP) obtained in both places and incidents as a reference metric. Table 14.5 reports experimental results and compares obtained models with the ones proposed in the original work. It shows how the obtained models produce comparable but slightly worse results both in incidents and places categories. The performance loss is probably due to the convolutional backbone (*MobileNet*), which is smaller and less complex than the proposed one (*Resnet50*). However, the gap can be considered negligible, given that the approach is more suitable for real-time utilization. Results also show the importance of the model pre-training on the *Places365* (Zhou et al. 2018) dataset: *mAP* in places is higher for the models that benefited from the preliminary training, even for the proposed *Resnet18* baseline trained with a standard binary Cross-Entropy loss, which had worse results in the incident classes. In these experiments, the pretraining on the *Places365* dataset has brought benefits, albeit not substantial, in terms of performance.

Multilabel predictions are also used to assign to each media a relevance score, which measures the relevance of the media to its associated event. The relevance score is evaluated by taking the *Multilabel* model classification output in the *incidents* classes for the given image and evaluating the scalar product with a previously defined weights vector, where each element represents how much an incident class is relevant for a given hazard type. The weights used in this application have been estimated as a similarity between the incident class and the hazard type, trying to prioritize the most

Table 14.5 Experimental results of multilabel model and comparison with reference work models.

CNN architecture	Class negative loss	Pretrained on places	Incidents mAP	Places mAP
Resnet18		✔	62.33	46.33
Resnet50	✔	✔	68.22	46.21
MobileNet (Ours)	✔		67.79	44.97
MobileNet (Ours)	✔	✔	68.00	45.95

Figure 14.8 Relevant media for a flood event. The media ranking system assigned it a score of 0.05.

similar incident classes for each hazard type. Given the flexibility of this technique, it is possible to easily change the media scoring criteria by tuning the weights according to what is considered a priority in the given application scenario. The two most relevant images captured by our system by, respectively, a flood and storm event are reported in Figures 14.8 and 14.9.

14.6 Event Detection

Every automation so far works on a single-document granularity. Each instance is either filtered, processed, classified, or ranked by a set of rules or a machine learning model. Considered singularly, the information contained in tweets is fragmented and disconnected, often lacking the complete picture or additional contextual data. An aggregation step is required to cope with the lack of a comprehensive output. In our case, this task goes under the name of event detection (ED), where tweets are incrementally clustered in real time to form groups of similar information bits, where the full picture of a given event can be observed. The following sections provide a detailed description of the underlying system, presenting similar approaches and the differences with our solution, the devised methodology and the experimental results on the available datasets in literature.

Figure 14.9 Relevant media for the storm event. The media ranking system assigned it a score of 0.05.

14.6.1 Related Work

Event detection algorithms applied to social media, and especially Twitter content, have been extensively studied and classified according to specific taxonomies (Farzindar and Khreich 2015), (Saeed et al. 2019). A first division occurs between the detection of *unspecified* and *specified events*. In the first case, no prior information or filtering is provided, often defining an event as a topic attracting the attention of a large number of users, causing a surge in volume for some theme or entity in the general discussion. In the second case, several attributes of the event of interest (current, past, or planned) are known beforehand, such as location, time, or involved entities. Disaster-related event detection systems often operate between these two extremes, analysing streams filtered to include only a set of topics or cover a specific geographical area of interest, but without any information on whether, when or where an event of some kind will occur, attributes that are often the target of the detection.

A second division is the one between document-pivot and feature-pivot techniques. In the former, the complete documents (i.e. the tweets) are clustered according to their content similarity. At the same time, in the latter, the volumetric evolution of some unique features (single keywords or entities, locations) is analysed and correlated.

Lastly, event detection systems may operate in near real time, attempting a New Event Detection (NED) as additional information is received from the streams, or a Retrospective Event Detection (RED), having available all the documents produced in the time frame of interest.

14.6.1.1 Feature-Pivot Methods Related to the problem of unspecified NED, (Fedoryszak et al. 2019) and (Long et al. 2011) extract cyclically relevant and bursting entities, hashtags or keywords, and place them as nodes in a clique graph where the edge weight is a measure of co-appearance of the two entities. The correlated entities are then clustered using Louvain community detection and divisive hierarchical clustering. The resulting groups are linked via Bipartite Graph Matching with those produced in the previous cycle, creating cluster chains. With a similar approach, Weng and Lee (2011) adopted wavelet transform to build entropy-based signals from the cyclical sampling of each word *DF-IDF*, filtering away trivial ones according to their autocorrelation. The graph built from the remaining signals' inter-correlation is then recursively partitioned to maximize modularity.

These methods perform well in detecting and linking generic trending entities and topics but cannot catch minor, temporally sparse events or distinguish similar but independent events.

Filtered-stream methods often exploit pure volumetric analysis: Sakaki et al. (2010) were able to detect earthquakes in Japan before the official announcement of the relevant agency by evaluating the number of tweets containing selected keywords and classified as event-relevant by an SVM-based model over time, using a probabilistic model to confirm spikes in tweet count. In addition, the epicentre is estimated from the available geolocated tweets using Kalman and particle filters. Leveraging only geolocated tweets (Dittrich and Lucas 2014), partition the area of interest into a grid and for every cell, a time-series anomaly detection is performed on the number of tweets localized inside of it, hypothesizing a normal distribution. Considering the tweets in each anomalous cell, the most represented words are extracted and matched to a multilingual list of disaster-related keywords. Same-event-type neighbouring cells are merged in a unique event. Similarly Thapen et al. (2016) detect geographically partitioned time-series volumetric anomalies, but the cells have an irregular shape and are defined with a previous clustering step, while the event topics are separated into independent streams beforehand. Considering independent tweet streams by location, Hossny and Mitchell (2018, Event Detection in Twitter: A Keyword Volume Approach) extracted the word pairs most associated with protests as a first model-building step. Then their daily count is used as input features to several classifiers to predict whether an event has occurred in the location in that day.

Volumetric-based methods are effective and reliable for significant events but less so for minor or temporally sparse ones and may be incapable or require tremendous resources to distinguish neighbouring same-type independent incidents.

14.6.1.2 Document-Pivot Methods Considering only textual information, (Hasan et al. 2016) apply a two-stage process looking for newsworthy events: in the first step an incoming tweet is compared against a buffer of recent ones, forwarding it to the second step only if it has at least one match with cosine similarity over a threshold. In the second stage, tweets are clustered incrementally and greedily, attaching a tweet to a cluster if its cosine similarity with the centroid is again over a threshold and merging clusters that match strongly with the same single tweet. Each stage has a lookup map between the most informative keywords (determined via an incremental TD-IDF) and the tweet or cluster match candidates, limiting the number of comparisons. A cluster is upgraded to event if it contains at least ten items and satisfies additional criteria, such as containing a link to a news portal. Petrovic et al. (2010) and McCreadie et al. (2013) followed a similar incremental clustering approach, leveraging Locality Sensitive Hashing to evaluate the novelty and group in the same bucket similar tweets. Liu et al. (2016) used a stack of filters based on fuzzy rules, topic modelling, and classification algorithms to isolate tweets regarding breaking news and perform a first aggregation to create unit clusters of three tweets that contain similar named entities, hashtags, and textual features. The unit clusters are further filtered by topic-level credibility and merged into compatible clusters contained in a 24-hour cache. Once generic tweet clusters are composed of unfiltered streams, the type of referenced event can be determined by evaluating and matching their keywords against those recurring in disasters, as performed by Angaramo and Rossi (2018), or more advanced classifiers.

Addressing specifically emergency events, (Klein et al. 2013) used a Named Entity Recognition (NER) model to extract from each tweet an emergency category given predefined keywords and a location or an event name (e.g. Hurricane Sandy). Tweets with the same tuple values are grouped, comparing locations along administrative hierarchies.

Incorporating geographical features in a RED clustering process, the approach followed by Ghaemi and Farnaghi (2019) is to apply and expand the VDBSCAN algorithm to group geolocated tweets, employing exponential spline interpolation to determine the different geographical search radiuses and imposing a minimum value of the textual similarity for geographically neighbouring elements to be clustered together. In a near real-time context, (Zhou and Chen 2014) propose LTT, an expansion of LDA taking in consideration beyond text content the location coordinates and time to represent a tweet as a probability distribution over a fixed number of topics (periodically refreshed), with the distance between messages computed as a combination of the bidirectional KL Divergence between the two distributions and the user link similarity. The clustering of similar elements is then accelerated through an optimized hash structure, considering hourly time slots, showing an application during Cyclone Ului and the Queensland floods. Lastly, Becker et al. (2010) show an application of single-pass threshold-based incremental clustering to group-related Flickr post using multiple and diverse features (textual content, tags, location, upload time), evaluating and discussing the performance of both ensemble and learned similarity measures.

Document-pivot methodologies can detect minor events but can be complex to scale and suffer from high noise and ambiguity in social media content. Considering geospatial features helps to disambiguate similar but independent events and to partition or index the items intuitively, leading to improved scaling.

14.6.2 Methodology

Several use-case-specific requirements drove the design of the Event Detection methodology that was developed, which is more extensively described in Salza et al. (2022):

- The system should be able to operate in real time but produce a meaningful result even in retrospective, with an unlimited number of simultaneous events. The set of tweets referencing an event should be part of the produced output and updated as new information becomes available.
- The system should detect both large- and small-scale events: while larger disasters often offer a continuous arrival of information, tweets relative to smaller incidents usually are much more temporally sparse.
- Events should be defined by their location, type, and time, and, when applicable, distinct events of the same type in the same or neighbouring locations should be detected separately. Event duration should be unbounded.
- The events should be able to span multiple locations, whether they are a large-scale disaster or a minor incident that is being reported in logically linked but different places (such as the actual location and its overarching area, e.g. county or province). The same tweet may reference more than one separate event.

The overall pipeline of the system is based on two main concepts: *cluster* and *event*. The former simply represents a collection of similar documents and is exploited as an internal data structure. The latter can be seen as a promoted (flagged) cluster that satisfies a set of required criteria and constitutes the system's output. The general schematic of the proposed system, parallel and independent for each language and hazard type combination, is reported in Figure 14.10, and follows in the first part a greedy incremental clustering approach: tweets, received from a streaming source or in batches, are individually pre-processed (**Tweet Pre-processing**) and attached to one or multiple clusters (**Incremental Clustering**). Cyclically, this process is suspended, and several cluster management operations are performed, such as **Cluster Defragmentation**, which finds and merges independent, active clusters that may describe the same event, having become similar or neighbouring over time. In these first two steps, all clusters, whether flagged as event or not, have no difference in behaviour. Subsequently, non-event clusters are evaluated for activation (**Event Activation**), after which their content is purged if expired and removed if they remain empty (**Tweet and cluster expiration**).

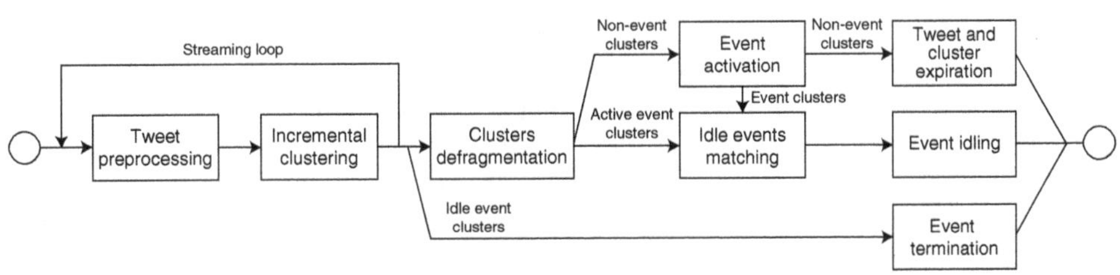

Figure 14.10 Event detection pipeline, highlighting the major processing steps.

Regarding events, the system keeps in memory two sets: active and idle events. While the first is full clusters that continue accumulating tweets in the incremental clustering and defragmentation phases, the second ones are excluded from these processes and keep a reduced representation in memory. Active events are converted into idle ones when they reach a certain age (**Event Idling**), and these last are similarly removed from memory after a timeout (**Event Termination**). Active events are periodically evaluated against idle ones, and when matching, they inherit their identifier, thus being set as a continuation of the same event (**Idle Events Matching**). After the completion of all cluster management tasks the system loops, resuming the tweet pre-processing and clustering operations.

14.6.2.1 Tweet Pre-Processing The input tweets are fetched from the backend and filtered by the informativeness classification of their textual content, discarding all non-informative tweets to remove noise and reduce volume. For all locations, a bounding box is considered rather than its centroid point since the latter does not allow computing distance correctly between neighbouring or overlapping vast areas. Tweets with no geographical information are ignored. The textual content is then tokenized, lemmatized, and POS-tagged, removing non-keyword tokens using a set of accepted POS tags rather than relying on a fixed list of stop words to remove. In addition to these, some words that, while part of the allowed POS, are not meaningful keywords useful to infer a distinction between different events (such as 'breaking news') are also removed. A single tweet is therefore represented as its tweet and author id, information and hazard type, a set of unique keyword lemmas and hashtags, a list of geographical bounding boxes and timestamps.

14.6.2.2 Incremental Clustering The clustering process is parallel and independent for every language and hazard type, and it follows the schematic of two-pass incremental clustering: an element is first assigned to a cluster with the highest similarity score that is over a predefined threshold. Otherwise, a new cluster is created. Then, recently modified clusters are periodically compared and merged to reduce fragmentation if their similarity is over a second threshold. The similarity score considers the textual keywords, geographical proximity, and hashtags. Since a single tweet may be about more than one independent event and thus point to more than one location, the first assignment is not performed directly on tweets, but rather once per named location: each surrogate document used in the comparison and then assigned, denominated *hint*, has the same identifier, text content, and hashtags of the source post, but only one of its geographic entities. Hints from the same tweet can be assigned to different clusters. For both hint assignation and cluster defragmentation, the candidates for comparison are selected and filtered by geographical proximity using an index, reducing the number of operations and improving performance. While all the recently modified clusters, flagged as event or not, are considered in the defragmentation process, only those activated are evaluated in the Idle Events Matching step. In this additional defragmentation operation, whenever the similarity between a currently active event and an idle one is sufficiently high, the active cluster inherits the identifier and starting time of the older one (and as such is marked as its continuation), while the older one is removed from memory, preventing further linking to the same event.

14.6.2.3 Event Activation All clusters modified during the current cycle and not flagged as events are evaluated and activated if several constraints are satisfied. These constraints take into account the number of unique tweet authors (to assess significance), the number of duplicate tweets (to counteract uninformative spam clusters), the distribution of hazard types (e.g. overabundance of *caution and advice* may be a sign of an incoming event, rather than an occurring one), the distribution of date entities (if a future or past date in frequently referenced, the event may be not a current one). A cluster can be evaluated for activation in an unlimited number of cycles, as long as it gets updated and modified by the inclusion of new tweets or the expiration of older ones.

14.6.2.4 Tweets and Cluster Expiration, Event Idling, and Termination Clusters must be activated and flagged as an event within a time limit from the timestamp of the earlier tweets in the cluster, otherwise all the relative hints are removed from the cluster, and their entities references are subtracted from the centroid. If, after this removal, the cluster remains without tweets, it is then completely removed from memory. Similarly, upon activation, the timestamp of the first tweet is set as the event start time, and the cluster lasts only a fixed duration time. After this time has elapsed, the event is subsequently removed from the list of active clusters and added to the idle ones. Since the continuation of the event is dependent on Idle Events Matching step and thus on the existence of a second and compatible cluster, to help the aggregation to a unique cluster of new incoming tweets related to the same event, a spinoff cluster is created. This cluster is equal to new empty clusters and is not linked in any way to the originating one, except that it is created with an already instantiated centroid, a reduced copy of the original one produced by the process shown in the algorithm, instead of an empty one. This idling with spinoff-matching and continuation allows the events to have an unbounded life span and to change over time while decreasing the probability of a cluster slowly becoming boundless and continuous in time. Finally, after a hard time limit from its start time, if the event is still present in the idle event list, it is terminated and removed from memory, setting as an end mark the time of the last tweet associated with it.

14.6.3 Evaluation of the Event Detection Pipeline

The module was implemented in Python script, leveraging the spaCy library to support the additional NLP tasks, such as POS tagging and lemmatization. At the same time, most operations on the location bounding boxes were performed using the Shapely library. As batch/cycle time, that is, the frequency at which incremental clustering is suspended and management operations are executed, five minutes was chosen as a trade-off between latency and compute time.

As a case study of the proposed system, tweets containing a set of predefined keywords covering multiple hazard types and the English, Spanish, and Italian languages were fetched and clustered in real time between 1 January and 31 August 2021. Each country was covered by exactly one language: in Spanish and Italian-speaking countries, only the events created from tweets in these languages were saved, while English covered the rest of the world. Other tweets collected between November and December 2020 were instead used to tune the various hyperparameters, manually grouping them to define the distribution of values of the similarity metrics among same-event and different-event tweets to define a first rough estimate and tweaking empirically the values with multiple runs of the system over the same period.

In the eight months of evaluation, 8.5 million out of 28 million were served as input to the clustering steps, a reduction due to the filtering by informativity and location presence, creating a total of 32 thousand events composed of more than 2 million tweets. The peak tweet count per cycle (of five minutes) was 5331 tweets the 30/07 in the morning when the Turkish wildfires gained international attention.

While 49% of events were composed of less than 10 tweets, 14 events contained more than 10 thousand, among which were the Turkish wildfires between the end of July and the beginning of August and several episodes of arson in Myanmar. Conversely, minor events were as expected relative to smaller and localized incidents; 88% of the events lasted less than one day, while the most prolonged events lasted up to 43 days.

To validate the quality of the system output, a total of 456 events, comprising 76876 tweets, were sampled randomly. Each cluster was then evaluated manually over multiple criteria (Table 14.6):

- *Content type*: whether the cluster is regarding a breaking disaster-type event and its developments (Event); is a generic discussion regarding a recent event or some aspects of it (Event-Relative); concerns an event of the past, such as in the case of

Table 14.6 Quality metrics concerning content type (A), Coherency (B), Hazard type (C) and Location correctness (D).

	Absolute	Relative (%)
A. Content type		
Event	239	52
Event-relative	83	18
Past event	11	2
Event alert	52	11
Not event	71	16
Total	456	
B. Coherency		
Coherent	397	87
Coherent with noise	39	9
Coherent with overlapping subevents	10	2
Not coherent	10	2
Total	456	
C. Hazard type		
Correct type	365	95
Wrong type	11	3
Other events	9	2
Total	385	
D. Location correctness		
Correct location	308	80
Imprecise location	30	8
Incorrect named entity extraction	12	3
Incorrect geocoding	35	9
Total	385	

an anniversary or breaking news due to a new development, such as a trial ruling (Past Event); is relative to a forecast, warning, or precautions against an expected or probable future event (Event Alert); is not a disaster-type event, but unrelated breaking news, discussed topic or generic noise (Not event).

- *Coherency*: independently from the content, if all the tweets in the cluster appear to be on the same topic (Coherent); if most of them are, and it is possible to understand a main discussion topic, but there are some evident outliers (Coherent with noise); if the cluster appears to be coherent at first, but closer inspection reveals multiple sub-topics, such as subsequent close shakes in an earthquake event or multiple neighbouring landslides or floodings during the same intense storm (Coherent with overlapping subevents); if the cluster topic is unclear or contains various independent events or topics (Not coherent).
- *Hazard/disaster type*: discarding all not-event clusters, whether the disaster type of the cluster is tagged correctly (Correct type); is not correct but is one of the other considered disaster types (Wrong type); is not correct and relative to a disaster type outside of the considered ones, such as a volcanic eruption, an explosion, or a plane crash (Other events).
- *Location correctness*: discarding all not-event clusters and considering as event location the most referenced location entity in the cluster and the corresponding bounding box as extracted by the pre-processing pipeline, it was evaluated if it was the correct one (Correct location); a location near or containing the correct one (Imprecise location); if the location was incorrect due an error of the NER (Incorrect named entity extraction); if the location name is correct, but the geocoding produced a homonym location in a different region, or even a completely unrelated result (Incorrect geocoding).

Additionally, to verify the ability of the system to detect events, the list of incidents in the considered period was extracted from the EM-DAT disaster database,[6] removing from the list those with an approximate start date or a disaster type not among the considered ones. Afterwards, Nominatim was used to geocode all the bounding boxes corresponding to the entities named in the 'Location' column, merging them in one target multi-polygon area for each event. It was then computed how many of the EM-DAT events matched to one or more activated clusters produced by our system, considering clusters with the same disaster type, a location intersecting the target one, and a starting date between the start date and up to 48 hours after the end date reported by EM-DAT. The process was then repeated for the disaster types reported in the associated disaster column. The resulting detection rates are presented in Table 14.7.

Table 14.7 Detection rates computed over the EM-DAT dataset.

Subset	Detected	Total	Percentage (%)
Global disasters	180	259	69
Disasters in English, Spanish, Italian speaking countries	55	77	71
Global associated disasters	53	97	55
Associated disasters in English, Spanish, Italian speaking countries	17	35	49

14.7 Impact Extraction

Once a hazardous event has been detected, it could be of interest to estimate, using social media data, the impact the event had on the territory in terms of damaged infrastructure and affected people. The impact estimation module described extensively also in (Blanco et al. 2022) accepts as input the list of tweets assigned to the events detected by the *Event Detection* module and for each event it generates, or updates as long as the event is updated, a comprehensive view of the situation, estimating the number of damaged infrastructure and affected people in a set of categories specifically designed for the problem and reported in Table 14.8.

It works in a two-stage fashion. The first one is responsible for extracting information about the disaster impact from each tweet by exploiting a multilingual rule-based algorithm (provided that the necessary rules have been written for the target languages that currently are English, Spanish, and Italian) and then associating each detected category with the numeric estimate reported in the text, validating this association thanks to a language model. In the second stage, the system aggregates the fine-grained information extracted from all event tweets to provide the final estimate associated with the whole event.

14.7.1 Related Work

Monitoring social media usage during disasters has been proven to be effective in estimating some important properties of the event, obtaining promising results in increasing situational awareness (Yin et al. 2012) and understanding its effects on

Table 14.8 Impact estimation taxonomy proposed. Natural disasters typically affect either population or infrastructures or both, so classes proposed in this taxonomy can be associated with two macro groups named *Infrastructures* and *Population*.

Infrastructure	Population
Road	Dead
Railway	Injured
Bridge	Missing
Port	Evacuated
Airport	Rescued
School	Infected
Hospital	Hospitalized
Residential	Recovered
Facility	
Power network	
Water network	
Cultural heritage	
Burned area (km^2)	
CO_2 emission (kg)	

society (Eyre et al. 2020; Dou et al. 2021). Provided that an accurate tweet geocoding mechanism is provided, it is possible to precisely locate the area affected by the event in a timely manner (Wang et al. 2019). The work of rapid assessment from social media (Kryvasheyeu et al. 2016) demonstrates how the activity of Twitter users was often positively correlated with the path of Hurricane Sandy. The more intense the social media activity of users affected by the storm, the higher their proximity to the hurricane's path. Similar outcomes can be provided for other hazard types, such as earthquakes: some studies (Liang et al. 2013) accurately estimate the epicentre of two earthquakes that happened in Japan and New Zealand, just by looking for the area with the highest density of earthquake-related tweets. The volume of disaster-related tweets produced in an area is shown to be strongly correlated also with the level of damage caused by the event itself, as demonstrated for hurricanes Sandy (Wu and Cui 2018) and Matthew (Yuan and Liu 2018, 2020).

Once a disaster event has been established, it is crucial to estimate its impact on the territory. Several works have demonstrated the potential and benefits of using social media streams for rapid damage assessment during and after a hazardous event. Other proposals (Resch et al. 2018) estimate earthquake footprint from information retrieved on Twitter and obtained reasonable results compared with *USGS*. Similar solutions (Shan et al. 2019) propose a disaster damage assessment model based on social media texts focusing on timeliness and disaster damage categories. They validate their findings on two disaster events that happened in China by comparing the results of Weibo (Chinese social media) analysis and the report published by official authorities obtaining reasonably reliable results in a much faster way. In Fohringer et al. (2015), the authors focused on flood events. They proposed an approach capable of providing inundation mapping with information about inundation extent and water depth leveraging images posted on social media. Li et al. (2021) propose a text-based system to classify the damage level, on a scale from 0 to 3, reported in damage-related tweets. A more formal and automated damage assessment method has been proposed in Mangalathu and Burton (2019), where a deep-learning-based text classifier has been used to estimate the damage level reported by buildings during an earthquake by analysing the messages posted on social media. The work follows the ATC-20[7] taxonomy, where a tag among red, yellow or green, is assigned to each building, indicating severe, moderate, or slight damage reported. One of the most comprehensive approaches found in literature is represented by the study of Imran et al. (2013): the authors presented a system leveraging a set of machine-learning-based text classifiers able to distinguish informative content from not informative one in social media, categorize the information type and extract from these tweets potentially useful information according to the information type, like the presence of affected people or buildings. Given the existing approaches, to the best of our knowledge, our impact estimation methodology can be seen as a practical and effective solution for two main reasons: it can be applied as a drop-in replacement for any hazardous event, given that it is not constrained to a specific hazard, and it aims at estimating a precise number of specific affected entities in a predefined taxonomy, reported in Table 14.8, instead of providing only large-scale information.

14.7.2 Methodology

The overall schema of the information extraction stage is presented in Figure 14.11: it comprises a series of modules that progressively elaborate the previous layer's output into something more semantically meaningful, structured, and, most importantly, machine readable. The process is applied independently to every single tweet belonging to the event currently under analysis. As the first operation, the input text is pre-processed to remove noisy content such as non-Unicode characters, tags, URLs, brackets, and retweet strings. The content of eventual hashtags is kept without the hash symbol and any number written in characters is converted into digits. Additional information is extracted in this phase thanks to a language model, namely the word dependencies graph, sentence boundaries, and named entities, which are instead removed from the text. Right after text pre-processing, the system identifies the mentioned categories of affected entities by exploiting a rule-based algorithm. To provide an estimate for each detected impacted category, the system retrieves, once per category, the sentence in which the affected entity was found together with any numeric expression in the sentence itself. Among these numbers, if any, we expect to find a numerical quantity describing an impact estimate for the observed category.

Moreover, the module also performs a check on each digit found to assess whether it is a good candidate estimate or not: we are interested in raw measurements, meaning that a good candidate estimate must not be part of either a date or a time string and it must not refer to scientific measurements or currencies. The system collects all the valid candidate estimates found in the sentence, keeping for each of them the impact category they are referred to, the value, the position in the input text, and the distance from the related mentioned entity. If no correct candidate estimate has been found in the sentence, a placeholder value of 0 is used. This value is used to describe the situation in which the system has detected the mention of an impact category but has not been able to find its associated estimate.

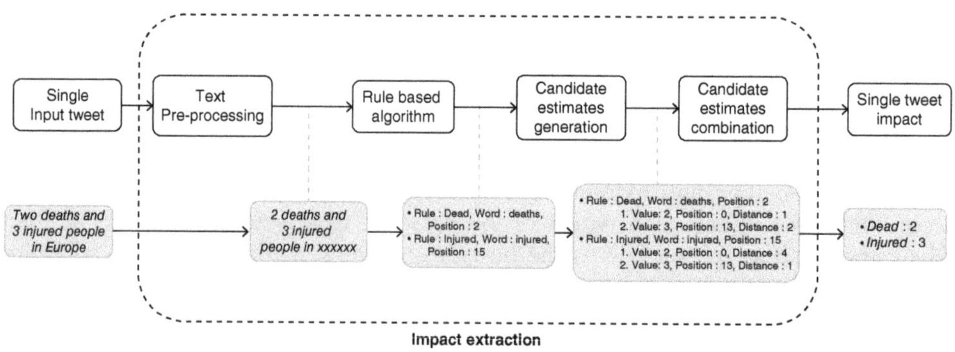

Figure 14.11 Impact extraction building blocks. The yellow blocks represent the intermediate result, for example, input text after each block.

After all the good candidate estimates have been collected, the system produces the final estimate for each mentioned category sequentially, following their appearance order in the text. Among all the good candidate estimates associated with each category, the first (in order of appearance) satisfying a set of conditions is selected as an estimate for the given category. The conditions are as follows:

- There is a path in words dependencies graph connecting mentioned category and candidate estimate.
- The position of the candidate estimate precedes any other mentioned category after the currently observed one.
- The position of the candidate estimate follows any other mentioned category preceding the current one (intuitively, it must reside between other categories, if present).
- The candidate estimate has not already been associated with any other impact category.

If no candidate estimate satisfies these conditions, the placeholder value of 0 is associated with the given impact category. The reasoning for these conditions stands in how a text reporting the number of damaged infrastructure or affected people is typically structured: in fact, it is often the case that the number of affected entities is close to the related entity (before or after) or at most in the same sentence, but it is not common to find this numerical quantity once another affected entity has been mentioned in the text, even less so to have the same number linked to two different entities.

At the end of this stage, the module has assigned an estimate for each detected impact category yielding the impact estimation information associated with the given tweet. A schematic view of the information extraction process and the results produced by each phase, decorated by an example input text, is reported in Figure 14.11.

14.7.3 Aggregating the Information

The information extracted in the first stage is organized according to the time of tweet creation: for each category reported in Table 14.8, the system stores all the impact measures extracted from event-related tweets sorted according to the creation time of the single tweet. After the first-stage completion, the impact aggregation phase combines the information extracted from each tweet to produce the overall estimation associated with the whole event, using the following simple and intuitive approach for each impact category: if there are no measures, it is established that the event has not provoked an impact for the given category. Otherwise, the event impact measure in the observed category is given by the most frequent value among the most recent 100 ones. If two measures appear with the same frequency, the biggest one is chosen as representative.

The described approach was motivated by empirical observations on a large sample of documents. The information about crisis events shared on social media platforms can easily be noisy and rapidly change over time: choosing the final value among the most recent ones helps in ignoring the initial estimates, which are usually inaccurate while choosing the most frequent value helps in ignoring eventual noisy estimations.

14.7.4 Evaluation Results

The proposed system has been realized using the Python programming language and a small set of third-party libraries. The impact extraction stage leverages two main modules: a language model and rule-based algorithms. The former comprises a text processing pipeline composed of a sentence parser (which finds sentence boundaries and captures existing dependencies between words) used by the system to create the words dependencies graph and an entity recognizer, which detects named entities in the text to remove them during text pre-processing phase.

The rule-based algorithm has been realized using the Yara Rules[8] tool. This framework allows the definition of rules in human-readable format, using simple strings of interest organized in different files. Each rule has a *strings* section in which it is possible to define all the required words for the given set in a regex-like fashion and the conditions that must be satisfied by the detected words to produce a match. For each impact category reported in Table 14.8 and for each supported language (currently English, Italian, and Spanish), a Yara rule has been manually defined. We consider a positive rule match as evidence that the related impact category has been mentioned in the text. The definition of Yara rules is quite similar among impact categories of the same type; however, rules for infrastructures and population categories have some minor differences. For the sake of completeness, we provide in **Annex 1** two sample rule definitions for English texts: one related to the population category and one specific to infrastructures.

The proposed system has been evaluated on a sample of 975 manually annotated English tweets posted during four different crisis events that happened in 2021: a flood in Nepal, a flood in Germany, a flood in North Carolina, storm in the Czech Republic. These tweets have been collected thanks to the processing pipeline explained earlier in the document. The events were carefully chosen to be representative on a global scale, considering a yearly scenario, as well as on an individual scale, selecting only crises big enough to contain impact information. Most of them are related to flood disasters since they are the most frequent natural disasters (Corban et al. 2017). The annotation task consisted of labelling for each tweet the impact information it is reporting, i.e. the number of entities affected for each category, if any.

This performance evaluation process aims to measure the system's ability to detect the presence of affected entities in an input text, specifically damaged infrastructures and affected people, and provide an estimate as accurately as possible for the detected entities. Unfortunately, tweet annotations do not cover all the classes reported in Table 14.8, so the evaluation has been done only for the mentioned categories. The number of tweets reporting an impact for each class is reported in the *Support* columns of Table 14.9.

The detection of the impact category in a text can be formalized as a classification problem. Therefore, the traditional metrics used for this task have been measured to

Table 14.9 Impact estimation classification results.

Category	Accuracy	Precision	Recall	F1-score	Support
Road	0.99	0.50	1	0.66	2
Bridge	1	1	1	1	3
Residential	0.98	0.92	0.99	0.95	197
Power network	1	1	1	1	8
Water network	1	1	1	1	1
Cultural heritage	1	1	1	1	4
Dead	0.99	0.99	0.99	0.99	714
Injured	1	1	1	1	283
Missing	0.99	0.99	1	0.99	357
Evacuated	1	1	1	1	1
Rescued	0.99	0.75	1	0.86	3
Hospitalized	0.99	1	0.5	0.67	2

characterize system performances in each impact category, whose results have been reported in Table 14.9. First, the table highlights the greater support of population-related categories concerning infrastructure in the validation set used in this experiment. High values in both precision and recall metrics show how the system most of the time correctly detects the presence of impact categories while producing false positive predictions (i.e. detecting an affected category when it is not) very rarely in almost all categories.

Once the system has detected the mention of an impact class, it is also asked to provide an estimate for the given category, which can be treated as a regression problem. For this reason, standard regression metrics such as *Mean Absolute Error* (*MAE*) and *Mean Squared Error* (*MSE*) have been measured to characterize system performances in this kind of task. The regression metrics reported in Table 14.10 have been evaluated considering only the estimates provided when the given category was mentioned in the text. Indeed, it would be inappropriate and unfair to evaluate the goodness of an estimate when it should not have been proposed in the first place. Results show an acceptable Mean Absolute Error (MSE) in each category, except in *Evacuated* and *Other* classes, where the few tweets mentioning these categories were poorly estimated. Given the low amount of validation data available, in turn, caused by these categories' peculiarity, it is impossible to validate such categories without further manual labelling.

Even though the system works at the event level, providing an estimate after aggregating information extracted from each tweet, the validation has only been done at the tweet level. The objective was to measure the system's ability to extract the correct information from each tweet given that it is the most critical phase of the process. Moreover, information associated with a single tweet is more straightforward for a human annotator. Working at the event level, instead, implies the necessity to perform the aggregation of tweet information, which can be done in several ways and is dependent on the results of the tweet extraction at the tweet level.

Table 14.10 Impact estimation regression results.

Category	MAE	MSE	Support
Road	0	0	2
Bridge	10	150	3
Residential	5.27	659.05	197
Power network	0	0	8
Water network	0	0	1
Cultural heritage	1	1	4
Dead	1.82	185.47	714
Injured	8.62	1033.97	283
Missing	3.27	73.90	357
Evacuated	140	19600	1
Rescued	0	0	3
Hospitalized	1.5	4.5	2

14.8 Annex 1: Definition of Yara Rules for Impact Estimation

We report in Table 14.11 an example of the Yara rule for the population, whose definition allows to capture the mention of dead people in an English text.

The *strings* section contains all the words that must be detected in the text because all of them could be used to describe the presence of dead people. The full word modifier at the end of each string definition is used to inform the Yara Rules tool that the full word must be reported in the text preventing partial words (e.g. *dece*) to be detected. The instruction *condition: 1 of them* informs the Yara Rules tool that any of these words is enough to produce a match and assess that the text reports information about dead people.

As previously mentioned, the definition of infrastructure rules is not as straightforward as for population ones. Finding the mention in the text of any infrastructure does not directly imply damage has happened to that infrastructure. For this reason, the detection of damaged infrastructure has been divided into two rules: one to detect the presence of any damage and one to detect the kind of damaged infrastructure. The presence of general damage in the text is captured via the rule displayed in Table 14.12.

The structure of this rule resembles population rules because the presence of any of the words specified in the strings section is enough to assess the text describing some damage. This general rule has defined a rule for each type of infrastructure of interest.

We report in Table 14.13 an example of the English Yara rule used to detect the mention of a damaged road.

The specified *condition* informs the Yara Rules tool it must produce a match for the road category if and only if the input text matches the Damage rule (i.e. the text reports information about damages). It contains one of the words in the *strings* section of this rule.

Table 14.11 Example of Yara rules for the population category.

```
rule DeadImpact : Dead
{
meta:
      name = "dead"
            description = "Rule description"
strings:
      $word1  = /(dead[s]?|death[s]?)/ fullword
      $word2  = /(die|died|dying)/ fullword
      $word3  = /deceas(e|es|ing|ed)/ fullword
      $word4  = /pass(ing|ed|es)? away/ fullword
      $word5  = /kill(s|ed|ing)?|executed/ fullword
      $word6  = /murder(s|ed|ing)?/ fullword
      $word7  = /assassinate(s)?/ fullword
      $word8  = /assassinat(ed|ing)/ fullword
      $word9  = /(lose(s)|lost|losing)\s((his|her|its|their)\s)?
(life|lives)/ fullword
      $word10 = /fatalit(y|ies)/ fullword
      $word11 = /casualt(y|ies)/ fullword
condition:
             1 of them
}
```

Table 14.12 Example of Yara rule for infrastructures and damages.

```
rule DamageImpact : Damage
{
meta:
    name = "damage"
    description = "Rule description"

strings:
    $word1 = /
  land(slide[s]?|slip[s]?)|avalanche(s)?|rock\s?slide(s)?|mudslide
  (s)?|subsidence(s)?/ fullword
    $word2 = /rockfall[s]?/ fullword
    $word3 = /collapse(d|s)?|collapsing|downfall(s)?/ fullword
    $word4 = /clos(e(s)?|ed|ure|ing)/ fullword
    $word5 = /(inter|dis)rupt(s?|ed|ion[s]?|ing)|suspen(ded|d(s)?|t
  |ding|sion(s)?)/ fullword
    $word6 = /(damag|compromis|degrad)(e[sd]?|ing)/ fullword
    $word7 = /destruct(ion(s)?|s|ed|ing)?/ fullword
    $word8 = /demoli(tion(s)?|shed|sh(es)?)/ fullword
    $word9 = /destroy(ed|ing|s)?/ fullword
    $word10 = /ruin(s|ed|ing)?/ fullword
    ellipse;
condition:
    1 of them
}
```

Table 14.13 Example of Yara rule to detect damaged roads.

```
rule RoadImpact : Road
{
meta:
    name = "road"
    description = "Rule description"
strings:
    $word1 = /road[s]?/ fullword
    $word2 = /arter(y|ies)/ fullword
    $word3 = /pathway[s]?/ fullword
    $word4 = /highway[s]?/ fullword
    $word5 = /roadway[s]?/ fullword
    $word6 = /route[s]?/ fullword
    $word7 = /itinerar(y|ies)/ fullword
    $word8 = /lane[s]?/ fullword
    $word9 = /avenue[s]?/ fullword
    $word10 = /street[s]?/ fullword
    $word11 = /motorway(\sexit[s]?)?/ fullword
condition:
    1 of them and DamageImpact
}
```

Funding

This work was supported by the SAFERS project (H2020 programme, GA n.869353).

Notes

1. https://www.internetlivestats.com/twitter-statistics
2. https://subredditstats.com
3. https://www.reddit.com/dev/api
4. https://archive.org/details/NudeNet_classifier_dataset_v1
5. https://cordis.europa.eu/project/id/700256/it
6. https://www.emdat.be
7. https://www.atcouncil.org
8. https://github.com/Yara-Rules/rules

References

Angaramo, F. and Rossi, C. (2018). Online clustering and classification for real-time event detection in Twitter. *Proceedings of the 15th International Conference on Information Systems for Crisis Response and Management* (ed. K. Boersma and B.M. Tomaszewski), Rochester, NY, USA (20–23 May 2018). ISCRAM Association. http://idl.iscram.org/files/federicoangaramo/2018/1630_FedericoAngaramo+ClaudioRossi2018.pdf.

Barbieri, F., Anke, L.E., and Camacho-Collados, J. (2022). XLM-T: multilingual language models in Twitter for sentiment analysis and beyond. *Proceedings of the Thirteenth Language Resources and Evaluation Conference, LREC 2022* (eds. N. Calzolari, F. Béchet, P. Blache et al.), pp. 258–266, Marseille, France (20–25 June 2022). European Language Resources Association. https://aclanthology.org/2022.lrec-1.27.

Becker, H., Naaman, M., and Gravano, L. (2010). Learning similarity metrics for event identification in social media. *Proceedings of the Third International Conference on Web Search and Web Data Mining, WSDM 2010* (eds. B.D. Davison, T. Suel, N. Craswell, and B. Liu), pp. 291–300, New York, NY, USA (4–6 February 2010). ACM. https://doi.org/10.1145/1718487.1718524.

Bellman, R. (1966). Dynamic programming. *Science* 153: 34–37.

Bishop, C.M. (2007). *Pattern Recognition and Machine Learning*, 5e. Springer. https://www.worldcat.org/oclc/71008143.

Blanco, G., Arnaudo, E., Salza, D., and Rossi, C. (2022). Impact estimation of emergency events using social media streams. *Impact Estimation of Emergency Events Using Social Media Streams*. IEEE. https://doi.org/10.1109/RTSI55261.2022.9905213.

Boulis, C. and Ostendorf, M. (2005). Text classification by augmenting the bag-of-words representation with redundancy-compensated bigrams. *Proc. of the International Workshop in Feature Selection in Data Mining*, Newport Beach, California (21 April 2005), pp. 9–16. https://www1.icsi.berkeley.edu/pubs/speech/bagofwords05.pdf.

Corban, C., Florczyk, A., Freire, S. et al. (2017). Atlas of the human planet 2017: global exposure to natural hazards. https://publications.jrc.ec.europa.eu/repository/bitstream/JRC106292/atlas2017_online.pdf (accessed November 2021).

Deng, J., Dong, W., Socher, R. et al. (2009). *ImageNet: A Large-Scale Hierarchical Image Database*, 248–255. Miami: IEEE https://doi.org/10.1109/CVPR.2009.5206848.

Devlin, J., Chang, M.-W., Lee, K., and Toutanova, K. (2019). BERT: pre-training of deep bidirectional transformers for language understanding. *Proceedings of the 2019 Conference of the North American Chapter of the Association for Computational Linguistics: Human Language Technologies, NAACL-HLT 2019* (eds. J. Burstein, C. Doran, and T. Solorio), pp. 4171–4186, Minneapolis, MN, USA (2–7 June 2019). Association for Computational Linguistics. Volume 1 (Long and Short Papers). https://doi.org/10.18653/v1/n19-1423.

Dittrich, A. and Lucas, C. (2014). Is this Twitter event a disaster?. *Proceedings of the The 17th AGILE International Conference on Geographic Information Science,* Castéllon, Spain (3–16 June 2014) (eds. J. Huerta, S. Schade, and C. Granell). https://agile-online.org/images/conferences/2014/documents/agile2014_97.pdf.

Dou, M., Wang, Y., Gu, Y. et al. (2021). Disaster damage assessment based on fine-grained topics in social media. *Computational Geosciences* 156: 104893. https://doi.org/10.1016/j.cageo.2021.104893.

Eyre, R., De Luca, F., and Simini, F. (2020). Social media usage reveals recovery of small businesses after natural hazard events. *Nature Communications* 11: 1–10. https://www.nature.com/articles/s41467-020-15405-7.

Farzindar, A. and Khreich, W. (2015). A survey of techniques for event detection in Twitter. *Computational Intelligence* 31: 132–164. https://doi.org/10.1111/coin.12017.

Fedoryszak, M., Frederick, B., Rajaram, V., and Zhong, C. (2019). Real-time event detection on social data streams. *Proceedings of the 25th ACM SIGKDD International Conference on Knowledge Discovery & Data Mining, KDD 2019* (eds. A. Teredesai, V. Kumar, Y. Li et al.), pp. 2774–2782, Anchorage, AK, USA (4–8 August 2019). ACM. https://doi.org/10.1145/3292500.3330689.

Fohringer, J., Dransch, D., Kreibich, H., and Schröter, K. (2015). Social media as an information source for rapid flood inundation mapping. *Natural Hazards and Earth System Sciences* 15: 2725–2738. https://doi.org/10.5194/nhess-15-2725-2015.

Forbes (2019). Is Twitter Really Faster Than The News? (forbes.com) (accessed November 2021).

Ghaemi, Z. and Farnaghi, M. (2019). A varied density-based clustering approach for event detection from heterogeneous Twitter data. *ISPRS International Journal of Geo-Information* 8: 82. https://doi.org/10.3390/ijgi8020082.

Hardoon, D.R., Szedmák, S., and Shawe-Taylor, J. (2004). Canonical correlation analysis: an overview with application to learning methods. *Neural Computation* 16: 2639–2664. https://doi.org/10.1162/0899766042321814.

Hasan, M., Orgun, M.A., and Schwitter, R. (2016). TwitterNews+: a framework for real time event detection from the Twitter data stream. *Social Informatics – 8th International Conference, SocInfo 2016* (eds. E.S. Spiro, and Y.-Y. Ahn), pp. 224–239, Bellevue, WA, USA (11–14 November 2016), Proceedings, Part I, 10046. https://doi.org/10.1007/978-3-319-47880-7_{1}{4}.

Henry, E.R. and Hofrichter, J. (1992). Singular value decomposition: application to analysis of experimental data – ScienceDirect. In: *Methods in enzymology*, vol. 210, 129–192. Elsevier https://doi.org/10.1016/0076-6879(92)10010-b.

Hossny, A.H. and Mitchell, L. (2018). Event detection in Twitter: a keyword volume approach. *2018 IEEE International Conference on Data Mining Workshops, ICDM Workshops* (eds. H. Tong, Z.J. Li, F. Zhu, and J. Yu), pp. 1200–1208, Singapore, Singapore (17–20 November 2018). IEEE. https://doi.org/10.1109/ICDMW.2018.00172.

Howard, A.G., Zhu, M., Chen, B. et al. (2017). MobileNets: efficient convolutional neural networks for mobile vision applications. *arXiv e-prints, arXiv* 1704: 04861. https://ui.adsabs.harvard.edu/abs/2017arXiv170404861H.

Imran, M., Elbassuoni, S., Castillo, C. et al. (2013). Extracting information nuggets from disaster-Related messages in social media. *Annual Conference on Information Systems for Crisis Response and Management (ISCRAM)* 201: 791–801. https://mimran.me/papers/imran_shady_carlos_fernando_patrick_iscram2013.pdf.

Kim, S.-B., Han, K.-S., Rim, H.-C., and Myaeng, S.-H. (2006). Some effective techniques for naive bayes text classification. *IEEE Transactions on Knowledge and Data Engineering* 18: 1457–1466. https://doi.org/10.1109/TKDE.2006.180.

Klein, B., Castanedo, F., Elejalde, I. et al. (2013). Emergency event detection in Twitter streams based on natural language processing. *Ubiquitous Computing and Ambient Intelligence. Context-Awareness and Context-Driven Interaction – 7th International Conference, UCAmI 2013* (eds. G. Urzaiz, S.F. Ochoa, J. Bravo et al.), pp. 239–246, Carrillo, Costa Rica (2–6 December 2013). Springer. Proceedings. 8276. https://doi.org/10.1007/978-3-319-03176-7_{3}{1}.

Kryvasheyeu, Y., Chen, H., Obradovich, N. et al. (2016). Rapid assessment of disaster damage using social media activity. *Science Advances* 2: https://doi.org/10.1126/sciadv.1500779.

Lai, S., Xu, L., Liu, K., and Zhao, J. (2015). Recurrent convolutional neural networks for text classification. *Proceedings of the Twenty-Ninth AAAI Conference on Artificial Intelligence* (eds. B. Bonet and S. Koenig), pp. 2267–2273, Austin, Texas, USA (25–30 January 2015). AAAI Press. http://www.aaai.org/ocs/index.php/AAAI/AAAI15/paper/view/9745.

Lample, G., Conneau, A., Ranzato, M. et al. (2018). Word translation without parallel data. *6th International Conference on Learning Representations, ICLR 2018,* Vancouver, BC, Canada (30 April–3 May 2018). OpenReview.net. Conference Track Proceedings. https://openreview.net/forum?id=H196sainb.

Li, L., Bensi, M., Cui, Q. et al. (2021). Social media crowdsourcing for rapid damage assessment following a sudden-onset natural hazard event. *International Journal of Information Management* 60: 102378. https://doi.org/10.1016/j.ijinfomgt.2021.102378.

Liang, Y., Caverlee, J., and Mander, J. (2013). Text vs. images: on the viability of social media to assess earthquake damage. *22nd International World Wide Web Conference, WWW'13* (eds. L. Carr, A. H. Laender, B. F. Lóscio et al.), pp. 1003–1006, Rio de Janeiro, Brazil (13–17 May 2013). International World Wide Web Conferences Steering Committee/ACM. Companion Volume. https://doi.org/10.1145/2487788.2488102.

Liu, X., Li, Q., Nourbakhsh, A. et al. (2016). Reuters tracer: a large scale system of detecting & verifying real-time news events from Twitter. *Proceedings of the 25th ACM International Conference on Information and Knowledge Management, CIKM* 2016 (eds. S. Mukhopadhyay, C. Zhai, E. Bertino et al.), pp. 207–216, Indianapolis, IN, USA (24–28 October 2016). ACM. https://doi.org/10.1145/2983323.2983363.

Liu, Y., Ott, M., Goyal, N. et al. (2020). Roberta: A robustly optimized bert pretraining approach. *Proceedings of the International Conference on Machine Learning,* Addis Ababa, Ethiopia (26 April–1 May 2020). https://openreview.net/attachment?id=SyxS0T4tvS&name=original_pdf.

Long, R., Wang, H., Chen, Y. et al. (2011). Towards effective event detection, tracking and summarization on microblog data. *Web-Age Information Management – 12th International Conference, WAIM* 2011 (eds. H. Wang, S. Li, S. Oyama et al.), pp. 652–663, Wuhan, China (14–16 September 2011). Springer. Proceedings. 6897. https://doi.org/10.1007/978-3-642-23535-1_{5}{5}.

Mangalathu, S. and Burton, H.V. (2019). Deep learning-based classification of earthquake-impacted buildings using textual damage descriptions. *International Journal of Disaster Risk Reduction* 36: 101111. https://doi.org/10.1016/j.ijdrr.2019.101111.

McCreadie, R., Macdonald, C., Ounis, I. et al. (2013). Scalable distributed event detection for Twitter. *2013 IEEE International Conference on Big Data (IEEE BigData 2013)* (eds. X. Hu, T.Y. Lin, V.V. Raghavan et al.), pp. 543–549, Santa Clara, CA, USA (6–9 October 2013). IEEE Computer Society. https://doi.org/10.1109/BigData.2013.6691620.

Mikolov, T., Karafiát, M., Burget, L. et al. (2010). Recurrent neural network based language model. *INTERSPEECH 2010, 11th Annual Conference of the International Speech Communication* Association (eds. T. Kobayashi, K. Hirose, and S. Nakamura), pp. 1045–1048, Makuhari, Chiba, Japan (26–30 September 2010). ISCA. http://www.isca-speech.org/archive/interspeech_2010/i10_1045.html.

Mikolov, T., Chen, K., Corrado, G., and Dean, J. (2013a). Efficient estimation of word representations in vector space. *1st International Conference on Learning Representations, ICLR* 2013 (eds. Y. Bengio and Y. LeCun), Scottsdale, Arizona, USA (2–4 May 2013). Workshop Track Proceedings. http://arxiv.org/abs/1301.3781.

Mikolov, T., Sutskever, I., Chen, K. et al. (2013b). Distributed representations of words and phrases and their compositionality. *Advances in Neural Information Processing Systems 26: 27th Annual Conference on Neural Information Processing Systems* 2013 (eds. C.J. Burges, L. Bottou, Z. Ghahramani, and K.Q. Weinberger), pp. 3111–3119, Lake Tahoe, Nevada, United States. https://proceedings.neurips.cc/paper/2013/hash/9aa42b31882ec039965f3c4923ce901b-Abstract.html.

Nguyen, D.T., Al-Mannai, K., Joty, S.R. et al. (2017). Robust classification of crisis-related data on social networks using convolutional neural networks. *Proceedings of the Eleventh International Conference on Web and Social Media, ICWSM 2017*, Montréal, Québec, Canada (15–18 May 2017), pp. 632–635. AAAI Press. https://aaai.org/ocs/index.php/ICWSM/ICWSM17/paper/view/15655.

Peng, F. and Schuurmans, D. (2003). Combining naive bayes and n-gram language models for text classification. *Advances in Information Retrieval, 25th European Conference on IR Research, ECIR 2003* (ed. F. Sebastiani), pp. 335–350, Pisa, Italy (14–16 April 2003). Springer. Proceedings, 2633. https://doi.org/10.1007/3-540-36618-0_{2}{4}.

Pennington, J., Socher, R., and Manning, C. D. (2014). Glove: global vectors for word representation. *Proceedings of the 2014 Conference on Empirical Methods in Natural Language Processing, EMNLP 2014* (eds. A. Moschitti, B. Pang, and W. Daelemans), pp. 1532–1543, Doha, Qatar (25–29 October 2014). ACL. A meeting of SIGDAT, a Special Interest Group of the ACL. https://doi.org/10.3115/v1/d14-1162.

Peters, M. E., Neumann, M., Iyyer, M. et al. (2018). Deep contextualized word representations. *Proceedings of the 2018 Conference of the North American Chapter of the Association for*

Computational Linguistics: Human Language Technologies, NAACL-HLT 2018 (eds. M.A. Walker, H. Ji, and A. Stent), pp. 2227–2237, New Orleans, Louisiana, USA (1–6 June 2018). Association for Computational Linguistics. Volume 1 (Long Papers). https://doi.org/10.18653/v1/n18-1202.

Petrovic, S., Osborne, M., and Lavrenko, V. (2010). Streaming first story detection with application to Twitter. *Human Language Technologies: Conference of the North American Chapter of the Association of Computational Linguistics, Proceedings*, pp. 181–189, Los Angeles, California, USA (2–4 June 2010). The Association for Computational Linguistics. https://aclanthology.org/N10-1021/.

Piscitelli, S., Arnaudo, E., and Rossi, C. (2021). Multilingual text classification from Twitter during emergencies. *IEEE International Conference on Consumer Electronics, ICCE 2021*, pp. 1–6, Las Vegas, NV, USA (10–12 January 2021). IEEE. https://doi.org/10.1109/ICCE50685.2021.9427581.

Radford, A., Narasimhan, K., Salimans, T. et al. (2018). Improving language understanding by generative pre-training. https://www.cs.ubc.ca/ amuham01/LING530/papers/radford2018improving.pdf (acessed November 2021).

Resch, B., Usländer, F., and Havas, C. (2018). Combining machine-learning topic models and spatiotemporal analysis of social media data for disaster footprint and damage assessment. *Cartography and Geographic Information Science* 45: 362–376. https://doi.org/10.1080/15230406.2017.1356242.

Saeed, Z., Abbasi, R.A., Maqbool, O. et al. (2019). What's happening around the world? a survey and framework on event detection techniques on Twitter. *Journal of Grid Computing* 17: 279–312. https://doi.org/10.1007/s10723-019-09482-2.

Sakaki, T., Okazaki, M., and Matsuo, Y. (2010). Earthquake shakes Twitter users: real-time event detection by social sensors. *Proceedings of the 19th international conference on World wide web*, New York, NY, USA (26–30 April 2010), pp. 851–860. Association for Computing Machinery. https://doi.org/10.1145/1772690.1772777

Saleem, H.M., Xu, Y., and Ruths, D. (2014). Effects of disaster characteristics on Twitter event signature. *Procedia Engineering* 78: 165–172. https://doi.org/10.1016/j.proeng.2014.07.053.

Salza, D., Arnaudo, E., Blanco, G., and Rossi, C. (2022). ISCRAM 2022 conference proceedings – 19th international conference on information systems for crisis response and management. *Annual Conference on Information Systems for Crisis Response and Management (ISCRAM)*, Tarbes, France (22–25 May 2022), pp. 570–583. http://idl.iscram.org/files/dariosalza/2022/2440_DarioSalza_etal2022.pdf.

Sanh, V., Debut, L., Chaumond, J., and Wolf, T. (2019). DistilBERT, a distilled version of BERT: smaller, faster, cheaper and lighter. *Proceedings of the Conference on Neural Information Processing (NEURIPS) Workshops*. https://arxiv.org/abs/1910.01108.

Settles, B. (2004). Biomedical named entity recognition using conditional random fields and rich feature sets. *Proceedings of the International Joint Workshop on Natural Language Processing in Biomedicine and its Applications, NLPBA/BioNLP* 2004 (eds. N. Collier, P. Ruch, and A. Nazarenko), Geneva, Switzerland (28–29 August 2004). https://aclanthology.org/W04-1221/.

Shan, S., Zhao, F., Wei, Y., and Liu, M. (2019). Disaster management 2.0: a real-time disaster damage assessment model based on mobile social media data – a case study of Weibo (Chinese Twitter). *Safety Science* 115: 393–413. https://doi.org/10.1016/j.ssci.2019.02.029.

Shi, M., Wang, K., and Li, C. (2019). A C-LSTM with word embedding model for news text classification. *18th IEEE/ACIS International Conference on Computer and Information Science, ICIS 2019*, Beijing, China (17–19 June 2019), pp. 253–257. IEEE. https://doi.org/10.1109/ICIS46139.2019.8940289.

Soumya George, K. and Joseph, S. (2014). Text classification by augmenting bag of words (BOW) representation with co-occurrence feature. *IOSR Journal of Computer Engineering* 16: 34–38. https://doi.org/10.9790/0661-16153438.

Tang, Z., Wan, B., and Yang, L. (2020). Word-character graph convolution network for chinese named entity recognition. *IEEE/ACM Transactions on Audio, Speech, and Language Processing* 28: 1520–1532. https://doi.org/10.1109/TASLP.2020.2994436.

Thapen, N.A., Simmie, D.S., and Hankin, C. (2016). The early bird catches the term: combining Twitter and news data for event detection and situational awareness. *Journal of Biomedical Semantics* 7: 61. https://doi.org/10.1186/s13326-016-0103-z.

Tong, S. and Koller, D. (2001). Support vector machine active learning with applications to text classification. *Journal of Machine Learning Research* 2: 45–66. http://jmlr.org/papers/v2/tong01a.html.

Vaswani, A., Shazeer, N., Parmar, N. et al. (2017). Attention is all you need. *Advances in Neural Information Processing Systems 30: Annual Conference on Neural Information Processing Systems 2017* (eds. I. Guyon, U. von Luxburg, S. Bengio et al.), pp. 5998–6008, Long Beach, CA, USA (4–9 December 2017). https://proceedings.neurips.cc/paper/2017/hash/3f5ee243547dee91fbd053c1c4a845aa-Abstract.html.

Wang, Y., Ruan, S., Wang, T., and Qiao, M. (2019). Rapid estimation of an earthquake impact area using a spatial logistic growth model based on social media data. *International Journal of Digital Earth* 12: 1265–1284. https://doi.org/10.1080/17538947.2018.1497100.

Weber, E., Marzo, N., Papadopoulos, D.P. et al. (2020). Detecting natural disasters, damage, and incidents in the wild. *European Conference on Computer Vision (ECCV)* (23–28 August). https://arxiv.org/pdf/2008.09188.pdf

Weng, J. and Lee, B.-S. (2011). Event detection in Twitter. *Proceedings of the Fifth International Conference on Weblogs and Social Media* (eds. L.A. Adamic, R. Baeza-Yates, and S. Counts), Barcelona, Catalonia, Spain (17–21 July 2011). The AAAI Press. http://www.aaai.org/ocs/index.php/ICWSM/ICWSM11/paper/view/2767.

Wu, D. and Cui, Y. (2018). Disaster early warning and damage assessment analysis using social media data and geo-location information. *Decision Support Systems* 111: 48–59. https://doi.org/10.1016/j.dss.2018.04.005.

Yin, J., Lampert, A., Cameron, M. et al. (2012). Using social media to enhance emergency situation awareness. *IEEE Intelligent Systems* 27 (6): 52–59. https://doi.org/10.1109/MIS.2012.6.

Yuan, F. and Liu, R. (2018). Feasibility study of using crowdsourcing to identify critical affected areas for rapid damage assessment: Hurricane Matthew case study. *International Journal of Disaster Risk Reduction* 28: 758–767. https://doi.org/10.1016/j.ijdrr.2018.02.003.

Yuan, F. and Liu, R. (2020). Mining social media data for rapid damage assessment during hurricane matthew: feasibility study. *Journal of Computing in Civil Engineering* 34: https://doi.org/10.1061/(asce)cp.1943-5487.0000877.

Zhang, X., Zhao, J.J., and LeCun, Y. (2015). Character-level convolutional networks for text classification. *Advances in Neural Information Processing Systems 28: Annual Conference on Neural Information Processing Systems 2015* (eds. C. Cortes, N.D. Lawrence, D.D. Lee et al.), pp. 649–657, Montreal, Quebec, Canada (7–12 December 2015). https://proceedings.neurips.cc/paper/2015/hash/250cf8b51c773f3f8dc8b4be867a9a02-Abstract.html.

Zhou, X. and Chen, L. (2014). Event detection over Twitter social media streams. *VLDB Journal* 23: 381–400. https://doi.org/10.1007/s00778-013-0320-3.

Zhou, B., Lapedriza, A., Khosla, A. et al. (2018). Places: a 10 million image database for scene recognition. *IEEE Transactions on Pattern Analysis and Machine Intelligence* 40 (6): 1452–1464. https://doi.org/10.1109/TPAMI.2017.2723009.

15

Conclusions and Perspectives

Philippe Quevauviller
European Commission, Vrije Universiteit Brussel, Brussels, Belgium

15.1 Introduction

This book provides an overview of three major projects funded under the EU Horizon 2020 Research and Innovation (R&I) programme (2014–2020), namely ANYWHERE, BeAWARE and I-REACT. Horizon 2020's successor is the Horizon Europe programme, which is running from 2021 to 2027. By coupling research and innovation, the EU R&I framework programmes are supporting excellent science and industrial leadership while tackling societal challenges. The goal is to ensure Europe produces world-class science, removes barriers to innovation and makes it easier for the public and private sectors to work together in delivering innovation. In the context of the Horizon 2020/Horizon Europe 'Civil Security for Society' programme (or Security Research Programme), the Disaster Resilient Societies (DRS) area supports the implementation of international policy initiatives (e.g. the Sendai Framework for Disaster Risk Reduction), EU Disaster Risk Reduction and climate change adaptation policies tackling natural and man-made threats (either accidental or intentional) and the Security Union Strategy (in particular for disasters linked to terrorism). The implementation of these policies calls for enhanced capacities in risk and resilience management and governance, including instruments for better prevention, mitigation, preparedness, monitoring, detection and response, technologies for practitioners and overall societal resilience. This covers specific areas of research, namely on extreme weather events

Responding to Extreme Weather Events, First Edition. Edited by Daniel Sempere-Torres, Anastasios Karakostas, Claudio Rossi, and Philippe Quevauviller.

(e.g. floods, heat and cold waves, storms, forest fires) resulting from anthropogenic climate change, geological hazards (earthquakes, tsunamis, volcanic eruptions), as well as anthropogenic threats (industrial accidents and terrorist attacks associated with chemical, biological, radiological, nuclear and explosive materials (CBRN-E)). It also covers research on health crisis management, in particular pandemic risks.

In this context, a wide range of research and technological developments, as well as capacity-building and training projects, has supported the development and implementation of the policies and strategies (see Section 15.2). Integrating needs are, however, overwhelming owing to the complexity of the policy framework and the high level of fragmentation of research and capacity-building initiatives. In addition, enhanced cooperation and involvement of different sectors and actors are essential, including policymakers, scientists, industry/small and medium enterprises (SMEs), public administration (both at national and regional/local level), scientists, financial institutions, practitioners, non-governmental organizations (NGOs), and civil-society organizations (CSOs), notwithstanding the citizen dimension.

When dealing with hydrometeorological extreme events, integrated approaches are thus essential to bridge different policy areas including civil protection, environment (including water, forestry, biodiversity/nature), and climate change adaptation and mitigation. They also require strengthening of transdisciplinary and transboundary cooperation with all the relevant actors. A huge body of knowledge and technology was developed in the Seventh Framework Programme (2007–2013) and Horizon2020 (2014–2020), which constituted a strong legacy for the current Horizon Europe (2021–2027) research in support of an enhanced resilience of European society to disasters of any kind.

Project outputs materialized in different ways, such as relevant scientific findings, the maturation of promising methods, solutions and technology areas, the operational validation of innovative concepts or the support to policy implementation. From the early 2010s onward, a strong research community integrated by highly committed stakeholders was consolidated. Policymakers, practitioners, industry, scientists and citizens are the pillars based on which a sound security research agenda can be built. The dialogue with those who define priorities and those who have the productive capacity guarantees not only that research addresses real needs but also that the investment in research will deliver tangible results. In the past years, this had a clear effect on an enhanced practitioner's participation in Security Research, promoting research results that are relevant to them, including the most promising tools that might have the potential to be taken up by them, and ensuring that their expertise is made available to policymakers. Building on several years of experience of community building, the Community of European Research and innovation for Security (CERIS) is now facilitating a more proactive participation of different actors (see section 4.3.2).

15.2 Policy Background

Many interrelated policies, legal-binding frameworks of actions across different sectors cover secure, safe and resilient societies, involving diverse EU directorates and intergovernmental agencies involved in the preparation and implementation of the legal instruments in the form of decisions, general frameworks, directives, communications, strategies and plans (European Commission 2020a). Crisis Management policies follow an integrated approach for the management of natural and human-made hazards (European Commission 2020b) focusing on disaster risk reduction (prevention and preparedness) and disaster response (recovery and learning). Figure 15.1 illustrates the relation between international frameworks, type of crises and legal instruments.

The Sendai Framework for Action 2015–2030 (UN-DRR 2015) 'Building the resilience of nations and communities to disasters' adopted by 168 UN member states sets a voluntary commitment towards the improvement of disaster resilience and disaster risk reduction as essential ingredients for the achievement of poverty reduction and sustainable development. The Sendai framework relates closely to the EU disaster risk management policies, in particular the EU Civil Protection Mechanism (UCPM, European Commission 2013a). Climate-related disasters are covered by environmental and climate policies (DG ENV through the Flood Directive (European Commission 2007a) and DG CLIMA through the EU strategy on adaptation to climate change, European Commission 2021). Further development and implementation of these policies represents a complex and ambitious challenge as they involve a wide variety of players. It involves member states (MS) with specific national approaches (national action plans) for dealing with crises. These are also differently organized in terms of disaster risk management capabilities to be adapted to diverse context. The EU framework represents a means and a real opportunity to discuss possible ways to improve coordination and knowledge sharing among the various national approaches. It enables the development of a common EU vision strengthened by a joint strategy and its implementation in this field.

15.2.1 Civil Protection Policies

The UCPM aims to facilitate reinforced cooperation and coordination between the EU and the MS in the field of civil protection to improve the effectiveness of systems for preventing, preparing for, and responding to natural and man-made disasters. It supports and complements the efforts of the MS for the protection, primarily of people but also of the environment and property, including cultural heritage, in the event of disasters. Built upon these policy instruments and on principles of solidarity, the UCPM targets the development of an integrated approach to disaster management.

The overall mechanism takes due to consideration of laws and international commitments, and exploits synergies with relevant union initiatives such as the European Earth Observation Programmes (Copernicus), the European Programme for Critical Infrastructure Protection (EPCIP) and the Common Information Sharing Environment (CISE). The UCPM is operationalized through the Emergency Response Coordination Centre (ERCC) and the European Emergency Response Capacity (EERC) in the form of a voluntary pool of pre-committed capacities from the MS, trained experts, a Common Emergency Communication and Information System (CECIS) managed by the Commission and contact points in the MS. It also recognizes the role of regional and local authorities in disaster management. Outside the union, disaster response is coordinated with the United Nations and other relevant international actors with

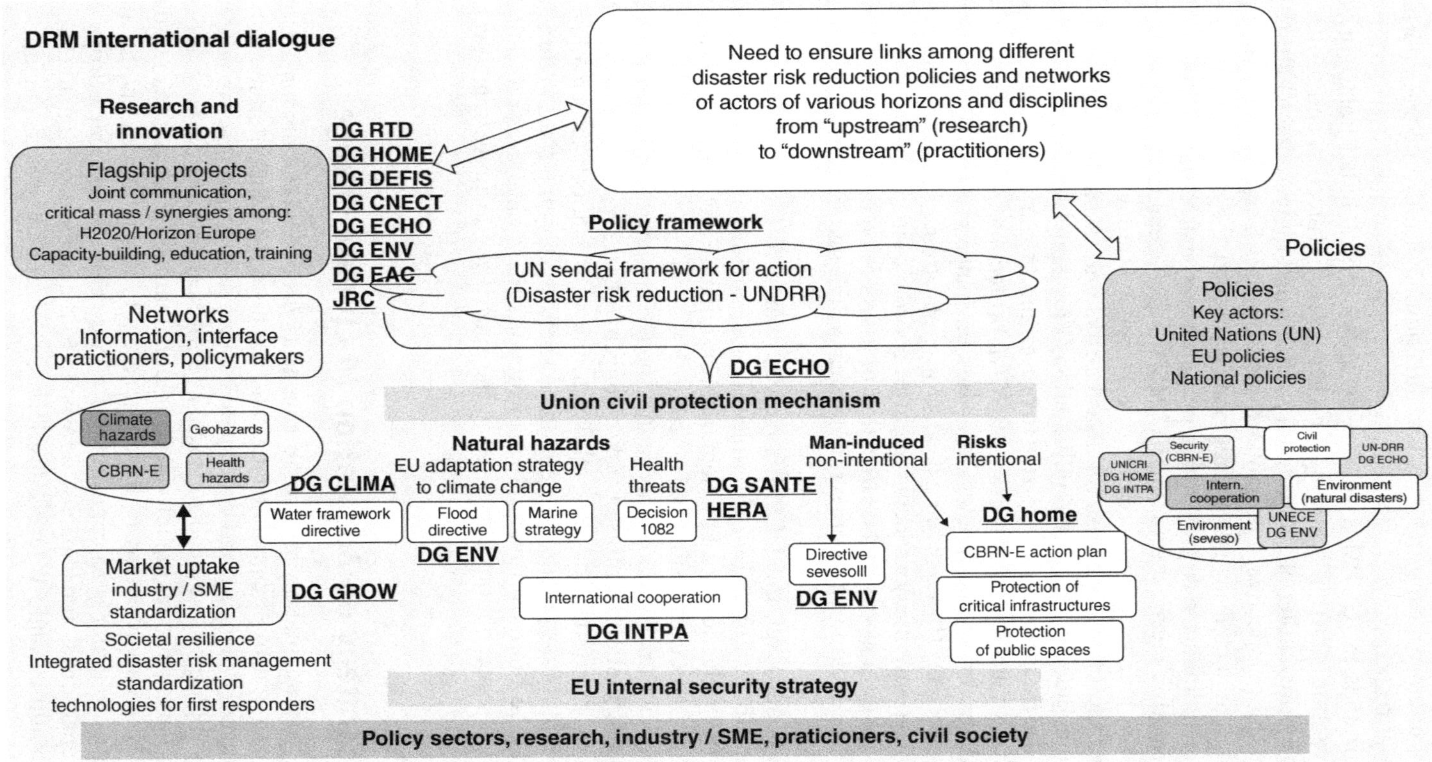

Figure 15.1 Policies related to Disaster Resilient Societies. Acronyms stand for different names of policy Directorate General (DG) of the European Commission, namely Research and Technological Developments (RTD) or Research & Innovation, Home Affairs and Migration (HOME), Defence and), Internal Market, Industry, Entrepreneurship and SMEs (GROW), Health and Food Safety (SANTE), Health Emergency, Preparedness and Response (HERA), International Partnerships (INTPA). United Nations (UN)-related acronyms stand for Disaster Risk Reduction (UN-DRR), Economic Commission for Europe (UNECE).

reference to Council Regulation concerning humanitarian aid (United Nations 1996). Finally, the use of military means under civilian direction may constitute an important resource to disaster response.

On technical grounds, the UCPM is working towards a general policy framework on disaster risk prevention aimed at achieving a higher level of protection and resilience against disasters by preventing or reducing their effects and by fostering a culture of prevention. From this perspective, it promotes the review of risk assessment, risk management planning conducted at national/regional level, and the development of an integrated approach, linking risk prevention, preparedness and response actions. On the basis of information received from the EU MS, the European Commission establishes and regularly updates a cross-sectoral risk overview. Among its priorities is the action to improve the knowledge base on disaster risks and facilitate the sharing of knowledge, recommended practices and information (European Commission 2013b).

The UCPM is also financing actions related to preventing, preparing for and responding to disasters. These include an important EU Civil protection training programme, regular large-scale exercises and modules exercises, exchange of experts, prevention and preparedness projects (through annual calls for applications), logistical and transport support for response missions, deployment of coordination, assessment or advisory missions, adaptation grants and certification of assets to be included in the Civil Protection Pool, the development of crisis response capacities under rescEU (European Commission 2021).

At the international level, the UCPM is closely related to the above-mentioned Sendai Framework for Action 2015–2030, which sets out an ambitious set of priorities to place disaster risk reduction as a key element of sustainable development efforts to define further steps to reduce existing and emerging risks and foster disaster resilience. As stressed in Council Conclusions on this matter, the EU supports a framework which strengthens the contribution of disaster risk management to smart, sustainable, and inclusive growth by promoting the use and development of innovative technologies and encouraging a more systematic and reinforced science-policy interface in disaster risk management. The Intergovernmental Panel supports these objectives on Climate Change (IPCC) recommendations expressed in the special report on extreme events (IPCC 2012).

15.2.2 EU Strategy on Adaptation to Climate Change

The EU Strategy on adaptation to climate change (European Commission 2021a) highlights the consequences of climate change and the need for adaptation measures. It gives priority to early, planned, and coordinated action rather than reactive adaptation. The strategy highlights the need for exchanges of best practice on how to best adapt to climate change, also taking into account global impacts that, e.g., lead to disruptions in the EU of supply chains, or impaired access to raw materials, energy, and food supplies. Its overall aim is to contribute to a more climate resilient Europe by enhancing the preparedness and capacity to respond to the impacts of climate change at local, regional, national, and EU levels, developing a coherent approach and improving coordination.

The requirement for 'climate-proofing' and mainstreaming of adaptation measures in various sectors also calls for strengthened preparedness and science-policy links. The strategy features a novel international component, aligning with the Paris Agreement and other relevant international policies. A close coordination between climate change adaptation and disaster risk management and policies is also required. Several of the actions proposed, including related to international aspects of adaptation, are of direct relevance and in full synergy with DRR.

15.2.3 Water Framework and Marine Policies

Specific policy instruments are in place in the water sector related to extreme hydrometeorological events such as floods and droughts. In the first place, complementing the Water Framework Directive (WFD, European Commission 2000) and its daughter Directives, namely the Directive on Environmental Quality Standards (European Commission 2008a) and the Groundwater Directive (European Commission 2006). The WFD has clear deadlines and milestones from 2000 until 2027. It provides a legal tool to address water scarcity,[1] by requiring long-term planning of water resources, economic tools to incentivize efficient use of water resources and specific measures on water abstraction and water efficiency. The strategic approach has been taken forward with the 2007 Commission Communication on water scarcity and droughts (European Commission 2007b) and its reviews (2008–2010). An in-depth assessment of EU policy on water scarcity and droughts was carried out for the 2012 Blueprint to Safeguard Europe's Water Resources, resulting in recommendations for follow-up actions (European Commission 2015). The WFD includes all elements needed to address adaptation to climate change. The definition of water status allows to cater for climate change impacts, and the cyclical approach to regularly update River Basin Management Plans provides the opportunity to constantly re-assess the actual status of water bodies and take the most effective actions to preserve them. The CIS Guidance No 24 on river basin management in the changing climate illustrates ways in which climate change impacts should be considered within river basin management cycles, including provision for floods and droughts.

More recently, the 2019 Fitness Check of EU water law (European Commission 2019) has shown that the Water Framework Directive is flexible enough to address also drought and water scarcity issues, as it requires Member States to achieve the overarching goal of good status of all their waters, and when needed drought management plans should be elaborated and implemented. Finally, the new EU Strategy on Adaptation to Climate Change (European Commission 2021b) calls for the wider use of drought management plans. The Environment Council of the EU adopted its conclusions on the Strategy on 10 June 2021, calling for integrated cross-sector and cross-border water management that contributes to increased resilience, the promotion of sustainable water use, the protection and conservation of aquatic ecosystems and the improvement of flood risk and water scarcity and drought management strategies, incorporating future floods and drought risk and arranging timely and reliable exchange of data and early warnings.

Flood prevention and management are tackled by the Flood Directive (European Commission 2007a), which requires EU Member States to assess and manage flood risks, with the aim of reducing adverse consequences for human health, the environment, cultural heritage and economic activity associated with floods in Europe. The Floods Directive, which coordinates its objectives with those of the WFD, has no deadlines (in recognition of the ever-changing nature of flooding risk), but it takes into account the likely impact of climate change on the occurrence of floods. It therefore provides a mechanism for assessing and monitoring increased risks of flooding, considering the possible impacts of climate change, and for developing appropriate adaptation approaches.

The Marine Strategy Framework Directive (European Commission 2008b), adopted in 2008, is the EU's main tool for marine environmental protection. It looks at all pressures on the seas and ocean and brings them together under one umbrella to tackle the cumulative impacts of human activities in one strategic framework. The Marine Strategy Framework Directive requires EU Member States to set up marine strategies in respect of the marine region of the Member States' marine waters to achieve 'Good Environmental Status'. One of the criteria for good environmental status is the level of chemical pollution in the marine environment. This includes acute pollution events, such as discharges, spills and emissions of hydrocarbons and other harmful chemicals from ships or offshore oil and gas installations. Shipping accidents involving loss of containers can also be a major source of marine pollution, notably from litter. The Directive require Member States to take the necessary measures to minimize the spatial extent and duration of significant acute pollution events, as well as to ensure that litter in the surface layer of the water column, and on the seabed are at levels that do not cause harm to the marine environment.

15.2.4 Links with Projects Subject to this Book

The ANYWHERE project targeted the empowerment exposed responder institutions and citizens to enhance their adaptation and proactive capacity when dealing with extreme and high-impact weather and climate events. The project implemented diverse activities such as workshops, table-top exercises, citizen science campaign to validate tools. It also delivered hands-on training that involved civil protection staff as well as representatives of the river basin agencies. In this respect, the project was at the crossroad of the three main policies described in the above sections (civil protection, adaptation to climate change, water framework and flood directives).

The I-REACT project was more about public administration authorities, private companies and citizens. The project used a gamification approach to promote education and boost citizen engagement and hence was closely supporting the Union Civil Protection Mechanism. Similarly, pilot demonstration by BeAWARE allow reviews with improved quality and speed of information sharing, cooperation between citizens and first responders, training and exercise programmes.

All projects delivered platforms supporting forecasting, early warning and real-time information sharing integrating data from different sources (such as Copernicus earth observation) and advance technologies (such as machine learning, drones). These platforms support local decisions and multimodal data management for coordination between first responders and authorities. Mobile applications included multi-language social media, visuals and routing for emergency managers to support and empower citizens to report and carry out initial actions.

15.3 Actor's Interactions and Community Building

15.3.1 Who are the Actors?

Disaster Risk/Crisis Management related to natural hazards (including hydrometeorological extreme events) is scattered into many different disciplines and sectors, e.g. various types of hazard-related communities are often working independently on somehow overlapping crisis situations (e.g. experts dealing with climate extreme events vs earthquake specialists), so definitions about 'Actors' may be prone to various interpretations. To simplify, we will distinguish five main categories: (i) Policymakers and Stakeholders; (ii) Scientists; (iii) Industry (including SMEs); (iv) training and operational units; and (v) general public:

a. ***Policymakers and stakeholders***
   - At EU level, main policy areas concerned with Crisis Management are DGs HOME (security), ECHO (civil protection), SANTE (health), ENV (water-related hazards and industrial accidents), CLIMA (climate-related hazards)
   - At Member State's level, Ministries of Defence, Interior, Foreign Affairs, Civil Protection, Ecology, Research and Industry), Agencies as well as Regional Authorities all benefit from research outputs.
   - Often working at the interface between policy and science, various stakeholders are involved in bridging interests of different communities, e.g. consultancy companies
b. ***Scientists***
   - Research on climate-related hazards involves a wide range of scientific disciplines which have to interact, ensuring complementarity and building interdisciplinary networks
   - Different types of scientists are to be considered (University, Research Institutes, research units linked to Defence/Interior ministries or agencies)
c. ***Industry (including SMEs)***
   - The sector involves many industry branches and stakeholders in the areas of climate adaptation, environment, civil protection etc. Research results can benefit most first responders
   - Different communication approaches to be followed towards large industries and SMEs often disconnected from discussions at EU level
d. ***Operational units***
   - First responders, emergency services, police, civil protection units, military units, laboratories, etc.
   - Training centres for first responders
   - Command control centres
   - Decision-makers (at national or regional levels)

c. ***General public***
   - NGOs
   - Public at large
   - Education (schools) and training

While some of the above actors are used to participate in international meetings, this is less frequent for SMEs and even less for actors of operational units and the general public. New ways have hence to be found to ensure that information may freely circulate 'horizontally' as well as 'vertically' in order to fertilize all project deliverables while maturing them to the final operational phase (also called 'usefulness & use') by end users and integrating them into appropriate EU strategy and security policy.

Transfer of knowledge has to be coordinated at various levels:

A. **Horizontally**
   - *Policy to policy*: while International and EU policies are developed in close consultation among different sectors, in practice few interactions take place at the implementation level among sectors within the Member States. This is partly due to insufficient sharing of information and joint actions.
   - *Science to science*: EU-funded projects respond to topics that are generally based on well-defined policy hooks. We might hence expect that projects supporting common policy goals will establish synergies, which is rarely the case without a push from the Commission owing to various considerations (IPR and classified information in particular). Here again, sharing information and developing interactions on a regular basis should become a practice that the Commission should ask of projects.
   - *Policy to science*: an essential component of the policy to science interaction is the capacity for policymakers to identify research needs on the short to long term and communicate these needs in anticipation to the research community so that programming, research development and implementation can match the policy timeline (e.g. access to the scientific state-of-the-art, short-term research/capacity building, longer term research goals, pre- and co-normative research).
   - *Science to policy*: this is obviously directly linked to the above, with the requirement for the scientific community to format/translate research information in a way which is tailor-made to policy applications, basically responding to well-specified technical challenges. This is the subject of the mapping described in this document.

B. **Vertically**
   - *International/EU to National*: at international/EU level, policies are elaborated by relevant organizations (e.g. UN for various conventions and European Commission for climate-related EU policies). The links to the National level take place through Committees represented by EU and Associated Member States (e.g. Civil Protection Committee, H2020 Committees). There is a need to

ensure that these Committees be informed on similar grounds about science & policy developments.

- *National to Regional/Local*: once Member State's Committee representatives are duly informed, it is to be expected that appropriate relays with regional/local implementers will then take place under the MS responsibility. This also requires a level of coordination that depends upon the willingness and capacity of each Member State. This level of interaction is less well defined than the EU level because of different settings within the Member States.

15.3.2 Community Building

15.3.2.1 A Community of Users The large span of projects leads to a huge dispersion of resources as mechanisms are hardly operational to date to establish a common platform to exchange information of public character, boost awareness and transfer of relevant research projects to relevant users and make them 'useful and used'. In addition, efforts are done to better address user's needs that may be reflected into possible inputs to research programming. Another aspect stems from contribution of scientific progress to policy formulation, implementation and review, which also require an enhanced coordination. In 2014, the European Commission (Security Research Programme) launched a Community of Users to boost information exchanges among various Disaster Risk Management (DRM) actors. Its development was based on two FP7 Integrated Projects (EDEN for CBRN-E, DRIVER for Crisis Management) that played the role of catalyst. Both projects inter alia demonstrated the added value of large-scale integration of security solutions (related to various risks) and the support to Member States' preparedness and response organizations in improving integration and information sharing in countering various threats. The Community of Users (CoU) hence provided an 'umbrella' at EU level, which provided an overall outlook (based on publicly available information) of science and policy developments in the relevant sectors (natural and man-made disaster risk management). Throughout the years, a strong security research community integrated by highly committed stakeholders has been consolidated. Policymakers, practitioners, industry, scientists and citizens are the pillars based on which a sound security research agenda is built, providing the basis for the formulation of relevant and targeted work programmes. The dialogue with those who define the security priorities and those who have the productive capacity guarantees not only that research addresses real needs, but also that the investment in research will deliver tangible results. Over the years 2014–2021, this informal platform currently involved a wide range of stakeholders (policy makers, end users, academia, industry and civil society) and regularly held thematic events with the security research community. The CoU enabled to gather policymakers, scientists, practitioners, industry/SMEs, and civil society organizations at different levels (International to Regional), creating dialogues around research in various security areas and 'bridges' among different sectors, which was an essential step. CoU dialogues and events had a clear effect on an enhanced practitioner's participation in Security Research in the H2020 framework, promoting research results that are relevant to them, including the most promising tools that might have the potential to be taken up by them, and ensuring that their expertise was made available to policymakers. The initiative also supported the competitiveness of EU industry in information sharing aiming to increase the visibility of the market for research results. Finally, the CoU stimulated the dialogue between the five pillars of security research as well as synergies between different research and capacity-building projects.

15.3.2.2 The Community of European Research and Innovation for Security With *Horizon Europe* (2021–2027) – the new framework programme for research and innovation – starting in 2021, the European Commission re-designed its instruments to facilitate necessary exchanges in all domains of security research. This further aligned research with the EU's main policies in all areas of security[2] and is contributing to an overall shift towards a forward-looking and anticipatory policymaking. Discussions held through Thematic Groups of the Community of Users (CoU) representing the main areas of Security Research were found to be effective but as such they appeared to lack a top-down direction in terms of objectives to be met and issues to be addressed, and a high-level planning, coordination and operational direction. They also did not consider any formal channel of communication between the CoU and receiving entities for any recommendation stemming from the discussions in the Thematic Groups. Finally, an overall working structure for the groups was missing. As a consequence, DG HOME, in consultation with other policy DGs[3] and Member States, proposed a possible governance structure aiming to organize activities in the context of Horizon Europe developments. Therefore, from 2021 onwards, the intention was to move to a more pro-active participation of experts, calling for inputs and defining what would be optimal outputs from the overall Community Building. In this respect, shifting from a user's perspective, and in the light of the forthcoming Horizon Europe developments in the years 2021–2027, the CoU enlarged its scope and became a forum on European research and innovation for security under the new name **Community of European Research and Innovation for Security (CERIS)**. While the developing activities keep alive elements that are already running, a strong shift has been done in the emphasis and modus operandi of the former CoU's work. Until 2020, discussions were mainly based on ad-hoc workshops taking place a few times per year (in a number of different sessions) in which representatives of the security research community discussed current trends and presented projects within their respective thematic areas. In 2020, Thematic Scoping Groups were created, which contributed to the planning, coordinating, and directing of the workshops in different areas of the Security Research Programme throughout the year. It was followed by the creation of a permanent CERIS Expert Group composed of Thematic Working Groups (ThWGs) related to four Thematic Areas (Destinations[4]) of Horizon Europe's Cluster for Civil Security for Society, with specific tasks and mandate. The aim of this new approach is to establish a structured bottom-up/top-down flow that will *inter alia* contribute to the development of a consolidated civil security research agenda.

15.4 Research Trends Related to Disaster Risks (Including Climate Extremes) in the Security Research Area

15.4.1 Societal Resilience

Trends in the area are covered by a topic that was opened in 2021 on 'Improved understanding of risk exposure and its public awareness in areas exposed to multi-hazards'. The awareness of multiple hazards (including hydrometeorological extreme events) and the understanding and the assessment of risks and their consequences is a critical and fundamental step towards the development of local, national and international policies and strategies within all phases of the disaster risk management cycle, in particular preparedness. The availability of reliable scientific data and information (including historical occurrences and climate projections) to anticipate future disaster events or crisis situations, considering uncertainties inherent to natural systems characterization, and effectively support decision-making processes at all levels represents a global challenge for both the research community and governance institutions. Actions at national/local and global/regional levels rely on knowledge of risks in all its dimension and changeable nature. A strengthened understanding of risks by the population (and decision-makers) is needed, based on both records of past events and forecasts and projections (with quantified uncertainties) that reflect consideration of evolving trends and dynamics over time and space. This is particularly acute in the case of multi-hazard risks, i.e. occurrences of several disasters either in cascade or at once. Moreover, the work needs to be complemented with improved knowledge on how risk awareness and actions are influenced and shaped by diverse aspects such as past events, cultures and traditions. The understanding of multiple disaster risks (and related awareness) relies on knowledge gained about historical data and information about past events and related lessons learned, as well as the ability to forecast and assess future risks under uncertainty (including impacts of pandemics, as well as global change, including climate trends and earth system and environment dynamics). These complex interactions between human decisions and multiple hazards require novel risk assessment approaches such as agent-based modelling and systems dynamics methods. This will result in improved preparedness actions built upon these analyses (e.g. defining evacuation routes, responsiveness of health services, etc.). Social media also plays a role in disaster analytics. For example, an increasing number of location-based social network services can provide time-stamped, geo-located data that opens new opportunities and solutions to a wide range of challenges by analysing the extracted public behaviour responses from social media before, during and after disaster events. When using social media data, the design for data collection and analysis has to respect fundamental rights, privacy and data protection and analyses have to take related societal effects in online and offline environments into account as well as possible disinformation and fake news. Also, risk awareness, understanding and preparedness are unequally distributed along a wide range of variables (socio-economic, cultural, regional etc.) that may generate drawbacks and conflicting issues with respect to groups' vulnerability.

Related to the above research area, further research will be undertaken from 2023 onward, issued from a topic opened in 2022 about 'Enhanced citizen preparedness in the event of a disaster or crisis-related emergency'. Improving societal resilience to disasters or crises relies on various features related to preparedness of citizens, communities, education systems, public administrations, business companies and practitioners. These concern, in particular, ways to react and informed decisions to take in case of an event. Individual, public and multi-level actions are needed in disaster risk management and they have huge implications on potentially reducing losses and increasing the operational capacity of responders, along with significant impacts on the emergency planning and management phases and its relation to continuous operations and existing safety management. In particular, the level of awareness of EU citizens of the risks in their region is an indicator for measuring progress in increasing public awareness and preparedness for disasters and in the implementation of the Union Civil Protection Mechanism legislation. Besides the required risk understanding, research is needed in several domains. With regard to public administrations, it is relevant to conceptualize how to increase risk awareness by building processes capable of fostering a long-lasting coalition with citizens around the objective of reducing vulnerability. This implies the definition of action protocols and models of responsibility that mobilize the intervention of individual employees of public administrations. With regard to business companies and practitioners, it is relevant to integrate their emergency activities in the local context. With regard to citizens and communities, it is necessary to design preparedness actions enabling an empowerment of citizens (including particularly vulnerable groups), their communities and NGOs through bottom-up participatory and learning processes. This includes school/university curricula and professional training and trust building among local actors, integrating relevant traditional knowledge, incorporating a gender perspective where relevant, best practices, guidelines and possible changes of regulations, to allow participatory actions. Difficulties in communication to the public in preparedness (and response) phases requires the consideration of legal aspects, along with investigations into innovative practices, forms and tools that will enable the more effective sharing of information, taking into account possible risks of disinformation and fake news. These will support citizens in acting efficiently by themselves, through enhanced collaboration and communication and improving information exchanges between local authorities (including first and second responders), vulnerable populations (e.g. socio-economic groups, ethnic groups, people with illnesses or disabilities, children, elderly, hospital patients), and the private sector. Moreover, recent crises have shown that there is a large sense of solidarity among the population during a disaster or crisis situation. Many citizens that were not involved in disaster relief organizations before the crisis want to offer support to their fellow citizens and the broader community in times of crises. These initiatives of 'spontaneous volunteers' are however most efficient if they are informed and trained and if their valuable contributions are coordinated with the authorities and first and second responders on the ground. Preparedness plans, tests and continued adaption on how best to manage spontaneous volunteers and integrate those into the response are needed.

Another axis of research, also launched in 2022, is focused on 'Better understanding of citizens' behavioural and psychological reactions in the event of a disaster or crisis situation'. Human actions and behaviour may strongly influence the effects and dynamics of a disaster or crisis situation and on the response, potentially modifying the vulnerability of the population. For example, inadequate design of technological systems may favour cascading consequences due to limited consideration of human performance, and insufficient planning. Linked to this, due to extreme time pressure, crisis managers are often forced to make decisions on the basis of inadequate information. The behaviour of the general public, mostly influenced by demographic factors (e.g. gender, age, income, education, risk-tolerance, social connectivity etc.) and the perception of risks, depends on the availability, form and access to information about the crisis and management of trade-offs (e.g. efficiency and thoroughness trade-offs). Social media play an important role here being a means of disinformation and misinformation. Recent disasters related either to natural causes (including climate-related and geological hazards), man-made causes (including industrial accidents or terrorist attacks) or the COVID-19 pandemic crisis have shown the lack of sufficient knowledge in the way citizens react in case of disasters or crisis situations, with implications on policy design and implementation for example in the form of preparedness plans. In this respect, taking into account the knowledge gathered by projects funded in Horizon 2020 and ensuring complementarity, behavioural and psychological research on how citizens behave in the event of a disaster or crisis situation is needed to better understand how to best raise awareness in the population and develop tools to facilitate this. It is hence necessary to better investigate how historical, cultural and emotional factors (e.g. anxiety, panic etc.) during a disaster or a crisis influence rational actions, evaluations of options and information seeking. In addition, the impact of disasters on health also requires looking into the short- and long-term consequences of exposure to high stress/threat levels bears, in particular for mental health.

15.4.2 Tools for Integrated Risk Reduction for Extreme Climate Events

Built up on a legacy of projects, in particular those described in the present book, research trends are being pursued in the area of 'Integrated Disaster Risk Reduction for extreme climate events: from early warning systems to long term adaptation and resilience building' (topic opened in 2022). In contemporary society, the capacity of communities and governments to manage expected and/or unexpected extreme climate events depends heavily on effective governance throughout the entire Disaster Risk Management cycle. This covers operational mechanisms ranging from short-term actions (e.g. early warning and forecast-based actions) to long-term adaptation strategies and resilience building, including nature-based solutions. A coherent integration between Disaster Risk Reduction, Climate Adaptation policies and Sustainable Development Goals as fostered by the European Green Deal and major UN initiatives should result in a comprehensive resilience framework while improving synergies and coherence among the institutions and international agencies involved. The effective implementation of global and European risk governance and policies to enable integrated disaster risk reduction for extreme climate events requires a collaborative involvement in risk assessment and information sharing across involved institutions, including the civil and private sector and the population. Cross-regional, cross-border and cross-sector agreements covering all phases of Disaster Risk Management can improve the knowledge about extreme climate events such as forest fires, droughts, floods, heatwaves, storms and storm surges. In addition, improving effective prevention, preparedness and response rely upon specific national or local expertise and experience. It is important to overcome silos between technical and political authorities at all levels and advocate integration among involved actors. Multi-risk governance frameworks related to climate extremes, shifting from single to multi-risk thinking in governmental agencies, represents the key challenge for the future, considering how measures to improve the resilience of the built environment and communities may provide effective solutions to strengthen adaptation measures. Creating an overview of existing knowledge, integrating tools and developing new ones for resilience and emergency management should include careful planning for interoperability amongst many actors. It is important that solutions pay attention to societal side-effects of integrating data about emergencies, for instance Apps, where persons concerned tend to share more willingly, but do not reflect consequences of that. Thus, the development of data management tools for emergencies need to respect fundamental rights, data protection and avoid function creep.

Another axis of research (also from a topic opened in 2022) looks at 'Enhanced assessment of disaster risks, adaptive capabilities and scenario building based on available historical data and projections'. The assessment of disaster risks requires different types of actions ranging from soft measures to technologies. Simulation-based risk and impact assessments represent an effective approach to make science understandable to decision-makers and streamline national to local mitigation/adaptation actions. This is especially the case if they are integrated with evaluation tools for cost-benefit/effectiveness and multi-criteria analyses, data-farming experiments, serious games, and are tailored to meet end user's needs, to assess the effectiveness of alternative options in different phases of the Disaster Risk Management cycle. Specific risk assessments should be decision- or demand-driven and informed by scientific evidence, and there is a clear need to translate the results to ensure they are relevant, usable, legitimate and credible from the perspectives of the users. Co-design, co-development, co-dissemination and co-evaluation engaging the intended end users represent in this sense key features of improved risk, resilience and impact assessments. In a first place, the acquisition of data is an essential feature and this requires innovative solutions for faster risk assessment and reduction. This includes the identification of precursors for different types of threats, supporting the design or improvement of risk-targeted monitoring programmes. In addition, risk assessments themselves are primarily designed to predict the likelihood of a specific event, whereas what is of primary concern is the impact of that event on society, infrastructure, governance, etc. Numerous experiences gathered in the natural hazards area showed that an enhanced assessment of risks and scenario building may be improved by taking into account reliable data (both quantitative and qualitative) and historical occurrences, when available, including disaster loss data (studies of past events in particular low-probability/long-time recurrence events). This includes for example a higher completeness of the historical-geological records of volcanic eruptions, major earthquakes, tsunamis etc. In the case of extreme climate events such as storms and related storm surges, or health crises (outbreaks, pandemics) the analysis should draw on the outputs of state-of-the-art climate projections, including by taking into account the uncertainties brought on by climate change and our state of knowledge of the key processes underpinning the functioning of the Earth system. In cases where there are not be enough historical data and a high level of uncertainty, assessments and decision-making will have to rely on qualitative data. The action should take into account disaster loss databases and risk data repositories in Member States and relevant hubs.

The above research trends are complemented by actions opened in the 2022 call related to 'Improved impact forecasting and early warning systems supporting the rapid deployment of first responders in vulnerable areas'. Enhanced risk and crisis assessment and preparedness to natural hazards rely on tools using different types of data, information and forecasts (e.g. meteorological data, physical data related to geohazards and climate projections etc.), which may enable to anticipate the occurrence of disasters. Based on the legacy of existing solutions, in particular in the area of extreme weather events, further developments are required to compare impact forecasting and early warning approaches at international level. The aim of such comparisons would be to design EU-wide decision-support and information systems supporting planning authorities and civil protection agencies in the rapid deployment of first responders and communication to citizens in vulnerable areas in the case of extreme climate events or geological disasters. This platform development might be prone to international cooperation, hence supporting the implementation of both EU policies and the UN Sendai Framework for Action. Innovation actions should improve measures and technologies that are needed to better plan for extreme climate events and geological disasters, reduce risks, as well as manage the immediate consequences of natural disasters, in particular regarding emergency responses. This should lead to sound and timely operational forecasts of severe (short-term focus) extreme weather events or geological hazards to aid planning authorities, civil protection agencies and first responders in their decision-making. Built up on developments from relevant H2020 projects, a European-scale multi-hazard platform should be designed, taking into account existing developments at EU level and available space information, in order to facilitate the identification of expected natural hazards with great specificity in time and space. The aim is to utilize largely existing capabilities and combine them into a single, user-friendly platform. In order to achieve the expected outcomes, international cooperation is encouraged, in particular with vulnerable countries, e.g. African and South Mediterranean members of the Union for the Mediterranean.

15.5 Conclusions, Gaps and Recommendations

Exchanges among DRM actors within various workshops, conferences and CERIS events led to the identification of gaps and to recommendations for future research programming. In the first place, the need for better data should be addressed through an EU framework for data collection and management, integrating existing private critical data and making them available for public entities responsible for crisis management across Europe following FAIR principles (findable, accessible, interoperable and re-usable). Another dimension to be considered (along the trends described in Section 15.4.1) is societal resilience, in particular making citizens the first beneficiary of research and innovation, enhancing citizens' knowledge, situational understanding related to risk and empowering them to act. This might be done better together with authorities and first responders acting in a systemic way in order to establish common governance and operational practices such as insurance adoption in disaster prone areas. Sharing knowledge, practices and lessons learnt to manage local and transnational emergencies is also a clear recommendation that was highlighted by DRM actors. This needs to address in particular management practitioners and public authorities.

Regarding operational needs, discussions highlighted the need to integrate existing emergency management volunteer organizations in the National Civil Protection system at every level (national to local). Volunteerism represents indeed a great added value to the Civil Protection system as it provides experienced human power and additional communication channels between and towards the population. Volunteers are an efficient means of bridging the gap between national authorities and citizens.

Finally, education on disaster risk reduction and initiatives aimed at engaging citizens in tailored emergency management tasks should be promoted: educational content for citizens should be created and distributed to citizen using online as well as offline resources to increase citizen awareness number of injured people as well as the death toll. Gamification approaches such as the one implemented by I-REACT are recommended to boost citizen engagement.

Notes

1. Water scarcity is a demand and supply issue, in which the increasing demand is greater than the available resources.
2. Supporting EU policies related to Disaster Risk Management, Critical Infrastructure Protection, Fight against Crime and Terrorism, and Border Management.
3. Notably DGs HOME, ECHO, ENV, DEVCO, RTD, TAXUD and JRC.
4. Namely Disaster-Resilient Societies (DRS), Fight Against Crime and Terrorism (FCT) along with Infrastructure Protection (INFRA), Border Management (BM) and Strengthened Security Research and Innovation (SSRI).

References

IPCC (2012). Field, C.B., V. Barros, T.F. Stocker, D. Qin, D.J. Dokken, K.L. Ebi, M.D. Mastrandrea, K.J. Mach, G.-K. Plattner, S.K. Allen, M. Tignor, and P.M. Midgley (Eds.) Available from Cambridge University Press, The Edinburgh Building, Shaftesbury Road, Cambridge CB2 8RU ENGLAND, 582 pp. Available from June 2012Managing the Risks of Extreme Events and Disasters to Advance Climate Change Adaptation (SREX) – https://www.ipcc.ch/report/managing-the-risks-of-extreme-events-and-disasters-to-advance-climate-change-adaptation/.

European Commission (2000). Directive 2000/60/EC.

European Commission (2006). Directive 2006/118/EC.
European Commission (2007a). Directive 2007/60/EC.
European Commission (2007b). COM(2007) 414 final.
European Commission (2008a). Directive 2008/105/EC.
European Commission (2008b). EU Marine Strategy Framework Directive, 2008/56/EC.
European Commission (2013a). Decision 1313/2013 on a Union Civil Protection Mechanism.
European Commission (2013b). Art.5.1(a), Council Decision No. 1313/2013/EU. *Official Journal of the European Union*, L347, 20.12.2013.
European Commission (2015). CIS guidance on ecological flows (endorsed in 2015), CIS Guidance on water balances (endorsed in 2015), the EU Regulation on water reuse (adopted by the Commission in 2018 and came into force in 2020)
European Commission (2019). https://ec.europa.eu/environment/water/fitness_check_of_the_eu_water_legislation/index_en.htm.
European Commission (2020a). Mapping Horizon 2020 projects and EU capacity Building Projects under 2016-2018 Programmes. https://tinyurl.com/ex9u789j.
European Commission (2020b). Overview of natural and man-made disaster risks the European Union may face. https://tinyurl.com/p2du9ktc.
European Commission (2021a). https://ec.europa.eu/echo/what/civil-protection/resceu_en.
European Commission (2021b). COM(2021) 82 final. Adopted by the Commission on 24 February 2021. https://eur-lex.europa.eu/legal-content/EN/TXT/?uri=COM:2021:82:FIN.
United Nations (1996). Council Regulation No 1257/96 concerning humanitarian aid.
UNDRR (2015). Sendai framework for disaster risk reduction. https://www.undrr.org/publication/sendai-framework-disaster-risk-reduction-2015-2030.

Index

Responding to Extreme Weather Events, First Edition. Edited by Daniel Sempere-Torres, Anastasios Karakostas, Claudio Rossi, and Philippe Quevauviller.

www.ingramcontent.com/pod-product-compliance
Lightning Source LLC
LaVergne TN
LVHW062205200926
848484LV00006B/1322
* 9 7 8 1 1 1 9 7 4 1 5 8 9 *